Field Manual
No. 7-15

Headquarters
Department of the Army
Washington, DC, 27 February 2009

The Army Universal Task List

Contents

Page

Distribution Restriction: Distribution Restriction: This manual is approved for public release; distribution is unlimited.

*This publication supersedes FM 7-15, 31 August 2003.

Contents

Figures

Tables

Preface

FM 7-15 describes the structure and content of the Army Universal Task List (AUTL). The AUTL is a comprehensive, but not all-inclusive listing of Army tasks, missions, and operations. Units and staffs perform these tasks, mission, and operations or capability at corps level and below. For each task, the AUTL provides a numeric reference hierarchy, a task title, a task description, a doctrine reference, and, in most cases, recommended measures of performance (measures) for training developers to develop training and evaluation outline evaluation criteria for supporting tasks. The task proponent is responsible for developing the training and evaluation outlines that supports each AUTL task. As a catalog, the AUTL captures doctrine as it existed on the date of its publication.

The AUTL can help commanders develop a mission-essential task list (METL). It (the AUTL) provides tasks, missions and operations or capabilities for a unit, company-sized and above, and staffs. Commanders should use the AUTL as a cross-reference for tasks. Commanders may use the AUTL to supplement their core training focused METL or the directed training focused METL as required. FM 7-0 and FM 7-1 discuss in detail METL development and requirements.

The primary source for standards for most Army units is their proponent-approved individual and collective tasks. Proponents revise standards when the factors of mission, enemy, terrain and weather, troops and support available, time available, civil considerations (METT-TC) significantly differ from those associated with a task training and evaluation outline. Significant differences in METT-TC may include new unit equipment; a table of organization; force packaging decisions during deployment; or new unit tasks.

Proponents and trainers will use the unit's assigned table of organization and equipment, as the basis for mission analysis during the analysis phase of the Systems Approach to Training process. Trainers may use the AUTL as a catalog of warfighting function tasks when developing collective tasks. The AUTL is not all-inclusive. If the proponent or school identifies or develops a new AUTL task requirement, the new task will be provided to the Collective Training Directorate for approval and the Combined Arms Doctrine Directorate for input to AUTL revision. Task proponents and schools write and define the conditions and standards (training and evaluation outlines) for individual and collective tasks which support the AUTL.

The AUTL does not include tasks Army forces perform as part of joint and multinational forces at the strategic and operational levels. Those tasks are included in the Universal Joint Task List (UJTL). The UJTL defines tasks and functions performed by Army elements operating at the operational and strategic levels of war. The UJTL provides an overall description of joint tasks to apply at the national strategic, theater strategic, operational, and tactical levels of command. The UJTL also provides a standard reference system used by United States Army Training and Doctrine Command (TRADOC) combat developers for analysis, such as front-end analysis of force element capabilities. Each military Service is required to publish its own tactical task list to supplement the UJTL. (The UJTL bibliography includes the other Services' task lists.) The AUTL is the Army supplement to the UJTL.

PURPOSE

The AUTL complements the UJTL by providing tactical-level of war, Army-specific tasks. The AUTL—

- Provides a common, doctrinal structure for collective tasks that support Army tactical missions and operations conducted by Army units and staffs.
- Articulates what tasks the Army performs to accomplish missions without describing how success occurs.
- Applies across the full spectrum operations (offensive, defensive, and stability or civil support).

- Provides standard definitions and helps establish a common language and reference system for all echelons (from company to corps) and staff sections.
- Uses approved definitions or derived definitions from current doctrine.
- Addresses each Army tactical task (ART) in only one location.
- Lists ARTs subordinate to each of the six warfighting functions (chapters 1-6) and tasks that support execution of the Army's missions (chapter 7).
- Provides a table with measures of performance that can be used to develop standards for each task in chapters 1-6 and generic measures of performance for tasks in chapter 7.

At the upper levels, the AUTL concisely identifies the major activities of a force. At lower levels, it provides increased detail on what the force must do to accomplish its mission.

SCOPE

AUTL tasks apply at the tactical level of war. Although the AUTL emphasizes tasks performed by Army units, the Army does not go to war alone. Chapters 1-6 detail the tasks of each warfighting function: movement and maneuver, intelligence, fires, sustainment, command and control, and protection. The warfighting functions group the related tasks according to battlefield use. In chapters 1-6, ARTs generally have recommended measures of performance associated with each warfighting function task, but commanders may develop their own. Even if a sample measure is selected, commanders must determine their criterion.

Chapter 7 captures tasks that describe results the commander wants to generate or create to accomplish the mission. These tasks are often given to small units as the tasks or purpose parts of their mission statement. Chapter 7 is not another warfighting function. The missions and operations described in this chapter are combined arms in nature and do not fall under the purview of any one warfighting function. Commanders, their staffs, combat developers, training developers, and doctrine analysts can use this chapter to determine what missions and operations a given tactical organization is designed or should be designed to accomplish. In some cases, ARTs in chapter 7 have specific recommended measures of performance to articulate mission accomplishment more clearly.

Throughout the manual, where no specific measures are developed, the generic measures for task performance include:

- Mission accomplishment occurs within the higher commander's intent of what the force must do and under what conditions.
- Mission accomplishment occurs within the higher commander's specified timeline and the risk assessment for fratricide avoidance and collateral damage.
- Mission accomplishment occurs with the minimum expenditure of resources.
- After mission accomplishment, the unit remains capable of executing assigned future missions and operations.

Trainers will use definitions in these chapters to describe specified and implied tasks of missions in common terms. However, ART definitions do not specify who or what type of unit performs the task, what means to use the task, when to perform the task, or how to perform a task. A complete mission statement provides those specifics. Trainers determine those specifics based on their unique circumstances since ARTs are independent of conditions.

Trainers use recommended measures of performance provided in chapters 1-6 as a basis from which to develop standards of performance for a specific unit under specific conditions. Examples of such standards are found in Proponent approved collective tasks delivered by the Digital Training Management System. When Army forces receive a mission, they use a much more focused analysis of the factors of the situation. At the tactical level, and at the operational level for major operations, Army leaders use METT-TC as a planning and execution tool.

For example, time (part of METT-TC) is a measure of performance for the displacement of a command post. However, a trainer may use a standard measuring in minutes for the displacement of a battalion-level tactical command post. A trainer may use a standard measuring in days for the displacement of a corps-level main command post. Measures of performance are neither directive nor all-inclusive. Trainers should use them as a guide and modify or expand them based on their experience and needs.

A *condition* consists of those variables of an operational environment or situation in which a unit, system, or individual is expected to operate and may affect performance (JP 1-02). Refer to enclosure C of the UJTL for descriptions of joint conditions.

APPLICABILITY

FM 7-15 applies to commanders and trainers at all echelons and to doctrine, combat, and training developers who develop doctrine, tactics, techniques, and procedures for the tactical level of war. This publication applies to the Active Army, the Army National Guard (ARNG)/the Army National Guard of the United States (ARNGUS), and the United States Army Reserves (USAR). It applies to both Soldiers and Department of the Army Civilians.

The AUTL provides a common language and reference system for doctrine, combat, and training developers. Proponents and schools use AUTL tasks, mission, and operation or capability along with their recommended measures of performance as the basis for establishing unit-specific, collective task training and evaluation outlines (T&EOs) based on the table of organization and equipment. Proponent T&EOs provide the measurable conditions and standards to be used in evaluating an organization and individuals abilities to execute these tasks.

The AUTL also provides a basis for establishing a unit-specific combined arms training strategy. It supplements the Digital Training Management System by providing a catalog of tasks to assist in identifying and developing a unit METL. The AUTL's link to the UJTL at the operational and strategic level aids analysts and planners in understanding the Army role and integrating joint operations.

ADMINISTRATIVE INSTRUCTIONS

In this manual, the term "forces" refers to Army combined arms organizations that apply the synchronized or simultaneous combat power of several arms and Services.

Opposing forces is used throughout the AUTL for opposition forces, enemies, and any other group or nation with whom the Nation's forces are in conflict.

This manual lists a single reference for each task after the task definition. It also gives the abbreviation for the proponent for the task after the manual reference. Proponents, as defined by AR 5-22 and TRADOC Regulation 350-70, are responsible for developing the training and evaluation outline that supports each AUTL task.

The Army operates in an environment of changing threats, technology, doctrine, and resources. As a result, the Army must update this manual periodically. Such updates draw heavily from experienced users in the field. Only then will the AUTL maintain utility to these users.

U.S. Army Training and Doctrine Command (TRADOC) is the proponent for this publication. The preparing agency is the Combined Arms Doctrine Directorate, U.S. Army Combined Arms Center. Task proponency is according to AR 5-22 and TRADOC Regulation 350-70. Send written comments and recommendations on DA Form 2028 (Recommended Changes to Publications and Blank Forms) directly to: Commander, U.S. Army Combined Arms Center and Fort Leavenworth, ATTN: ATZL-CD (FM 7-15), 201 Reynolds Avenue, Fort Leavenworth, KS 66027-2337. Send comments and recommendations by e-mail to leav-cadd-web-cadd@conus.army.mil. Follow the DA Form 2028 format or submit an electronic DA Form 2028.

Introduction

Army forces integrate their efforts with other Services to achieve the joint force commander's intent. The primary functions of the Army, as outlined in Department of Defense (DOD) Directive 5100.1, are to organize, equip, and train forces for the conduct of prompt and sustained combat operations on land. Accordingly, the Army possesses the capability to defeat enemy land forces and to seize, occupy, and defend land areas. Additionally, it can conduct air and missile defense, space and space control operations, and joint amphibious and airborne operations.

The Army Universal Task List (AUTL) describes what well-trained, superbly led, and well-equipped Soldiers do for the Nation. While focused on the land dimension, abilities of Army forces complement abilities of other Services. The ability of Army forces to perform tasks builds the credible land power necessary for joint force commanders to preclude and deter enemy action, win decisively if deterrence fails, and establish a rapid return to sustained stability. Thus, Army forces expand a joint force commander's military options in full spectrum operations.

Joint tasks describe the current and potential capabilities of the U.S. forces in broad terms. Joint force commanders assign these tasks to joint staffs and integrated service components. The AUTL does not include tasks Army forces perform as part of joint and multinational forces at the strategic level of war. Those tasks are included in the Universal Joint Task List (UJTL). UJTL tasks, when associated with conditions and standards, describe a required capability without describing the means necessary to fulfill a requirement.

The seven operational-level UJTL task areas do not reflect how the Army has traditionally organized its physical means (Soldiers, organizations, and equipment) to accomplish missions. The Army organizes the Army tactical tasks (ARTs) under the six warfighting functions. A warfighting function does not represent an Army branch or proponent. Any Army organization, regardless of branch or echelon, performs tasks related to one or more of the warfighting functions. The introductory figure shows the links between the six warfighting functions and six of the operational level UJTL functional task areas.

Note: The introductory figure does not link any Army warfighting function to the joint operational task area of counter chemical, biological, radiological, nuclear, and high-yield explosives (CBRNE) weapons. The protection warfighting function includes CBRNE. Outside United States Army Training and Doctrine Command combat development-related activities, the Army regards the presence of CBRNE effects as another environment just as it regards the desert, jungle, mountain, and urban terrain as environments. The physical environment forms part of the conditions under which tasks are performed. Different physical environments may require different standards. They do not require different tasks.

Army Warfighting Functions	Joint Functions (JP 3-0)	UJTL Operational Task Areas
Movement and Maneuver (ART 1.0)	Movement and Maneuver	Movement and Maneuver
Intelligence (ART 2.0)	Intelligence	Intelligence, Surveillance, and Reconnaissance
Fires (ART 3.0)	Fires	Fires
Sustainment (ART 4.0)	Sustainment	Perform Logistics and CSS
Command and Control (ART 5.0)	Command and Control	Command and Control
Protection (ART 6.0)	Protection	Force Protection Counter CBRNE Weapons

ART	Army tactical task	CSS	combat service support
CBRNE	chemical, biological, radiological, nuclear, and high-yield explosives	JP	joint publication
		UJTL	Universal Joint Task List

Introductory figure. Links between Army warfighting functions and UJTL task areas

The AUTL divides the warfighting functions into ARTs. Almost any ART can be the "what" of a unit mission statement. Most ARTs can subdivide into subordinate ARTs. Subordinate ARTs can break down until they are no longer tasks, missions, or capabilities. At that level, tasks, missions, and operations or capabilities become individual tasks addressed in Soldier training publications. ART definitions at a lower level elaborate on higher-level ART definitions.

The AUTL numbering system provides a standard reference for addressing and reporting requirements, capabilities, or issues. Diagrams in each chapter show how each ART relates to the warfighting function that contains it and to the doctrinal mission hierarchy. The horizontal links of ARTs from different warfighting functions require synchronizing their performance in space and time based on the concept of operations. The position of any Army tactical mission or task within the AUTL structure has no relationship to its importance. That importance always depends on the mission. Likewise, the position of an Army tactical mission or task does not imply either command or staff oversight.

Each ART appears only once in the AUTL. Doctrine determines the subordination of ARTs. If several logical locations for an ART exist, then it appears where it depicts its most common relationships. While some warfighting functions resemble one another, their definitions clearly distinguish them.

Many ARTs have parallel tasks at the other levels of war, which are addressed in the UJTL. For example, ARTs associated with preparations for overseas movement link vertically to operational-level force projection tasks. Vertical task links provide connectivity among tactical, operational, and strategic activities. However, contributions of tactical land power to joint military power are unique in some cases and do not necessarily link directly to operational- and strategic-level UJTL tasks.

In applying the AUTL to the Army training process, a number of basic terms apply. The introductory table shows their definitions and proponents. Understanding the relationships of Army tactical mission tasks, operations, and missions makes using the AUTL to establish training requirements more successful.

Introductory table. Basic terms, their definitions, and their proponents

Term	Definition
collective task	A unit of work or action requiring interaction between two or more individuals for its accomplishment. It may also be a mission requirement which can be broken down into individual tasks. (TRADOC Pamphlet 350-70-1)
condition	Those variables of an operational environment or situation in which a unit, system, or individual is expected to operate and may affect performance. (JP 1-02)
core mission-essential task list	A list of a unit's core capability mission-essential tasks and general mission-essential tasks. (FM 7-0)
criterion	The minimum acceptable level of performance associated with a particular measure of task performance. (CJCSM 3500.04E)
directed mission-essential task list	A list of mission-essential tasks that must be performed to accomplish a directed mission. (FM 7-0)
measure	A parameter that provides the basis for describing varying levels of performance of a task. (CJCSM 3500.04E)
measure of performance	A criterion used to assess friendly actions that are tied to measuring task accomplishment. (JP 3-0)
mission	The task, together with the purpose, that clearly indicates the action to be taken and the reason therefor. (JP 1-02)
mission-essential task	A collective task a unit must be able to perform successfully in order to accomplish its doctrinal or directed mission. (FM 7-0)
mission-essential task list	A compilation of mission-essential tasks that an organization must perform successfully to accomplish its doctrinal or directed missions. (FM 7-0)
operation	1. A military action or the carrying out of a strategic, operational, tactical, service, training, or administrative military mission. 2. The process of carrying on combat, including movement, supply, attack, defense, and maneuvers needed to gain the objectives of any battle or campaign. (JP 1-02)
standard	A quantitative or qualitative measure and criterion for specifying the levels of performance of a task. (FM 7-0)
task	A clearly defined and measurable activity accomplished by individuals and organizations. (FM 7-0)
warfighting function	A group of tasks and systems (people, organizations, information, and processes) united by a common purpose that commanders use to accomplish missions and training objectives. (FM 3-0)

The primary source for tasks and training and evaluation outlines is the Digital Training Management Systems. Proponents approve tasks and training and evaluation outlines. Commanders request from the task proponent, new or revised standards when the factors of METT-TC significantly differ from those associated with a task identified in a training and evaluation outline. Significant METT-TC differences may result from new unit equipment or force package decisions in support of directed mission. An example of this would be a cannon equipped field artillery battery providing installation security.

Chapter 1

ART 1.0: The Movement and Maneuver Warfighting Function

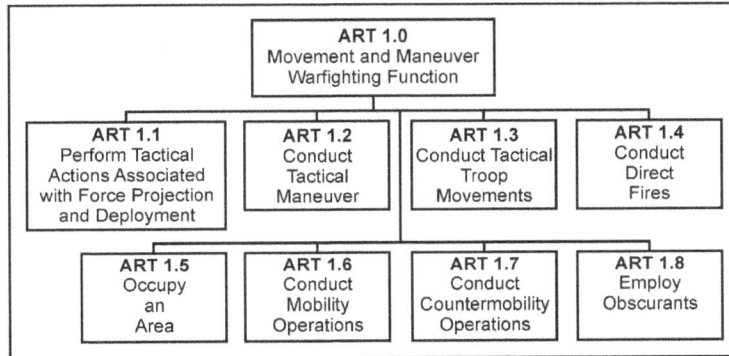

```
                         ART 1.0
                 Movement and Maneuver
                   Warfighting Function

   ART 1.1          ART 1.2        ART 1.3        ART 1.4
 Perform Tactical   Conduct      Conduct Tactical Conduct
Actions Associated  Tactical         Troop        Direct
with Force Projection Maneuver     Movements       Fires
 and Deployment

                    ART 1.5        ART 1.6        ART 1.7        ART 1.8
                    Occupy         Conduct        Conduct        Employ
                      an           Mobility     Countermobility  Obscurants
                     Area        Operations      Operations
```

The *movement and maneuver warfighting function* is the related tasks and systems that move forces to achieve a position of advantage in relation to the enemy. Direct fire is inherent in maneuver, as is close combat. The function includes tasks associated with force projection related to gaining a positional advantage over an enemy. One example is moving forces to execute a large-scale air or airborne assault. Another is deploying forces to intermediate staging bases in preparation for an offensive. *Maneuver* is the employment of forces in the operational area through movement in combination with fires to achieve a position of advantage in respect to the enemy in order to accomplish the mission. Maneuver is the means by which commanders mass the effects of combat power to achieve surprise, shock, and momentum. Effective maneuver requires close coordination with fires. Movement is necessary to disperse and displace the force as a whole or in part when maneuvering. Both tactical and operational maneuver require logistic support. (FM 3-0) (USACAC)

SECTION I – ART 1.1: PERFORM TACTICAL ACTIONS ASSOCIATED WITH FORCE PROJECTION AND DEPLOYMENT

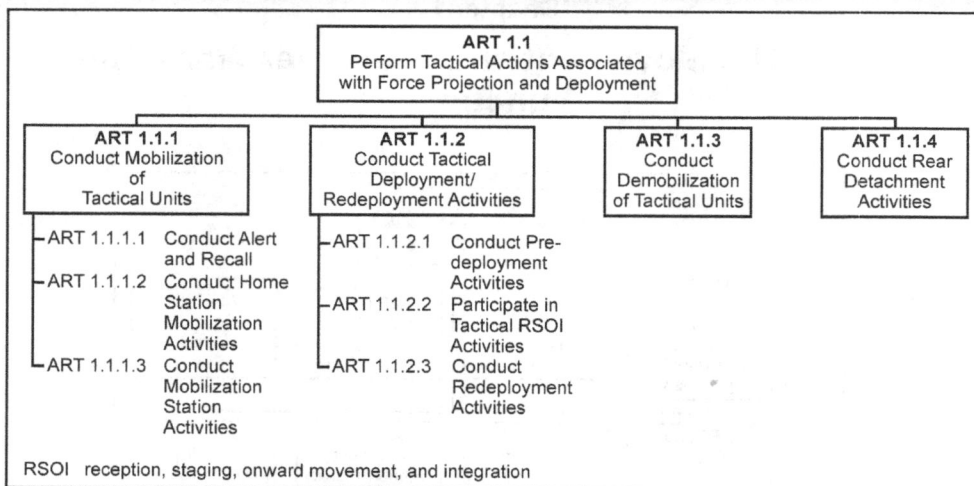

```
                            ART 1.1
                  Perform Tactical Actions Associated
                  with Force Projection and Deployment

        ART 1.1.1          ART 1.1.2          ART 1.1.3          ART 1.1.4
     Conduct Mobilization  Conduct Tactical    Conduct           Conduct Rear
           of              Deployment/        Demobilization     Detachment
      Tactical Units       Redeployment Activities of Tactical Units  Activities

   ART 1.1.1.1  Conduct Alert   ART 1.1.2.1  Conduct Pre-
                and Recall                   deployment
   ART 1.1.1.2  Conduct Home                 Activities
                Station        ART 1.1.2.2  Participate in
                Mobilization                 Tactical RSOI
                Activities                   Activities
   ART 1.1.1.3  Conduct        ART 1.1.2.3  Conduct
                Mobilization                 Redeployment
                Station                      Activities
                Activities

  RSOI  reception, staging, onward movement, and integration
```

1-1. Force projection is the military instrument of national power that systemically and rapidly moves military forces in response to requirements across the spectrum of conflict. It is a demonstrated ability to alert, mobilize, rapidly deploy, and operate effectively anywhere. The Army, as a key member of the joint team, must be ready for global force projection with an appropriate mix of combat and support forces. The world situation demands that the Army project its power at an unprecedented pace. Therefore, the Army must be able to defuse crises early to prevent escalation by using flexible, rapidly deployable forces with sufficient depth and strength to sustain multiple, simultaneous operations. (FMI 3-35) (CASCOM)

No.	Scale	Measure
01	Yes/No	Unit deployed from its current location to the area of operations per the time-phased force and deployment list.
02	Time	To complete unit mobilization.
03	Time	To complete required training before employment in a theater of operations.
04	Time	To determine available transportation infrastructure and resources.
05	Time	To deploy unit from home or mobilization station to a theater of operations.
06	Time	To redeploy unit from theater of operations to home station or another theater.
07	Percent	Of combat effectiveness of unit when employed in a theater of operations.

ART 1.1.1 CONDUCT MOBILIZATION OF TACTICAL UNITS

1-2. Mobilization is the process by which Army tactical forces or part of them is brought to a state of readiness for war or other national emergency. (See JP 1-02 for a complete definition.) It includes activating all or part of the Reserve Components as well as assembling and organizing personnel, supplies, and materiel. (FMI 3-35) (CASCOM)

No.	Scale	Measure
01	Yes/No	Unit was brought to its planned state of readiness in specified time.
02	Yes/No	Unit mobilization progress monitored by higher headquarters.
03	Time	Between planned and actual unit arrival time at mobilization station.

No.	Scale	Measure
04	Time	To process Reserve Components through their mobilization station or continental United States replacement center.
05	Percent	Of key personnel reporting within planning timelines.
06	Percent	Of alert and activation messages dispatched within timelines.
07	Percent	Of required initial mobilization reports submitted on time.

ART 1.1.1.1 CONDUCT ALERT AND RECALL

1-3. This task involves activities required when units and individuals receive mobilization and alert orders. Individuals assigned to the unit are notified of the situation. All individuals report to the designated location at the designated time with designated personal items. (FMI 3-35) (CASCOM)

No.	Scale	Measure
01	Yes/No	Unit mustered its assigned personnel at the designated location within the designated time.
02	Yes/No	Unit alert and recall progress reported to and monitored by higher headquarters.
03	Time	For notified units to identify and report preliminary list of deployable and nondeployable personnel.
04	Percent	Of key personnel reported within planning timelines.
05	Percent	Of alert messages dispatched within timelines.
06	Percent	Of alert messages returned for incomplete or inaccurate addresses.
07	Percent	Of notified units able to alert all their personnel within 24 hours.

ART 1.1.1.2 CONDUCT HOME STATION MOBILIZATION ACTIVITIES

1-4. This task involves activities of Reserve Components at home station after receiving a mobilization order followed by entry onto Federal active duty or other command and control changes. It includes taking action to speed transition to active duty status, such as identifying shortages of critical personnel and equipment. Task activities include inventorying unit property, dispatching an advance party to the mobilization station, and loading the unit on organic equipment or designated carriers. Movement is coordinated with state area commands defense movement coordinator, unit movement coordinator, installation transportation offices, and Military Surface Deployment and Distribution Command. (FMI 3-35) (CASCOM)

No.	Scale	Measure
01	Yes/No	Unit deployed to its mobilization station according to its mobilization timetable.
02	Yes/No	Unit home station mobilization activities reported to and monitored by higher headquarters.
03	Time	To activate key unit personnel.
04	Time	To conduct preparatory administrative, logistic, medical, and readiness activities.
05	Time	To submit initial mobilization reports.
06	Percent	Of key personnel reported within planning timelines.
07	Percent	Of initial mobilization reports submitted on time.

ART 1.1.1.3 CONDUCT MOBILIZATION STATION ACTIVITIES

1-5. This task encompasses actions required to meet deployment or other unit validation criteria. It results in assuring the unit's mission capability. Members of Reserve Components transition to active duty. Commanders conduct necessary individual and collective training that may vary as evaluations and circumstances dictate. Personnel complete preparation for overseas movement. Soldiers prepare equipment for deployment. Mobilization station commanders verify training and processing completed at home station to preclude it being repeated. Depending on the situation, units may move through a mobilization site rather than a mobilization station. (FMI 3-35) (CASCOM)

No.	Scale	Measure
01	Yes/No	Unit met deployment or other unit validation criteria within established mobilization timetables.
02	Yes/No	Unit home station mobilization activities reported to and monitored by higher headquarters.
03	Time	To assemble unit and report status.
04	Time	To conduct specified training.
05	Time	To requisition mobilization station training and support requirements.
06	Time	To transfer home station property and prepare to move to the mobilization station.
07	Percent	Of specified training completed.
08	Percent	Of required mobilization station training and support requirements requisitioned.
09	Percent	Of home station property transferred to the appropriate agency.

ART 1.1.2 CONDUCT TACTICAL DEPLOYMENT/REDEPLOYMENT ACTIVITIES

1-6. Deployment is composed of activities required to: prepare and move forces, move sustainment equipment, and move supplies within a theater of operations. This task involves the force as it task organizes, echelons and tailors itself for movement based on the mission, concept of operations, available lift, and other resources. Redeployment is transferring forces and materiel to support another joint force commander's operational requirements, or to return personnel, equipment, and materiel to the home and/or demobilization stations for reintegration and/or out-processing. Redeployment optimizes readiness of redeploying forces and materiel to meet new contingencies or crises. (FMI 3-35) (CASCOM)

Note: Three phases of deployment/redeployment are tasks addressed elsewhere in the AUTL:

- ART 1.3 (Conduct Tactical Troop Movements) addresses onward movement.
- ART 1.6.5 (Conduct Nontactical Movements) addresses movement of deploying/redeploying units to air/sea ports of embarkation.
- ART 4.1.2.3.2 (Move by Air) addresses strategic lift.

No.	Scale	Measure
01	Yes/No	Unit loaded its designated operational and strategic lift systems per the force deployment plans.
02	Yes/No	Unit conducted a loading exercise of its vehicles and equipment to include containers and garrison close out procedures (if applicable) before developing and inputting data into automated deployment systems. These systems can include Transportation Coordinator's Automated Command and Control Information System, Transportation Coordinator's Automated Information for Movements System II, automated unit equipment list, and time-phased force and deployment data database of Global Command and Control System-Army.
03	Yes/No	Deployment and redeployment activities were coordinated with all required agencies.
04	Time	To task-organize the unit based on mission, concept of operations, available lift, and other resources.
05	Time	To echelon the unit based on mission, concept of operations, available lift, and other resources.
06	Time	To tailor the unit based on mission, concept of operations, available lift, and other resources.
07	Percent	Of available lift resources lost awaiting arrival and loading of unit.

No.	Scale	Measure
08	Percent	Of unit not closed on port of debarkation by scheduled date and time.
09	Percent	Of available lift needed to move unit configured for tactical application not required to move unit administratively.

ART 1.1.2.1 CONDUCT PREDEPLOYMENT ACTIVITIES

1-7. Predeployment activities include planning and preparing for deployment. They include updating unit deployment information for input into command and control and standard Army multi-command management information system including Global Command and Control System-Army (GCCS-A), the automated unit equipment list (AUEL), the Transportation Coordinator's Automated Command and Control Information System (TC-ACCIS), and the Transportation Coordinator's Automated Information for Movements System II (TC-AIMS II). Units update and obtain maps and update databases, organizational structures, and networks to support the Army Battle Command System. Updated information affects the Joint Operation Planning and Execution System (JOPES), the time-phased force and deployment data (TPFDD), and the time-phased force and deployment list. Tasks include maintaining the AUEL; updating AUEL data to become the deployment equipment list; and writing radio frequency tags for unit vehicles, containers, and other critical pieces of equipment to achieve in transit visibility of these items. (FMI 3-35) (CASCOM)

No.	Scale	Measure
01	Yes/No	Unit loaded its designated operational and strategic lift systems per the force deployment plans.
02	Yes/No	Unit conducted a loading exercise of its vehicles and equipment to include containers and garrison close out procedures (if applicable) before developing and inputting data into automated deployment systems. These systems can include TC-ACCIS, TC-AIMS II, AUEL, and TPFDD database of GCCS-A.
03	Time	To cross level and redistribute personnel and equipment.
04	Time	To train to minimum operationally ready status.
05	Time	To evaluate unit for deployment.
06	Time	To secure clearance for a nonvalidated unit before its deployment.
07	Time	To integrate unit movement information into automated transportation information system and other information systems, such as TC-AIMS II and JOPES.
08	Percent	Of required base and operations support, such as training areas, ranges, and ammunition received by deploying unit.
09	Percent	Of mission-essential and other required tasks performed to standard.

ART 1.1.2.2 PARTICIPATE IN TACTICAL RECEPTION, STAGING, ONWARD MOVEMENT, AND INTEGRATION ACTIVITIES

1-8. This task involves unit reception at the port of debarkation (POD). It includes drawing equipment from pre-positioned stocks. This task begins when the first strategic lift system of the main body arrives at the POD. It ends when adequate equipment and supplies are discharged and issued, unit tactical information systems are operational, units have moved from the port to tactical assembly areas, and units are combat ready. (FMI 3-35) (CASCOM)

Note: This task describes the Army's contribution toward the Universal Joint Task List (UJTL) task of Conduct Joint Reception, Staging, Onward Movement, and Integration (JRSOI) in the Joint Operations Area (JOA) (OP 1.1.3).

No.	Scale	Measure
01	Yes/No	Unit was combat ready and in a tactical assembly area per deployment plans.
02	Yes/No	PODs were used efficiently and effectively with no noticeable delay in the throughput of forces into theater.
03	Yes/No	The commander directed that contingency plans be developed for reception, staging, onward movement, and integration activities.
04	Time	To discharge cargo from ship in a logistics over-the-shore situation.
05	Time	To clear aerial port of debarkation (APOD) of aircraft cargo.
06	Time	To clear APOD of chalk's personnel.
07	Time	To clear seaport of debarkation (SPOD) of ship's cargo.
08	Time	To clear SPOD of personnel landing administratively.
09	Time	That ship remains in vicinity of port awaiting discharge of cargo.
10	Time	To clear frustrated cargo from POD.
11	Time	To match personnel arriving by air with equipment arriving by sea.
12	Time	To match personnel arriving by air with pre-positioned equipment.
13	Time	To begin unloading ships on arrival in theater.
14	Time	To accomplish linkup of personnel and equipment.
15	Time	For unit to be en route to final destination from staging area or POD.
16	Percent	Of throughput actually achieved.
17	Percent	Of POD capability within the theater used for the reception of forces.
18	Percent	Of transportation assets available for onward movement to staging area or destination.
19	Percent	Of time airfield is in the maximum on ground category.
20	Percent	Of unit personnel and equipment assembled when transferred to tactical commander.
21	Number	Of personnel per day moved by host-nation support to marshaling areas.

ART 1.1.2.3 CONDUCT REDEPLOYMENT ACTIVITIES

1-9.   This task involves the unit moving to assembly areas and beginning recovery and reconstitution. The commander rebuilds unit integrity and accounts for personnel, equipment, and supplies. The unit develops movement data, washes equipment, completes customs and Department of Agriculture inspections, prepares documentation, and plans movement. This task also includes environmental considerations. At the port, the unit processes personnel and equipment for strategic lift. (FMI 3-35) (CASCOM)

No.	Scale	Measure
01	Yes/No	Unit loaded its personnel and equipment on strategic and operational lift systems per the redeployment plan.
02	Time	To determine lift and security requirements.
03	Time	To select routes and assembly areas.
04	Time	To deploy transportation and security forces.
05	Time	To prepare movement orders.
06	Time	To close unit into designated assembly areas.
07	Time	To integrate unit movement information into automated transportation information systems and other information systems, such as Transportation Coordinator's Automated Information for Movements System II and Joint Operation Planning and Execution System.
08	Time	For unit to prepare equipment for shipment back to home station, depot, or future duty location.
09	Time	For unit to prepare personnel for return to home station or future duty assignment to include the conduct of individual health assessments.
10	Time	To pass command authority of in-transit forces.

No.	Scale	Measure
11	Percent	Of movement orders requiring revision.
12	Percent	Of unit personnel and equipment that move as scheduled to designated ports of embarkation.
13	Percent	Of unit equipment and supplies remaining in theater properly accounted for.
14	Percent	Of available transportation systems used.

ART 1.1.3 CONDUCT DEMOBILIZATION OF TACTICAL UNITS

1-10. Demobilization is the act of returning the force and materiel to a premobilization posture or to some other approved posture. Demobilization actions occur in the area of operations, between the port of embarkation (POE) and demobilization station or POE and continental United States (CONUS) demobilization center, at the demobilization station and CONUS demobilization center, and at home station or home of record. This task also includes environmental considerations. (FMI 3-35) (CASCOM)

No.	Scale	Measure
01	Yes/No	Unit demobilized personnel, equipment, and supplies per plans.
02	Yes/No	Unit maintained accountability of property and personnel.
03	Time	To prepare unit equipment and supplies for movement (includes cleaning, maintaining, and configuring vehicles for movement by removing canvas tarps and folding down windows).
04	Time	To move demobilizing organization from its current location to demobilization station.
05	Time	Between planned and actual unit arrival at home station or demobilization station.
06	Time	To process Reserve Components through the demobilization station.
07	Time	To prepare to transfer unit supplies and equipment listed in table of organization and equipment (TOE) to appropriate storage location.
08	Time	To conduct individual demobilization administrative, logistic, medical, and financial management activities.
09	Time	To transfer home station property back to the unit.
10	Time	To terminate outstanding contracts.
11	Percent	Of unit supplies and equipment listed in TOE transferred to temporary or permanent storage locations.
12	Percent	Of home station property transferred back to the unit.
13	Percent	Of outstanding contracts to terminate.
14	Percent	Of individual demobilization administrative, logistic, medical, and financial management activities completed.

ART 1.1.4 CONDUCT REAR DETACHMENT ACTIVITIES

1-11. Rear detachment activities use nondeployable and other personnel to maintain facilities and equipment at home station when the deployed force is expected to return. Such activities include supporting families of deployed personnel. This task includes turning over residual equipment, supplies, and facilities to appropriate authorities (for example, the mobilization station commander) if the unit does not return to the mobilization station. (FM 1-0) (USAAGS)

No.	Scale	Measure
01	Yes/No	Unit family members continued to receive authorized support, assistance, and counseling during operational deployments of Soldiers.
02	Yes/No	Rear detachment maintained and accounted for unit installation property and equipment throughout the unit's deployment.
03	Yes/No	Rear detachment established rear detachment program before beginning deployment activities.

No.	Scale	Measure
04	Yes/No	Environmental considerations associated with departing units have been identified and appropriate actions taken.
05	Time	To turn over residual equipment, supplies, and facilities to appropriate authorities when the unit will not return to home or mobilization station.
06	Time	To conduct rear detachment administrative and logistic activities.
07	Time	To establish a functioning family support group.
08	Time	To provide quality and meaningful assistance to unit family members during times of need and support.
09	Percent	Of rear detachment administrative and logistic reports submitted on time.
10	Percent	Of residual equipment, supplies, and facilities turned over to appropriate authorities when unit returns to home or mobilization station.

SECTION II – ART 1.2: CONDUCT TACTICAL MANEUVER

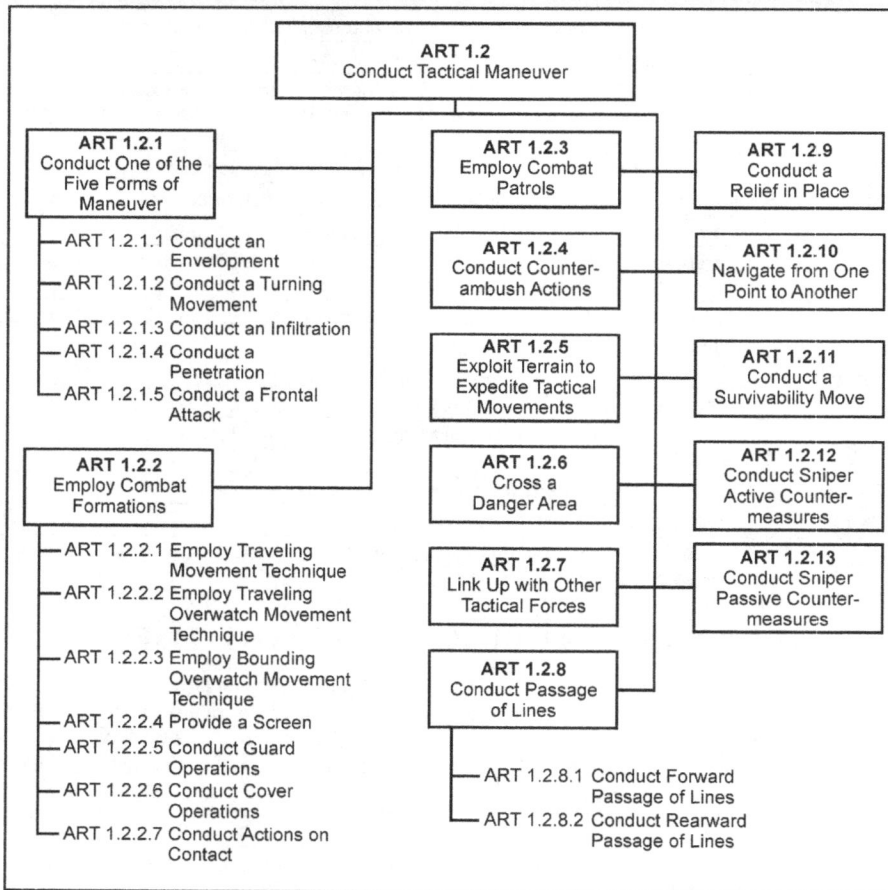

1-12. Maneuver is the means by which commanders mass the effects of combat power to achieve surprise, shock, momentum, and dominance. Commanders take full advantage of terrain and combat formations when maneuvering their forces. (FM 3-90) (USACAC)

No.	Scale	Measure
01	Yes/No	Unit achieved a position of advantage with respect to the enemy.
02	Yes/No	Subordinate commanders used their initiative to achieve the commander's intent when the plan no longer applied.
03	Yes/No	The commander maintained control of associated forces.
04	Time	To initiate maneuver.
05	Time	To complete maneuver.
06	Percent	Of forces achieving position of advantage after executing the plan.
07	Percent	Of enemy force destroyed and neutralized by the maneuver of the friendly force.
08	kph	Rate of movement across the battlefield.

ART 1.2.1 CONDUCT ONE OF THE FIVE FORMS OF MANEUVER

1-13. The forms of maneuver are envelopment, turning movement, infiltration, penetration, and frontal attack. While normally combined, each form of maneuver attacks the enemy differently. A single operation may contain several forms of offensive maneuver. For example, a frontal attack to clear a security area may be followed by a penetration to create a gap in the defenses. This could be followed by an envelopment to destroy the first line of defense. Each form of maneuver poses different challenges for attackers and different dangers for defenders. Commanders determine the form of maneuver to use by analyzing the factors of mission, enemy, terrain and weather, troops and support available, time available, civil considerations. (FM 3-90) (USACAC)

ART 1.2.1.1 CONDUCT AN ENVELOPMENT

1-14. An envelopment is a form of maneuver in which an attacking force seeks to avoid the principal enemy defenses by seizing objectives to the enemy rear to destroy the enemy in his current positions. At the tactical level, envelopments focus on seizing terrain, destroying specific enemy forces, and interdicting enemy withdrawal routes. Envelopments avoid the enemy front where enemy forces are protected and can easily concentrate fires. Single envelopments maneuver against one enemy flank; double envelopments maneuver against both. A vertical envelopment (an air assault or airborne operation) creates an assailable flank by arriving from an unexpected direction. An envelopment may result in an encirclement. (FM 3-90) (USACAC)

> *Note:* Airborne, air assault, and amphibious operations are tactical aspects of forcible entry operations that involve seizing and holding a military lodgment in the face of armed opposition. Joint force commanders conduct forcible entry operations at the operational level. (See JP 3-18.) Forcible entry operation tasks are addressed in the Universal Joint Task List (UJTL).

ART 1.2.1.1.1 Conduct an Airborne Assault into Area of Operations

1-15. An airborne assault secures a defendable airhead from which to conduct lodgment activities or to seize key terrain to block or deny its use by the enemy. This mission begins when paratroopers and equipment exit the aircraft by parachute or are air landed. The mission ends when all elements of the relevant airborne echelon are delivered to the objective area and the assault objective has been seized. (FM 3-90) (USACAC)

> *Note:* Marshaling forces for airborne operations is addressed as ART 1.3.1.3 (Marshal Forces to Conduct an Airborne Assault).
> The air movement phase of airborne operations is addressed as ART 4.1.2.3.2 (Move by Air).
> Tasks for the ground tactical-phase of airborne operations duplicate missions are contained elsewhere in this chapter.

No.	Scale	Measure
01	Yes/No	Enemy situation in the area of operations known and disseminated to assault force.
02	Yes/No	Drop zone/landing zone marked by pathfinders.
03	Yes/No	En route communications established between assault force and force headquarters.
04	Yes/No	Preassault fires executed before H-hour.
05	Yes/No	Assault force establishes contact with force headquarters.
06	Percent	Of friendly casualties occurring during the airborne assault.

ART 1.2.1.1.2 Conduct an Air Assault

1-16. An air assault secures a defendable airhead or other key terrain from which to conduct lodgment activities or follow-on tactical operations. The mission ends when all elements of the relevant air assault echelon are delivered to the objective area and the assault objective has been seized. (FM 90-4) (USAIS)

Note: ART 1.2.1.1.2 does not include air traffic control. Air traffic control is addressed under ART 5.1.3.1.7 (Control Tactical Airspace).

ART 1.2.1.1.2 does not include efforts to improve the landing zone. Landing zone improvement is addressed under ART 1.6.2 (Enhance Movement and Maneuver).

Pickup zone operations are addressed as ART 1.3.1.4 (Conduct Pickup Zone Operations).

The air movement phase of air assault operations is addressed as ART 4.1.2.3.2 (Move by Air).

Tasks for the ground tactical phase of air assault operations duplicate missions contained elsewhere in this chapter.

Scale and measures assigned to ART 1.2.1.1.1 apply.

ART 1.2.1.1.3 Conduct an Amphibious Assault

1-17. The Army conducts an amphibious assault as a member of the landing force in conducting one of four forcible entry operations. It may also conduct forcible entry operations via airborne, air assault or a combination of any or all of these forcible entry techniques. An amphibious assault secures a defendable beachhead from which to conduct lodgment activities. This mission begins when Soldiers and equipment leave the ships that bring them to the amphibious objective area and transfer to the landing craft that will deliver them to the beach. The mission ends when all elements of the amphibious force are delivered onto the beachhead and the assault objective has been seized. (JP 3-02) (USJFCOM JWFC)

Note: Army forces follow joint doctrine and Marine Corps doctrinal publications when tasked to conduct amphibious operations.

Scale and measures assigned to ART 1.2.1.1.1 apply.

ART 1.2.1.1.4 Conduct an Encirclement

1-18. Encirclements are operations where one force loses its freedom of maneuver because an opposing force is able to isolate it by controlling all ground lines of communications. (FM 3-90) (USACAC)

ART 1.2.1.2 CONDUCT A TURNING MOVEMENT

1-19. A turning movement is a maneuver in which an attacking force seeks to avoid the enemy's principal defensive positions by seizing objectives to the enemy rear and causing the enemy to move out of his current positions or divert major forces to meet the threat. (FM 3-90) (USACAC)

ART 1.2.1.3 CONDUCT AN INFILTRATION

1-20. An infiltration is a maneuver in which an attacking force conducts undetected movement through or into an area occupied by enemy forces to occupy a position of advantage while exposing only small elements to enemy defensive fires. (FM 3-90) (USACAC)

Note: Infiltration is also a tactical march technique used within a friendly territory to move forces in small groups at extended or irregular intervals. (See FM 3-90.)

ART 1.2.1.4 CONDUCT A PENETRATION

1-21. A penetration is a maneuver in which an attacking force seeks to rupture enemy defenses on a narrow front to disrupt the defensive system. Commanders direct penetrations when enemy flanks are not assailable or time does not permit another form of maneuver. Successful penetrations create assailable flanks and provide access to enemy sustainment areas. (FM 3-90) (USACAC)

ART 1.2.1.5 CONDUCT A FRONTAL ATTACK

1-22. A frontal attack is a maneuver in which the attacking force seeks to destroy a weaker enemy force or fix a larger enemy force in place over a broad front. At the tactical level, an attacking force can use a frontal attack to overrun a weaker enemy force rapidly. A frontal attack strikes the enemy across a wide front and over the most direct approaches. Commanders normally use it when they possess overwhelming combat power and the enemy is at a clear disadvantage. (FM 3-90) (USACAC)

ART 1.2.2 EMPLOY COMBAT FORMATIONS

1-23. Use an ordered arrangement of troops and vehicles for a specific purpose. Commanders use one of seven different combat formations: column, line, echelon (left or right), box, diamond, wedge, and vee. Their use depends on the factors of mission, enemy, terrain and weather, troops and support available, time available, civil considerations (METT-TC). (FM 3-90) (USACAC)

No.	Scale	Measure
01	Yes/No	Combat formation reflected the existing factors of METT-TC.
02	Yes/No	Unit complied with all graphic control measures.
03	Yes/No	Unit employed the movement formation ordered by the leader.
04	Time	To plan and prepare operation order.
05	Time	To initiate movement.
06	Time	To complete movement.
07	Percent	Of area of operations observed during reconnaissance conducted before starting operations.
08	Percent	Of allocated forces in place at the start of the plan.
09	Percent	Of unit in designated combat formation throughout the movement.
10	Percent	Of unit moving on the specified route or axis.
11	Percent	Of casualties to the force that occurred during the operation.
12	kph	Rate of movement.

ART 1.2.2.1 EMPLOY TRAVELING MOVEMENT TECHNIQUE

1-24. All unit elements move simultaneously. (FM 3-90) (USACAC)

No.	Scale	Measure
01	Yes/No	Combat formation reflected the existing factors of mission, enemy, terrain and weather, troops and support available, time available, civil considerations.
02	Yes/No	Unit complied with all graphic control measures.
03	Yes/No	Unit employed the movement formation ordered by the leader.
04	Yes/No	Unit commander did not expect to encounter the enemy.
05	Yes/No	All subordinate elements of the unit assigned areas of operations for observation during the movement.
06	Yes/No	Unit leaders ensured 360-degree coverage of the unit for observation and fire, to include air guard.
07	Yes/No	Route of movement offered concealment from enemy ground and air observation, avoided skylining, avoided moving directly forward from firing positions, crossed open areas quickly, avoided possible kill zones, avoided wide open spaces (especially where high ground dominates or where the terrain covers and conceals the enemy), and avoided obvious avenues of approach.
08	Time	To plan and prepare operation order.
09	Time	To initiate movement.
10	Time	To complete movement.
11	Percent	Of allocated forces in place at the start of the plan.
12	Percent	Of unit in designated combat formation throughout the movement.
13	Percent	Of unit moving on the specified route or axis.
14	Percent	Of casualties to the force occurring during the operation.

ART 1.2.2.2 EMPLOY TRAVELING OVERWATCH MOVEMENT TECHNIQUE

1-25. The lead element moves continuously. Trailing elements move at varying speeds, sometimes pausing to overwatch movement of the lead element. (FM 3-90) (USACAC)

No.	Scale	Measure
01	Yes/No	Combat formation reflected the existing factors of mission, enemy, terrain and weather, troops and support available, time available, civil considerations.
02	Yes/No	Unit complied with all graphic control measures.
03	Yes/No	Unit employed the movement formation ordered by the leader.
04	Yes/No	Unit commander determined that enemy contact is possible, but speed is important.
05	Yes/No	All subordinate elements assigned area of operations for observation during the movement.
06	Yes/No	Unit leaders ensured that 360-degree coverage exists for observation and fire, to include air guard.
07	Yes/No	Route of movement offered concealment from enemy ground and air observation, avoided skylining, avoided moving directly forward from firing positions, crossed open areas quickly, avoided possible kill zones, avoided wide open spaces (especially where high ground dominates or where the terrain covers and conceals the enemy), and avoided obvious avenues of approach.
08	Time	To plan and prepare operation order.
09	Time	To initiate movement.
10	Time	To complete movement.
11	Percent	Of allocated forces in place at the start of the plan.
12	Percent	Of unit in designated combat formation throughout the movement.
13	Percent	Of unit moving on the specified route or axis.
14	Percent	Of casualties to the force that occurred during the operation.
15	kph	Rate of movement.

ART 1.2.2.3 EMPLOY BOUNDING OVERWATCH MOVEMENT TECHNIQUE

1-26. Using successive or alternate bounds, overwatching elements cover bounding elements from covered, concealed positions with good observation of, and fields of fire against, possible enemy positions. (FM 3-90) (USACAC)

No.	Scale	Measure
01	Yes/No	Combat formation reflected the existing factors of mission, enemy, terrain and weather, troops and support available, time available, civil considerations (METT-TC).
02	Yes/No	Unit complied with all graphic control measures.
03	Yes/No	Unit employed the movement formation ordered by the leader.
04	Yes/No	Unit commander expected to make enemy contact.
05	Yes/No	All subordinate elements of the unit assigned an area of operations for observation during the movement.
06	Yes/No	Unit leaders ensured 360-degree coverage exists for observation and fire, to include air guard.
07	Yes/No	Length of bounds, selected overwatch positions, and variation of techniques—use of alternate or successive bounds—reflected existing factors of METT-TC.
08	Yes/No	Route of movement offered concealment from enemy ground and air observation, avoided skylining, avoided moving directly forward from firing positions, crossed open areas quickly, avoided possible kill zones, avoided wide open spaces (especially where high ground dominates or where the terrain covers and conceals the enemy), and avoided obvious avenues of approach.
09	Time	To initiate movement.
10	Time	To complete movement.
11	Percent	Of allocated forces in place at the start of the execution of the plan.
12	Percent	Of unit in designated combat formation throughout the movement.
13	Percent	Of unit moving on the specified route or axis.
14	Percent	Of casualties to the force occurring during the operation.
15	kph	Rate of movement.

ART 1.2.2.4 PROVIDE A SCREEN

1-27. A screen is a security operation that primarily provides early warning to the protected force. The unit executing a screen observes, identifies, and reports enemy actions. Generally, a screening force, augmented by indirect fires, engages and destroys enemy reconnaissance elements within its capabilities, but otherwise fights only in self-defense. (FM 3-90) (USACAC)

No.	Scale	Measure
01	Yes/No	The screening force's operations provided the protected force or installation with sufficient reaction time and maneuver space to conduct defensive operations.
02	Yes/No	Screening force was in place not later than time specified in operation order.
03	Yes/No	Screening force prevented enemy ground observation of protected force or installation.
04	Yes/No	Collateral damage from the screening force's activities was within acceptable limits.
05	Yes/No	Screening force provided early and accurate warning of enemy approach.
06	Yes/No	Screening force oriented its operations of the force or facility to be secured.
07	Yes/No	Screening force performed continuous reconnaissance.
08	Yes/No	Screening force maintained contact with enemy forces.
09	Yes/No	Commander developed criteria for ending the screening operation.
10	Time	To conduct reconnaissance of the area surrounding the secured force or installation.
11	Time	To plan the screen.
12	Time	To prepare for the screen including movement into security area.

No.	Scale	Measure
13	Time	To execute the screen.
14	Time	To report enemy activities to appropriate headquarters.
15	Time	Of warning that the screening force gives to the secured unit or installation before the secured unit or installation makes contact with the enemy.
16	Percent	Of screening force casualties during the screen.
17	Percent	Of secured force or installation casualties during the conduct of the screen.
18	Percent	Of unit combat power used to provide a screen.
19	Percent	Of enemy reconnaissance elements destroyed or repelled by the screening force.
20	Percent	Of friendly operations judged as not compromised before or during execution.
21	Percent	Of operations not compromised (based on enemy prisoner of war interrogations or captured documents).
22	Percent	Of the area of operations (AO) or security area that can be observed at any given time by the screen force using visual observation and sensors.
23	Number	Of incidents where enemy forces affect the security of friendly units and facilities.
24	Number	Of incidents where enemy reconnaissance forces compromise friendly courses of action causing them to be delayed, disrupted, canceled, or modified.
25	Number	Of screening force casualties during the security operation.
26	Number	Of secured force or installation casualties during the conduct of the security operation.
27	Number	Of enemy reconnaissance elements destroyed during security operation.
28	Number	Of mobility corridors or avenues of approach that the screening force can observe.
29	Number	Of observation posts that the screening force can establish.
30	Square Kilometers	Size of security area or AO.

ART 1.2.2.5 CONDUCT GUARD OPERATIONS

1-28. Guard is a security operation. Its primary task is to protect the main body. It gains time by fighting. It also observes and reports information while preventing enemy ground observation of and direct fire against the main body. A guard differs from a screen in that a guard force contains sufficient combat power to defeat, repel, or fix the lead elements of an enemy ground force before it can engage the main body with direct fires. The guard force operates within the range of the main body's fire support weapons, deploying over a narrower front than a comparable-sized screening force to permit concentrating combat power. (FM 3-90) (USACAC)

No.	Scale	Measure
01	Yes/No	Guard force's operations provided the protected force or installation with sufficient reaction time and maneuver space to conduct defensive operations.
02	Yes/No	Guard force was in place not later than time specified in operation order.
03	Yes/No	Guard force prevented enemy ground observation of protected force or installation.
04	Yes/No	Collateral damage from the guard force's activities was within acceptable limits.
05	Yes/No	Guard force provided early and accurate warning of enemy approach.
06	Yes/No	Guard force oriented its operations of the force or facility to be secured.
07	Yes/No	Guard force performed continuous reconnaissance.
08	Yes/No	Guard force maintained contact with enemy forces.
09	Yes/No	Commander developed criteria for ending the guard operation.
10	Yes/No	Guard force caused the enemy main body to deploy.
11	Yes/No	Guard force impeded and harassed the enemy within its capabilities while displacing.
12	Time	To conduct reconnaissance of the area surrounding the secured force or installation.
13	Time	To plan the guard operation.

No.	Scale	Measure
14	Time	To prepare for the guard operation to include movement into security area.
15	Time	To execute the guard operation.
16	Time	To report enemy activities to appropriate headquarters.
17	Time	Of warning that the guard force gives to the secured unit or installation before the secured unit or installation makes contact with the enemy.
18	Percent	Of guard force casualties during the guard operation.
19	Percent	Of secured force or installation casualties during the guard operation.
20	Percent	Of unit combat power used to provide the guard force.
21	Percent	Of enemy reconnaissance elements destroyed or repelled by the guard force.
22	Percent	Of friendly operations judged as not compromised prior to or during execution.
23	Percent	Of operations not compromised (based on enemy prisoner of war interrogations or captured documents).
24	Percent	Of the area of operations or security area that can be observed at any given time by the guard force using visual observation and sensors.
25	Number	Of incidents where enemy forces affect the security of the secured force or facilities.
26	Number	Of incidents where enemy reconnaissance or advance guard forces compromise friendly courses of action.
27	Number	Of guard force casualties during the guard operation.
28	Number	Of secured force or installation casualties during the guard operation.
29	Number	Of enemy reconnaissance and advance guard elements destroyed during the guard operation.
30	Number	Of mobility corridors or avenues of approach that the guard force can observe.
31	Number	Of observation posts that the guard force can establish.
32	Square Kilometers	Size of security area or area of operations.

ART 1.2.2.6 CONDUCT COVER OPERATIONS

1-29. Cover is a security operation. Its primary task is to protect the main body. It gains time by fighting. It also observes and reports information while preventing enemy ground observation of and direct fire against the main body. A covering force operates outside supporting range of the main body. It promotes early situational development as it deceives the enemy about the location of the main battle area while disrupting and destroying enemy forces. Cover operations provide the main body with the maximum early warning and reaction time. (FM 3-90) (USACAC)

No.	Scale	Measure
01	Yes/No	Covering force's operations provided the protected force with sufficient reaction time and maneuver space.
02	Yes/No	Covering force was in place not later than time specified in operation order.
03	Yes/No	Covering force prevented enemy ground observation of protected force or installation.
04	Yes/No	Collateral damage from the covering force's activities was within acceptable limits.
05	Yes/No	Covering force provided early and accurate warning of enemy approach.
06	Yes/No	Covering force oriented its operations on the secured force.
07	Yes/No	Covering force performed continuous reconnaissance.
08	Yes/No	Covering force maintained contact with enemy forces.
09	Yes/No	Commander developed criteria for ending the covering operation.
10	Yes/No	Covering force caused the enemy main body to deploy.
11	Yes/No	Covering force defeated or repelled enemy forces as directed by the higher commander.

No.	Scale	Measure
12	Yes/No	During an offensive cover operation, the covering force penetrated the enemy's security area and located the enemy's main defensive positions.
13	Yes/No	During an offensive cover operation, the covering force determined enemy strengths and dispositions and located gaps or weak points within the enemy's defensive scheme.
14	Yes/No	During an offensive cover operation, the covering force deceived the enemy into thinking the main body had been committed.
15	Yes/No	During an offensive cover operation, the covering force fixed enemy forces in current positions to allow the main body to maneuver against them.
16	Yes/No	During a defensive cover operation, the covering force avoided being bypassed by attacking enemy forces.
17	Time	To conduct zone reconnaissance of the area surrounding the secured force.
18	Time	To plan the cover operation.
19	Time	To prepare for the cover operation to include movement to security area.
20	Time	To execute the cover operation.
21	Time	To report enemy activities to appropriate headquarters.
22	Time	Of warning that the covering force gives to the secured unit or installation before the secured unit or installation makes contact with the enemy.
23	Percent	Of covering force casualties during the cover operation.
24	Percent	Of secured force casualties during the cover operation.
25	Percent	Of unit combat power used to provide the covering force.
26	Percent	Of enemy reconnaissance, advance guard, and main body elements destroyed or repelled by the covering force.
27	Percent	Of friendly operations judged as not compromised prior to or during execution.
28	Percent	Of operations not compromised (based on enemy prisoner of war interrogations or captured documents).
29	Percent	Of the security area that can be observed at any given time by the covering force using visual observation and sensors.
30	Percent	Of area of operations cleared of enemy forces by an offensive covering force.
31	Percent	Of enemy forces in an area of operations bypassed by an offensive covering force.
32	Number	Of incidents where enemy forces affect the security of the secured force.
33	Number	Of incidents where enemy reconnaissance, advance guard, or first echelon forces compromise friendly courses of action.
34	Number	Of covering force casualties during the cover operation.
35	Number	Of secured force casualties during the cover operation.
36	Number	Of enemy reconnaissance, advance guard, and main body first echelon elements destroyed during the security operation.
37	Number	Of mobility corridors or avenues of approach that the covering force can observe.
38	Number	Of observation posts that the covering force can establish.
39	Square Kilometers	Size of security area or area of operations.

ART 1.2.2.7 CONDUCT ACTIONS ON CONTACT

1-30. Develop the situation once contact is made, concentrate combat power, and transition to a hasty attack or defense. Actions on contact include all forms of contact: sensor; direct and indirect lethal and nonlethal fires; air; obstacle or device; electronic warfare; and chemical, biological, radiological, and nuclear. Whether attacking or defending, commanders generate and sustain overwhelming combat power at the point combat forces collide to defeat the enemy rapidly. (FM 3-90) (USACAC)

No.	Scale	Measure
01	Yes/No	Unit generated and sustained overwhelming combat power at the point of contact if the element that made contact was able to defeat the enemy unassisted.
02	Yes/No	The generation of overwhelming combat power was the product of the recommended course of action to the higher commander.
03	Yes/No	Intelligence, surveillance, and reconnaissance assets were used to develop situation without main body being in contact with the enemy.
04	Time	To deploy and report.
05	Time	To evaluate and develop the situation.
06	Time	To choose a course of action (COA).
07	Time	To execute a selected COA.
08	Time	To recommend a COA to the higher commander.
09	Time	To return to previous mission.
10	Percent	Of friendly forces available to continue previous mission.
11	Percent	Of combat effectiveness of enemy force that made contact.

ART 1.2.3 EMPLOY COMBAT PATROLS

1-31. Use ground and air detachments to provide security and harass, destroy, or capture enemy troops, equipment, or installations. (FM 3-21.10) (USAIS)

No.	Scale	Measure
01	Yes/No	Combat patrols provided security and harassed, destroyed, or captured enemy troops, equipment, or installations per the commander's intent.
02	Time	To prepare patrol plan.
03	Time	To conduct rehearsals.
04	Time	To conduct the combat patrol within time higher headquarters allocates.
05	Percent	Of assigned area covered during the patrol.
06	Percent	Of friendly casualties received during the combat patrol.
07	Percent	Of encountered enemy troops and equipment destroyed or captured.
08	Percent	Of information requirements achieved.

ART 1.2.4 CONDUCT COUNTERAMBUSH ACTIONS

1-32. Execute immediate action against near and far ambushes to minimize casualties, exit the enemy engagement area, inflict casualties on the enemy ambush force, and continue the mission. (FM 3-21.10) (USAIS)

No.	Scale	Measure
01	Yes/No	Unit continued its mission after exiting the enemy engagement area.
02	Yes/No	Unit security element detected the ambush.
03	Yes/No	Unit prevented the enemy from gaining intelligence.
04	Yes/No	Unit security element prevented the enemy from engaging the unit main body.
05	Yes/No	Unit bypassed the ambush kill zone and the enemy's associated security positions.
06	Yes/No	Unit attacked and defeated the enemy ambush force before the enemy initiated the ambush.
07	Yes/No	Unit disengaged its elements in the kill zone before destroying all elements in the kill zone.
08	Yes/No	Unit engaged and fixed enemy forces to prevent their withdrawal.
09	Number	Of enemy casualties.

ART 1.2.5 EXPLOIT TERRAIN TO EXPEDITE TACTICAL MOVEMENTS

1-33. Use terrain as a combat equalizer or multiplier by positioning or maneuvering forces to outfight the enemy. Reinforce natural terrain advantages through mining, barriers, and other obstacles. (FM 3-90) (USACAC)

No.	Scale	Measure
01	Yes/No	Unit used terrain to provide concealment from enemy ground and air observation.
02	Yes/No	Unit avoided skylining vehicles, and enemy Soldiers, fighting positions, and survivability positions.
03	Yes/No	Unit avoided moving directly forward from firing positions toward the enemy.
04	Yes/No	Unit crossed open areas quickly.
05	Yes/No	Unit avoided possible enemy kill zones.
06	Yes/No	Unit avoided open spaces where the enemy can detect and engage in it at long ranges, especially where high ground dominates or where the terrain covers and conceals the enemy.
07	Yes/No	Unit avoided obvious avenues of approach into enemy positions.

ART 1.2.6 CROSS A DANGER AREA

1-34. Move forces rapidly across potential enemy engagement areas without detection by the enemy and without exposing the entire force. (FM 7-85) (USAIS)

No.	Scale	Measure
01	Yes/No	Unit prevented decisive engagement by the enemy.
02	Yes/No	Unit prevented the enemy from surprising the main body.
03	Time	To submit reports of the danger area to higher headquarters per unit standing operating procedures.
04	Time	For unit to cross danger area.
05	Percent	Of unit personnel that became casualties while crossing the area.
06	Percent	Of unit equipment that was damaged or immobilized while crossing the area.
07	Percent	Of unit personnel and equipment that crossed the danger area.

ART 1.2.7 LINK UP WITH OTHER TACTICAL FORCES

1-35. A *linkup* is a meeting of friendly ground forces that occurs in many circumstances. It happens when an advancing force reaches an objective area previously seized by an airborne or air assault force; when an encircled element breaks out to rejoin friendly forces or a force comes to the relief of an encircled force; and when converging maneuver forces meet. Forces may be moving toward each other, or one may be stationary. (FM 3-90) (USACAC)

No.	Scale	Measure
01	Yes/No	Units made physical contact with each other while accomplishing their assigned missions.
02	Yes/No	Main bodies of units linking up were not surprised by the enemy.
03	Yes/No	Higher headquarters directing linkup established control measures that protected both forces from fratricide and expedited execution of the linkup.
04	Yes/No	Higher headquarters issued instructions and control measures with adequate time for affected units to plan and prepare for linkup.
05	Time	To complete link-up plan.
06	Time	Between planned and actual link-up times.

No.	Scale	Measure
07	Time	For units linking up to establish a consolidated chain of command.
08	Distance	Between planned and actual link-up locations.
09	Number	Of instances of fratricide occurring during the linkup.

ART 1.2.8 CONDUCT PASSAGE OF LINES

1-36. A *passage of lines* is a tactical enabling operation in which one unit moves through another unit's positions with the intent of moving into or out of enemy contact. (FM 3-90) (USACAC)

No.	Scale	Measure
01	Yes/No	The unit moving in or out of contact accomplished its mission after passing through the stationary unit.
02	Yes/No	The enemy surprised neither the stationary nor the passing unit main body.
03	Yes/No	Higher headquarters directing the passage designated subsequent missions for both forces, when and under what conditions passage of command took place, start and finish times for the passage, contact points between the units involved, and common maneuver control measures and graphics.
04	Yes/No	Stationary unit provided guides and other assistance to the passing unit per the operation order directing the passage.
05	Yes/No	All personnel and equipment moved through the stationary unit by time specified in operation or fragmentary order.
06	Time	To pass through the lines.
07	Time	Between planning contact and making contact.
08	Time	Between planning the transfer of responsibility for the area of operations and when it actually occurs.
09	Percent	Of support (such as fires, maintenance, medical) the passing unit requests and the stationary unit provides.
10	Distance	Between planned and actual locations of contact points.
11	Distance	Between planned and actual locations where the passage of lines takes place.
12	Number	Of casualties from friendly fire or obstacles in either the stationary or the passing unit.

ART 1.2.8.1 CONDUCT FORWARD PASSAGE OF LINES

1-37. A *forward passage of lines* occurs when a unit passes through another unit's positions while moving toward the enemy. (FM 3-90) (USACAC)

No.	Scale	Measure
01	Yes/No	The unit conducting the forward passage of lines accomplished its mission after passing through the stationary unit.
02	Yes/No	The enemy surprised neither the stationary nor the passing unit main body.
03	Yes/No	Higher headquarters directing the passage designated subsequent missions for both forces, when and under what conditions passage of command took place, start and finish times for the passage, contact points between the units involved, and common maneuver control measures and graphics.
04	Yes/No	Stationary unit provided guides and other assistance to the passing unit per the operation order directing the passage.
05	Yes/No	All personnel and equipment moved through the stationary unit by time specified in operation order.
06	Time	To complete the forward passage of lines.
07	Time	Between planning the contact and making contact.
08	Time	Between planning the transfer of responsibility and when it actually occurs.

No.	Scale	Measure
09	Percent	Of support (such as fires, maintenance, and medical) the passing unit requests and the stationary unit provides.
10	Distance	Between planned and actual locations of contact point.
11	Distance	Between planned and actual locations where the passage of lines takes place.
12	Number	Of casualties from friendly fire or obstacles in either the stationary or the passing unit.

ART 1.2.8.2 CONDUCT REARWARD PASSAGE OF LINES

1-38. A *rearward passage of lines* occurs when a unit passes through another unit's positions while moving away from the enemy. (FM 3-90) (USACAC)

No.	Scale	Measure
01	Yes/No	Unit conducting the rearward passage of lines accomplished its mission after passing through the stationary unit.
02	Yes/No	The enemy surprised neither the stationary nor the passing unit main body.
03	Yes/No	Higher headquarters directing the passage designated subsequent missions for both forces, when and under what conditions passage of command took place, start and finish times for the passage, contact points between the units involved, and common maneuver control measures and graphics.
04	Yes/No	Stationary unit provided guides and other assistance to the passing unit per the operation order directing the passage.
05	Yes/No	Higher headquarters issued instructions and control measures with adequate time for affected units to plan and prepare for linkup.
06	Yes/No	All personnel and equipment moved through the stationary unit by time specified in operation order.
07	Time	To complete the rearward passage of lines.
08	Time	Between planning contact and making contact.
09	Time	Between planning the transfer of responsibility for the area of operations and when it actually occurs.
10	Percent	Of support (such as fires, maintenance, and medical) the passing unit requests and the stationary unit provides.
11	Distance	Between planned and actual locations of contact point.
12	Distance	Between planned and actual locations where the passage of lines takes place.
13	Number	Of casualties from friendly fire or obstacles in either the stationary or the passing unit.

ART 1.2.9 CONDUCT A RELIEF IN PLACE

1-39. A *relief in place* is a tactical enabling operation in which, by the direction of higher authority, all or part of a unit is replaced in an area by the incoming unit. The responsibilities of the replaced elements for the mission and the assigned zone of operations are transferred to the incoming unit. The incoming unit continues the operation as ordered. The relieving unit usually assumes the same responsibilities and initially deploys in the same configuration as the outgoing unit. Relief in place is executed for a number of reasons including introducing a new unit into combat, changing a unit's mission, relieving a depleted unit in contact, retaining a unit, relieving the stress of prolonged operations in adverse conditions, resting a unit after long periods in a mission-oriented protective posture, decontaminating a unit, and avoiding excessive radiation exposure. A relief in place may be hasty or deliberate. This task may contain significant environmental considerations, especially if the relief in place occurs at a base camp or other similar place. (FM 3-90) (USACAC)

No.	Scale	Measure
01	Yes/No	The relieving unit assumed command of the area of operations after the previously designated trigger event occurred.
02	Yes/No	The relieved unit started its next mission per operation order.
03	Yes/No	Higher headquarters directing the relief designated subsequent missions for both forces, when and under what conditions passage of command took place, start and finish times for the relief, contact points between the units involved, and common maneuver control measures and graphics.
04	Yes/No	Enemy did not detect the relief.
05	Yes/No	Enemy attacked during the relief was defeated.
06	Time	To complete the plan for conducting the relief in place.
07	Time	Of starting the relief is by time specified in operation order.
08	Time	Of completing the relief is by time specified in operation order.
09	Percent	Of designated supplies and equipment left in position.
10	Percent	Of relieved unit's fire, weapons plans, and range cards passing to relieving unit.
11	Percent	Of lanes marked and guides in place prior to initiating the relief.
12	Percent	Of friendly casualties resulting from an enemy attack during the relief.
13	Number	Of casualties from friendly fire or obstacles in either the relieving or relieved unit.

ART 1.2.10 NAVIGATE FROM ONE POINT TO ANOTHER

1-40. Use navigational aids, such as maps, compasses, charts, stars, dead reckoning, and global positioning system receivers. ART 1.2.10 includes determining distance; direction; location; elevation and altitude; route; and data for navigational aids, orientation, and rate of movement. (FM 3-25.26) (USAIS)

No.	Scale	Measure
01	Yes/No	The unit was in the correct position at the correct time.
02	Time	To plan the movement of the individual, unit, vehicle, ship, or aircraft.
03	Time	Of force delays due to navigational error.
04	Percent	Of force no longer mission capable due to navigational error accidents.
05	Percent	Of force that arrives at the correct destination at the planned time.

ART 1.2.11 CONDUCT A SURVIVABILITY MOVE

1-41. Rapidly displace a unit, command post, or facility in response to enemy direct and indirect fires, the approach of an enemy unit, or as a proactive measure based on intelligence preparation of the battlefield and risk analysis. (FM 3-90) (USACAC)

No.	Scale	Measure
01	Yes/No	Unit prevented the enemy from engaging the main body.
02	Yes/No	Unit prevented decisive engagement by the enemy.
03	Time	To report survivability move to higher headquarters per unit standing operating procedures.
04	Time	For unit to displace from area projected to be crossed by approaching enemy unit.
05	Time	For unit to become operational again after survivability move.
06	Percent	Of unit casualties while displacing from the area projected to be traversed by the approaching enemy unit.
07	Percent	Of unit equipment damaged or immobilized while displacing from the area projected to be traversed by the approaching enemy unit.
08	Percent	Of unit personnel and equipment that displaces before the enemy unit arrives.

ART 1.2.12 CONDUCT SNIPER ACTIVE COUNTERMEASURES

1-42. Use active sniper countermeasures to detect and destroy the sniper before the sniper can fire, or engage and neutralize the sniper after the sniper fires. (FM 3-06.11) (USAIS)

No.	Scale	Measure
01	Yes/No	Unit maintained 360-degree security.
02	Yes/No	Unit did not set patterns.
03	Yes/No	Unit used observation posts and aerial observers.
04	Yes/No	Unit used magnified optics to scan for snipers.
05	Yes/No	Unit used small reconnaissance and security patrols.
06	Yes/No	Unit identified sniper location and returned fire.
07	Yes/No	Unit attacked by maneuvering on enemy location and killing the enemy.
08	Yes/No	Unit reorganized and evaluated casualties.

ART 1.2.13 CONDUCT SNIPER PASSIVE COUNTERMEASURES

1-43. Use passive sniper countermeasures to prevent the sniper from acquiring a clear target or prevent sniper fires from inflicting casualties. (FM 3-06.11) (USAIS)

No.	Scale	Measure
01	Yes/No	Unit planned and rehearsed actions on sniper control.
02	Yes/No	Unit maintained 360-degree security.
03	Yes/No	Unit limited exposure of personnel and equipment.
04	Yes/No	Soldiers stuck to the shadows, used cover, and used concealment.
05	Yes/No	Soldiers removed rank insignia and did not salute in the field.
06	Yes/No	Leaders blended into element.
07	Yes/No	Soldiers wore protective armor.
08	Yes/No	Soldiers used armored vehicles.
09	Yes/No	Unit erected screens and shields for concealment.
10	Yes/No	Unit denied enemy the ability to overwatch terrain.
11	Yes/No	Unit used smoke hazes, smoke screens, and/or nonlethal weapons to distract, disorient, degrade, and/or obscure the sniper's field of view.
12	Yes/No	Unit used smoke hazes, smoke screens, and/or nonlethal weapons to limit the effectiveness of the sniper's fire.

SECTION III – ART 1.3: CONDUCT TACTICAL TROOP MOVEMENTS

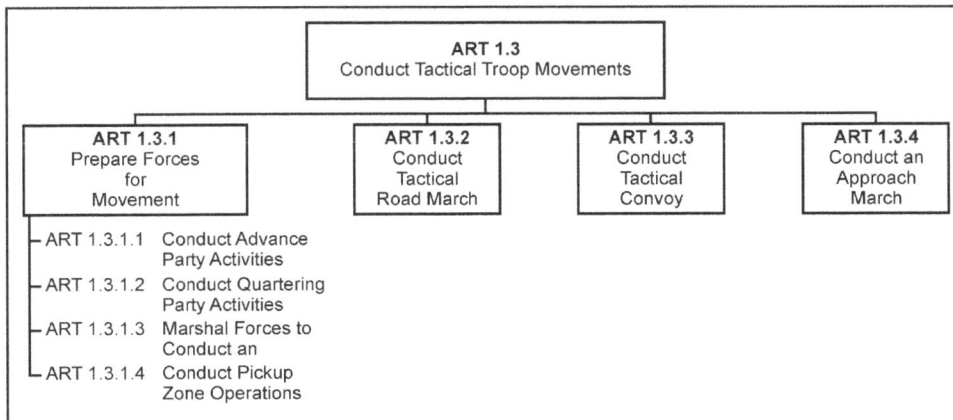

```
                          ┌─────────────────────────────┐
                          │          ART 1.3            │
                          │ Conduct Tactical Troop Movements │
                          └─────────────────────────────┘
        ┌───────────────┬──────────────┴──────────────┬───────────────┐
 ┌──────────────┐ ┌──────────────┐ ┌──────────────┐ ┌──────────────┐
 │   ART 1.3.1  │ │   ART 1.3.2  │ │   ART 1.3.3  │ │   ART 1.3.4  │
 │ Prepare Forces│ │   Conduct   │ │   Conduct   │ │  Conduct an  │
 │     for      │ │  Tactical   │ │  Tactical   │ │   Approach   │
 │   Movement   │ │ Road March  │ │   Convoy    │ │    March     │
 └──────────────┘ └──────────────┘ └──────────────┘ └──────────────┘
```

- ART 1.3.1.1 Conduct Advance Party Activities
- ART 1.3.1.2 Conduct Quartering Party Activities
- ART 1.3.1.3 Marshal Forces to Conduct an
- ART 1.3.1.4 Conduct Pickup Zone Operations

1-44. Relocate or move by any means or mode of transportation preparatory to deploying into combat formations to support tactical commander and joint force commander plans. Positioning and repositioning must support the commander's intent and concept of operation. ART 1.3 includes generating and dispersing tactical forces. It also includes moving units by military, host-nation, or contracted vehicles. (FM 3-90) (USACAC)

No.	Scale	Measure
01	Yes/No	Unit followed the prescribed route at the prescribed speed without deviating unless required otherwise by enemy action or on orders from higher headquarters.
02	Yes/No	Unit crossed and cleared start point and release point at designated time.
03	Yes/No	Enemy did not surprise the unit's main body because of a failure to conduct security operations.
04	Percent	Of casualties sustained during the movement.
05	kph	Rate of movement.

ART 1.3.1 PREPARE FORCES FOR MOVEMENT

1-45. Assemble, inspect, and load personnel, equipment, and supplies to prepare for a tactical movement. (FM 3-90) (USACAC)

No.	Scale	Measure
01	Yes/No	Unit prepared to move at the appointed time and place.
02	Time	To load forces and equipment.
03	Time	To prepare movement orders.
04	Time	To deploy transportation and security forces.
05	Percent	Of forces and equipment loaded within established time requirements.

ART 1.3.1.1 CONDUCT ADVANCE PARTY ACTIVITIES

1-46. Send a detachment ahead of the main body to establish conditions for the main body's arrival. Conditions include administrative and logistic actions. (FMI 3-35) (CASCOM)

No.	Scale	Measure
01	Yes/No	The advance party established necessary conditions for the unit main body to conduct and complete the reception, staging, onward movement and integration by time specified.
02	Yes/No	Advance party arranged to receive the main body.
03	Yes/No	Advance party assisted point of debarkation with discharge operations.
04	Yes/No	Advance party consisted of battery teams, fuel handlers, drivers, and property book and supply personnel.
05	Yes/No	Advance party deployed sufficiently in advance of the main body to accomplish its assigned responsibilities.
06	Yes/No	Unit requested sufficient external transportation support to accomplish assigned responsibilities.

ART 1.3.1.2 CONDUCT QUARTERING PARTY ACTIVITIES

1-47. Secure, reconnoiter, and organize an area for the main body's arrival and occupation. (FMI 3-35) (CASCOM)

No.	Scale	Measure
01	Yes/No	Quartering party deployed sufficiently in advance of the main body to accomplish its assigned responsibilities.
02	Yes/No	Quartering party guided unit main body into position from the release point to precise locations within the assembly area.
03	Yes/No	Quartering party secured the designated assembly area.
04	Yes/No	Quartering party conducted an area reconnaissance of the designated assembly area per environmental considerations.
05	Distance	Of assembly area (and positions in it) changed from tentative locations selected by unit commander based on a map reconnaissance.

ART 1.3.1.3 MARSHAL FORCES TO CONDUCT AN AIRBORNE ASSAULT

1-48. Marshaling airborne forces involves conducting the planning, rehearsals, and briefbacks addressed in ART 5.0. It involves assembling and preparing paratroopers, equipment, and supplies for the jump. It includes airborne-specific briefings; prejump training; the actual movement of paratroopers, equipment, and supplies to departure airfields; and loading them into aircraft. (FM 90-26) (USAIS)

No.	Scale	Measure
01	Yes/No	Unit completed all preparations—such as assembling, organizing, marking, and rigging unit equipment, rations, ammunition, water, and other supplies—by time specified in operation order. Preparations also include, but are not limited to, ensuring the cross loading of personnel and key weapon systems, conducting prejump refresher training and mission rehearsals, and moving to the departure airfield by time specified in operation order.
02	Yes/No	Unit commander issued warning order and operation order.
03	Yes/No	Unit met station time.
04	Yes/No	Unit met load time.
05	Time	From receiving warning order to completing preparations for airborne operation.

ART 1.3.1.4 CONDUCT PICKUP ZONE OPERATIONS

1-49. Pickup zone operations involve assembling and preparing Soldiers, equipment, and supplies for an air assault. ART 1.3.1.4 includes conducting air assault-specific briefings and training; moving Soldiers, equipment, and supplies to pickup zones; and loading into rotary aircraft. (FM 90-4) (USAIS)

No.	Scale	Measure
01	Yes/No	Unit commander issued warning order and operation order.
02	Yes/No	Unit adjusted load plan and ground tactical plan to account for less than scheduled amount of aircraft.
03	Yes/No	Unit met load time.
04	Yes/No	Unit maintained local security during loading.
05	Yes/No	Unit completed all preparations—such as assembling, organizing, marking, and rigging unit equipment, rations, ammunition, water, and other supplies—by time specified in operation order. Preparations also include, but are not limited to, ensuring the cross loading of personnel and key weapon systems, conducting air assault refresher training and mission rehearsals, conducting an air mission brief, and moving to the departure airfield by time specified in operation order.
06	Yes/No	Unit released available attack and reconnaissance assets from pickup zone security to perform air route reconnaissance and to establish mobile flank screens for air movement to landing zones.
07	Time	From receiving warning order to completing preparations for air assault operation.
08	Percent	Of changes in numbers and types of rotary-wing aircraft.

ART 1.3.2 CONDUCT A TACTICAL ROAD MARCH

1-50. A tactical road march is a rapid movement used to relocate units within an area of operations in order to prepare for combat operations. Commanders arrange troops and vehicles to expedite their movement and conserve time, energy, and unit integrity. They anticipate no interference except by enemy forces, or sympathizers. Units conducting tactical road marches employ three tactical march techniques: open column, close column, and infiltration. (FM 3-90) (USACAC)

No.	Scale	Measure
01	Yes/No	Unit crossed and cleared start point and release point at designated times.
02	Yes/No	Unit followed the prescribed route without deviation unless required otherwise by enemy action or on orders from higher headquarters.
03	Yes/No	Enemy did not surprise the unit main body through a failure to conduct appropriate security operations.
04	Yes/No	Unit employed appropriate tactical road march technique (open column, close column, or infiltration).
05	Yes/No	Unit task-organized properly for tactical road march.
06	Time	To initiate movement.
07	Time	To complete movement.
08	Percent	Of force completing the movement.
09	Percent	Of unit casualties.

ART 1.3.3 CONDUCT A TACTICAL CONVOY

1-51. Conduct tactical convoys by employing one or a combination of three types of column formations: open, close, and infiltration. Tactical convoys are combat operations in which forces and materiel are moved overland from one location on the battlefield to another while maintaining the ability to aggressively respond to enemy attempts to impede, disrupt, or destroy elements of the convoy. (FM 55-30) (CASCOM)

No.	Scale	Measure
01	Yes/No	Unit conducted precombat checks and inspections to include unit test fire of all weapons.
02	Yes/No	Unit was equipped with primary and alternate communications.
03	Yes/No	Unit had frequencies for close air support, fires, and medical evacuation and duress frequencies.

No.	Scale	Measure
04	Yes/No	Unit crossed start point by time specified in operation order.
05	Yes/No	Unit reported crossing start point to higher headquarters.
06	Yes/No	Unit had constant and uninterrupted internal and external communications throughout the convoy.
07	Yes/No	Unit maintained 360-degree security and situational awareness during the convoy.
08	Yes/No	Unit conducted rehearsals to react to enemy attack per operation order, battle drill, and standing operating procedures.
09	Yes/No	Unit rehearsed actions to deal with noncombatants who attempt to pilfer for personal reasons; unit was able to respond passively to these disruptions.
10	Yes/No	Unit identified checkpoints en route and reported crossing checkpoints.
11	Yes/No	Unit maintained visual contact and proper intervals between vehicles.
12	Yes/No	Forward security element provided the convoy with sufficient reaction time and maneuver space to avoid or react to enemy contact.
13	Yes/No	Unit crossed the release point by time specified in operation order.
14	Yes/No	Unit reported arrival at release point to higher headquarters.
15	Yes/No	Unit rehearsed special teams, actions on enemy contact, and improvised explosive devices to include medical evacuations and vehicle recovery.
16	Time	To designate and position security teams throughout the convoy.
17	Time	To designate quick reaction force, special teams for aid and litter, combat lifesaver, and recovery and landing zones.
18	Time	To provide situational reports to higher headquarters.
19	Number	Of types of enemy forces active within area of influence (convoy route).
20	Number	Of enemy actions in the last 30 days (or other selected period).

ART 1.3.4 CONDUCT AN APPROACH MARCH

1-52. Conduct an advance of a combat unit when direct contact with the enemy is intended. An approach march emphasizes speed over tactical deployment. It is used when the enemy's approximate location is known and allows the attacking force to move with greater speed and less physical security or dispersion. An approach march ends in an attack position, assembly area, or assault position; or it transitions to an attack. (FM 3-90) (USACAC)

No.	Scale	Measure
01	Yes/No	Using the approach march allowed the force to move quickly to the area where it expected to make contact with the enemy and transition to an appropriate combat formation.
02	Yes/No	Unit task-organized properly for approach march.
03	Yes/No	Unit used established control measures and control graphics.
04	Time	To initiate the approach march.
05	Time	To complete the approach march.
06	Time	Between planned and actual unit arrival at checkpoints.
07	Percent	Of force completing the approach march.
08	Percent	Of force maintaining correct interval between units.
09	Percent	Of force using correct movement techniques (traveling, traveling overwatch, and bounding overwatch).
10	kph	Rate of movement.

SECTION IV – ART 1.4: CONDUCT DIRECT FIRES

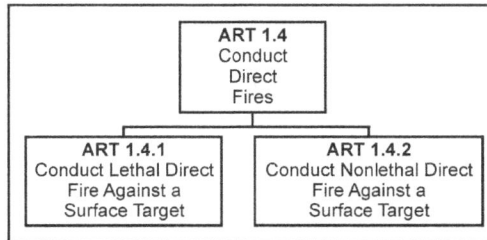

```
                    ┌─────────────────┐
                    │     ART 1.4      │
                    │     Conduct      │
                    │     Direct       │
                    │     Fires        │
                    └─────────────────┘
                 ┌───────────┴───────────┐
    ┌────────────────────────┐  ┌────────────────────────┐
    │       ART 1.4.1         │  │       ART 1.4.2         │
    │ Conduct Lethal Direct   │  │ Conduct Nonlethal Direct│
    │    Fire Against a        │  │    Fire Against a        │
    │    Surface Target        │  │    Surface Target        │
    └────────────────────────┘  └────────────────────────┘
```

1-53. Conduct lethal and nonlethal direct fires. Examples of direct-fire systems include small arms, tanks, antitank weapons, automatic weapons, directed energy, optical, acoustic, and blunt trauma weapons. ART 1.4 includes attack helicopter fires and direct fire tied directly to battlefield movement. (FM 3-90) (USACAC)

Note: ART 6.1.2 (Destroy Aerial Platforms) includes the attack of aerial targets.

ART 1.7.1 (Site Obstacles) addresses the elements of direct fire planning, such as the integration of indirect fires, obstacles and terrain, and air and ground assets with control measures designed to mass fires.

No.	Scale	Measure
01	Yes/No	Unit direct fires contributed to accomplishing unit mission.
02	Yes/No	Conduct of direct fires done was done per established rules of engagement.
03	Yes/No	Unit used correct weapon to engage target.
04	Time	To get complete attack on direct fire target after detecting and identifying target.
05	Time	To suppress targets.
06	Percent	Of probability of suppressing a target.
07	Percent	Of probability of a hit.
08	Percent	Of probability of a kill given a hit.
09	Percent	Of missions flown and fired to achieve desired target damage.
10	Percent	Of available direct fire weapon systems engaging direct fire targets.
11	Percent	Of direct fire targets not engaged.
12	Percent	Of enemy performance degraded due to direct fire attack.
13	Percent	Of direct fire attacks that result in collateral damage.
14	Percent	Of direct fire attacks that result in friendly or neutral casualties.
15	Number	Of direct fire attacks that result in collateral damage.
16	Number	Of direct fire attacks that result in friendly or neutral casualties.

ART 1.4.1 CONDUCT LETHAL DIRECT FIRE AGAINST A SURFACE TARGET

1-54. Engage enemy equipment and materiel, personnel, fortifications, and facilities with direct fire designed to destroy the target. These direct fires may be from fixed- or rotary-wing systems. (FM 3-90) (USACAC)

No.	Scale	Measure
01	Yes/No	Direct fires contributed to accomplishing unit mission.
02	Yes/No	Direct fire attack was conducted per established rules of engagement.
03	Yes/No	Unit used correct weapon to engage target.
04	Time	To get complete attack on direct fire target after detecting and identifying target.
05	Time	To suppress targets.
06	Percent	Of probability of suppressing a target.
07	Percent	Of probability of a hit.
08	Percent	Of probability of a kill given a hit.
09	Percent	Of missions flown and fired to achieve desired target damage.
10	Percent	Of available direct fire weapon systems engaging direct fire targets.
11	Percent	Of direct fire targets not engaged.
12	Percent	Of enemy performance degraded due to lethal direct fire attack.
13	Percent	Of lethal direct fire attacks that result in collateral damage.
14	Percent	Of lethal direct fire attacks that result in friendly or neutral casualties.
15	Number	Of lethal direct fire attacks that result in collateral damage.
16	Number	Of lethal direct fire attacks that result in friendly or neutral casualties.

ART 1.4.2 CONDUCT NONLETHAL DIRECT FIRE AGAINST A SURFACE TARGET

1-55. Employ nonlethal direct fires to incapacitate threat personnel and materiel without causing permanent injury or destruction. ART 1.4.2 includes using nonlethal weapons–such as directed energy, blunt trauma, riot control agents, vehicle and vessel arresting devices, and water cannons. (FM 3-19.15) (USAMPS)

No.	Scale	Measure
01	Yes/No	Direct fires contributed to accomplishing the unit mission.
02	Yes/No	Conduct of nonlethal direct fires against surface target was done per rules of engagement, to include receipt of the approval of weapons released from a competent authority.
03	Time	To develop nonlethal direct fire options after receiving warning order.
04	Time	To complete nonlethal direct fire attack on target (after initiation).
05	Percent	Of all targets evaluated as candidates for nonlethal direct fire attack.
06	Percent	Of nonlethal direct fire attacks on selected targets that achieve desired effect.
07	Percent	Of nonlethal direct fire attacks without lethal results.
08	Percent	Of nonlethal direct fire attacks that require lethal fires to achieve desired operational effects.
09	Percent	Of nonlethal direct fire attacks that result in result in collateral damage.
10	Percent	Of threat actions that are denied, stopped, moved, diverted, suppressed, and/or disabled due to nonlethal direct fire attack.
11	Percent	Of nonlethal direct fire attacks that result in friendly or neutral casualties.
12	Number	Of nonlethal direct fire attacks that result in friendly or neutral casualties.
13	Number	Of nonlethal direct fire attacks that result in collateral damage.

SECTION V – ART 1.5: OCCUPY AN AREA

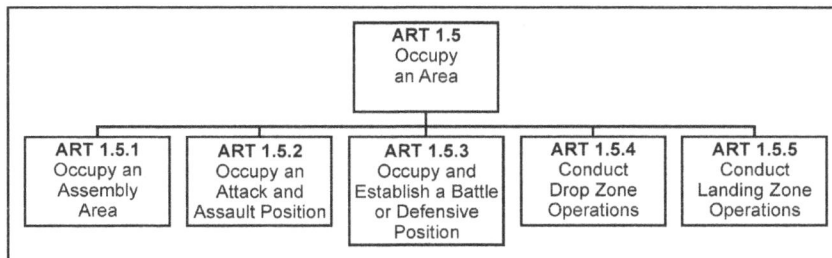

```
                          ART 1.5
                          Occupy
                          an Area

  ART 1.5.1      ART 1.5.2      ART 1.5.3      ART 1.5.4      ART 1.5.5
  Occupy an      Occupy an      Occupy and     Conduct        Conduct
  Assembly       Attack and     Establish a Battle   Drop Zone      Landing Zone
  Area           Assault Position   or Defensive    Operations     Operations
                                 Position
```

1-56. Move forces into and secure an area from which to conduct future operations. This task includes occupying assembly areas; occupying attack or assault positions; and establishing and occupying defensive positions, including the five types of battle positions: primary, alternate, supplementary, subsequent, and strong point. (FM 3-90) (USACAC)

No.	Scale	Measure
01	Yes/No	Unit controlled the area so the enemy could not use the area.
02	Yes/No	Personnel and essential equipment were in assigned positions by time specified in operation order.

ART 1.5.1 OCCUPY AN ASSEMBLY AREA

1-57. Move forces into and occupy an assembly area in which to assemble and prepare for further action. Actions include resupplying and organizing forces for future operations. (FM 3-90) (USACAC)

No.	Scale	Measure
01	Yes/No	Unit moved into and occupied an assembly area in which it assembles and prepares for further action by time specified in operation order.
02	Yes/No	Unit forced enemy reconnaissance elements to withdraw without allowing penetration of the assembly area perimeter.
03	Yes/No	The enemy did not surprise the unit main body through a failure to conduct security operations.
04	Yes/No	Unit dispersed its forces appropriately; used cover and concealment; and designated entrances, exits, and internal routes per the factors of mission, enemy, terrain and weather, troops and support available, time available, civil considerations to include drainage and soil conditions per environmental considerations.
05	Yes/No	All personnel and essential equipment moved into assigned initial defensive positions by time specified in operation order.
06	Yes/No	Unit completed preparations for next operation per commander's intent by time specified in operation order.

ART 1.5.2 OCCUPY AN ATTACK AND ASSAULT POSITION

1-58. As part of an offensive operation, move tactical forces into and through these positions to prepare for further action or support actions of another force. Activities include making last-minute coordination and tactical adjustments, preparing specialized equipment for immediate use, and protecting the occupying force until supporting fire is lifted or shifted. This task includes the use of attack-by-fire and support-by-fire positions and holding areas by attack helicopters. (FM 3-90) (USACAC)

No.	Scale	Measure
01	Yes/No	Unit occupied attack and assault positions only as necessary to ensure the attack's success.
02	Yes/No	Unit cleared these positions of enemy forces.
03	Yes/No	The enemy did not surprise the unit main body.
04	Yes/No	All personnel and essential equipment moved into assigned positions by time specified in operation order.
05	Yes/No	Unit completed attack and assault preparations per commander's intent and the factors of mission, enemy, terrain and weather, troops and support available, time available, civil considerations.

ART 1.5.3 OCCUPY AND ESTABLISH A BATTLE OR DEFENSIVE POSITION

1-59. As part of a defensive operation, move tactical forces into positions to prepare for further action. A *battle position* is a defensive location oriented on a likely enemy avenue of approach. Five kinds of battle positions exist: primary, alternate, supplementary, subsequent, and strongpoint. The positions may be located on any type of land and terrain, such as urban, natural, mountainous, piedmont, steppe, delta, desert, jungle, and arctic. (See CJCSM 3500.04E condition C 1.0 for factors that describe the physical environment.) (FM 3-90) (USACAC)

Note: ART 1.7.1 (Site Obstacles) addresses defensive planning.

No.	Scale	Measure
01	Yes/No	Unit can conduct a coherent defense from its positions.
02	Yes/No	Unit cleared enemy forces from the defended area.
03	Yes/No	The enemy did not surprise the unit main body.
04	Yes/No	All personnel and essential equipment moved into assigned positions by time specified in operation order.
05	Yes/No	Unit was prepared to defend by the time specified in operation order.
06	Percent	Of acceptable friendly losses.
07	Number	Of casualties from friendly fire.

ART 1.5.4 CONDUCT DROP ZONE OPERATIONS

1-60. ART 1.5.4 begins when paratroopers and equipment exit the aircraft by parachute or airland. ART 1.5.4 ends when all elements of the relevant airborne echelon arrive in the objective area. (FM 90-26) (USAIS)

No.	Scale	Measure
01	Yes/No	Unit assembled according to its landing plan (on the objective, on the drop zone, or in an assembly area adjacent to the drop zone) and began to execute the ground tactical plan by time specified in operation order.
02	Yes/No	Enemy forces were unable to engage forces landing on the drop zone.
03	Yes/No	Security positions were positioned around drop zone until completion of vertical envelopment.
04	Yes/No	Unit cleared drop zone of equipment and debris for use by follow-on forces or future airland operations.
05	Time	To conduct map or physical reconnaissance of site to ensure that the drop zone supports operational requirements, such as acceptable degree of slope, surface conditions, appropriate size, and free of obstacles.

No.	Scale	Measure
06	Time	To clear or mark obstacles such as stumps, fences, and barbed wire located on the drop zone.
07	Time	To ensure drop zone approach and exit paths are free of obstructions.
08	Time	To mark the drop zone.
09	Time	For pathfinder elements to establish communications with follow-on aircraft.
10	Time	For pathfinder elements to confirm or determine drop heading with aircrew.
11	Percent	Of dropped and airlanded aircraft loads under control of a ground station located on or near the drop zone.
12	Percent	Of airborne unit personnel and cargo drops landing in the drop zone.
13	Number	Of personnel landing in the drop zone.
14	Number	Of tons and types of cargo landing in the drop zone or that the airborne unit can recover.
15	Number	Of casualties from accidents caused by conditions on the drop zone, such as wind speed, obstacles, and surface conditions.

ART 1.5.5 CONDUCT LANDING ZONE OPERATIONS

1-61. ART 1.5.5 begins when Soldiers and equipment exit the helicopters. It does not include air traffic control or efforts to improve the landing zone. ART 1.5.5 ends when all elements of the relevant air assault echelon arrive in the objective area. (FM 90-4) (USAIS)

No.	Scale	Measure
01	Yes/No	Unit assembled according to its landing plan (on the objective, on the landing zone, or in an assembly area adjacent the landing zone) and began to execute the ground tactical plan by time specified in operation order.
02	Yes/No	Security positioned around landing zone occupied zone until completion of the vertical envelopment process.
03	Yes/No	Unit cleared landing zone of equipment and debris for use by follow-on forces or future airland operations.
04	Time	To conduct map or physical reconnaissance of site to ensure that landing zone supports operational requirements, such as acceptable degree of slope, surface conditions, appropriate size to accommodate the helicopters delivering the force, and free of obstacles.
05	Time	To clear or mark obstacles such as stumps and fences located on the landing zone.
06	Time	To ensure landing zone approach and exit paths are free of obstructions.
07	Time	To mark the landing zone.
08	Percent	Of air assault unit personnel and cargo landing in the landing zone.
09	Number	Of personnel landing on the landing zone.
10	Number	Of tons and types of cargo landing on the landing zone.
11	Number	Of casualties from accidents caused by landing zone conditions, such as foreign object damage and hidden obstacles.

SECTION VI – ART 1.6: CONDUCT MOBILITY OPERATIONS

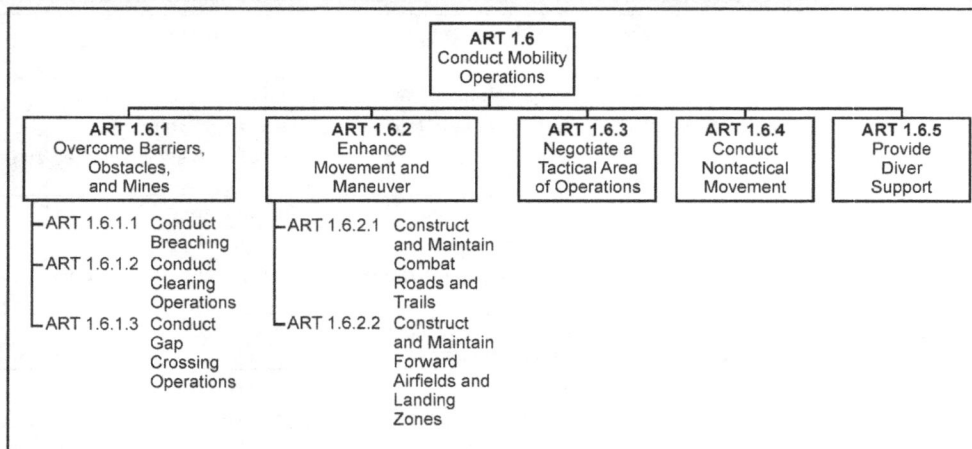

```
                            ART 1.6
                       Conduct Mobility
                          Operations
   ┌────────────┬────────────┬────────────┬────────────┐
ART 1.6.1     ART 1.6.2     ART 1.6.3     ART 1.6.4     ART 1.6.5
Overcome Barriers,  Enhance    Negotiate a    Conduct      Provide
Obstacles,      Movement and  Tactical Area  Nontactical    Diver
and Mines        Maneuver   of Operations   Movement     Support
```

- ART 1.6.1.1 Conduct Breaching
- ART 1.6.1.2 Conduct Clearing Operations
- ART 1.6.1.3 Conduct Gap Crossing Operations

- ART 1.6.2.1 Construct and Maintain Combat Roads and Trails
- ART 1.6.2.2 Construct and Maintain Forward Airfields and Landing Zones

1-62. Maintain freedom of movement for personnel and equipment in an area of operations without delays due to terrain or barriers, obstacles, and mines. Mobility operations preserve the freedom of maneuver of friendly forces. Mobility tasks include breaching, clearing, or crossing obstacles; increasing battlefield circulation; improving or building roads; providing bridge and raft support; and identifying routes around contaminated areas. Countermobility denies mobility to enemy forces. It limits the maneuver of enemy forces and enhances the effectiveness of fires. (FM 3-90.12) (USAES)

Note: The term "breaching system" used in this section includes both manual and mechanical means.

No.	Scale	Measure
01	Yes/No	Terrain, barriers, and obstacles were overcome within the period the operation order specifies.
02	Yes/No	Commander coordinated with higher, adjacent, supported, and supporting units to maintain freedom of movement in the area of operations.
03	Time	That terrain, barriers, obstacles, and mines delay movement of friendly forces.
04	Time	To conduct route, zone, and area reconnaissance to determine terrain trafficability and the location and boundaries of barriers, obstacles, and minefields.
05	Time	After discovery for staff to disseminate terrain trafficability and barrier, obstacles, and mine data to higher headquarters, laterally, and subordinate units.
06	Time	To conduct successful execution of breach fundamentals—suppress, obscure, secure, reduce, and assault—at the obstacle.
07	Time	To complete mobility activities that improve the unit's capability to cross the terrain, such as applying a rock layer to a combat road and cutting down trees to make a trail.
08	Time	To reduce lanes through obstacles.
09	Time	To complete minefield reduction.
10	Time	To move breaching equipment to breach site.
11	Percent	Of obstacles in the area of operations that have been breached.
12	Percent	Of breaching systems that are mission capable.
13	Percent	Of completed engineer efforts designed to enhance the unit's capability to cross terrain.

No.	Scale	Measure
14	Number	Of breaching systems that are mission capable.
15	Number	Of friendly and neutral casualties during mobility enhancing activities.

ART 1.6.1 OVERCOME BARRIERS, OBSTACLES, AND MINES

1-63. Enable a force to maintain its mobility by reducing, bypassing, or clearing obstacles. An *obstacle* is any obstruction designed or employed to disrupt, fix, turn, or block the movement of an opposing force, and to impose additional losses in personnel, time, and equipment on the opposing force. Obstacles can be natural or man-made, or a combination of both. Naturally existing obstacles can include rivers, mountains, barrier reefs, and cities. Man-made or reinforcing obstacles can include minefields and antitank ditches. A complex obstacle is a combination of different types of individual obstacles that requires more than one reduction technique (explosive, mechanical, or manual) to create a lane through the obstacle. A reinforcing obstacle is an obstacle that is specifically constructed, emplaced, or detonated through military effort. (FM 3-90.12) (USAES)

No.	Scale	Measure
01	Yes/No	Unit overcame obstacles and barriers by time specified in operation order.
02	Time	That enemy-emplaced obstacles delay friendly force movement.
03	Time	For staff to disseminate barrier, obstacle, and mine data to subordinate units, higher headquarters, and laterally after discovery.
04	Time	To conduct reconnaissance of obstacle focused on answering obstacle intelligence information requirements—obstacle location, length, width, and depth; obstacle composition (such as wire and mines by type); soil conditions; locations of lanes and bypasses; and the location of enemy direct fire systems.
05	Time	To conduct successful execution of breach fundamentals—suppress, obscure, secure, reduce, and assault—at the obstacle.
06	Time	To reduce lane through obstacles.
07	Time	To complete mine clearing.
08	Time	To move breaching equipment to breach site.
09	Time	To reduce underwater obstacles at crossing sites.
10	Percent	Of obstacles in the area of operations that have been breached.
11	Percent	Of breaching systems that are mission capable.
12	Number	Of breaching systems that are mission capable.
13	Number	Of friendly and neutral casualties caused by detonation of mines or explosives.

ART 1.6.1.1 CONDUCT BREACHING OPERATIONS

1-64. Conduct a combined arms operation to project combat power to the far side of an obstacle. Breaching tenets include intelligence, synchronization, mass, breach fundamentals (suppress, obscure, secure, reduce, and assault), and breach organization (support, assault, and breach forces). ART 1.6.1.1 includes the reduction of minefields and other obstacles. Reduction is the creation of lanes through or over an obstacle to allow an attacking force to pass. The number and width of lanes created varies with the enemy situation, the assault force's size and composition, and the concept of operations. The lanes must allow the assault force to rapidly pass through the obstacle. The breach force will reduce, proof (if required), mark, and report lane locations and the land-marking method to higher headquarters. Follow-on units will further reduce or clear the obstacle when required. (FM 3-90.12) (USAES)

No.	Scale	Measure
01	Yes/No	Unit completed breaching operation by time specified in the operation order.
02	Time	For staff to disseminate barrier, obstacle, and mine data to subordinate units, higher headquarters, and laterally after discovery.

No.	Scale	Measure
03	Time	That enemy-emplaced obstacles delay friendly force movement.
04	Time	To conduct reconnaissance of barriers, obstacles, and minefields.
05	Time	To conduct successful execution of breach fundamentals—suppress, obscure, secure, reduce, and assault—at the obstacle.
06	Time	To reduce lane through obstacles (one lane per assault company, two lanes per task force).
07	Time	To move breaching equipment to breach site.
08	Percent	Of obstacles in the area of operations that have been breached.
09	Percent	Of breaching systems that are mission capable.
10	Number	Of lanes opened by the breaching operation.
11	Number	Of breaching systems that are mission capable.
12	Number	Of friendly and neutral casualties caused by detonation of explosives.

ART 1.6.1.2 CONDUCT CLEARING OPERATIONS

1-65. Clearing operations (area or route clearance) are conducted to enable the use of a designated area or route. Clearing is the total elimination or neutralization of an obstacle (to include explosives hazard) or portions of an obstacle. Clearing operations are typically not conducted under fire and may be performed after a breaching operation where an obstacle is a hazard or hinders friendly movement or occupation of a location. ART 1.6.1.2.1 (Conduct Area Clearance) focuses on obstacle clearance of a designated area and is typically not a combined arms operation. ART 1.6.1.2.2 (Conduct Route Clearance)focuses on obstacle clearance along a specific route, typically conducted as a combined arms operation, and may be performed in situations where enemy contact is likely. (FM 3-90.12) (USAES)

No.	Scale	Measure
01	Yes/No	Unit completed obstacle clearance mission by time specified in operation order.
02	Yes/No	Unit conducted emergency de-mining and unexploded explosive ordnance removal.
03	Yes/No	Unit conducted mapping and survey exercises of mined areas.
04	Yes/No	Unit marked minefields.
05	Yes/No	Unit identified and coordinated emergency de-mining and unexploded explosive ordnance removal requirements.
06	Yes/No	Unit established priorities and conducted de-mining operations.
07	Yes/No	Unit initiated large-scale de-mining and unexploded explosive ordnance removal operations.
08	Time	For staff to disseminate obstacle data to subordinate units, higher headquarters, and laterally after discovery.
09	Time	To conduct reconnaissance of obstacle focused on answering obstacle intelligence information requirements—obstacle location, length, width, and depth; obstacle composition (such as wire, mines by type); soil conditions; locations of lanes and bypasses; and the location of enemy direct fire systems.
10	Time	To plan how to clear the obstacle.
11	Time	To clear the obstacles.
12	Time	To move equipment to the area where the clearance mission takes place.
13	Percent	Of obstacle that has been removed or neutralized.
14	Percent	Of systems committed to the clearance mission that are mission capable.
15	Number	Of lanes opened by the reducing operation.
16	Number	Of systems that are mission capable.
17	Number	Of friendly and neutral casualties during the clearance mission.

ART 1.6.1.2.1 Conduct Area Clearance

1-66. Area clearance is the total elimination or neutralization of an obstacle or portions of an obstacle in a designated area. (FM 3-90.12) (USAES)

No.	Scale	Measure
01	Yes/No	Unit completed area clearance mission by time specified in operation order.
02	Time	That obstacles delay friendly force movement.
03	Time	For staff to disseminate obstacle data to subordinate units, to higher headquarters, and laterally after discovery.
04	Time	To conduct area reconnaissance.
05	Time	To plan how to clear the area.
06	Time	To clear the area.
07	Time	To move equipment to the area where the clearance mission takes place.
08	Percent	Of area that has been cleared.
09	Percent	Of systems committed to the clearance mission that are mission capable.
10	Number	Of area clearance systems that are mission capable.
11	Number	Of friendly and neutral casualties during the area clearance mission.

ART 1.6.1.2.2 Conduct Route Clearance

1-67. A route clearance is a combined arms operation conducted to remove mines and other obstacles along preexisting roads and trails. (FM 3-90.12) (USAES)

No.	Scale	Measure
01	Yes/No	Unit accomplished route clearance by time specified in operation order.
02	Yes/No	Unit dismantled roadblocks and established checkpoints.
03	Time	That obstacles along the route delay the friendly force movement.
04	Time	For staff to disseminate obstacle data to subordinate units, higher headquarters, and laterally after discovery.
05	Time	To conduct route reconnaissance.
06	Time	To plan how to clear the route.
07	Time	To clear the route.
08	Time	To move equipment from its current location to the route where the clearance mission takes place.
09	Time	To establish security along portion of the route being cleared.
10	Percent	Of route cleared by time specified in operation order.
11	Percent	Of systems committed to route clearance that are mission capable.
12	Percent	Of increase in transportation and maneuver efficiency due to the completion of route clearance mission.
13	Number	Of obstacles along the route that have been cleared.
14	Number	Of breaching systems that are mission capable.
15	Number	Of friendly and neutral casualties during the route clearance mission.

ART 1.6.1.3 CONDUCT GAP CROSSING OPERATIONS

1-68. A gap crossing operation is a combined arms operation to project combat power across a linear obstacle. The obstacle is linear in that it creates a line crossing all or a significant portion of the area of operations. The obstacle can be wet gap (water obstacle) or dry gap that is too wide to overcome by self-bridging. The nature of the obstacle differentiates a gap crossing from a breaching operation. A wet gap crossing (river crossing) is also unique because the water obstacle is significantly large enough to prevent normal ground maneuver. A gap crossing generally requires special planning and support. The factors of

METT-TC—mission, enemy, terrain and weather, troops and support available, time available, civil considerations—dictate the type of crossing (hasty, deliberate, or retrograde). Gap crossing generally includes preparing access and egress routes, completing a hydrographic survey (underwater obstacle detection or reduction), employing crossing means (bridging and rafts), and operating an engineer regulating point if required. This task is measured against a river crossing, the most difficult standard of gap crossing operation. Crossing fundamentals include surprise, extensive preparation, a flexible plan, traffic control, organization, and speed. Gap crossings may be conducted in support of combat maneuver or in support of lines of communication. (FM 3-90.12) (USAES)

Note: The engineer bridge, raft, and assault boat systems percentages in this task apply to the individual ribbon bridge bays and rafts, and to individual assault boats and not to a ribbon or assault bridge set as a whole.

No.	Scale	Measure
01	Yes/No	Unit accomplished gap crossing by time specified in operation order.
02	Time	That the gap or obstacle delays friendly force movement.
03	Time	To conduct area reconnaissance of the terrain surrounding the gap.
04	Time	To plan the gap crossing.
05	Time	For staff to disseminate data concerning the gap to subordinate units, higher headquarters, and laterally after determination.
06	Time	For underwater reconnaissance to be performed by dive team.
07	Time	To move engineer bridging equipment to the crossing site.
08	Time	To establish conditions necessary for success, such as suppressing enemy systems overwatching the river, breaching minefields and other obstacles barring access to the riverbanks, and preparing access and egress routes.
09	Time	To emplace and construct crossing assets.
10	Time	To complete gap crossing.
11	Percent	Of crossing unit that has moved to the far shore of the gap.
12	Percent	Of engineer bridge, raft, and assault boat systems that are mission capable.
13	Percent	Of crossing area seeded with obstacles if conducting a retrograde crossing.
14	Number	Of bridges and crossing sites established.
15	Number	Of engineer bridge, raft, and assault boat systems that are mission capable.
16	Number	Of friendly casualties due to accidents and enemy action during the river crossing.
17	Number	Of obstacles emplaced in the crossing area, if conducting a retrograde crossing.
18	Rate	Per hour that personnel, tactical, and combat vehicles can cross the river.

ART 1.6.1.3.1 Conduct Gap Crossing in Support of Combat Maneuver

1-69. Conduct gap crossing in support of combat maneuver includes both hasty and deliberate gap crossings and the majority of river crossing operations. It includes both those operations conducted primarily at the brigade combat team level and those conducted by the division or corps level organization. Those gap crossings conducted as a reduction method within a combined arms breaching operation are also included in this art, but since the primary focus of planning and preparation is on the breaching operation they are typically discussed as a part of the breaching operation rather than as a separate gap crossing operation in that context. (FM 3-90.12) (USAES)

No	Scale	Measure
01	Yes/No	Unit accomplished gap crossing by time specified in operation order.
02	Time	That the gap or obstacle delays friendly force movement.
03	Time	To conduct area reconnaissance of the terrain surrounding the gap.
04	Time	To plan the gap crossing.

No	Scale	Measure
05	Time	For staff to disseminate data concerning the gap to subordinate units, to higher headquarters, and laterally after determination.
06	Time	To move engineer bridging equipment to the crossing site.
07	Time	To establish conditions necessary for success, such as suppressing enemy systems over watching the river, breaching minefields and other obstacles barring access to the riverbanks, and preparing access and egress routes.
08	Time	To emplace and construct crossing assets.
09	Time	To complete gap crossing.
10	Percent	Of crossing unit that has moved to the far side of the gap.
11	Percent	Of engineer bridge, raft, and assault boat systems that are mission capable.
12	Percent	Of crossing area seeded with obstacles if conducting a retrograde crossing.
13	Number	Of crossing area seeded with obstacles if conducting a retrograde crossing.
14	Number	Of engineer bridge, raft, and assault boat systems that are mission capable.
15	Number	Of friendly casualties due to accidents and enemy action during the gap crossing.

ART 1.6.1.3.2 Conduct Line of Communications Gap Crossing Support

1-70. Conduct line of communications (LOC) gap crossing support is not tactically focused, although it may clearly affect tactical operations. This support may provide the means for combat maneuver forces to move, but it is not directly in support of combat maneuver. As the title implies this is focused on ultimately using nonstandard bridging. Both assault and tactical bridging is designed to support the flow of traffic requirements (number of passes) of LOCs. (FM 3-90.12) (USAES)

Note: For construction and maintenance or roads and highways, see ART 1.6.2.1 (Construct and Maintain Combat Roads and Trails).

No	Scale	Measure
01	Yes/No	Unit constructed or maintained adequate bridging for a given LOC road within the timeframe of the construction directive without degrading or delaying movement along the LOC.
02	Yes/No	Unit developed detailed plans for all necessary gap crossings.
03	Yes/No	Unit inspected project for quality control and ensured gap crossings were completed on time and to appropriate standards.
04	Time	To conduct reconnaissance to determine how the local environment will affect the bridging.
05	Time	To conduct underwater inspection to support the bridging for a wet gap crossing.
06	Time	To review available information in construction directive, intelligence reports, and site investigation to develop an operation plan or operation order.
07	Time	To plan the bridging requirements including construction estimate, construction directive, and quality control.
08	Time	To prepare a bridging estimate.
09	Time	To prepare a bridging construction directive and issue it to the construction units.
10	Time	To coordinate additional personnel, equipment, and critical items.
11	Time	To monitor construction and conduct quality assurance inspections.
12	Time	To perform final inspection of finished bridging and turn it over to the user.
13	Time	To construct and maintain bridging.
14	Time	That scheduled arrivals in area of operations (AO) are delayed on the average due to interruptions in roads and highways by combat actions or natural disasters.

No	Scale	Measure
15	Percent	Difference between planned and actual requirements for bridging construction and maintenance requirements.
16	Percent	Of force becoming casualties due to enemy action or accidents during bridging construction and repair.
17	Percent	Increase in the carrying capability of a road or highway due to bridging construction and maintenance.
18	Percent	Of planned bridging construction and maintenance capability achieved in AO.
19	Percent	Of personnel in AO required to construct and maintain bridging.
20	Percent	Of bridging construction and repair capability provided by host nation.
21	Percent	Of existing bridging in AO improved.
22	Percent	Of bridging in AO that can be used in their current condition by military load classification.
23	Percent	Of unit operations degraded, delayed, or modified in AO due to bridge or gap impassability.
24	Number	Of bridges in the AO damaged by enemy fire or natural disaster.
25	Number	Of bridges in the AO requiring construction and maintenance in AO.
26	Number	Of bridges constructed and improved in AO.
27	Number	Of meters of bridging constructed and improved in AO in a specified time.
28	Number	Of instances of delays in scheduled arrivals due to interruption of bridging in AO by combat actions or natural disaster.
29	Number	Of instances in which troop movement or sustaining operations were prevented due to bridge or gap impassability.
30	Number	Of bridging maintenance inspections conducted per month in AO.

ART 1.6.2 ENHANCE MOVEMENT AND MANEUVER

1-71. Enhance force mobility in the forward area by constructing or repairing combat roads, trails, and forward airfields and landing zones to facilitate the movement of personnel, equipment, and supplies. Apply environmental considerations as appropriate. (FM 3-34) (USAES)

Note: Mobility enhancing systems referred to in this task include, but are not limited to, bulldozers, road graders, tanks with mine plows, armored combat earthmovers, dump trucks, cranes, scoop loaders, and explosives used to remove obstacles.

No.	Scale	Measure
01	Yes/No	Unit completed mobility enhancing activity by time specified in operation order.
02	Time	To respond to an event (natural disaster or combat activity) that impacts the unit's movement and maneuver.
03	Time	That the preparation and execution of unit operations are delayed due to a natural disaster or combat activity that impacts the unit's movement and maneuver.
04	Time	To conduct a route or area reconnaissance of location where mobility enhancing activity is required.
05	Time	For staff to disseminate event data to subordinate units, to higher headquarters, and laterally after discovery.
06	Time	To plan for the mobility enhancement effort.
07	Time	To move mobility enhancing systems to work site.
08	Time	To complete mobility enhancing activity.
09	Time	Of movement between given points reduced due to the mobility enhancing activity.

No.	Scale	Measure
10	Time	To establish conditions necessary for the success of the mobility enhancement effort, such as establishing security, gaining permission from local authorities for construction, and obtaining supplies—gravel, sand, airfield mats, soil stabilization systems—necessary for construction.
11	Percent	Of mobility enhancing activity completed.
12	Percent	Of mobility enhancing systems available to the commander that are committed to the task.
13	Percent	Of increase in unit mobility and maneuver due to completion of the mobility enhancing activity.
14	Number	Of mobility enhancing systems that are mission capable.
15	Number	Of necessary and unnecessary environmentally harmful incidents, such as petroleum spills in watersheds and soil spills into fish habitats.
16	Number	Of friendly and neutral casualties during the mobility enhancing activity.

ART 1.6.2.1 CONSTRUCT AND MAINTAIN COMBAT ROADS AND TRAILS

1-72. Prepare and maintain routes for equipment and personnel. ART 1.6.2.1 includes delineating routes, conducting reconnaissance, clearing ground cover, performing earthwork, providing drainage, stabilizing soil, and preparing the road surface for transit by Army combat and tactical vehicles. (FM 3-90.12) (USAES)

No.	Scale	Measure
01	Yes/No	Unit completed combat road and trail construction or maintenance operation by time specified in operation order.
02	Time	To respond to an event (natural disaster or combat activity) that impacts existing combat roads and trails.
03	Time	That the preparation and execution of unit operations are delayed due to a natural disaster or combat activity that impacts the unit's capability to use a combat road or trail.
04	Time	To conduct area reconnaissance of location where the construction and repair of combat roads and trails will take place.
05	Time	For staff to disseminate reconnaissance results to subordinate units, to higher headquarters, and laterally after discovery.
06	Time	To plan the construction and maintenance of combat roads and trails.
07	Time	To establish the conditions necessary for success of the construction and maintenance effort, such as establishing security, gaining permission from local authorities for construction, and obtaining supplies—gravel, sand, and soil stabilization systems.
08	Time	To move mobility enhancing systems to the work site.
09	Time	To construct, improve, and repair the required combat roads and trails.
10	Time	Of movement between given points reduced due to the construction and maintenance of combat roads and trails.
11	Percent	Of combat roads and trails construction and maintenance operation completed.
12	Percent	Of reduction in speed of vehicles traversing existing combat roads and trails due to existing environmental conditions, such as snow, ice, and grade.
13	Percent	Of mobility enhancing systems available to the tactical force commander that are committed to the task.
14	Percent	Of increase in movement time during the actual repair of combat roads and trails.
15	Percent	Of decreased movement time due to construction of combat roads and trails.
16	Number	Of mobility enhancing systems that are mission capable.
17	Number	Of friendly and neutral casualties during the combat roads and trails construction and maintenance operation.
18	Number	And type of vehicles unable to traverse existing terrain, combat roads, and trails.

No.	Scale	Measure
19	Number	And type of vehicles able to traverse combat roads and trails after their construction and maintenance.
20	Number	Of necessary and unnecessary environmentally harmful incidents, such as petroleum spills in watersheds and soil spills into fish habitats.

ART 1.6.2.2 CONSTRUCT AND MAINTAIN FORWARD AIRFIELDS AND LANDING ZONES

1-73. Prepare and maintain landing zones and landing strips to support Army and joint aviation ground facility requirements. (FM 3-90.12) (USAES)

No.	Scale	Measure
01	Yes/No	Unit completed forward airfield and landing zone construction and maintenance effort by time specified in operation order.
02	Time	To respond to an event (natural disaster or combat activity) that negatively impacts the capability of existing forward airfields and landing zones.
03	Time	That the preparation and execution of unit operations are delayed due to a natural disaster or combat activity that negatively impacts a unit's use of existing forward airfields and landing zones.
04	Time	To conduct an area reconnaissance of the current location of forward airfields and landing zones that have been negatively impacted due to an event or of proposed locations for forward airfields and landing zones.
05	Time	For the staff to format and disseminate information obtained by the area reconnaissance to subordinate units, to higher headquarters, and laterally.
06	Time	To plan for the construction and repair of forward airfields and landing zones.
07	Time	To establish the conditions necessary for success of the construction and repair effort, such as establishing security, gaining permission from local authorities, and obtaining supplies—gravel, sand, airfield mats, and soil stabilization systems.
08	Time	To move mobility enhancing systems to the work site.
09	Time	To complete construction and repair of the forward airfield or landing zone.
10	Percent	Of reduction in forward airfield and landing zone capacity due to existing environmental conditions, such as snow, ice, and fog.
11	Percent	Of forward airfield and landing zone construction and repair completed.
12	Percent	Of mobility enhancing systems available to the tactical force commander that are committed to the task.
13	Percent	Of increase in capacity of unit forward airfields and landing zones due to completion of the construction and repair effort.
14	Percent	Of forward airfields and landing zones in the area of operations (AO) with approaches compatible with Army fixed-wing operational support aircraft.
15	Percent	Of forward airfields and landing zones in the AO with navigational aids allowing for landings in bad weather.
16	Distance	Between existing and proposed forward airfields and landing zones.
17	Number	Of mobility enhancing systems that are mission capable.
18	Number	And types of aircraft unable to use existing forward airfields and landing zones.
19	Number	And types of aircraft able to use forward airfields and landing zones after the construction and maintenance of those zones.
20	Number	And types of aircraft able to use forward airfields and landing zones simultaneously—maximum on ground—after the construction and repair of those zones.
21	Number	Of friendly and neutral casualties during the construction and repair of forward airfields and landing zones due to accidents and enemy actions.
22	Number	Of necessary and unnecessary environmentally harmful incidents, such as petroleum spills in watersheds and soil and spills into fish habitats.

ART 1.6.3 NEGOTIATE A TACTICAL AREA OF OPERATIONS

1-74. Overcome the challenges presented by the trafficability or configuration of the ground, air, or sea environment through the inherent characteristics of personnel or their equipment. This task involves overcoming aspects of the physical environment—such as high winds and rain—and the presence of chemical, biological, radiological, and nuclear agents. It includes crossing or bypassing contaminated areas. (FM 3-11) (USACBRNS)

Note: ART 1.6.3 differs from ART 1.3.2 (Conduct Tactical Road March) and ART 1.3.3 (Conduct Tactical Convoy) by the environment in which it takes place. ARTs 1.3.3 and 1.3.4 involve only the act of moving units. ART 1.2.2.2 (Employ Traveling Overwatch Movement Technique) includes crossing and bypassing contaminated areas.

No.	Scale	Measure
01	Yes/No	The unit was in the correct position at the correct time.
02	Time	Of force delays due to poor trafficability or environmental conditions.
03	Time	Of force delays to assume appropriate mission-oriented protective posture.
04	Percent	Of decrease in rate of movement resulting from actual terrain trafficability differing from that in the plan.
05	Percent	Of force no longer fully mission capable resulting from terrain accidents.
06	Percent	Of force delayed due to terrain conditions.
07	kph	Rate of movement.

ART 1.6.4 PROVIDE DIVER SUPPORT

1-75. Provide diving equipment and personnel to conduct underwater operations. Engineer dive teams provide underwater reconnaissance, salvage, recovery, construction, demolition, repairs, inspections, and hydrographic survey operations in support of the full spectrum operations. (FM 3-34.280) (USAES)

No.	Scale	Measure
01	Yes/No	Engineer diving support increased available berthing positions to on load and off-load cargo from ships.
02	Yes/No	Engineer diving team conducted hydrographic surveys and established navigational waterways.
03	Yes/No	Engineer diving teams eliminated underwater obstacles in support of bridging operations and ships traffic.
04	Yes/No	Engineer diving teams inspected underwater structures to aid in military load class analysis.
05	Yes/No	Engineer diving teams supported mobility and countermobility along inland waterways, ports, and harbors.
06	Yes/No	Engineer diving teams located and recovered submerged personnel, equipment, weapon systems, or all of these.
07	Yes/No	Engineer dive teams conducted damage survey and repair operations.
08	Yes/No	Unit conducted joint logistics over-the-shore operations by time specified by higher headquarters.
09	Time	To repair underwater portions of waterfront facilities.
10	Time	To conduct hydrographic survey of 1,000 square meters.
11	Time	To remove obstacles to navigation and bridging operations.
12	Time	To plan and inspect underwater structures.
13	Time	To clear debris and wreckage from underwater structures.
14	Time	To repair underwater structures.

No.	Scale	Measure
15	Time	To install physical security systems.
16	Percent	Of underwater pipelines inspected, maintained, or repaired.
17	Percent	Of channels and waterways surveyed in an area of operations.
18	Percent	Of identified obstacles emplaced or reduced.
19	Percent	Of port facilities open, construction completed, and rehabilitation completed.
20	Percent	Of vessel underwater hulls inspected during security swim operations.
21	Percent	Of vessels with current in-water hull inspections completed.
22	Percent	Of mooring systems inspected and repaired in a specified area.
23	Percent	Of offshore petroleum distribution pipeline and components inspected and maintained in a specified area.
24	Percent	Of construction completed on underwater structures.
25	Number	Of fording sites identified and reported to higher headquarters.
26	Number	Of rafting sites identified and reported to higher headquarters.
27	Number	Of wet gap crossing operations conducted by a specified time.
28	Number	Of bridge inspection and repairs conducted by a specified time.
29	Number	Of hydrographic survey products rendered in a specified time.
30	Number	Of salvage operations completed in a specified time.
31	Number	Of search and recovery missions completed in a specified time.
32	Number	Of security inspection missions of bridges, ports, locks, and dams missions completed during rotation.
33	Number	Of ships husbandry missions completed in a specified time.
34	Number	Of in-water maintenance missions conducted in a port facility.

ART 1.6.5 CONDUCT NONTACTICAL MOVEMENTS

1-76. Execute a movement in which troops and vehicles are arranged to expedite their movement and conserve time and energy when no enemy interference, except by air, is anticipated. Environmental considerations are applied as appropriate. (FM 3-90) (USACAC)

No.	Scale	Measure
01	Yes/No	Completed unit movement by time specified in operation order.
02	Yes/No	Used transportation resources and assets efficiently.
03	Time	To initiate movement.
04	Time	To complete movement.
05	Percent	Of force completing the movement.
06	kph	Rate of movement.

SECTION VII – ART 1.7: CONDUCT COUNTERMOBILITY OPERATIONS

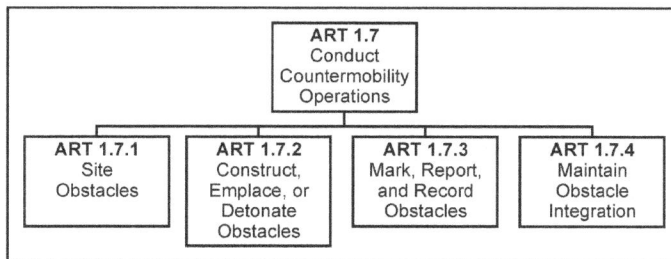

```
                              ART 1.7
                              Conduct
                           Countermobility
                             Operations
    ┌───────────────┬───────────┴──────────┬────────────────┐
  ART 1.7.1       ART 1.7.2           ART 1.7.3         ART 1.7.4
   Site           Construct,          Mark, Report,      Maintain
  Obstacles       Emplace, or         and Record         Obstacle
                  Detonate            Obstacles          Integration
                  Obstacles
```

1-77. Countermobility operations involve constructing reinforcing obstacles integrated with fires to inhibit the maneuver of an enemy force, increase time for target acquisition, and increase weapon effectiveness. Commanders integrate obstacle planning into the military decisionmaking process (see FM 5-0); integrate obstacles into the concept of operations (primarily through proper siting); and maintain integration through obstacle turnover, protection, and tracking. The force constructs, emplaces, or detonates tactical and protective obstacles to reinforce existing obstacles. Tactical obstacles are designed and integrated with fires to achieve a tactical effect—disrupt, fix, turn, or block. The three types of tactical obstacles are directed obstacles, situational obstacles, and reserve obstacles. They are distinguished by the differences in execution criteria. Protective obstacles are a key component of survivability operations. (See ART 6.7, Conduct Survivability Operations.) The force may employ any type of individual obstacle as a tactical obstacle. (FM 5-102) (USAES)

Note: The term "countermobility systems" used in this section is generic in nature and includes both manual and mechanical means, such as sapper units, cratering demolition kits, and mine dispensing systems.

No.	Scale	Measure
01	Yes/No	Friendly obstacle effect accomplished commander's guidance (block, disrupt, fix, or turn).
02	Yes/No	Unit emplaced obstacles per restrictions established by higher headquarters, to include obstacle control measure graphics and rules of engagement.
03	Yes/No	Unit integrated fires (direct and indirect, lethal and nonlethal) in the obstacle plan and were effective when required.
04	Time	That enemy forces are delayed in the conduct of their movement and maneuver due to friendly obstacles. (Delay time depends on type of effect.)
05	Time	Required by the enemy to repair and replace facilities (such as bridges, railroad switching yards, dockyard cranes, and airfield runways) damaged and destroyed by friendly countermobility efforts.
06	Time	To conduct area reconnaissance of proposed locations of obstacle complexes.
07	Time	For staff to format and disseminate information obtained by the area reconnaissance to subordinate units, to higher headquarters, and laterally.
08	Time	To conduct terrain analysis to assist in selecting obstacle locations.
09	Time	To plan construction of the obstacle effort.
10	Time	To establish conditions necessary for construction of obstacles, such as establishing security and moving class IV and class V material to obstacle locations.
11	Time	To move countermobility systems to work site.
12	Time	To emplace obstacles.

No.	Scale	Measure
13	Time	To employ appropriate lethal and nonlethal fires when enemy is engaged in friendly obstacles.
14	Percent	Of countermobility effort completed.
15	Percent	Of total available countermobility effort in a given time not used because of poor management.
16	Percent	Of enemy forces unable to reach their objective due to obstacles.
17	Percent	Of mobility corridors and avenues of approach closed to enemy maneuver by friendly obstacles.
18	Percent	Of enemy casualties inflicted by friendly obstacles.
19	Percent	Of available countermobility assets that are mission capable.
20	Percent	Of enemy sustainment capability interdicted by friendly obstacle efforts.
21	Percent	Of enemy engineering capability devoted toward enhancing enemy mobility and maneuver.
22	Percent	Of friendly capability devoted to conducting countermobility operations.
23	Percent	Of friendly fires systems used to emplace field artillery- and air-delivered obstacles.
24	Number	Of enemy main supply routes and lines of communication interdicted by friendly obstacles.
25	Number	Of friendly and civilian casualties during the conduct of countermobility operations.
26	Number	Of available countermobility assets that are mission capable.
27	Number	Of potential enemy courses of action no longer feasible due to friendly countermobility efforts.

ART 1.7.1 SITE OBSTACLES

1-78. Determine the location of individual obstacles based on the enemy force (target), desired location of massed fires, tentative weapon system positions, and the intended effect (disrupt, fix, turn, or block). ART 1.7.1 includes verifying that the obstacle is covered by fires, noting locations of fire control measures and obstacles, and recording the appropriate data on range cards. (FM 90-7) (USAES)

No.	Scale	Measure
01	Yes/No	Obstacle location accomplished intended effect when integrated with available fires.
02	Yes/No	Unit emplaced planned obstacles per restrictions established by higher headquarters, to include obstacle control measure graphics, applying environmental considerations, and rules of engagement.
03	Yes/No	Planned obstacles had a reasonable probability of being emplaced, given protected availability of countermobility systems, supplies, manpower, and time.
04	Yes/No	Unit coordinated with supported unit to ensure obstacle coverage by fires.
05	Time	To conduct terrain analysis to support selecting locations for obstacles.
06	Time	To conduct area reconnaissance of proposed obstacle locations applying environmental considerations.
07	Time	For the staff to format and disseminate information obtained by the area reconnaissance to subordinate units, to higher headquarters, and laterally.
08	Time	To plan the countermobility effort.
09	Percent	Of mobility corridors and avenues of approach that will be closed to enemy maneuver by friendly obstacles once they are emplaced.
10	Number	Of enemy main supply routes and lines of communications that will be interdicted by friendly obstacles.
11	Number	Of potential enemy courses of action that are no longer feasible due to friendly countermobility efforts.

ART 1.7.2 CONSTRUCT, EMPLACE, OR DETONATE OBSTACLES

1-79. Reinforce the terrain and combine obstacles with fires to disrupt, fix, turn, or block an enemy force. ART 1.7.2 includes emplacing special purpose munitions; constructing wire obstacles, antitank ditches, tetrahedrons and log obstacles; and detonating explosives to create road craters, destroy bridges, and construct abatises. (FM 90-7) (USAES)

No.	Scale	Measure
01	Yes/No	Friendly obstacle effect accomplished the commander's intent (block, disrupt, fix, and turn).
02	Yes/No	Unit emplaced obstacles per restrictions established by higher headquarters, to include obstacle control measure graphics, environmental considerations, and rules of engagement.
03	Time	Required to conduct area reconnaissance of proposed obstacle locations, applying environmental considerations.
04	Time	For the staff to format and disseminate information from the area reconnaissance to subordinate units, to higher headquarters, and laterally.
05	Time	To plan the design of individual obstacles and obstacle complexes.
06	Time	To establish the conditions necessary for obstacle construction, such as establishing security and moving class IV and class V materials to obstacle locations.
07	Time	That the obstacle construction effort is delayed due to insufficient engineer support.
08	Time	To move countermobility systems to the work site.
09	Time	To construct, emplace, or detonate (underwater) obstacles.
10	Percent	Of obstacle effort completed.
11	Percent	Of total available countermobility effort in a given time not used because of poor management.
12	Percent	Of available countermobility assets that are mission capable.
13	Percent	Of friendly fires systems used to emplace field artillery- and air-delivered obstacles.
14	Number	Of available countermobility assets that are mission capable.
15	Number	Of friendly and civilian casualties during the construction, emplacement, or detonation of obstacles.

ART 1.7.3 MARK, REPORT, AND RECORD OBSTACLES

1-80. Mark all obstacles to aid in fratricide prevention. Report the intention to emplace obstacles (if required), initiation of construction and emplacement, and completion and execution of obstacles. As a minimum, record the obstacle location, type, and (if applicable) number and types of mines, placement of mines, use of antihandling devices, location of lanes and gaps, and description of marking. (FM 90-7) (USAES)

No.	Scale	Measure
01	Yes/No	Unit marked, reported, and recorded all obstacles per standing operating procedures.
02	Time	To identify and determine the limits of minefields and other obstacles.
03	Time	To mark obstacle limits per doctrine and international agreements.
04	Time	To transmit obstacle information to appropriate agencies in area of operations.
05	Percent	Of obstacle location and composition information correctly recorded in the unit database.
06	Percent	Of obstacle location and composition information correctly transmitted to appropriate agencies in area of operations.
07	Number	Of friendly and neutral casualties resulting from improperly marked obstacles.

ART 1.7.4 MAINTAIN OBSTACLE INTEGRATION

1-81. Ensure emplaced obstacles remain integrated into the concept of operations. ART 1.7.4 includes turnover and transfer, protection, repair, and tracking of obstacles. Obstacle protection focuses on two tasks: counterreconnaissance to prevent the enemy from gathering obstacle intelligence and enemy mobility asset destruction to ensure maximum effectiveness of obstacles. Obstacle tracking includes supervising achievement of key milestones as part of the unit's timeline (class IV and V forward, initiate engagement area development, siting complete), collation, and dissemination of obstacle information, and maintenance of records. (FM 90-7) (USAES)

No.	Scale	Measure
01	Yes/No	Obstacle turnover and transfer occurred per doctrinal guidance and international standardization agreements.
02	Yes/No	Friendly unit was able to prevent enemy reconnaissance elements from gaining information on the obstacle.
03	Yes/No	Obstacle tracking occurred within an acceptable level of accuracy, as determined by the unit commander.
04	Time	To plan and coordinate obstacle turnover and transfer.
05	Time	To restore a partially reduced obstacle.
06	Time	To conduct obstacle tracking.
07	Percent	Of enemy reconnaissance assets destroyed while maintaining obstacle integration.
08	Percent	Of enemy mobility assets destroyed before they could reduce friendly obstacles.
09	Number	Of enemy reconnaissance assets destroyed while maintaining obstacle integration.
10	Number	Of enemy mobility assets destroyed before they could reduce friendly obstacles.
11	Number	Of friendly and civilian casualties during the maintenance of obstacle integration.

SECTION VIII – ART 1.8: EMPLOY OBSCURANTS

1-82. Use obscurants to conceal friendly positions and screen maneuvering forces from enemy observation. An obscurant is a chemical agent that decreases the level of energy available for the functions of seekers, trackers, and vision enhancement devices. ART 1.8. includes obscuring and screening in full spectrum operations. Apply environmental considerations as appropriate. (FM 3-11.50) (USACBRNS)

No.	Scale	Measure
01	Yes/No	Use of obscurants improved unit survivability and maneuverability.
02	Yes/No	Use of obscurants compromised unit course of action.
03	Time	To assess unit concealment requirements beyond that provided by camouflage systems.
04	Time	To employ obscurants to screen personnel, major combat equipment, bridge sites, and obstacles in an area of operations (AO).
05	Percent	Of unit commanders and planners able to effectively plan the use of obscurants to protect friendly personnel, equipment, and positions from enemy direct fire, observation, and surveillance for deception operations.
06	Percent	Of units, installations, and facilities in the AO employing obscurants.
07	Percent	Of increased time to conduct operations in limited visibility conditions due to the use of obscurants.

Chapter 2

ART 2.0: The Intelligence Warfighting Function

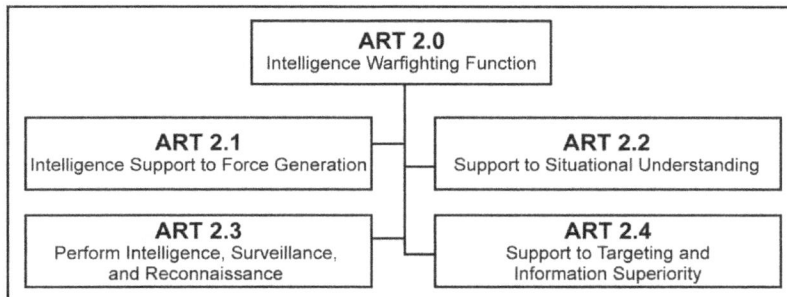

The *intelligence warfighting function* is the related tasks and systems that facilitate understanding of the operational environment, enemy, terrain, and civil considerations. It includes tasks associated with intelligence, surveillance, and reconnaissance operations, and is driven by the commander. Intelligence is more than just collection. As a continuous process, it involves analyzing information from all sources (human, imagery, measurement and signature, signal) and conducting operations to develop the situation. (FM 3-0) (USACAC)

SECTION I – ART 2.1: INTELLIGENCE SUPPORT TO FORCE GENERATION

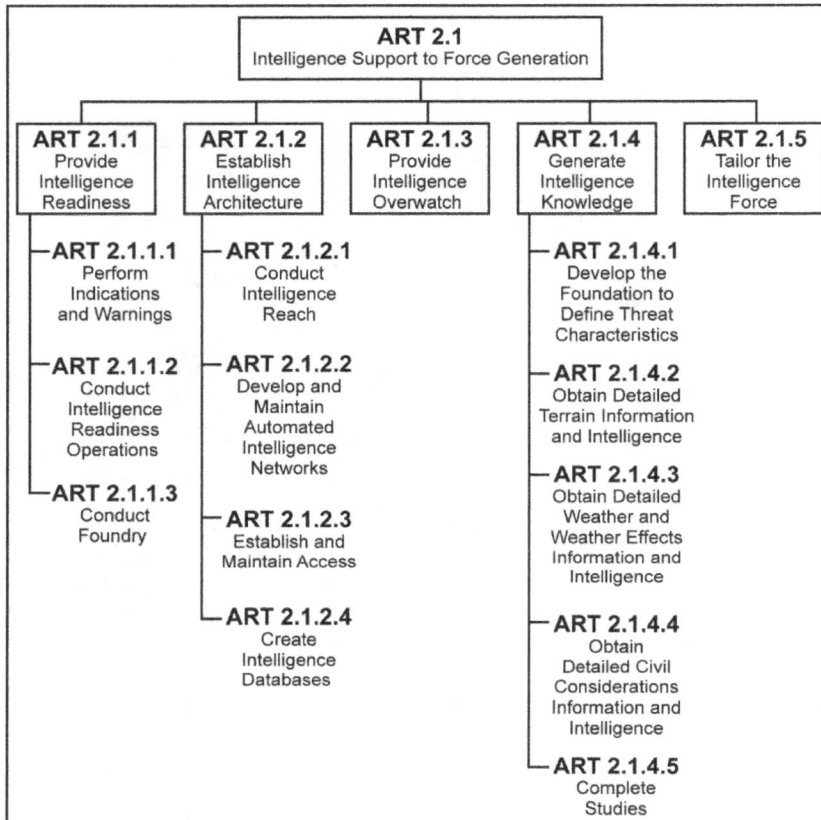

```
                        ┌────────────────────────────┐
                        │          ART 2.1           │
                        │ Intelligence Support to     │
                        │ Force Generation            │
                        └────────────────────────────┘
```

ART 2.1.1 Provide Intelligence Readiness	ART 2.1.2 Establish Intelligence Architecture	ART 2.1.3 Provide Intelligence Overwatch	ART 2.1.4 Generate Intelligence Knowledge	ART 2.1.5 Tailor the Intelligence Force

ART 2.1.1
- ART 2.1.1.1 Perform Indications and Warnings
- ART 2.1.1.2 Conduct Intelligence Readiness Operations
- ART 2.1.1.3 Conduct Foundry

ART 2.1.2
- ART 2.1.2.1 Conduct Intelligence Reach
- ART 2.1.2.2 Develop and Maintain Automated Intelligence Networks
- ART 2.1.2.3 Establish and Maintain Access
- ART 2.1.2.4 Create Intelligence Databases

ART 2.1.4
- ART 2.1.4.1 Develop the Foundation to Define Threat Characteristics
- ART 2.1.4.2 Obtain Detailed Terrain Information and Intelligence
- ART 2.1.4.3 Obtain Detailed Weather and Weather Effects Information and Intelligence
- ART 2.1.4.4 Obtain Detailed Civil Considerations Information and Intelligence
- ART 2.1.4.5 Complete Studies

2-1.   Intelligence support to force generation is the task of generating intelligence knowledge concerning an area of interest, facilitating future intelligence operations, and tailoring the force. It includes establishing intelligence communication and knowledge management architectures. These architectures enable collaboration among strategic, operational, and tactical intelligence organizations in the following areas—intelligence reach; collaborative analysis; data storage; processing and analysis; and intelligence production support to force generation consists of five tasks. (FM 2-0) (USAICoE)

ART 2.1.1 PROVIDE INTELLIGENCE READINESS

2-2.   Intelligence readiness operations develop baseline knowledge of multiple potential threats and operational environments. These operations support ongoing operations, contingency planning, and operational preparation. These operations and related intelligence training activities enable the intelligence warfighting function to support the commander's intelligence requirements effectively. (FM 2-0) (USAICoE)

No.	Scale	Measure
01	Yes/No	Unit provided indications and warning reports.
02	Yes/No	Unit provided intelligence readiness.
03	Yes/No	Unit conducted foundry training.
04	Percent	Of intelligence relating to indications and warning prior to incident occurring.
05	Percent	Of relevant data, information, intelligence, and products received on the operational environment.
06	Percent	Of relevant training received for the conduct of operations.

ART 2.1.1.1 PERFORM INDICATIONS AND WARNING

2-3. This task provides the commander with advance warning of threat actions or intentions. The intelligence officer develops indications and warnings to rapidly alert the commander of events or activities that would change the basic nature of the operation. It enables the commander to quickly reorient the force to unexpected contingencies and shape the operational environment. (FM 2-0) (USAICoE)

No.	Scale	Measure
01	Yes/No	Unit issued indications and warnings in sufficient time to prevent the enemy or threat from surprising the targeted friendly unit or installation.
02	Yes/No	Unit disseminated indications and warnings after development and compared to other information and intelligence to ensure accuracy.
03	Yes/No	Intelligence officer monitored event template and matrix to determine if the enemy or threat was performing a predicted course of action (COA).
04	Yes/No	Unit identified enemy and potential threats in the operational environment.
05	Yes/No	Unit updated enemy and threat identification, tactics, techniques, and procedures.
06	Time	Between receipt of significant information and intelligence and updates of indications and warnings conditions.
07	Time	To conduct predictive analysis in support of COA development.
08	Time	To predict significant changes in enemy or threat activities.
09	Time	Required to disseminate indications and warnings to appropriate echelons, agencies, and organizations.
10	Time	To submit intelligence portion of commander's situation report.
11	Percent	Of indications and warnings of threat actions reported that will impact friendly forces operations.
12	Percent	Of commander's threat conditions and attack warnings issued and disseminated.
13	Percent	Of threat indicators maintained and evaluated.

ART 2.1.1.2 CONDUCT INTELLIGENCE READINESS OPERATIONS

2-4. Intelligence readiness operations support contingency planning and preparation by developing baseline knowledge of multiple potential threats and operational environments. This information and training enables a collaborative effort and environment to provide the best possible initial threat understanding. (FM 2-0) (USAICoE)

No.	Scale	Measure
01	Yes/No	Unit provided information on the civil considerations using the factors of areas, structures, capabilities, organizations, people and events affecting the force.
02	Percent	Of intelligence relating to the enemy force.
03	Percent	Of intelligence relating to potential terrorist acts against U.S. forces.
04	Percent	Of intelligence relating to potential insurgent, guerrilla, or criminal groups.
05	Percent	Of intelligence relating to using chemical, biological, radiological, nuclear, and high-yield explosives (CBRNE) weapons.

No.	Scale	Measure
06	Percent	Of information and intelligence that facilitates a cultural understanding of the operational environment.
07	Percent	Of intelligence requirements to support the other warfighting functions or Army programs.

ART 2.1.1.3 CONDUCT FOUNDRY

2-5. Foundry is a training program designed to sustain critical intelligence capabilities and perishable intelligence skills, and provide regional focus, technical training and functional expertise to the tactical military intelligence (MI) force through home-station training platforms, mobile training teams, and live environment training opportunities. Foundry creates a single hub for advanced skills training across the Active Army, Army National Guard, and the Army Reserve MI force. It also provides training to leaders who supervise MI missions and Soldiers who perform MI activities. (FM 2-0) (USAICoE)

No.	Scale	Measure
01	Yes/No	Unit identified specific intelligence skills by each military occupational specialty needed to deploy to a specific theater.
02	Yes/No	Unit identified intelligence training requirements and skills needed for MI sustainment and the intelligence overwatch program.
03	Yes/No	Unit identified intelligence training resources available to conduct foundry training in order to meet theater specific mission and deployment standards.
04	Yes/No	Unit prepared foundry prioritized quarterly training requests for funding.
05	Yes/No	MI soldiers and units actively participated in foundry training opportunities to include live environment training.
06	Yes/No	Unit established unit and functional partnerships in support of MI training and readiness effort.
07	Yes/No	Unit used Project Foundry to develop MI adjunct faculty (subject matter expert) in support of unit and home station training efforts.
08	Yes/No	Unit conducted MI periodic training councils to share training resources and assets.
09	Yes/No	Units used Project Foundry to conduct unit collective training and for unit exercises.
10	Yes/No	Units reported tactics, techniques, and procedures (TTP) and lessons learned for integration into training programs, doctrine, and the Project Foundry.
11	Yes/No	Unit conducted an annual assessment of Project Foundry and MI training successes and needs.
12	Percent	Of unfunded quarterly foundry training requests.
13	Number	Of skills identified for deployment used in theater.
14	Number	Of skills not initially identified but were needed in theater.
15	Number	Of available MI soldiers requiring training who participate annually in foundry training.
16	Number	Of available MI units requiring training who participate annually in foundry training.
17	Number	Of units that utilize foundry for collective training.
18	Number	Of units that utilize foundry for unit exercises.
19	Number	Of units that report TTP and lessons learned for integration into training programs, doctrine, and the Project Foundry.
20	Number	Of items identified in the annual assessment needed to improve Project Foundry and MI training.

ART 2.1.2 ESTABLISH INTELLIGENCE ARCHITECTURE

2-6. Establishing an intelligence architecture includes complex and technical issues that include hardware, software, communications, communications security materials, network classification, technicians, database access, liaison officers, training, and funding. Well-defined and well-designed intelligence architecture can offset or mitigate structural, organizational, or personnel limitations. This architecture provides the best possible understanding of the threat, terrain, and weather, and civil considerations understanding. (FM 2-0) (USAICoE)

No.	Scale	Measure
01	Yes/No	Unit identified national agencies with relevant information, databases, and systems to form a collaborative environment.
02	Yes/No	Unit identified Department of Defense organizations with relevant information, databases, and systems to form a collaborative environment.
03	Yes/No	Unit identified multinational organizations and elements with relevant data, information, databases, and systems to form a collaborative environment.
04	Yes/No	Unit identified Army organizations with relevant information, databases, and systems to form a collaborative environment.
05	Yes/No	Unit achieved enemy and threat understanding through the intelligence architecture.
06	Percent	Of analysis, training, and intelligence production achieved through collaborative intelligence architecture.
07	Percent	Of structural, organizational, or personnel limitations offset or mitigated.

ART 2.1.2.1 CONDUCT INTELLIGENCE REACH

2-7.   Intelligence reach is a process by which intelligence organizations proactively and rapidly access information from, receive support from, and conduct direct collaboration and information sharing with other units and agencies, both within and outside the area of operations, unconstrained by geographic proximity, echelon, or command. Intelligence obtained through intelligence reach helps the staff plan and prepare for operations and answer commander's critical information requirements without the need for the information to pass through a formal hierarchy. (FM 2-0) (USAICoE)

No.	Scale	Measure
01	Yes/No	Unit established unit procedures or processes by which to conduct intelligence reach.
02	Yes/No	Soldiers trained, maintained, exercised, and sustained their intelligence reach procedures or processes.
03	Yes/No	Unit received information or intelligence through intelligence reach relevant to operations.
04	Yes/No	Unit used intelligence reach in intelligence, surveillance, and reconnaissance (ISR) planning.
05	Time	Required to update procedures and processes for intelligence reach sustainment.
06	Percent	Of trained personnel on intelligence reach procedures.
07	Percent	Of relevant information obtained.
08	Percent	Of information acquired through organic ISR taskings available through intelligence reach.
09	Percent	Of ISR tasks developed through information obtained by intelligence reach.
10	Percent	Of ISR tasks dynamically retasked due to information obtained by intelligence reach.

ART 2.1.2.2 DEVELOP AND MAINTAIN AUTOMATED INTELLIGENCE NETWORKS

2-8.   This task entails providing information systems that connect unique assets, units, echelons, agencies, and multinational partners for intelligence, collaborative analysis and production, dissemination, and intelligence reach. It uses existing automated information systems, such as the distributed common ground system-Army (known as DCGS-A), and, when necessary, creates operationally specific networks. In either case, these networks allow access to unclassified and classified means, and interoperability across the area of operations. This task includes identifying deficiencies in systems or networks, Service procedures, system administration procedures, security procedures, alternate power plan, redundancy capability, system backups, and update procedures. (FM 2-0) (USAICoE)

No.	Scale	Measure
01	Yes/No	Unit established intelligence specific networks (classified and unclassified).
02	Yes/No	The appropriate unique assets, units, echelons, agencies, and multinational partners were included in the network.

No.	Scale	Measure
03	Yes/No	The intelligence network was used for collaborative analysis and production, dissemination, and intelligence reach.
04	Yes/No	The intelligence network was adequate for collaborative analysis and production, dissemination, and intelligence reach.
05	Yes/No	The intelligence network was adequate for interoperability across the area of operations to include subordinate elements collaborative analysis and production, dissemination, and intelligence reach.
06	Yes/No	Policies and procedures were in place for identifying deficiencies in the systems or networks.
07	Yes/No	Policies and procedures were in place for implementing service procedures and an alternate power plan.
08	Yes/No	Policies and procedures were in place for network security.
09	Yes/No	Policies and procedures were in place for system backup and update.
10	Yes/No	Units used policies and procedures for system administration.
11	Yes/No	Adequate redundancy was built into the network.
12	Time	The network was down or unavailable due to service, system administration, or system update.
13	Time	Between the primary system failing and the redundancy capability initialized.
14	Percent	Of network used for collaborative analysis and production, dissemination, and intelligence reach.
15	Percent	Of time the redundancy capability was used in place of the primary system.

ART 2.1.2.3 ESTABLISH AND MAINTAIN ACCESS

2-9. Establish and provide access to classified and unclassified programs, databases, networks, systems, and other Web-based collaborative environments for Army and multinational organizations to facilitate intelligence reporting, production, dissemination, sustainment, and intelligence reach. This task also includes establishing access to classified and unclassified programs, databases, networks, systems, and other Web-based collaborative environments with joint forces and national agencies to facilitate a multilevel collaborative information environment. This task entails establishing, providing, and maintaining access to classified and unclassified programs, databases, networks, systems, and other Web-based collaborative environments for Army forces, joint forces, national agencies, and multinational organizations. Its purpose is to facilitate intelligence reporting, production, dissemination, and sustainment; intelligence reach; and a multilevel collaborative information environment. (FM 2-0) (USAICoE)

No.	Scale	Measure
01	Yes/No	Unit established or provided access to classified and unclassified programs, databases, networks, systems, and other Web-based collaborative environments for Army and multinational organizations.
02	Yes/No	Unit provided access to classified and unclassified programs, databases, networks, systems, and other Web-based collaborative environments with joint and interagency organizations to facilitate a multilevel collaborative environment.
03	Yes/No	Unit established individual user accounts per applicable policies and regulations and any applicable prerequisite training.
04	Yes/No	Unit established local standing operating procedures (SOPs) for individual user accounts.
05	Yes/No	Unit maintained individual user accounts per the SOPs, policies, regulations, and any applicable recertification or revalidation training.
06	Yes/No	Unit established system accounts per policies, regulations, and any prerequisite training.
07	Yes/No	Unit established local SOPs for system accreditation.
08	Yes/No	Unit established joint, interagency, and multinational accounts per policies, regulations, and any prerequisite training.
09	Yes/No	Unit established local SOPs for joint, interagency, and multinational accounts establishment and maintenance.

No.	Scale	Measure
10	Yes/No	Unit maintained joint, interagency, and multinational accounts per policies, regulations, and any prerequisite training.
11	Yes/No	Unit designated a security officer to ensure local SOPs meet appropriate standards for Army, joint, interagency, and multinational policies and regulations.
12	Yes/No	Unit designated a security officer to establish an oversight or inspection program to enforce Army, joint, interagency, and multinational policies and regulations.
13	Yes/No	Unit designated a security officer to serve as the liaison for higher headquarters, joint, and interagency oversight and inspection teams.

ART 2.1.2.4 CREATE INTELLIGENCE DATABASES

2-10. This task entails creating and maintaining unclassified and classified databases. Its purpose is to establish interoperable and collaborative environments for Army forces, joint forces, national agencies, and multinational organizations. This task facilitates intelligence analysis, reporting, production, dissemination, sustainment, and intelligence reach. It also includes the requirements for formatting and standardization, indexing and correlation, normalization, storage, security protocols, and associated applications. The following must be addressed in database development, management, and maintenance: data sources; information redundancy; import and export standards; data management and standards; update and backup procedures; and data mining, query, and search protocols. (FM 2-0) (USAICoE)

No.	Scale	Measure
01	Yes/No	Unit established and maintained classified and unclassified databases at the appropriate echelons and organizations.
02	Yes/No	Unit established prerequisite training for users on database use; data mining, searches, and queries; and update and backup procedures.
03	Yes/No	Unit established local standing operating procedures (SOPs) for database use; data mining, searches, and queries; and update and backup procedures.
04	Yes/No	The intelligence database was used for collaborative analysis and production, dissemination, and intelligence reach.
05	Yes/No	The intelligence database was adequate for collaborative analysis and production, dissemination, and intelligence reach.
06	Yes/No	The intelligence database was adequate for interoperability across the area of operations to include subordinate elements' collaborative analysis and production, dissemination, and intelligence reach.
07	Yes/No	Policies and procedures were in place for reporting issues with database security and database deficiencies or corruption.
08	Yes/No	Adequate redundancy was built into the database.
09	Yes/No	Policies and procedures were provided for database administration.
10	Yes/No	Unit designated a primary and alternate database manager to ensure local SOPs meet appropriate standards for Army, joint, interagency, and multinational policies and regulations.
11	Yes/No	Unit established training program for the primary and alternate database managers.
12	Yes/No	Policies and procedures were in place for implementing Service procedures and a data redundancy plan.
13	Yes/No	Policies and procedures were in place for database security.
14	Yes/No	Policies and procedures were in place for a database backup and update.
15	Time	That database was down or unavailable due to service, administration, or update.
16	Time	Between the primary database failing and the redundancy capability was initialized.
17	Time	That primary or alternate database manager spent on correcting or resolving deficiencies or corruptions with the primary or redundant database.
18	Percent	Of database used for collaborative analysis and production, dissemination, and intelligence reach.
19	Percent	Of time the redundancy capability was used in place of the primary database.

ART 2.1.3 PROVIDE INTELLIGENCE OVERWATCH

2-11. Intelligence overwatch is creating standing, fixed analytical intelligence capabilities that provide dedicated intelligence support to committed maneuver units. The overwatch cell is connected via a shared intelligence network that can extract information from multiple sources and provide succinct answers (vice megabytes of information) directly to supported units when time is critical. (FM 2-0) (USAICoE)

Note: This task branch supports ART 6.7 (Conduct Survivability Operations).

No.	Scale	Measure
01	Yes/No	Selected intelligence Soldiers provided a foundation of regional and subject matter expertise.
02	Yes/No	New intelligence technology was incorporated into the intelligence system.

ART 2.1.4 GENERATE INTELLIGENCE KNOWLEDGE

2-12. Generate intelligence knowledge is a continuous, user-defined task driven by the commander. It begins before mission receipt and provides the relevant knowledge required regarding the operational environment for the conduct of operations. As soon as the intelligence officer and other staff sections begin to collect data on the operational environment, they should organize the data into databases that meet the commander's visualization requirements. The execution of this task must follow all applicable policies and regulations on information collection and operations security. The information and intelligence obtained are refined into knowledge for use in mission analysis through functional analysis. Information is obtained through intelligence reach; research; data mining; database access; academic studies, products, or materials; intelligence archives; open-source intelligence; and other information sources. Generate intelligence knowledge is the foundation for performing intelligence preparation of the battlefield and mission analysis. The primary product of the generate intelligence knowledge task is the initial data files and intelligence survey. (FM 2-0) (USAICoE)

No.	Scale	Measure
01	Yes/No	Unit obtained detailed information on threat characteristics (composition, disposition, tactics, training, logistics, operational effectiveness, fires, communications, personality, intelligence, reach, agencies, nongovernmental organizations, and other threats) in and affecting the operational environment.
02	Yes/No	Unit obtained detailed information on the types of environment, such as desert, urban, jungle, arctic, and mountain, and the military aspects of terrain.
03	Yes/No	Unit obtained detailed information on recent and historical weather trends, seasonal patterns, aspects of weather, and weather zones in the operational environment. Unit effectively used and integrated the information to determine how the weather potentially affects friendly and enemy forces and operations in the operational environment.
04	Yes/No	Unit obtained detailed information on civil considerations within and affecting the operational environment.
05	Yes/No	Unit used the analytic functions and functional analysis.
06	Yes/No	Unit followed all applicable policies and regulations when collecting information and operations security.

ART 2.1.4.1 DEVELOP THE FOUNDATION TO DEFINE THREAT CHARACTERISTICS

2-13. Obtain detailed information and intelligence concerning threat characteristics (formerly order of battle) affecting the conduct of operations. Obtain this information from sources that include intelligence reach; research; data mining; database access; academic studies, products, or materials; intelligence archives; and open-source intelligence. This task develops specific, detailed information for each threat characteristic. The information, intelligence, products, and materials obtained are refined for use in mission analysis, intelligence preparation of the battlefield (IPB), and other planning tasks. This refinement occurs through functional analysis and other analytic techniques. (FM 2-0) (USAICoE)

No.	Scale	Measure
01	Yes/No	Unit obtained detailed information on composition, disposition, tactics, training, logistics, operational effectiveness, fires, communications, personality, intelligence, reach, agencies, nongovernmental organizations, and other threats in and affecting the operational environment.
02	Yes/No	Information obtained from external agencies, echelons, and partners was adequate for use in mission analysis, IPB, and planning.
03	Yes/No	Information obtained was used and integrated effectively to determine the effects on operations.
04	Percent	Of relevant information obtained from national agencies.
05	Percent	Of relevant information obtained from joint echelons.
06	Percent	Of relevant information obtained from multinational partners.
07	Percent	Of relevant information obtained from higher Army echelons.
08	Percent	Of relevant information obtained from lateral and subordinate echelons.

ART 2.1.4.2 Obtain Detailed Terrain Information and Intelligence

2-14. Obtain detailed information and intelligence about the terrain of the expected area of interest from sources that include intelligence reach; research; data mining; database access; academic studies, products, or materials; intelligence archives; and open-source intelligence. The information, intelligence, products, and material obtained are refined for use in mission analysis, intelligence preparation of the battlefield (IPB), and other planning tasks through functional analysis. This task encompasses the types of environments and the military aspects of terrain. (FM 2-0) (USAICoE)

No.	Scale	Measure
01	Yes/No	Unit obtained detailed information on the types of environment, such as desert, urban, jungle, arctic, and mountain.
02	Yes/No	Information obtained from external agencies, echelons, and partners was adequate for use in mission analysis, IPB, and planning.
03	Yes/No	Information obtained was used and integrated effectively into mission analysis, IPB, and planning.
04	Yes/No	Unit obtained detailed information on the military aspects of terrain.
05	Percent	Of relevant information obtained from national agencies.
06	Percent	Of relevant information obtained from joint echelons.
07	Percent	Of relevant information obtained from multinational partners.
08	Percent	Of relevant information obtained from higher Army echelons.
09	Percent	Of relevant information obtained from lateral and subordinate echelons.

ART 2.1.4.3 Obtain Detailed Weather and Weather Effects Information and Intelligence

2-15. Obtain detailed information and intelligence regarding recent and historical weather trends, seasonal patterns, aspects of weather, and weather zones. Obtain information on how the weather affects friendly and enemy forces and operations in the area of interest Sources of information include intelligence reach; research; data mining; database access via the digital topographic support system; academic studies, products, or materials; intelligence archives; and open-source intelligence. This task requires specific and detailed information for each weather factor. The Integrated Meteorological System (accessed through distributed common groundsystem-Army) provides commanders at all echelons of command with an automated weather system. This system receives, processes, and disseminates weather observations, forecasts, and decision aids for weather and environmental effects to all warfighting functions. The information, intelligence, products, and materials obtained are refined for use in mission analysis, intelligence preparation of the battlefield (IPB), and other planning tasks through functional analysis. (FM 2-0) (USAICoE)

No.	Scale	Measure
01	Yes/No	Unit obtained detailed information the recent and historical weather trends, seasonal patterns, aspects of weather, and weather zones in the operational environment.
02	Yes/No	Information obtained from external agencies, echelons, and partners was adequate for use in mission analysis, IPB, and planning.
03	Yes/No	Information obtained was used and integrated effectively into mission analysis, IPB, and planning.
04	Yes/No	Unit obtained detailed information on each weather factor.
05	Yes/No	Information obtained was used and integrated effectively to determine how the weather affected friendly and enemy forces and operations in the operational environment.
06	Percent	Of relevant information obtained from national agencies.
07	Percent	Of relevant information obtained from joint echelons.
08	Percent	Of relevant information obtained from multinational partners.
09	Percent	Of relevant information obtained from higher Army echelons.
10	Percent	Of relevant information obtained from lateral and subordinate echelons.

ART 2.1.4.4 OBTAIN DETAILED CIVIL CONSIDERATIONS INFORMATION AND INTELLIGENCE

2-16. Obtain detailed information and intelligence concerning the civil considerations—areas, structures, capabilities, organizations, people, events (ASCOPE)—within or affecting the expected operational environment through reach; research; data mining; database access; academic studies, products, or materials; intelligence archives; open-source intelligence; or other information sources to support operations, planning, execution, and commander's decisions. The data, information, intelligence, products, and material obtained are refined for use in mission analysis, intelligence preparation of the battlefield (IPB), and planning through functional analysis. This task requires specific and detailed information for each ASCOPE factor. This task entails obtaining detailed information and intelligence concerning the civil considerations within or affecting the expected area of interest. The intelligence section obtains this information from sources that include intelligence reach; research; data mining; database access; academic studies, products, or materials; intelligence archives; and open-source intelligence. The data, information, intelligence, products, and materials obtained are refined for use in mission analysis, IPB, and other planning tasks through functional analysis. This task develops specific and detailed information for each characteristic of civil considerations. (FM 2-0) (USAICoE)

Note: This task is supported by ART 2.2.1.5 (Conduct Geospatial Engineering Operations and Functions) and ART 2.2.5 (Conduct Police Intelligence Operations).

No.	Scale	Measure
01	Yes/No	Unit obtained detailed information on areas and structures within and affecting the operational environment.
02	Yes/No	Information obtained from external agencies, echelons, and partners was adequate for use in mission analysis, IPB, and planning.
03	Yes/No	Information obtained was used and integrated effectively into mission analysis, IPB, and planning.
04	Yes/No	Unit obtained detailed information on the capabilities within and affecting the operational environment.
05	Yes/No	Unit obtained detailed information on the organizations, people, and events within and affecting the operational environment.
06	Percent	Of relevant information obtained from national agencies.
07	Percent	Of relevant information obtained from joint echelons.
08	Percent	Of relevant information obtained from multinational partners.

No.	Scale	Measure
09	Percent	Of relevant information obtained from higher Army echelons.
10	Percent	Of relevant information obtained from lateral and subordinate echelons.

ART 2.1.4.5 COMPLETE STUDIES

2-17. To assist in achieving goals and objectives, this task entails providing the requesting command or organization with detailed information, assessments, and conclusions about the area of operations and area of interest. A study can be a systems or functional analysis product. It should be as detailed and in-depth as time allows. Studies provide knowledge that supports understanding of the local populations; cultures and caste system; societal systems or organizations; political systems and structures; religions practiced and their impacts; moral beliefs and their impacts; civil authority considerations; military organizations, structure, and equipment; and attitudes toward U.S., multinational, or host-nation forces. Studies can also include the views and attitudes of multinational and host-nation forces towards these factors. (FM 2-0) (USAICoE)

Note: This task can be conducted in support of ART 2.2 (Support to Situational Understanding).

No.	Scale	Measure
01	Yes/No	Study and conclusions provided commander and staff with necessary information for mission analysis and planning.
02	Yes/No	Unit completed study in time to support mission analysis and planning.
03	Yes/No	Study provided relevant information for mission analysis and planning.
04	Yes/No	Study did not reproduce data contained in readily available publications.
05	Yes/No	Study used maps, charts, and tables to portray data in easily understandable and retrievable formats.
06	Yes/No	Study included, as appropriate, areas (such as historical, religious, or culturally important buildings or landmarks) that could cause a negative view of U.S. forces if attacked or targeted.
07	Yes/No	Study included, as appropriate, industrial infrastructure, natural resources, and areas that may pose environmental or health threat, if attacked.
08	Yes/No	Information for the study was compiled per applicable regulations, policies, and procedures.
09	Yes/No	Unit completed study per applicable regulations, policies, and procedures.
10	Yes/No	Unit spent appropriate amount of time to complete the study.
11	Time	To identify shortfalls and gaps in available data to complete study.
12	Time	To request required data not available from available resources.
13	Percent	Of information requested and received from outside sources.
14	Percent	Of study information, as appropriate, accurate on or concerning the individual topics. Topics can include geography, history, population, culture and social structure, languages, religion, U.S. interests, civil defense, labor, legal, public administration, public education, public finance, public health, public safety, public welfare, civilian supply, civilian economics and commerce, food and agriculture, property control, public communications, public transportation, public works and utilities, arts, monuments, archives, civil information, cultural affairs, dislocated civilians, and host-nation support.

ART 2.1.4.5.1 Conduct Area, Regional, or Country Study of a Foreign Country

2-18. Study and provide mission-focused knowledge of the terrain and weather, civil considerations, and threat characteristics for a specified area or region of a foreign country— including the attitudes of the populace and leaders toward joint, multinational, or host-nation forces—to assist in achieving goals and objectives. Studies can also include the views and attitudes of multinational and host-nation forces. Studies provide detailed information, assessments, and conclusions on the areas of interest of the requesting command or organization. Studies should be as detailed as time allows. (FM 2-0) (USAICoE)

No.	Scale	Measure
01	Yes/No	Study provided commander and staff with necessary information for mission analysis and planning.
02	Yes/No	Unit completed study in time to support mission analysis and planning.
03	Yes/No	Study provided relevant information for mission analysis and planning.
04	Yes/No	Study did not reproduce data contained in readily available publications.
05	Yes/No	Study used maps, charts, and tables to portray data in easily understandable and retrievable formats.
06	Yes/No	Study included, as appropriate, areas (such as historical, religious, or culturally important buildings or landmarks) that could cause a negative view of U.S. forces if attacked or targeted.
07	Yes/No	Study included, as appropriate, industrial infrastructure, natural resources, and areas that may pose environmental or health threat, if attacked.
08	Yes/No	Unit spent appropriate time and resources to complete the study.
09	Yes/No	If study was requested from external organization, the request was made to the appropriate external organization.
10	Yes/No	If study was requested from external organization, the study was received in time to support mission analysis and planning.
11	Yes/No	If study was requested from external organization, the study provided relevant information for mission analysis and planning.
12	Yes/No	If study was requested from external organization, the study used maps, charts, and tables to portray data in easily understandable and retrievable formats.
13	Yes/No	If study was requested from external organization, amount and type of information adequate and accurate for the area, region, or country.
14	Yes/No	The initial study was maintained and updated.
15	Yes/No	Information for the study was compiled per applicable regulations, policies, and procedures.
16	Yes/No	Study was completed per applicable regulations, policies, and procedures.
17	Time	To identify shortfalls in available data to complete study.
18	Time	To request required data not available from available resources.
19	Time	Allocated in advance for a study by an external organization prior to mission analysis and planning.
20	Time	Prior to mission analysis and planning that the requested external agency used to deliver the study.
21	Time	From receipt of tasking until study was complete.
22	Percent	Of time available used toward completion of the study.
23	Percent	Of resources available used toward completion of the study.
24	Percent	Of information requested from outside sources provided by those outside sources.

ART 2.1.4.5.2 Conduct Specified Study

2-19. Study and provide focused knowledge of the terrain and weather, civil considerations, and threat characteristics for a specified topic or requirement. Studies provide the requesting command or organization with detailed information, assessments, and conclusions on the area of interest. Studies should be as detailed and in-depth as time allows. (FM 2-0) (USAICoE)

No.	Scale	Measure
01	Yes/No	Study provided commander and staff with the specified information for mission analysis and planning.
02	Yes/No	Unit completed study in time to support mission analysis and planning.

No.	Scale	Measure
03	Yes/No	Study provided relevant information for mission analysis and planning.
04	Yes/No	Study did not reproduce data contained in readily available publications.
05	Yes/No	Study used maps, charts, and tables to portray data in easily understandable and retrievable formats.
06	Yes/No	Study included, as appropriate, areas (such as historical, religious, or culturally important buildings or landmarks) that could cause a negative view of U.S. forces.
07	Yes/No	Study included, as appropriate, current industrial infrastructure, natural resources, and areas that may pose environmental or health threat.
08	Yes/No	Unit spent appropriate time and resources to complete the study.
09	Yes/No	If study was requested from external organization, the request was made to the appropriate external organization.
10	Yes/No	If study was requested from external organization, the study was received in time to support mission analysis and planning.
11	Yes/No	If study was requested from external organization, the study used maps, charts, and tables to portray data in easily understandable and retrievable formats.
12	Yes/No	If study was requested from external organization, amount and type of information was adequate and accurate for the area, region, or country.
13	Yes/No	The initial study was maintained and updated.
14	Yes/No	Information for the study was compiled per applicable regulations, policies, and procedures.
15	Yes/No	Study was completed per applicable regulations, policies, and procedures.
16	Time	To identify shortfalls in available data to complete study.
17	Time	To request required data not available from available resources.
18	Time	Allocated in advance for a study by an external organization prior to mission analysis and planning.
19	Time	Prior to mission analysis and planning that the requested external agency used to deliver the study.
20	Time	From receipt of tasking until study was complete.
21	Percent	Of time available used toward completion of the study.
22	Percent	Of resources available used toward completion of the study.
23	Percent	Of information requested from outside sources provided by those outside sources.

ART 2.1.5 TAILOR THE INTELLIGENCE FORCE

2-20. The generating force uses mission analysis to focus the allocation of intelligence resources for use by a joint task force or combatant commander as well as to support strategic objectives, the Army's mission, and operations at each echelon. Based on their own mission analysis, the staffs at each echelon allocate intelligence resources obtained through the generating force to support the commander's intent, guidance, and mission objectives. (FM 2-0) (USAICoE)

Note: This task supports ART 2.3.1 (Perform Intelligence, Surveillance, and Reconnaissance Synchronization).

No.	Scale	Measure
01	Yes/No	Unit used results of generate knowledge to determine the correct amount intelligence assets required to accomplish the mission.
02	Yes/No	Unit used results of generate knowledge to determine what existing national intelligence requirements need to be reallocated to Army Service component command (ASCC) intelligence services or organizations.

No.	Scale	Measure
03	Yes/No	If the national intelligence requirements were not reallocated, unit requested external intelligence support to accomplish the mission.
04	Yes/No	Unit kept intelligence assets at the ASCC level based on strategic requirements.
05	Yes/No	Unit reallocated intelligence assets to subordinate commanders based on the mission variables.
06	Yes/No	Unit reallocated intelligence assets to subordinate commanders based on operational requirements.
07	Yes/No	Unit reallocated intelligence assets to subordinate echelons as required by the ASCC and corps based on the results from generate knowledge.
08	Yes/No	Unit kept intelligence assets at the corps level based on operational requirements.
09	Yes/No	Intelligence assets allocated to the corps were adequate.
10	Yes/No	Minimum intelligence assets required for the corps mission were reevaluated periodically.
11	Yes/No	Unit reallocated intelligence assets to subordinate commanders based on the current operational environment and requirements.
12	Yes/No	Corps results from generate knowledge were used to determine the minimum intelligence assets required to accomplish the tactical mission.
13	Yes/No	Unit reallocated intelligence assets to subordinate echelons as required based on the corps results from generate knowledge.
14	Yes/No	Intelligence assets were kept at the tactical level commands—divisions, brigades, and battalions—based on tactical requirements.
15	Yes/No	Unit reallocated intelligence assets based on the subordinate commander's area of operation.
16	Yes/No	Intelligence assets allocated to the tactical level commands—divisions, brigades, and battalions—were adequate.
17	Yes/No	Minimum intelligence assets required for the mission were reevaluated periodically.
18	Percent	Of intelligence assets allocated based on the results of generate knowledge.
19	Percent	Of surplus national intelligence requirements reallocated to other intelligence services or organizations.
20	Percent	Of intelligence assets kept at the ASCC level that satisfy only the strategic requirements.
21	Percent	Of intelligence assets reallocated based on operational environment.
22	Percent	Of intelligence assets reallocated based on operational need.
23	Percent	Of intelligence assets allocated to the corps based on the ASCC results from generate knowledge.
24	Percent	Of intelligence assets kept at the corps level that satisfied only the operational requirements.
25	Percent	Of intelligence assets reallocated based on the current operational environment and requirements.
26	Percent	Of intelligence assets reallocated to subordinate units based on tactical need.
27	Percent	Of external intelligence assets received to accomplish the tactical mission.
28	Percent	Of intelligence assets allocated to the tactical level based on the results of corps generate knowledge.
29	Percent	Of intelligence assets were kept at the tactical level commands—divisions, brigades, and battalions—which satisfied only that command's requirements.
30	Percent	Of intelligence assets reallocated based on the subordinate commander's area of operation.
31	Percent	Of intelligence assets reallocated to subordinate units based on tactical requirements.
32	Percent	Of external intelligence assets received to accomplish the mission.

SECTION II – ART 2.2: SUPPORT TO SITUATIONAL UNDERSTANDING

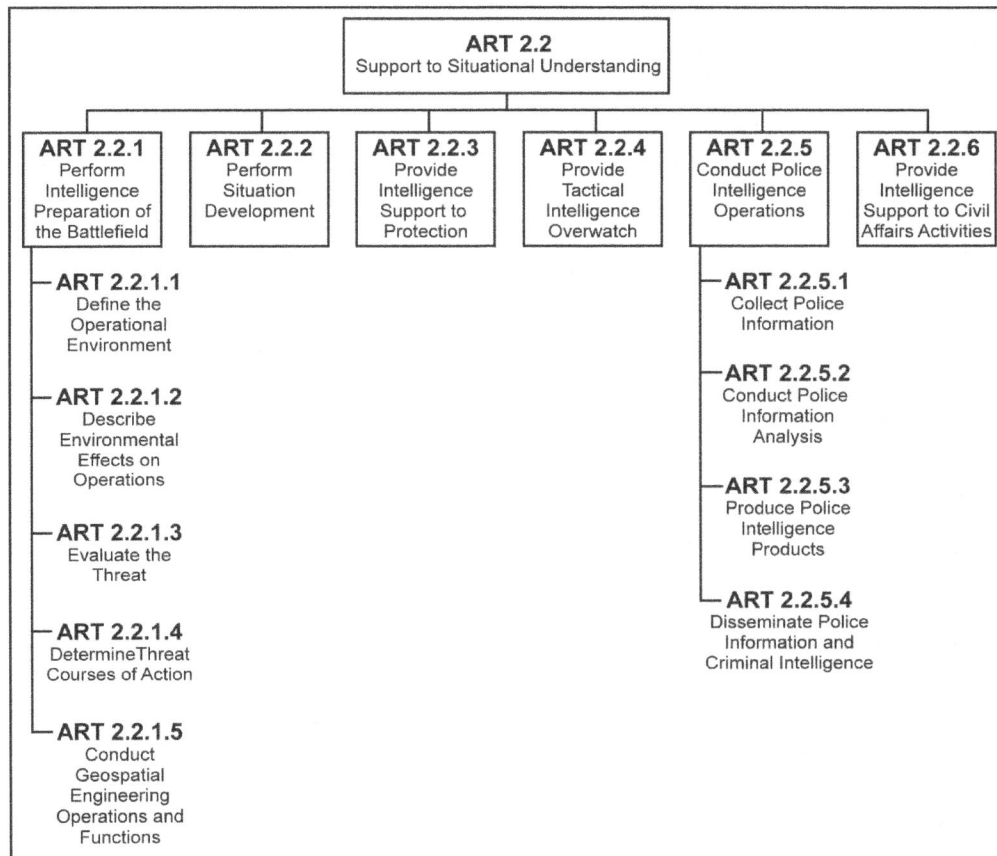

```
                              ART 2.2
                   Support to Situational Understanding

   ART 2.2.1      ART 2.2.2      ART 2.2.3      ART 2.2.4      ART 2.2.5      ART 2.2.6
   Perform        Perform        Provide        Provide        Conduct Police  Provide
   Intelligence   Situation      Intelligence   Tactical       Intelligence    Intelligence
   Preparation of Development     Support to     Intelligence   Operations      Support to Civil
   the Battlefield               Protection     Overwatch                      Affairs Activities

   ART 2.2.1.1                                                  ART 2.2.5.1
   Define the                                                   Collect Police
   Operational                                                  Information
   Environment

   ART 2.2.1.2                                                  ART 2.2.5.2
   Describe                                                     Conduct Police
   Environmental                                                Information
   Effects on                                                   Analysis
   Operations

   ART 2.2.1.3                                                  ART 2.2.5.3
   Evaluate the                                                 Produce Police
   Threat                                                       Intelligence
                                                                Products

   ART 2.2.1.4                                                  ART 2.2.5.4
   DetermineThreat                                              Disseminate Police
   Courses of Action                                            Information and
                                                                Criminal Intelligence

   ART 2.2.1.5
   Conduct
   Geospatial
   Engineering
   Operations and
   Functions
```

2-21. Support to situational understanding is the task of providing information and intelligence to commanders to assist them in achieving a clear understanding of the force's current state with relation to the threat and other aspects of the AO. It supports the commander's ability to make sound decisions. (FM 2-0) (USAICoE)

No.	Scale	Measure
01	Yes/No	Unit supported the commander's visualization of the battlefield and situational understanding of the threat.
02	Time	Required to provide intelligence products that facilitate the commander's visualization and situational understanding of the threat.
03	Percent	Of information and intelligence accurate in light of events.
04	Percent	Of information and intelligence requested by commander completed by latest time information and intelligence is of value.
05	Percent	Of produced intelligence judged relevant to military situations.
06	Percent	Of produced intelligence judged timely by users.
07	Percent	Of produced intelligence judged useable by users.

ART 2.2.1 PERFORM INTELLIGENCE PREPARATION OF THE BATTLEFIELD

2-22. Intelligence preparation of the battlefield (IPB) is a systematic process of analyzing and visualizing the portions of the mission variables of threat, terrain and weather, and civil considerations in a specific area of interest and for a specific mission. By applying IPB, commanders gain the information necessary to selectively apply and maximize operational effectiveness at critical points in time and space. IPB is a continuous staff planning activity undertaken by the entire staff. The staff aims to understand the operational environment and the options it presents to friendly and threat forces. (FM 2-01.3) (USAICoE)

No.	Scale	Measure
01	Yes/No	Products of the IPB process supported the preparation of running estimates and the military decisionmaking process by the supported commander and staff.
02	Yes/No	Unit intelligence staff, with the support of the entire staff, identified characteristics of the area of operations that will influence friendly and threat operations including terrain, civil considerations, weather, and threat considerations.
03	Yes/No	The entire staff, led by the intelligence staff, established limits to the area of interest following the commander's guidance.
04	Yes/No	Unit intelligence staff identified gaps in current intelligence holdings, identified information requirements, and recommended commander's critical information requirements.
05	Yes/No	Unit intelligence staff, with the support of the entire staff, identified characteristics of the battlefield that will influence friendly and threat operations.
06	Yes/No	Unit intelligence staff, with the support of the entire staff, described effects that military actions will have on future operations in the area of operations.
07	Yes/No	Unit intelligence staff, with the support of the entire staff, evaluated the threat.
08	Yes/No	Unit intelligence staff determined the threat course of action.
09	Yes/No	Unit intelligence staff validated templates with updated information.
10	Yes/No	Commander and other unit staff elements in addition to the intelligence staff participated in the process.
11	Yes/No	IPB products assisted the commander's and staff's visualization and decisionmaking.
12	Time	Since IPB products have been updated.
13	Time	To disseminate updated IPB products.
14	Percent	Of produced intelligence judged to be timely (latest time information and intelligence is of value) by users.
15	Percent	Of produced intelligence judged to be accurate in light of events.
16	Percent	Of produced intelligence judged to be useable by users.
17	Percent	Of produced intelligence judged to be complete based upon requests for clarification or expansion.
18	Percent	Of produced intelligence judged to be relevant to the military situation.

ART 2.2.1.1 DEFINE THE OPERATIONAL ENVIRONMENT

2-23. Initially examine the area of operations (AO), define the area of interest, identify other characteristics—such as the role of nongovernmental and international organizations—in the AO that influence friendly and threat operations, and identify gaps in current intelligence holdings. (FM 2-01.3) (USAICoE)

Note: The term "operational" in the title of this task does not refer to the operational level of war.

No.	Scale	Measure
01	Yes/No	Unit operations were delayed, disrupted, or canceled because the staff failed to identify environmental characteristics of the AO.
02	Yes/No	Commander, with the assistance of intelligence staff officer, identified the area of interest.
03	Yes/No	Unit staff identified gaps in current information databases.
04	Yes/No	Public affairs elements accompanied the maneuver brigades and established field media centers to take maximum advantage of these resources.
05	Yes/No	The entire staff identified significant characteristics of the operational environment, to include the status-of-forces agreement, rules of engagement, and other constraints on unit operations.
06	Yes/No	Staff collected and refined their input such as religious, health threat, terrain, meteorological, and hydrological data.
07	Time	To establish or obtain a database that defines the operational environment.
08	Time	Since intelligence preparation of the battlefield products addressing the operational environment have been updated.
09	Time	Required to determine threat characteristics and determine patterns of operations.
10	Percent	Of terrain, meteorological, and hydrological products produced and issued on time to assigned and gained units.
11	Percent	Of information such as religious, health threat, terrain, meteorological, and hydrological data received from higher headquarters or other organizations.
12	Percent	Of accuracy of threat characteristics, to include doctrine, tactics, techniques, and procedures data.
13	Percent	Of accuracy of operational forecasts and products, to include weather effects and terrain trafficability matrices, tide forecasts, light data, and tactical decision aids.
14	Percent	Of hazards in the AO identified and reported to appropriate headquarters.
15	Percent	Of noncombatants in the AO whose location is accurately reported to appropriate headquarters.
16	Percent	Of press coverage and threat propaganda that addresses friendly activities in an AO.
17	Percent	Of the local legal and illegal economy correctly identified in area studies.
18	Percent	Of local decisionmakers and centers of influence correctly identified in area studies. This includes official and unofficial leaders.

ART 2.2.1.2 DESCRIBE ENVIRONMENTAL EFFECTS ON OPERATIONS

2-24. The activity to evaluate all aspects of the area of operations (AO) with which all forces involved —enemy, friendly, and neutral—must contend with during the conduct of full spectrum operations. This includes the portions of the mission variables of terrain, civil considerations, weather, and threat considerations of the AO and the area of interest (FM 2-01.3) (USAICoE)

No.	Scale	Measure
01	Yes/No	Unit operations were delayed, disrupted, or canceled because the staff failed to identify environmental characteristics of the AO.
02	Yes/No	Unit staff identified gaps in current databases.
03	Yes/No	Staff identified and evaluated how terrain affects military operations and the corresponding positive or negative impacts of the operation on the terrain.
04	Yes/No	Staff identified and evaluated how weather affects military operations.
05	Yes/No	Products assisted the commanders and staff's visualization and decisionmaking.
06	Time	To establish or obtain initial operational environment (terrain, civil considerations, weather, and threat considerations) database.
07	Time	Required to evaluate the impact of combat operations and weather on trafficability of the terrain.
08	Time	Since intelligence preparation of the battlefield (IPB) products have been updated.

No.	Scale	Measure
09	Time	To disseminate updated IPB products.
10	Percent	Of analytical products produced and issued on time to assigned and task-organized units.
11	Percent	Of analytical products received from higher headquarters.
12	Percent	Of accuracy of products provided to unit elements.
13	Percent	Of entire staff effort devoted to updating initial operational environment (terrain, civil considerations, weather, and threat considerations) database.
14	Number	Of restrictions on friendly operations resulting from the presence and movement of noncombatants in the AO.

ART 2.2.1.3 EVALUATE THE THREAT

2-25. This activity analyzes current intelligence to determine how the threat normally organizes for combat and conducts operations. The evaluation includes each threat operating system as well as potential criminal organizations, factions, guerrillas, or insurgents. This step focuses on creating threat models and templates that depict how the threat operates when unconstrained by effects of the environment (FM 2-01.3) (USAICoE)

No.	Scale	Measure
01	Yes/No	The threat's capabilities were stated in the intelligence preparation of the battlefield process and accounted for in the military decisionmaking process.
02	Yes/No	Unit used pattern analysis, event analysis, and intelligence from higher headquarters and other organizations to create threat templates, models, and methods of operation and identify high-payoff targets and high-value targets.
03	Yes/No	Unit used Red Team concept to confirm or deny estimates and assessments about threat intentions.
04	Time	Required to incorporate new intelligence data and products into ongoing threat evaluations.
05	Time	To identify threat capabilities and limitations.
06	Time	To update or create threat templates, models, and methods of operation.
07	Time	To disseminate updated threat templates, models, and methods of operation.
08	Percent	Of correctly identified threat templates, models, and methods of operations; capabilities and limitations; high-payoff targets; high-value targets; and threat models.
09	Percent	Of new, processed intelligence integrated to update broad courses of action.

ART 2.2.1.4 DETERMINE THREAT COURSES OF ACTION

2-26. The activity to determine possible threat courses of action (COAs), describe threat COAs, rank COAs in probable order of adoption, and at a minimum identify the most probable and the most dangerous COAs. (FM 2-01.3) (USAICoE)

No.	Scale	Measure
01	Yes/No	The entire staff, under the direction of the intelligence staff, assessed the effects of friendly actions on threat COAs.
02	Yes/No	The threat's likely objectives and desired end state were identified, beginning with the threat command level at one echelon above the friendly unit and ending the process at two echelons below.
03	Yes/No	Unit staff identified opportunities and constraints that the operational environment offers or affords to threat and friendly forces.
04	Yes/No	Unit staff assessed effects of friendly actions on threat COAs.
05	Yes/No	Units considered threat capabilities, effects of the operational environment, and the threat's preference in operations.
06	Yes/No	Units determined most probable and most dangerous COAs and other threat COAs to a micro level of detail as time permitted.

No.	Scale	Measure
07	Yes/No	Units disseminated threat COAs to lower, adjacent, and next higher echelon.
08	Yes/No	Units delivered threat COAs in time to be of value for developing friendly COAs.
09	Yes/No	The G-3 or S-3 led staff war-gaming with full staff participation to validate and update assessments.
10	Time	Required to identify likely threat objectives and desired end states at different threat echelons of command.
11	Time	To identify and analyze the feasibility of each threat COA in terms of time, space, resources, and force ratios required to accomplish its objective.
12	Time	To evaluate and prioritize each identified threat COA.
13	Percent	Of threat branches and sequels correctly identified during planning.
14	Percent	Of new intelligence integrated to update threat COAs.
15	Percent	Of forecasted significant threat actions correctly identified during planning.
16	Percent	Of correctly identified threat high-value targets and tactical centers of mass (before selecting COAs).

ART 2.2.1.5 CONDUCT GEOSPATIAL ENGINEERING OPERATIONS AND FUNCTIONS

2-27. Geospatial engineer operations include functions of terrain analysis, data collection, data generation, database management, data manipulation and exploitation, cartographic production and reproductions, and geodetic survey. Geospatial engineering operations focus on data generation, data management, terrain analysis, and the presentation of their results to the commander. The functions are all interdependent so to prepare a geospatially accurate and timely enabled common operational picture for the commander. (FM 3-34.230) (USAES)

> *Note:* ART 5.3.3 (Facilitate Situational Understanding Through Knowledge Management) includes the processing of sensor data, the interpretation of data into intelligent information, fusion and integration of separate source data, management of the data to include accuracy and data topology, and dissemination of tactical data information. The inclusion of this task does not change the steps of the intelligence preparation of the battlefield process described in FM 2-01.3.

No.	Scale	Measure
01	Yes/No	The availability of accurate geospatial products allowed the commander to deploy and employ the weapon systems effectively.
02	Yes/No	The availability of accurate geospatial products allowed supported commanders and staffs to visualize their areas of operations, interest, and influence.
03	Yes/No	The availability of accurate geospatial products allowed supported commanders and staffs to target enemy systems effectively.
04	Yes/No	The availability of accurate geospatial products allowed supported commanders and staffs to plan air and ground missions efficiently.
05	Yes/No	The availability of accurate geospatial products allowed supported commanders and staffs to counter enemy weapons and intelligence-collection capabilities.
06	Yes/No	The availability of accurate geospatial products supported the commander's and staff's plans to avoid areas.
07	Time	To complete terrain analysis of an area of operations and prepare products supporting intelligence preparation of the battlefield.
08	Time	To complete geodetic survey in the area of operations.
09	Time	To produce and reproduce geospatial information in sufficient quantities to meet supported unit demand.
10	Number	And types of engineer topographic elements available to support unit operations.

ART 2.2.2 PERFORM SITUATION DEVELOPMENT

2-28. Situation development is a process for analyzing information and producing current intelligence concerning the portions of the mission variables of enemy, terrain, and weather, and civil considerations within the area of operations before and during operations. The process helps the intelligence officer recognize and interpret indicators of threat intentions, and objectives. Situation development confirms or denies threat courses of actions (COAs), provides threat locations, explains what the threat is doing in relation to the friendly force commander's intent, and provides an estimate of threat combat effectiveness. The locations and actions of noncombatant elements and nongovernmental organizations in the area of operations that may impact operations should also be considered. Through situation development, the intelligence officer quickly identifies information gaps, explains threat activities in relation to the unit's operations, and assists the commander in gaining and maintaining situational understanding. Situation development helps the commander make decisions, including when to execute branches and sequels. (FM 2-0) (USAICoE)

No.	Scale	Measure
01	Yes/No	The COA executed by threat forces was predicted during the intelligence preparation of the battlefield process and accounted for in the military decisionmaking process.
02	Yes/No	Threat situation development provided information that helps the commander make decisions to execute branches and sequels.
03	Yes/No	Unit disseminated situation development intelligence to friendly forces.
04	Time	To disseminate situation development intelligence.
05	Time	To produce an updated situation template.
06	Time	To confirm or deny the existing estimate of the threat COA and update the estimate based on current terrain, civil considerations, weather, and threat characteristics.
07	Percent	Of produced intelligence judged accurate based on threat activity.
08	Percent	Of produced intelligence judged complete based on request for clarification or additional information.
09	Percent	Of produced intelligence judged useable to the current situation.
10	Percent	Of produced intelligence judged timely by consumers and users.

ART 2.2.3 PROVIDE INTELLIGENCE SUPPORT TO PROTECTION

2-29. This task includes providing intelligence that supports measures the command takes to remain viable and functional by protecting itself from the effects of threat activities. It also provides intelligence that supports recovery from threat actions. This task supports the protection warfighting function and is linked to antiterrorism and homeland security. (FM 2-0) (USAICoE)

Note: This task branch supports ART 6.7 (Conduct Survivability Operations) and ART 6.6.1 (Identify Potential Terrorist Threats and Other Threat Activities, ART 6.6 Apply Antiterrorism Measures, ART 6.2.5 Support Homeland Security Personnel Recovery Operations).

No.	Scale	Measure
01	Yes/No	Unit established force protection plan with commander's critical information requirements and reporting channels and requirements.
02	Yes/No	Unit provided information on incidents by threat forces affecting security of the force.
03	Yes/No	Unit provided intelligence support to the antiterrorism program and force protection mission.
04	Yes/No	Intelligence provided to support the antiterrorism program and force protection mission.
05	Percent	Of intelligence relating to potential hostile acts against U.S. forces or installations.
06	Percent	Of intelligence relating to potential criminal acts against U.S. forces or installations.
07	Percent	Of intelligence relating to chemical, biological, radiological, nuclear, and high-yield explosives employment and deployment.

No.	Scale	Measure
08	Percent	Of intelligence to support theater missile defense operations.
09	Percent	Of intelligence relating to threat surveillance and reconnaissance elements.
10	Percent	Of intelligence provided for combatant command to support homeland security or defense.

ART 2.2.4 PROVIDE TACTICAL INTELLIGENCE OVERWATCH

2-30. Tactical intelligence overwatch is creating standing, fixed analytical intelligence capabilities that provide dedicated intelligence support to committed maneuver units. The tactical intelligence overwatch element is connected through a shared intelligence network that can extract information from multiple sources and provide succinct answers directly to supported units when time is critical. (FM 2-0) (USAICoE)

Note: This task branch supports ART 6.7 (Conduct Survivability Operations).

No.	Scale	Measure
01	Yes/No	The tactical intelligence overwatch cell provided succinct information directly to the supported units.
02	Yes/No	Tactical intelligence forces provided a foundation of regional and subject matter expertise.

ART 2.2.5 CONDUCT POLICE INTELLIGENCE OPERATIONS

2-31. Police intelligence operations (PIO) is a military police (MP) function, integrated within all MP operations, which supports the operations process through analysis, production, and dissemination of information collected as a result of police activities to enhance situational understanding, protection, civil control, and law enforcement. Upon analysis, this information may contribute to commander's critical information requirements, intelligence-led, time-sensitive operations, or policing strategies necessary to forecast, anticipate, and preempt crime or related disruptive activities to maintain order. Police intelligence results from the application of systems, technologies, and processes that analyze applicable data and information necessary for situational understanding and focusing policing activities to maintain law and order. (FM 3-19.50) (USAMPS)

Notes: PIO function is not an intelligence discipline; it is a law enforcement function. However, it is within the critical intelligence task, "support situational understanding" that PIO best support the Army operations process and informs the intelligence process. PIO are essential to this task, particularly where irregular threats (criminal, terrorist, and insurgents) threaten the security of U.S. forces and military operations. This function supports and enhances the commander's situational awareness and common operational picture through collection, analysis, and appropriate dissemination of relevant criminal, police information, and police intelligence. PIO are a vital tool of law enforcement and criminal investigators that distributes and focuses military police and criminal investigations assets. U.S. Codes, executive orders, Department of Defense (DOD) directives, and Army regulations contain specific guidance regarding the prohibition of intelligence personnel from collecting intelligence on U.S. citizens, U.S. corporations, and resident aliens. Any access by the intelligence community to information or products resulting from PIO directed against U.S. citizens should undergo competent legal review.

No.	Scale	Measure
01	Yes/No	The conduct of PIO did not violate U.S. Code and applicable DOD and Army regulations against collecting intelligence on U.S. citizens.
02	Yes/No	PIO contributed to reducing criminal activity in the area of operations (AO).
03	Yes/No	Unit updated police intelligence products as additional police and criminal information and related data was collected.

No.	Scale	Measure
04	Yes/No	Unit obtained legal coordination and authorization before using technical listening equipment and technical surveillance equipment.
05	Yes/No	Commander provided list of critical assets to be protected to provost marshal.
06	Yes/No	Unit identified patterns or trends relevant to proactive law and order operations.
07	Yes/No	Military law enforcement received police and criminal information from the host nation.
08	Yes/No	Unit maintained information on known or suspected criminals per regulatory and legal guidance.
09	Yes/No	Unit established activity files, when applicable.
10	Yes/No	Unit established area files, when applicable.
11	Yes/No	Unit safeguarded juvenile records from unauthorized disclosure.
12	Yes/No	Unit established and maintained liaison with host-nation authorities, military and civilian law enforcement agencies, and other organizations as required by the factors of mission, enemy, terrain and weather, troops and support available, time available, civil considerations.
13	Yes/No	Unit provided tactical intelligence obtained to echelon intelligence staff.
14	Yes/No	Unit collected police and criminal information per the intelligence, surveillance, and reconnaissance plan.
15	Yes/No	Unit disseminated police and criminal information to appropriate agencies.
16	Yes/No	Police intelligence requirements and integration were incorporated into the operation plan or order.
17	Yes/No	Police and criminal databases were updated as unit received police and criminal information and data.
18	Time	To determine susceptibility of critical facilities to criminal threat.
19	Time	To identify military resources susceptible to theft and diversion.
20	Time	To identify criminal trends and patterns based upon PIO.
21	Time	To identify trends and patterns by continued association with identified offenders.
22	Time	To develop tactics, techniques, and procedures of operations to assist in eliminating or reducing vulnerability to criminal threat activities.
23	Time	To coordinate with the host nation for police and criminal information.
24	Time	To review internally created police and criminal information to see if information can answer police and criminal information requirements.
25	Time	To coordinate police and criminal information collection by organic assets.
26	Time	To coordinate with staff intelligence officer for required counterintelligence support.
27	Time	To recruit and develop police informants.
28	Percent	Of identified offenders linked to criminal trends.
29	Percent	Of available U.S. police intelligence resources in the AO identified by military law enforcement.
30	Percent	Of criminal-related activities reported by informants prior to their occurrence.
31	Percent	Of terrorist-related activities reported by informants prior to their occurrence.
33	Percent	Of criminal and other potentially disruptive elements in an AO identified before their committing hostile acts against U.S. interests and host-nation elements.

ART 2.2.5.1 COLLECT POLICE INFORMATION

2-32. Collection of police information is a continuous activity. Military police identify gaps in existing police information and develop intelligence requirements. This collection can be completed through several means; military police patrols, police engagement, criminal investigations, collected evidence, database queries and the use of reachback centers. Collection efforts also assist in enhancing protection operations and antiterrorism through identifying potential criminal threats and other threat activities. (FM 3-19.50) (USAMPS)

No.	Scale	Measure
01	Yes/No	The collection of police and criminal information did not violate U.S. Code and applicable Department of Defense and Army regulations against collecting intelligence on U.S. citizens.
02	Yes/No	Police and criminal information collected allowed the unit to prevent or reduce criminal activity in its area of operations (AO).
03	Yes/No	Unit updated police and criminal intelligence products as additional police and criminal information and data was collected.
04	Yes/No	Unit obtained legal coordination and authorization before using technical listening equipment and technical surveillance equipment.
05	Yes/No	Military law enforcement received police and criminal information from the host nation.
06	Yes/No	Unit maintained liaison with host-nation authorities, military and civilian police agencies, and other organizations as required by the factors of mission, enemy, terrain and weather, troops and support available, time available, civil considerations.
07	Yes/No	Unit disseminated police and criminal information to appropriate headquarters and agencies.
08	Yes/No	Unit recruited and developed police informants in the AO.
09	Time	To coordinate with the host nation for police and criminal information.
10	Time	To coordinate police and criminal information collection by organic assets.
11	Time	To coordinate with staff intelligence officer for required counterintelligence support.
12	Time	To develop police and criminal information requirements.
13	Time	To perform law enforcement patrols designed to obtain police and criminal information.
14	Percent	Of known terrorist- or criminal-related activities reported by informants before their occurrence.
15	Percent	Of identified individuals engaged in terrorist- or criminal-related activities reported by informants after their occurrence.
16	Percent	Of criminal and other potentially disruptive elements in an AO identified before they commit hostile acts against U.S. interests and host-nation elements.
17	Percent	Of criminal activities (such as smuggling, vice, counterfeiting, narcotics, extortion, rape, murder, robbery) occurring in an AO identified and reported.
18	Percent	Of AO covered by law enforcement patrols (mounted and dismounted).

ART 2.2.5.2 CONDUCT POLICE INFORMATION ANALYSIS

2-33. Conducting police information analysis is the process that organizes, analyzes, and interprets police information to police intelligence. Police information and intelligence contributes to all-source analysis and the Army operations process. It also assesses police and criminal information to identify trends and patterns of criminal activity. (FM 3-19.50) (USAMPS)

No.	Scale	Measure
01	Yes/No	The police information analysis did not violate U.S. Code and applicable Department of Defense and Army regulations against collecting intelligence on U.S. citizens.
02	Yes/No	The police information analysis allowed the unit to reduce criminal activity in its area of operations (AO).
03	Yes/No	Unit coordinated funds to establish and maintain a police informant operation.
04	Yes/No	Unit identified patterns or trends relevant to proactive law and order operations.
05	Yes/No	Unit analyzed police information and produced actionable criminal intelligence.
06	Yes/No	Unit recruited and developed police informants in the AO.
07	Time	To prepare annex K for provost marshal operation plan and order as required.
08	Time	To identify and assess latest criminal information collected.
09	Time	To identify criminal information resources in the AO.
10	Time	To identify criminal trends and patterns developed in the AO.

No.	Scale	Measure
11	Time	To analyze external criminal information reports.
12	Time	To assess internally created police information.
13	Time	To produce criminal information bulletins and alert notices.
14	Percent	Of criminal priority intelligence requirement collection efforts directed toward subordinate units.
15	Percent	Of available criminal intelligence resources in the AO.
16	Percent	Of known terrorist- and criminal-related activities reported by informants before their occurrence.
17	Percent	Of identified individuals engaged in terrorist- and criminal-related activities reported by informants after their occurrence.
18	Percent	Of accurate reported terrorist- and criminal-related activities.

ART 2.2.5.3 PRODUCE POLICE INTELLIGENCE PRODUCTS

2-34. Police intelligence operations use the intelligence process to produce police intelligence products used by military police leaders. These products focus police operations and contribute to the Army operations process. These products include standardized police information formatted for rapid dissemination, criminal threat assessments based on police information analysis, and assessment and analysis of police capability and capacity. (FM 3-19.50) (USAMPS)

No.	Scale	Measure
01	Yes/No	Developing the police intelligence products did not violate U.S. Code and applicable Department of Defense and Army regulations against collecting intelligence on U.S. citizens.
02	Yes/No	The police intelligence products developed enabled the unit to reduce criminal activity in its area of operations (AO).
03	Yes/No	Unit developed police intelligence products based on police and criminal information collected from external and internal sources.
04	Yes/No	Unit identified patterns or trends relevant to the conduct of proactive law and order operations.
05	Yes/No	Unit produced actionable police intelligence products.
06	Yes/No	Unit disseminated relevant police and criminal intelligence through military law enforcement and civilian and host-nation networks.
07	Yes/No	Products produced were relevant and answered identified police intelligence requirements.
08	Yes/No	Unit provided police information and police intelligence to the military intelligence community within applicable legal constraints.
09	Yes/No	Unit provided support to the police intelligence fusion cell.
10	Time	To identify police and criminal information resources in the AO.
11	Time	To identify criminal trends and patterns developed in the AO.
12	Time	To analyze external police and criminal information reports.
13	Time	To assess internally created police and criminal information.
14	Time	To produce police and criminal information bulletins and alert notices.

ART 2.2.5.4 DISSEMINATE POLICE INFORMATION AND CRIMINAL INTELLIGENCE

2-35. Police information and criminal intelligence is disseminated for use by law enforcement to focus policing activities. Police intelligence products are disseminated and integrated within the operations process enhancing situational understanding, mission planning, and execution at every echelon. These products may be disseminated in support of host-nation law enforcement in combating crime and neutralizing criminal threats to military operations based on trend and pattern analysis and shared with other law enforcement agencies. When legally allowable, these products are provided to the military intelligence community for fusion and incorporation to the all-source intelligence effort, contributing to a more complete intelligence picture. (FM 3-39) (USAMPS)

Note: The dissemination of police information and criminal intelligence is included in ART 5.8.1 (Conduct Law and Order Operations), ART 5.8.1.1 (Perform Law Enforcement), ART 5.8.1.2 (Conduct Criminal Investigations), ART 5.8.1.6 (Provide Customs Support), ART 4.4 (Conduct Internment and Resettlement Operations), and ART 7.3.2.3 (Perform Host-Nation Police Training and Support). Joint, interagency, and multination coordination is included in ART 5.2.2.1 (Prepare the Command Post for Displacement). (FM 3-39 and FM 3-19.50) (USAMPS)

No.	Scale	Measure
01	Yes/No	Law enforcement agency personnel investigated offenses against Army forces or property committed by persons subject to military law.
02	Yes/No	Law enforcement agency personnel referred offenses against Army forces or property committed by persons subject to military law.
03	Yes/No	Law enforcement agency personnel monitored all ongoing investigations in the theater of operation.
04	Yes/No	Law enforcement agency personnel received final reports from subordinate elements.
05	Yes/No	Law enforcement agency personnel coordinated with Army Criminal Investigation Command for investigation of all major incidents (deaths, serious bodily injury, and war crimes).
06	Yes/No	Law enforcement agency personnel took control of crime scene.
07	Yes/No	Law enforcement agency personnel identified personnel involved in the crime.
08	Yes/No	A senior law enforcement agency individual formulated an investigative plan.
09	Yes/No	Law enforcement agency personnel processed the crime scene. Processing includes recording investigative notes and sketches, photographing crime scene as necessary, and collecting physical evidence for evaluation.
10	Yes/No	Law enforcement agency personnel released the crime scene to appropriate individuals.
11	Yes/No	Law enforcement agency personnel pursued immediate leads. Pursuing leads includes: interviewing victims and witnesses, obtaining written statements (if appropriate), advising suspects of legal rights, obtaining written statements from suspects, collecting related documents necessary to support specific investigations, and coordinating with the judge advocate office as necessary.
12	Yes/No	Investigators law enforcement agency personnel modified standard interview techniques to overcome any language barriers and cultural differences.
13	Yes/No	Investigators maintained a disciplined and systematic approach in their questioning when doing a long series of interviews on the same basic topic.
14	Yes/No	Law enforcement agency personnel continued the investigation as necessary. Tasks include completing evidence documentation, releasing evidence to evidence custodian, identifying need for crime lab analysis of evidence, obtaining other needed specialized investigative support (polygraph or technical listening equipment), gathering related police intelligence operations from other sources, performing surveillance, submitting status reports, and coordinating investigative efforts and findings with supporting Staff Judge Advocate office.
15	Yes/No	Law enforcement agency personnel closed the case by preparing final report.
16	Time	To complete crime analysis.
17	Percent	Of investigations of minor reported incidents.
18	Percent	Of investigations conducted and reported per AR 195-2.
19	Percent	Of returns on deficient reports of investigations for corrective action or for further investigative activity.
20	Percent	Of crime analysis performed correctly.

No.	Scale	Measure
21	Percent	Of case documents and required reports prepared per legal, regulatory, and standing operating procedure guidance.
22	Number	Of hotline complaints referred to criminal investigation division or military police investigation.

ART 2.2.6 PROVIDE INTELLIGENCE SUPPORT TO CIVIL AFFAIRS ACTIVITIES

2-36. Military intelligence organizations performing this task collect and provide information and intelligence products concerning civil considerations in support of civil affairs activities. (FM 2-0) (USAICoE)

No.	Scale	Measure
01	Yes/No	Intelligence support required for civil affairs identified through intelligence preparation of the battlefield.
02	Yes/No	Unit provided intelligence information to identify pertinent civil considerations that affect civil affairs operations.
03	Yes/No	Unit provided intelligence support to identify populace perceptions, sentiments, attitudes, mores, allegiances, alliances, and behaviors.
04	Yes/No	Unit provided intelligence support to identify indigenous population and institutions, nongovernmental, and intergovernmental organizations in the area of operations (AO).
05	Yes/No	Unit identified all relevant government agencies, organizations, or departments that effect civil affairs activities.
06	Yes/No	Unit provided intelligence support to identify trends reflected by the national and international media.
07	Yes/No	Unit provided support for to identify resources and capabilities of host-nation population and institutions, nongovernmental, and intergovernmental organizations in the AO.
08	Yes/No	Unit provided intelligence and information to civil affairs within all applicable regulations, policies, and laws.
09	Time	To identify pertinent civil considerations which affect civil affairs activities.
10	Time	To identify populace perceptions, sentiments, attitudes, mores, allegiances, alliances, and behaviors.
11	Time	To identify nongovernmental and international organizations in the AO.
12	Time	To identify all relevant government agencies, organizations, or departments that affect civil affairs activities.
13	Time	To identify trends reflected by the national and international media.
14	Time	To identify resources and capabilities of nongovernmental and international organizations in the AO.
15	Percent	Of accurate information and intelligence provided to civil affairs for civil considerations that affect civil affairs activities.

SECTION III – ART 2.3: PERFORM INTELLIGENCE, SURVEILLANCE, AND RECONNAISSANCE

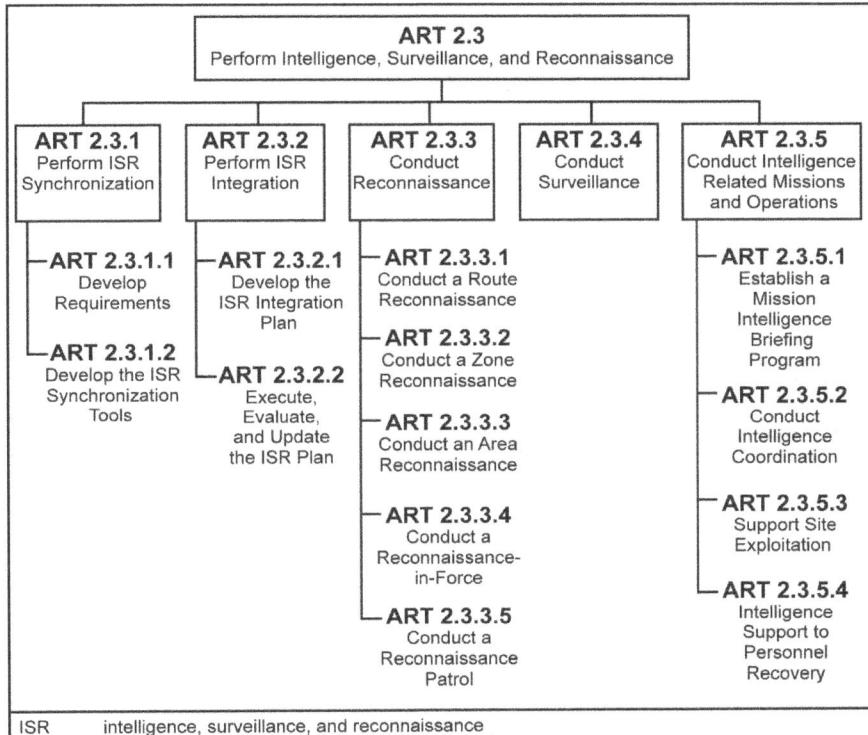

2-37. *Intelligence, surveillance, and reconnaissance* is an activity that synchronizes and integrates the planning and operation of sensors, assets, and processing, exploitation, and dissemination systems in direct support of current and future operations. This is an integrated intelligence and operations function. For Army forces, this combined arms operation focuses on priority intelligence requirements while answering the commander's critical information requirements (CCIRs). Through intelligence, surveillance, and reconnaissance (ISR), commanders and staffs continuously plan, task, and employ collection assets and forces. These forces collect, process, and disseminate timely and accurate information, combat information, and intelligence to satisfy the CCIRs and other intelligence requirements. When necessary, ISR assets focus on special requirements, such as personnel recovery. (FM 3-90) (USACAC)

No.	Scale	Measure
01	Yes/No	Unit developed a strategy to answer each CCIR to accomplish ISR integration.
02	Yes/No	Unit developed requirements that supported the commander's decisionmaking.
03	Yes/No	Unit planned scheme of support including fires, routes of movement and maneuver, air corridors, medical and casualty evacuation, and sustainment.
04	Yes/No	Unit planned for unique support requirements for ISR assets to include maintenance, crew training, downlink nodes, access, and connectivity.
05	Yes/No	Unit developed a detailed ISR plan through a staff process.
06	Yes/No	Unit task-organized ISR forces to perform tasks and accomplish missions.

No.	Scale	Measure
07	Yes/No	Unit developed a debriefing program to capture information observed or gathered by Soldiers (Soldier surveillance and reconnaissance) conducting the ISR operations.
08	Yes/No	Unit satisfied CCIRs using intelligence reach.
09	Yes/No	Unit submitted requests for information to satisfy CCIRs that the commander lacks the organic assets to answer.
10	Yes/No	The operations officer, with the intelligence officer, tasked and directed the available ISR assets to answer the CCIRs.
11	Yes/No	Unit received additional requests from higher headquarters to collect, produce, or disseminate combat information or intelligence products.
12	Yes/No	Unit evaluated each requirement for completion based on reported information.
13	Yes/No	Unit redirected or retasked ISR assets based on reported information.
14	Time	To redirect or retask ISR assets.
15	Percent	Of commander's requirements answered through intelligence reach.
16	Percent	Of commander's requirements answered through requests for information.

ART 2.3.1 PERFORM INTELLIGENCE, SURVEILLANCE, AND RECONNAISSANCE SYNCHRONIZATION

2-38. *Intelligence, surveillance, and reconnaissance synchronization* is the task that accomplishes the following: analyzes information requirements and intelligence gaps; evaluates available assets internal and external to the organization; determines gaps in the use of those assets; recommends intelligence, surveillance, and reconnaissance (ISR) assets controlled by the organization to collect on the commander's critical information requirements; and submits requests for information for adjacent and higher collection support. (FM 3-0) (USACAC)

Note: This task ensures that ISR, intelligence reach, and requests for information successfully report, produce, and disseminate information, combat information, and intelligence to support decision making. The intelligence officer, in coordination with the operations officer and other staff elements as required, synchronizes the entire collection effort. This effort includes assets the commander controls and those of adjacent and higher echelon units and organizations. It also uses intelligence reach to answer the commander's critical information requirements (CCIRs) and other requirements.

No.	Scale	Measure
01	Yes/No	Unit accomplished ISR synchronization by developing a strategy to answer each priority intelligence requirement (PIR).
02	Yes/No	Unit developed requirements that supported the commander's decisionmaking.
03	Yes/No	Unit determined unique support requirements for ISR assets to include maintenance, crew training, downlink nodes, access, and connectivity.
04	Yes/No	Unit determined scheme of support requirements including fires, routes of movement and maneuver, medical and casualty evacuation, and sustainment.
05	Yes/No	Unit developed and provided ISR task organization recommendations to the operations officer.
06	Yes/No	Unit satisfied CCIRs using intelligence reach and through the request for information (RFI) process.
07	Yes/No	Unit submitted requests for information to satisfy CCIRs that the commander lacks the organic assets to answer.
08	Yes/No	Unit requested collection, products, or dissemination from higher headquarters, adjacent units, or multinational partners to facilitate ISR operations.
09	Yes/No	Unit evaluated each requirement for completion based on reported information.

No.	Scale	Measure
10	Yes/No	Unit delivered the ISR synchronization plan to the operations officer.
11	Yes/No	Unit accomplished ISR synchronization by developing a strategy to answer each PIR.
12	Yes/No	Unit developed requirements that supported the commander's decisionmaking.
13	Percent	Of commander's requirements answered through intelligence reach and through the RFI process.
14	Percent	Of commander's requirements answered through requests for information and through collection, production, or dissemination from higher headquarters.

ART 2.3.1.1 DEVELOP REQUIREMENTS

2-39. The intelligence staff develops a prioritized list focusing on what information it needs to collect to produce intelligence. Additionally, the intelligence staff dynamically updates and adjusts the requirements in response to mission adjustments and changes. Each requirement is assigned a latest time information is of value to meet operational requirements. (FM 2-0) (USAICoE)

No.	Scale	Measure
01	Yes/No	Staff analyzed commander's critical information requirements (CCIRs), including requests for information from lower echelons, adjacent units and organizations, and orders from higher echelons.
02	Yes/No	Unit identified initial intelligence requirements during intelligence preparation of the battlefield.
03	Yes/No	Unit prioritized and validated all information and intelligence requirements by the staff.
04	Yes/No	Commander approved the prioritized and validated requirements developed by the staff.
05	Yes/No	Unit identified mission-specific priority intelligence requirements (PIRs) during the military decisionmaking process course of action (COA) development.
06	Yes/No	Unit performed wargaming for each COA.
07	Yes/No	Unit identified intelligence gaps.
08	Yes/No	Unit developed indicators for each PIR and friendly force information requirements.
09	Yes/No	Unit developed a specific information requirement for each PIR and friendly force information requirement.
10	Time	To validate and incorporate PIR from higher, lower, and adjacent units.
11	Time	Before next phase of an operation when the PIR are validated or updated.
12	Time	Required for the commander to approve the updated PIR.
13	Time	To identify intelligence requirements and if necessary dynamically update or revise in advance of collection.
14	Time	To prioritize information and intelligence collection requirements.
15	Percent	Of PIRs addressed in intelligence update.
16	Percent	Of subordinate echelon requirements supported by the echelon's information and intelligence requirements.
17	Percent	Of invalidated requirements returned to or resubmitted by originating agency or office.
18	Percent	Of validated requirements have ongoing collection efforts directed towards answering requirements.
19	Percent	Of requirements submitted by multiple organizations.
20	Percent	Of PIRs tied to commander's decision points.
21	Percent	Of PIRs linked to specific information requirements.
22	Number	Of open CCIRs and PIRs at any one time.
23	Number	Of requirements identified after collection begins.

ART 2.3.1.2 Develop Intelligence, Surveillance, and Reconnaissance Synchronization Tools

2-40. The entire unit staff develops their information requirements and determines how best to satisfy them. The staff uses surveillance and reconnaissance assets to collect information. There are at least three intelligence, surveillance, and reconnaissance (ISR) synchronization tools: the requirements management matrix, ISR synchronization matrix, and ISR overlay. ISR synchronization tools address all assets the operations officer can task or request as well as the coordinating mechanisms needed to ensure adequate coverage of the area of interest. ISR tools are used to develop the ISR plan. (FM 2-0) (USAICoE)

Note: This task supports ART 2.3.2.1 (Develop the Intelligence, Surveillance, and Reconnaissance Integration Plan)

No.	Scale	Measure
01	Yes/No	Intelligence officer, with staff participation, used the priority intelligence requirements (PIRs) to begin ISR synchronization.
02	Yes/No	Unit identified all organic, adjacent, and higher echelon ISR assets.
03	Yes/No	Intelligence officer and operations officer, with staff participation, determined asset availability.
04	Yes/No	Unit determined unique support requirements for ISR assets to include maintenance, crew training, downlink nodes, access, and connectivity.
05	Yes/No	Unit developed recommended scheme of support including fires, routes of movement and maneuver, medical and casualty evacuation, and sustainment.
06	Yes/No	Intelligence officer, with staff participation, determined availability and capabilities of assets from higher echelons.
07	Yes/No	Unit identified the reporting criteria, capabilities, limitations, and latest time information is of value for all ISR assets.
08	Yes/No	Unit possessed the right mix of ISR assets for the area of operations.
09	Yes/No	Unit requested additional ISR assets or support when organic assets and attachments cannot provide coverage.
10	Yes/No	ISR synchronization plan addressed redundancy, mix, and cueing.
11	Yes/No	ISR synchronization plan developed and linked to information and intelligence requirements.
12	Yes/No	ISR synchronization plan included the required action or production action needed once requirements were answered.
13	Yes/No	ISR synchronization plan included the reporting procedures and channels for each requirement.
14	Yes/No	Intelligence production answered requirements (specifically PIRs).
15	Yes/No	Unit disseminated ISR synchronization plan to operations officer in time to produce orders and perform troop leading procedures.
16	Time	To determine availability of ISR assets.
17	Time	To develop ISR synchronization plan.
18	Percent	Of requirements identified that is addressed in the ISR synchronization plan.
19	Percent	Of requirements identified that has to be supported by higher echelons.
20	Percent	Of available ISR assets identified.
21	Percent	Of requirements analyzed, validated, and prioritized.

ART 2.3.2 PERFORM INTELLIGENCE, SURVEILLANCE, AND RECONNAISSANCE INTEGRATION

2-41. *Intelligence, surveillance, and reconnaissance integration* is the task of assigning and controlling a unit's intelligence, surveillance, and reconnaissance assets (in terms of space, time, and purpose) to collect and report information as a concerted and integrated portion of operation plans and orders. Intelligence, surveillance, and reconnaissance (ISR) integration assigns and controls a unit's ISR assets (in terms of space, time, and purpose) to collect and report information as a concerted and integrated portion of operation plans and orders. The operations officer integrates the best ISR assets through a deliberate and coordinated effort of the entire staff across all warfighting functions into the operation. Specific information requirements facilitate tasking by matching requirements to assets. Intelligence requirements are identified, prioritized, and validated. ISR integration is vital in controlling limited ISR assets. During ISR integration, the staff recommends redundancy, mix, and cue as appropriate. The result of ISR synchronization and integration is an effort focused on answering the commander's requirements. (FM 3-90) (USACAC)

No.	Scale	Measure
01	Yes/No	ISR plan focused on the commander's critical information requirements.
02	Yes/No	Staff determined unique support requirements for ISR assets to include maintenance, crew training, downlink nodes, access, and connectivity.
03	Yes/No	Staff determined scheme of support including fires, routes of movement and maneuver, medical and casualty evacuation, and sustainment.
04	Yes/No	ISR plan included other staff information requirements.
05	Yes/No	ISR plan provided information and intelligence in time for the commander's decisionmaking.
06	Yes/No	ISR plan included a dissemination plan, to include information tasked or requested from external organizations.
07	Yes/No	Operations officer disseminated ISR plan in time to produce orders and perform troop leading procedures.
08	Yes/No	Staff executed ISR plan.
09	Yes/No	Unit identified indicators of previously unanticipated threat patterns or activities.
10	Yes/No	Unit identified new requirements based on reporting and ISR operations.
11	Yes/No	ISR plan updated and disseminated in time to produce orders.
12	Time	To produce and disseminate initial ISR plan.
13	Time	To evaluate reporting and ISR operations.
14	Time	To determine threat patterns or activities from reporting and ISR operations.
15	Time	To identify indicators of previously unanticipated threat patterns or activities.
16	Time	To review existing information on threat capabilities.
17	Time	To consider threat doctrine.
18	Time	To determine new information and intelligence requirements.
19	Time	To develop new ISR tasks and requests for information.
20	Time	To update and disseminate the ISR plan.
21	Percent	Of specific information and intelligence requirements correctly answered.
22	Percent	Of reporting requirements correctly answered.

ART 2.3.2.1 DEVELOP THE INTELLIGENCE, SURVEILLANCE, AND RECONNAISSANCE INTEGRATION PLAN

2-42. The operations officer develops the intelligence, surveillance, and reconnaissance (ISR) plan. The entire unit staff analyzes each requirement to determine how best it is to be satisfied. The staff receives ISR tasks and requests for information from subordinate and adjacent units and higher headquarters. The ISR plan includes all assets that the operations officer can task or request and coordinating mechanisms to ensure adequate coverage of the area of interest. (FM 3-90) (USACAC)

No.	Scale	Measure
01	Yes/No	Intelligence staff provided the operations officer with the ISR synchronization plan.
02	Yes/No	Intelligence staff and operations officer determined asset availability in coordination with the rest of staff.

No.	Scale	Measure
03	Yes/No	All organic, adjacent, and higher ISR assets were tasked against requirements.
04	Yes/No	Staff developed and linked ISR plan to commander and staff requirements.
05	Yes/No	Unit had the right mix of ISR assets for the area of operations.
06	Yes/No	ISR plan addressed redundancy, mix, and cueing.
07	Yes/No	Unit determined unique support requirements for ISR assets to include maintenance, crew training, and connectivity.
08	Yes/No	Unit determined a scheme of support including fires, routes of movement and maneuver, medical and casualty evacuation, and sustainment.
09	Yes/No	Unit identified the reporting criteria, capabilities, and limitations of all ISR assets.
10	Yes/No	Unit linked ISR plan specific information requirements to specific reporting criteria and latest time information is of value.
11	Yes/No	ISR plan provided information and intelligence in sufficient time for the commander's decisionmaking.
12	Yes/No	ISR plan required that ISR forces receive mission intelligence briefing before mission.
13	Yes/No	ISR plan required that ISR assets prepare mission report.
14	Yes/No	ISR plan required that ISR forces are debriefed.
15	Yes/No	ISR plan included a dissemination plan, to include information tasked or requested from external organizations.
16	Yes/No	Operations officer disseminated ISR plan in time to produce operation orders and perform troop leading procedures by subordinates.
17	Yes/No	ISR plan supported the intelligence production and provided answers to requirements.
18	Time	To develop ISR plan.
19	Time	To disseminate the ISR plan.
20	Time	To determine availability of ISR assets.
21	Percent	Of available ISR assets identified.

ART 2.3.2.2 EXECUTE, EVALUATE, AND UPDATE THE INTELLIGENCE, SURVEILLANCE, AND RECONNAISSANCE PLAN

2-43. The operations officer updates the intelligence, surveillance, and reconnaissance (ISR) plan based on information received from the intelligence officer. The operations officer is the integrator and manager of the ISR effort through an integrated staff process and procedures. As priority intelligence requirements (PIRs) are answered and new information requirements arise, the intelligence officer updates intelligence synchronization requirements and provides the new input to the operations officer who updates the ISR plan. The intelligence and operations officers work closely with all staff elements to ensure the unit's organic collectors receive appropriate taskings. This ISR plan reflects an integrated collection strategy and employment, production, and dissemination scheme that will effectively answer the commander's PIR. This task includes updating ISR operations through dynamic retasking and periodic updates of the ISR plan. (FM 3-90) (USACAC)

No.	Scale	Measure
01	Yes/No	Operations officer disseminated ISR plan in time to produce operation orders and perform troop leading procedures.
02	Yes/No	ISR plan provided information and intelligence in time for the commander to use it in the conduct of the operation.
03	Yes/No	Unit executed ISR plan.
04	Yes/No	ISR assets met latest time information is of value for ISR tasks.
05	Yes/No	ISR assets reported activity or lack of activity in the correct formats and through the required reporting channels.
06	Yes/No	The staff monitored and began evaluation of ISR operations.

No.	Scale	Measure
07	Yes/No	The entire staff conducted analysis of ISR reporting.
08	Yes/No	Unit evaluated reports and identified which requirements had been answered and identified which requirements (PIRs and commander's critical information requirements) to remove.
09	Yes/No	Unit evaluated reports and identified which commander and staff requirements had been satisfied.
10	Yes/No	Unit evaluated reports and identified new requirements.
11	Yes/No	Unit identified gaps or seams within the ISR plan (a continuous measure).
12	Yes/No	Technical authority and channels provided refined information or data to ISR assets.
13	Yes/No	Technical authority and channels provided new technical data to ISR assets.
14	Yes/No	Unit identified indicators of previously unanticipated threat patterns or activities.
15	Yes/No	Unit identified new requirements based on reporting and ISR operations.
16	Yes/No	Unit developed new ISR tasks and requests for information based on new specific information requirements.
17	Yes/No	Dynamic retasking refocused the ISR operation.
18	Yes/No	Operations officer updated and disseminated ISR plan periodically in time to produce operation orders and lead troop movements.
19	Time	To evaluate reporting and ISR operations.
20	Time	To determine threat patterns or activities from reporting and ISR operations.
21	Time	To evaluate and identify gaps or seams within the ISR plan.
22	Time	To identify indicators of previously unanticipated threat patterns or activities.
23	Time	To review existing information on threat capabilities.
24	Time	To consider threat doctrine.
25	Time	To determine new information and intelligence requirements.
26	Time	To develop new ISR tasks and requests for information.
27	Time	To generate and disseminate new ISR tasks as a result of dynamic retasking.
28	Time	To update and disseminate the ISR plan.
29	Percent	Of specific information and intelligence requirements correctly answered.
30	Percent	Of reporting requirements correctly answered.

ART 2.3.3 CONDUCT RECONNAISSANCE

2-44. 2-44. Reconnaissance is a mission undertaken to obtain, by visual observation or other detection methods, information about activities and resources of an enemy or potential enemy and about the meteorological, hydrographic, or geographic characteristics of an area of operations. Other detection methods include signals, imagery, measurement of signature, or other technical characteristics. This task includes performing chemical, biological, radiological, and nuclear reconnaissance; engineer reconnaissance (to include infrastructure reconnaissance and environmental reconnaissance). (FM 3-34.170) (USAIC&FH)

Note: This task branch includes techniques by which ART 6.9.6 may be performed.

No.	Scale	Measure
01	Yes/No	The specific information requirement that prompted the conduct of reconnaissance was answered.
02	Yes/No	Reconnaissance system or force oriented on the intelligence, surveillance, and reconnaissance (ISR) objectives.

No.	Scale	Measure
03	Yes/No	Human intelligence (HUMINT), signals intelligence (SIGINT), imagery intelligence (IMINT), measurement and signature intelligence (MASINT), technical intelligence (TECHINT), and counterintelligence (CI) assets were available for ISR tasking.
04	Yes/No	Reconnaissance system or force reported all information in a timely and accurate manner.
05	Yes/No	Reconnaissance mission completed no later than time specified in the order.
06	Yes/No	Identified support requirements for each reconnaissance asset were sufficient.
07	Yes/No	Available HUMINT, SIGINT, IMINT, MASINT, TECHINT, and CI assets deployed to maximize their capabilities.
08	Yes/No	Unit maintained continuous reconnaissance by employing appropriate asset mix and cueing.
09	Yes/No	Reconnaissance asset was dynamically retasked.
10	Yes/No	Technical authority and channels provided information or data that redirected reconnaissance assets on the ISR objective.
11	Yes/No	Units practiced tactical patience when using HUMINT, SIGINT, IMINT, MASINT, TECHINT, and CI assets.
12	Yes/No	Information observed or gained from contact with the local population (Soldier surveillance and reconnaissance) by units conducting reconnaissance missions was collected by debriefing Soldiers or from unit reports.
13	Yes/No	New tasking received as a result of dynamic retasking.
14	Time	From receipt of tasking until reconnaissance assets are in place and operational.
15	Time	To report information to requesting unit or agency to satisfy priority intelligence requirements (PIR) or information requirements.
16	Time	To redirect or reorient reconnaissance assets to meet new collection requirements.
17	Percent	Of ISR assets collecting against assigned named areas of interest and target areas of interest.
18	Percent	Of PIRs tied to commander's decision points.
19	Percent	Of specific information requirements developed from PIRs.
20	Percent	Of collection requirements fulfilled by reconnaissance assets.
21	Percent	Of accuracy of data provided.
22	Percent	Of reconnaissance mission-capable assets.
23	Percent	Of reconnaissance assets becoming casualties during the mission.

ART 2.3.3.1 CONDUCT A ROUTE RECONNAISSANCE

2-45. Conduct a reconnaissance operation focused along a specific route—such as a road, railway, or waterway—to provide new or updated information on route conditions and activities. (FM 3-90) (USACAC)

No.	Scale	Measure
01	Yes/No	Route reconnaissance accomplished its task or mission, such as determining the trafficability of the route and finding, reporting, and clearing within force capabilities all enemy forces that can influence movement along the route.
02	Yes/No	Reconnaissance force oriented on the reconnaissance objective.
03	Yes/No	Reconnaissance force reported all information rapidly and accurately.
04	Yes/No	Reconnaissance force retained its freedom to maneuver.
05	Yes/No	Reconnaissance force gained and maintained enemy contact.
06	Yes/No	Reconnaissance force rapidly developed the situation.
07	Yes/No	Force reported route reconnaissance critical tasks not performed to higher headquarters.
08	Yes/No	Force accomplished route reconnaissance mission by time specified in the order.

No.	Scale	Measure
09	Yes/No	Force collected information during the route reconnaissance and made it available to the commander.
10	Yes/No	Information observed or gained from contact with the local population (Soldier surveillance and reconnaissance) by units conducting route reconnaissance was collected by debriefing Soldiers or from unit reports.
11	Time	From receiving task until unit reconnaissance assets are in place.
12	Time	To provide collected route data to tasking agency analysts.
13	Time	To redirect reconnaissance assets to meet new requirements.
14	Time	From receiving task until completing route reconnaissance.
15	Percent	Of route reconnaissance critical tasks fulfilled by reconnaissance assets.
16	Percent	Of tactical-level requirements satisfied by higher or adjacent units' existing intelligence, surveillance, and reconnaissance assets on a noninterference basis.
17	Percent	Of accuracy of data provided.
18	Percent	Of operational assets committed to the route reconnaissance mission.
19	Percent	Of reconnaissance assets becoming casualties during the mission.

ART 2.3.3.2 CONDUCT A ZONE RECONNAISSANCE

2-46. Conduct a directed effort to obtain detailed information about all routes, obstacles (to include chemical, biological, radiological, and nuclear contamination), terrain, and enemy forces in an area defined by boundaries. The commander normally assigns a zone reconnaissance when the enemy situation is vague or when information concerning cross-country trafficability is desired. (FM 3-90) (USACAC)

No.	Scale	Measure
01	Yes/No	Zone reconnaissance accomplished its task or mission, such as finding and reporting all enemy forces in the designation area and clearing all enemy forces in the designated area of operations (AO).
02	Yes/No	Human intelligence (HUMINT), signals intelligence (SIGINT), imagery intelligence (IMINT), measurement and signature intelligence (MASINT), technical intelligence (TECHINT), and counterintelligence (CI) assets were available for intelligence, surveillance, and reconnaissance (ISR) tasking.
03	Yes/No	Reconnaissance force was oriented on the reconnaissance objective.
04	Yes/No	Reconnaissance force reported all information rapidly and accurately.
05	Yes/No	Reconnaissance force retained its freedom to maneuver.
06	Yes/No	Reconnaissance force gained and maintained enemy contact.
07	Yes/No	Reconnaissance force rapidly developed the situation.
08	Yes/No	Reported zone reconnaissance tasks performed to higher headquarters.
09	Yes/No	Unit cleared the AO that defines the zone of all enemy forces within the capability of the unit performing the zone reconnaissance.
10	Yes/No	Unit accomplished zone reconnaissance mission by time specified in the operation order.
11	Yes/No	Unit deployed available HUMINT, SIGINT, IMINT, MASINT, TECHINT, and CI assets to maximize their capabilities.
12	Yes/No	Unit practiced tactical patience when using HUMINT, SIGINT, IMINT, MASINT, TECHINT, and CI assets.
13	Yes/No	Information observed or gained from contact with the local population (Soldier surveillance and reconnaissance) by units conducting a zone reconnaissance was collected by debriefing Soldiers or from unit reports.
14	Time	From receiving task until unit reconnaissance assets are in place.
15	Time	To provide collected data to tasking agency analysts.
16	Time	To redirect reconnaissance assets to meet new requirements.
17	Time	From receiving task until completing zone reconnaissance.

No.	Scale	Measure
18	Percent	Of assigned zone reconnaissance tasks fulfilled by reconnaissance assets.
19	Percent	Of tactical-level requirements satisfied by higher or adjacent units' existing ISR assets on a noninterference basis.
20	Percent	Of accuracy of data provided.
21	Percent	Of zone reconnaissance mission-capable assets.
22	Percent	Of reconnaissance assets becoming casualties during the mission.

ART 2.3.3.3 CONDUCT AN AREA RECONNAISSANCE

2-47. Conduct a directed effort to obtain detailed information concerning the terrain or enemy activity within a prescribed area not defined by boundaries, such as a town, ridgeline, woods, or other feature critical to operations to include underwater reconnaissance, search, or recovery. (FM 3-90) (USACAC)

No.	Scale	Measure
01	Yes/No	Area reconnaissance accomplished its task or mission, such as finding and reporting all enemy forces within the designation area and clearing all enemy forces in the designated area of operations (AO) within the capability of the unit conducting reconnaissance.
02	Yes/No	Human intelligence (HUMINT), signals intelligence (SIGINT), imagery intelligence (IMINT), measurement and signature intelligence (MASINT), technical intelligence (TECHINT), and counterintelligence (CI) assets were available for intelligence, surveillance, and reconnaissance (ISR) tasking.
03	Yes/No	Reconnaissance force oriented on the reconnaissance objective.
04	Yes/No	Reconnaissance force reported all information rapidly and accurately.
05	Yes/No	Reconnaissance force retained its freedom to maneuver.
06	Yes/No	Reconnaissance force gained and maintained enemy contact.
07	Yes/No	Reconnaissance force rapidly developed the situation.
08	Yes/No	Reported area reconnaissance tasks not performed to higher headquarters.
09	Yes/No	Unit cleared AO that defined the area of all enemy forces within the capability of the unit performing the area reconnaissance.
10	Yes/No	Unit accomplished area reconnaissance mission by time specified in the order.
11	Yes/No	Unit deployed available HUMINT, SIGINT, IMINT, MASINT, TECHINT, and CI assets in a manner to maximize their capabilities.
12	Yes/No	Unit practiced tactical patience when using HUMINT, SIGINT, IMINT, MASINT, TECHINT, and CI assets.
13	Yes/No	Information observed or gained from contact with the local population (Soldier surveillance and reconnaissance) by units conducting an area reconnaissance was collected by debriefing Soldiers or from unit reports.
14	Time	From receiving task until unit reconnaissance assets are in place.
15	Time	To provide collected data to tasking agency analysts.
16	Time	To redirect reconnaissance assets to meet new requirements.
17	Time	From receiving task until completing area reconnaissance.
18	Percent	Of assigned area reconnaissance tasks fulfilled by reconnaissance assets.
19	Percent	Of tactical-level requirements satisfied by higher or adjacent units' existing ISR assets on a noninterference basis.
20	Percent	Of accuracy of data provided.
21	Percent	Of area reconnaissance mission-capable assets.
22	Percent	Of reconnaissance assets becoming casualties during the mission.

ART 2.3.3.4 CONDUCT A RECONNAISSANCE-IN-FORCE

2-48. A reconnaissance in force is a deliberate combat operation designed to discover or test the enemy's strength, dispositions, and reactions or to obtain other information. (FM 3-90) (USACAC)

No.	Scale	Measure
01	Yes/No	Reconnaissance-in-force accomplished its task or mission, such as penetrating the enemy's security area, determining its size and depth, and determining the location and disposition of enemy main positions.
02	Yes/No	Unit conducting the reconnaissance-in-force remained oriented on the reconnaissance objective.
03	Yes/No	Unit conducting the reconnaissance-in-force reported all information rapidly and accurately to its higher headquarters.
04	Yes/No	Unit conducting the reconnaissance-in-force retained its freedom to maneuver.
05	Yes/No	Unit conducting the reconnaissance-in-force gained and maintained enemy contact.
06	Yes/No	Unit conducting the reconnaissance-in-force rapidly developed the situation.
07	Yes/No	Unit conducting the reconnaissance-in-force reported reconnaissance-in-force tasks not performed to higher headquarters.
08	Yes/No	Unit completed the reconnaissance-in-force by time specified in the order.
09	Yes/No	Unit collected desired information during the reconnaissance-in-force and made it available to the commander.
10	Yes/No	Information observed or gained from contact with the local population (Soldier surveillance and reconnaissance) by units conducting a reconnaissance in force was collected by debriefing Soldiers or from unit reports.
11	Time	From receiving task until unit reconnaissance assets are in place.
12	Time	To provide collected data to tasking agency analysts.
13	Time	To redirect reconnaissance assets to meet new requirements.
14	Time	From receiving task until completing reconnaissance-in-force.
15	Percent	Of doctrinal reconnaissance-in-force tasks fulfilled by reconnaissance assets.
16	Percent	Of tactical-level requirements satisfied by higher or adjacent units' existing intelligence, surveillance, and reconnaissance assets on a noninterference basis.
17	Percent	Of accuracy of data provided.
18	Percent	Of unit assets mission capable at different points during the reconnaissance-in-force mission—beginning, end, and critical times.
19	Percent	Of unit Soldiers becoming casualties during the mission.

ART 2.3.3.5 CONDUCT A RECONNAISSANCE PATROL

2-49. Use a detachment of ground, sea, or air forces to gather information about the enemy, terrain, or civil environment. (FM 3-90) (USACAC)

No.	Scale	Measure
01	Yes/No	Patrol collected the information required.
02	Yes/No	Unit prepared patrol plan within time allowed.
03	Yes/No	Unit conducted rehearsals to standard within the time allowed.
04	Yes/No	Patrol used tactically appropriate reconnaissance tactic—fan, converging route, or successive bounds.
05	Yes/No	Enemy detected reconnaissance patrol.
06	Yes/No	Information observed or gained from contact with the local population (Soldier surveillance and reconnaissance) by units conducting a reconnaissance patrol was collected by debriefing Soldiers or from unit reports.
07	Time	To conduct the reconnaissance patrol by time allocated by higher headquarters.
08	Percent	Of assigned area covered during the patrol.

No.	Scale	Measure
09	Percent	Of tactical-level requirements satisfied by higher or adjacent units' existing intelligence, surveillance, and reconnaissance assets on a noninterference basis.
10	Percent	Of accuracy of data provided.
11	Percent	Of friendly casualties received during the combat patrol.
12	Percent	Of information requirements achieved.

ART 2.3.4 CONDUCT SURVEILLANCE

2-50. *Surveillance* is the systematic observation of aerospace, surface, or subsurface areas, places, persons, or things by visual, aural, electronic, photographic, or other means. (JP 3-0) Other means may include but are not limited to space-based systems, and special chemical, biological, radiological, and nuclear; artillery; engineer; special operations forces; and air defense equipment. Surveillance involves observing an area to collect information. (FM 2-0) (USAICoE)

No.	Scale	Measure
01	Yes/No	Surveillance assets collected required information.
02	Yes/No	Identified support requirements for each surveillance asset was sufficient.
03	Yes/No	Human intelligence (HUMINT), signals intelligence (SIGINT), imagery intelligence (IMINT), measurement and signature intelligence (MASINT), technical intelligence (TECHINT), and counterintelligence (CI) assets were available for intelligence, surveillance, and reconnaissance tasking.
04	Yes/No	Unit deployed available HUMINT, SIGINT, IMINT, MASINT, TECHINT, and CI assets to maximize their capabilities.
05	Yes/No	Enemy forces detected surveillance asset.
06	Yes/No	Surveillance system or force reported all information in a timely and accurate manner.
07	Yes/No	Unit maintained continuous surveillance by using appropriate asset mix and cueing.
08	Yes/No	Unit practiced tactical patience when using HUMINT, SIGINT, IMINT, MASINT, TECHINT, and CI assets.
09	Yes/No	Information observed or gained from contact with the local population (Soldier surveillance and reconnaissance) by units conducting a surveillance mission was collected by debriefing Soldiers or from unit reports.
10	Time	From receipt of tasking until surveillance assets are in place and operational.
11	Time	To respond to new taskings.
12	Time	To provide collected data to tasking agency analysts.
13	Time	To report information to requesting unit or agency to satisfy priority intelligence requirements or information requirements.
14	Percent	Of time able to respond to collection requirements.
15	Percent	Of collection requirements fulfilled by surveillance assets.
16	Percent	Of surveillance requirements satisfied using intelligence reach.
17	Percent	Of surveillance assets mission capable.
18	Percent	Of surveillance assets becoming casualties during the mission.
19	Percent	Of assets collecting against assigned named areas of interest and target areas of interest.

ART 2.3.5 CONDUCT INTELLIGENCE RELATED MISSIONS AND OPERATIONS

2-51. The associated intelligence tasks (mission and debriefing program, intelligence coordination, technical authority, and intelligence support to personnel recovery) that facilitate the conduct of reconnaissance and surveillance. It also includes specialized missions (such as sensitive site exploitation), that provide intelligence and information outside of the traditional ISR construct. (FM 2-0) (USAICoE)

ART 2.3.5.1 ESTABLISH A MISSION INTELLIGENCE BRIEFING AND DEBRIEFING PROGRAM

2-52. Commanders establish, support, and allocate appropriate resources for a mission briefing and debriefing program. Conducting battle updates and after action reviews are separate tasks from the mission briefing and debriefing program. The intelligence officer develops a mission intelligence briefing plan and complementary debriefing plan to support the commander's program. Soldiers receive a mission intelligence briefing before executing a patrol or similar operation. The briefing sensitizes Soldiers to specific information and reporting requirements, information gaps, and unique mission requirements. The mission intelligence briefings and debriefings generally follow the format of a mission briefing: review the route traveled, collection objectives of the patrol, and methods employed. The debriefing program captures the specific information requirements the patrol was to collect and any additional information and observations the patrol made concerning the operational environment. It also collects any fliers, pamphlets, media, or pictures the patrol found or obtained. (FM 2-0) (USAICoE)

No.	Scale	Measure
01	Yes/No	Command established and supported a mission briefing and debriefing program.
02	Yes/No	Command allocated appropriate resources (time, personnel, and a designated area) to support the mission briefing and debriefing program.
03	Yes/No	Units performed battle update briefing and after action reviews separately from the mission briefing and debriefing program.
04	Yes/No	G-2 or S-2 developed a mission intelligence plan.
05	Yes/No	G-2 or S-2 allocated resources for the mission briefing.
06	Yes/No	G-2 or S-2 developed a complementary debriefing plan.
07	Yes/No	G-2 or S-2 allocated resources for the debriefing.
08	Time	That command allocated to mission briefings and debriefings.
09	Time	That G-2 or S-2 allocated for mission intelligence briefing.
10	Time	That G-2 or S-2 allocated to conduct debriefings.

ART 2.3.5.1.1 Establish a Mission Intelligence Briefing Plan

2-53. The intelligence section develops a mission intelligence briefing plan. The mission intelligence briefing plan identifies information Soldiers executing patrols should be seeking. It ensures that all Soldiers conducting patrols, tactical movements, and nontactical movements are sensitized to specific information and reporting requirements, information gaps, and unique mission requirements. The intelligence mission briefing and debriefing generally follow the format of a mission briefing: review the route traveled, collection objectives of the patrol, and methods employed. (FM 2-0) (USAICoE)

No.	Scale	Measure
01	Yes/No	All patrols, tactical movements, and nontactical movements received mission intelligence briefing.
02	Yes/No	Mission intelligence briefing contained terrain impacts update.
03	Yes/No	Mission intelligence briefing contained civil considerations update.
04	Yes/No	Mission intelligence briefing contained weather effects update.
05	Yes/No	Mission intelligence briefing contained threat update.
06	Yes/No	Mission intelligence briefing contained route update.
07	Yes/No	Mission intelligence briefing contained focus areas for observation.
08	Yes/No	Mission intelligence briefing contained updated collection requirements.
09	Yes/No	Mission intelligence briefing provided criteria for immediate reporting requirements.
10	Yes/No	Mission intelligence briefing provided reporting requirements for nonpriority reporting.
11	Yes/No	Mission briefing contained reporting requirements for unusual activity or complete lack of activity by the local population.
12	Yes/No	Mission briefing contained requirements for handling and disposing collected documents.

No.	Scale	Measure
13	Yes/No	Mission briefing contained requirements for enemy prisoner of war and detainee and captured enemy document handling and disposition.
14	Yes/No	Mission intelligence briefing contained requirements for use of digital photography.

ART 2.3.5.1.2 Establish a Debriefing Plan

2-54. The intelligence section develops a complementary debriefing plan. The debriefing plan captures information related to the specific information requirements the patrol was to collect and any additional information and observations the patrol made concerning the operational environment. It also collects any fliers, pamphlets, media, or pictures the patrol found or obtained. The plan should include all returning patrols, leaders who have traveled to meetings, returning human intelligence collection teams, aircrews, and others who may have obtained information of intelligence value. The intelligence section debriefs personnel. Debriefers then write and submit a report or reports information verbally, as appropriate. The requirement for a debriefing by the intelligence section following each mission should be a part of the intelligence mission briefing. Leaders should not consider the mission complete and release the personnel until the reporting and debriefings are done. (FM 2-0) (USAICoE)

No.	Scale	Measure
01	Yes/No	Units attended and participated in all patrols, tactical movements, and nontactical movements in postmission intelligence debriefing upon return.
02	Yes/No	Units completed and submitted mission reports for all patrols, tactical movements, and nontactical movements upon return.
03	Yes/No	Debriefing conducted in a logical and organized manner.
04	Yes/No	Debriefing conducted with all members of the patrols, tactical movements, and nontactical movements.
05	Yes/No	Debriefing addressed information pertaining to the tasked collection requirements.
06	Yes/No	Debriefing addressed additional details on information provided through immediate reporting during the mission or movement.
07	Yes/No	Debriefing addressed information provided through nonpriority reporting during the mission or movement.
08	Yes/No	Debriefing addressed unusual activity or complete lack of activity by the local population observed during the mission or movement.
09	Yes/No	Debriefing addressed documents found or photographed during the mission or movement.
10	Yes/No	Debriefing addressed enemy prisoner of war and detainee encountered during the mission.
11	Yes/No	Debriefing addressed any conversations unit members had with the local populace.

ART 2.3.5.2 CONDUCT INTELLIGENCE COORDINATION

2-55. Conduct intelligence coordination is the task carried out by the intelligence section to facilitate active collaboration, laterally and vertically. It includes establishing and maintaining technical channels to refine and focus the intelligence disciplines in intelligence, surveillance, and reconnaissance (ISR) tasks. It also properly coordinates the discipline assets when operating in another unit's area of operations (AO). (FM 2-0) (USAICoE)

No.	Scale	Measure
01	Yes/No	Technical authority and channels capability and expertise resided in the G-2 or S-2.
02	Yes/No	Units used technical authority and channels to train assets before the mission.
03	Yes/No	Units used technical authority and channels to identify, define clearly, and disseminate the legal limits of how to use specific ISR assets.
04	Yes/No	Units handled technical information and guidance to support operations following all applicable regulations, policies, and procedures.
05	Yes/No	Intelligence assets and units coordinated with friendly units when entering, moving through, or departing friendly lines or areas of operations.

No.	Scale	Measure
06	Yes/No	Units monitored the technical authority and channels of operations to ensure missions conducted within applicable regulations, policies, and procedures; provided control measures for ISR synchronization effort; and ensured a collaborative collection and information environment.
07	Yes/No	Intelligence assets and units identified specific offices for liaison to exchange relevant information and intelligence with the friendly units.

ART 2.3.5.2.1 Establish and Maintain Technical Authority and Channels

2-56. Intelligence commanders and intelligence staffs maintain control of each intelligence discipline during operations through technical channels to ensure adherence to applicable laws and policies, ensure proper use of doctrinal techniques, and provide technical support and guidance. Applicable laws and policies include all relevant U.S. law, the law of war, international law, directives, DOD instructions, and orders. Commanders direct operations but often rely on the intelligence section's technical expertise to conduct portions of the unit's collection effort. Technical channels also involve translating ISR tasks into the specific parameters used to focus highly technical or legally sensitive aspects of the intelligence, surveillance, and reconnaissance (ISR) effort. Technical channels include but are not limited to defining, managing, or guiding the use of specific ISR assets; identifying critical technical collection criteria, such as technical indicators; recommending collection techniques, procedures, or assets; coordinating operations; and directing specialized training for specific MI personnel or units. (FM 2-0) (USAICoE)

Note. In specific cases, regulatory authority is granted to specific national and DOD intelligence agencies for specific intelligence discipline collection and is passed through technical channels.

No.	Scale	Measure
01	Yes/No	Technical authority and channels capability and expertise resided within the G-2 or S-2.
02	Yes/No	Unit requested technical authority and channel capability as well as resident expertise if it was not resident within the G-2 or S-2.
03	Yes/No	Unit used technical authority and channels to train assets before the mission.
04	Yes/No	Unit conducted operations per all applicable regulations, policies, and procedures, to include classification and security considerations.
05	Yes/No	Unit used technical authority and channels to identify, define clearly, and disseminate the legal limits of the use of specific ISR assets.
06	Yes/No	Unit handled technical information and guidance to support operations per all applicable regulations, policies, and procedures.
07	Yes/No	Unit monitored technical authority and channels to ensure missions conducted within applicable regulations, policies, and procedures.
08	Yes/No	Technical authority and channels reported operations and missions that violated applicable regulations, policies, and procedures.
09	Yes/No	Unit used technical authority and channels to determine what existing national intelligence assets or higher echelon Army assets were satisfying requirements that allow organic assets to be reallocated for other missions.
10	Yes/No	Unit used technical authority and channels to determine the minimum intelligence assets required for the mission periodically reevaluated.
11	Yes/No	Technical authority and channels provided technical guidance and control measures to the G-2 or S-2 for ISR synchronization efforts.
12	Yes/No	Technical channels provided the commander information and intelligence to support lethal and nonlethal targeting of the threat.
13	Yes/No	Technical authority and channels monitored and identified all technical control measures to ensure unit met reporting criteria.
14	Yes/No	Technical authority and channels ensured a collaborative collection and information environment for all available intelligence disciplines.

No.	Scale	Measure
15	Yes/No	Technical controls ensured continuous technical interface between all applicable ISR assets.
16	Yes/No	G-2 or S-2 section monitored ISR collection and compared it to the event template and matrix to determine if the technical authority and channels were meeting predicted course of action requirements.
17	Time	To identify, define, and disseminate the legal limits using technical authority and channels.
18	Time	To provide information and guidance to support operations using technical authority and channels.
19	Percent	Of technical guidance and control measures given to the G-2 or S-2 for ISR synchronization efforts.

ART 2.3.5.2.2 Conduct Deconfliction and Coordination

2-57. Deconfliction and coordination consists of a series of related activities that facilitate operations in another unit's area of operations (AO). These activities facilitate successful intelligence, surveillance, and reconnaissance collection, support of the operation, and fratricide avoidance. Military intelligence organizations may be used in general support for coverage of an AO or in direct support to support a specific unit. Military intelligence organizations operating in general support should coordinate with unit commanders when operating in that unit's AO. At a minimum, the military intelligence organizations announce their presence and request information on any conditions or ongoing situations that may affect how they conduct their mission—organizations should conduct thorough face-to-face coordination. A military intelligence organization operating in direct support of a specific unit coordinates with the unit for augmentation to conduct operations in accordance with force protection requirements. The military intelligence organization's leader also coordinates with the supported unit's intelligence section for debriefings of returning members, convoy leaders, and others. (FM 2-0) (USAICoE)

No.	Scale	Measure
01	Yes/No	Intelligence assets and units conducted coordination with friendly units when entering, moving through, or departing friendly lines or AOs.
02	Yes/No	Intelligence assets and units provided a liaison with the friendly forward unit, when available.
03	Yes/No	Intelligence assets and units coordinated with the friendly forward unit within whose AO they would be operating for exchanged graphics and overlays; fire support (en route, at mission location, and return); casualty evacuation procedures; passwords (running, forward of friendly lines); recognition signals, call signs, and frequencies; approved routes for movement; off-limits or restricted areas; maintenance and resupply points; reporting criteria to share relevant information and intelligence; and key information concerning the operational environment.
04	Yes/No	Intelligence assets and units exchanged relevant information and intelligence with the friendly forward unit.

ART 2.3.5.3 SUPPORT SITE EXPLOITATION

2-58. *Site exploitation* consists of a related series of activities inside a captured sensitive site to exploit personnel, documents, electronic data, and material captured at the site, while neutralizing any threat posed by the site or its contents. A sensitive site is a designated, geographically limited area with special diplomatic, informational, military, and economic sensitivity for the United States. This includes factories with technical data on enemy weapon systems, war crimes sites, critical hostile government facilities, areas suspected of containing persons of high rank in a hostile government or organization, terrorist money laundering areas, and document storage areas for secret police forces. While the physical process of exploiting the sensitive site begins at the site itself, full exploitation may involve teams of experts located around the world. (FM 3-90.15) (USACAC)

No.	Scale	Measure
01	Yes/No	Exploitation of the sensitive site supported U.S. operational or strategic diplomatic, informational, military, and economic goals.
02	Yes/No	Unit supporting the exploitation of the sensitive site had isolated, seized, secured, and cleared the site or relieved a unit that accomplished those tactical missions prior to exploiting the site under the technical direction of subject matter experts.
03	Yes/No	Enemy or adversary knew that U.S. forces had exploited the site.
04	Yes/No	Tactical unit task organized itself appropriately to accomplish the mission of supporting the exploitation of the sensitive site and compensated for losses.
05	Yes/No	Enemy or adversary failed to exfiltrate sensitive equipment or materiel from the sensitive site.
06	Yes/No	Unit supporting the sensitive site exploitation did not suffer casualties because of a failure to manage risks associated with the site properly.
07	Yes/No	Enemy or adversary failed to destroy sensitive equipment, materiel, and documents or to purge computers of sensitive information before securing the site.
08	Yes/No	Commander of the unit supporting the exploitation of the sensitive site maintained situational understanding throughout the operation.
09	Yes/No	Members of the unit supporting the exploitation had access to a high fidelity common operational picture throughout the operation consistent with operations security.
10	Yes/No	Leaders of the unit supporting the exploitation of the sensitive site used the military decisionmaking process or troop leading procedures correctly to include identifying search locations, security positions, boundaries, and fire support coordination measures. They coordinated and synchronized warfighting functions.
11	Yes/No	Unit conducted operations per established rules of engagement and consideration for the nature of the sensitive site.
12	Yes/No	Unit recorded results of the search and appropriately disseminated results.
13	Yes/No	Unit supporting the exploitation of the site appropriately killed, captured, or detained enemy soldiers, adversaries, sensitive individuals, and neutrals in the site.
14	Time	Necessary to isolate the sensitive site and forces and individuals located in the site from outside physical, informational, and psychological support.
15	Time	Necessary to seize the sensitive site.
16	Time	Necessary to secure the sensitive site.
17	Time	Necessary to search the sensitive site.
18	Time	Necessary to establish liaison with and deploy technical experts to the sensitive site.
19	Percent	Of potential sensitive personnel, documents, electronic data, and materiel located in the site discovered during the search.

ART 2.3.5.4 INTELLIGENCE SUPPORT TO PERSONNEL RECOVERY

2-59. Support to personnel recovery (PR) consists of intelligence activities and capabilities focused on gathering information to recover and return U.S. personnel—whether Soldier, Army civilian, selected Department of Defense contractors, or other personnel as determined by the Secretary of Defense—who are isolated, missing, detained, or captured (IMDC) in a specific area of operation. This support also includes detailed analysis, developing detailed products, and estimates to support operations undertaken to recover IMDC personnel. (FM 3-50.1) (USACAC)

No.	Scale	Measure
01	Yes/No	Unit conducted PR-focused mission analysis.
02	Yes/No	Unit conducted PR-focused intelligence preparation of the battlefield.
03	Yes/No	Unit conducted support to PR planning, preparation, and execution.
04	Yes/No	Unit incorporated PR into intelligence, surveillance, and reconnaissance plan.

No.	Scale	Measure
05	Yes/No	Unit established PR intelligence architecture and network.
06	Yes/No	Unit established PR communications architecture and network.

ART 2.3.5.4.1 Conduct Personnel Recovery-Focused Intelligence Preparation of the Battlefield

2-60. Although the steps in the intelligence preparation of the battlefield process remain the same for all operations, analysts realize that personnel recovery operations require additional considerations. Every echelon and unit must be prepared to support personnel recovery in their area of operations and for all missions and operations executed. (FM 3-50.1) (USACAC)

No.	Scale	Measure
01	Yes/No	Unit provided weather conditions, forecasts, and updates (advisories, watches, and warnings) that provided coverage during mission period and included data pertaining to starting point, recovery site, routes to and from recovery site, and effects on the threat or local population.
02	Yes/No	Unit provided effects of weather on survivability for isolated, missing, detained, or captured personnel.
03	Yes/No	Unit queried geospatial information and services archives and available databases for information pertaining to terrain analysis of recovery site, building plans, and layout of complexes, slope, and gradient analysis of recovery site and route (to include ingress and egress routes).
04	Yes/No	If lacking sufficient geospatial information and services information, unit consulted civil affairs, psychological operations, human intelligence, and other pertinent staff or files that might have pertinent information.
05	Yes/No	Unit determined production strategy to fill geospatial data shortfalls after receiving National Geospatial-Intelligence Agency's (known as NGA) initial assessment of product and data availability and suitability.
06	Yes/No	Unit developed modified combined obstacle overlays threat, situational, and event templates.
07	Yes/No	Intelligence personnel identified potential pickup zones, landing zones, and drop zones for feasibility, defense, threats; ground and air routes for conditions, detours, hazards and obstacles; and named areas of interest and target areas of interest in vicinity of recovery site to include areas for possible suppression of enemy air defense.
08	Yes/No	Intelligence personnel identified cultural considerations of local populations in vicinity of recovery site; population safe and support status; local customs and values that impact the recovery; social and human factors that impact the recovery; political parties and factions impacts; and economic impacts.
09	Yes/No	Intelligence personnel identified known or suspected threat forces in the area (military, paramilitary, guerilla forces, insurgents, and terrorists); enemy capabilities to counter combat search and rescue (CSAR); potential enemy captivity and interrogation procedures; and special capabilities (counterintelligence, electronic measures, ground surveillance radar, night vision devices, aerial and satellite surveillance capabilities, and suspected interrogation techniques).
10	Yes/No	Intelligence personnel identified known or suspected criminal groups, gangs, and organizations that pose a risk to CSAR operations.
11	Yes/No	Unit developed an intelligence overlay of pertinent threat characteristics.
12	Yes/No	Unit coordinated with the rest of the staff to identify and post friendly units (to include multinational partners), elements, or groups in the area of operations to include coordination lines; applicable local law enforcement agencies; and political and local leaders or local populations.

ART 2.3.5.4.2 Conduct Support to Personnel Recovery Planning, Preparation, and Execution

2-61. To accomplish the mission, the command and staff ensure resources are available to collect, analyze, and produce intelligence for the planning and execution of personnel recovery operations. The personnel recovery (PR) intelligence support team immediately identifies the appropriate assets to best satisfy the commander's critical information requirements for PR from the available resources. To accomplish this, the team is familiar with every sensor's capabilities and limitations so they can adjust sensor taskings to ensure the optimal resource usage in real time. The intelligence, surveillance, and reconnaissance (ISR) support starts with receiving and analyzing information from the unit reporting the isolated, missing, detained, or captured (IMDC). Then the effort focuses on receiving, processing, fusing, and analyzing information originating from organic collectors and theater, joint, and national agencies and organizations. It is critical that coordination with higher headquarters is established to assist in PR operations. (FM 3-50.1) (USACAC)

No.	Scale	Measure
01	Yes/No	The command and staff ensured resources were made available to collect, analyze, and produce intelligence for planning and executing PR operations to retrieve IMDC personnel (to include contractors and Army civilians).
02	Yes/No	The staff developed a PR plan and included PR in the ISR plan.
03	Yes/No	The staff consisted of adequately trained PR personnel.
04	Yes/No	The staff tasked identified resources in accordance with the ISR annex and PR annex of the operation order.
05	Yes/No	The available assets best satisfied the requirements identified.
06	Yes/No	The available assets best satisfied the requirements tasked in the order.
07	Yes/No	The intelligence assets allocated to PR were adequate for mission accomplishment.
08	Yes/No	Unit received report of IMDC personnel.
09	Yes/No	Unit standing operating procedures IMDC report standards were met when notified of IMDC personnel.
10	Yes/No	The information on the IMDC incident was sufficient to begin PR.
11	Yes/No	Units executed their be-prepared and on-order PR tasks per the operation order.
12	Yes/No	Unit established PR intelligence and communications architecture and network.
13	Yes/No	PR intelligence and communications architecture and network were adequate.
14	Yes/No	The ISR plan required significant adjustment.
15	Yes/No	Unit completed after action review of PR plan and operation.
16	Yes/No	Unit implemented the lessons learned from the after action review into future PR planning.
17	Yes/No	The PR plan was adequate to accomplish the mission.
18	Yes/No	Higher headquarters assets were required to cover gaps in the PR plan.
19	Time	To report IMDC personnel.
20	Time	For subordinate units to reorient, move into position, or reposition to execute be-prepared and on-order PR tasks.
21	Time	To request and task assets (higher or adjacent) to reorient, move into position, or reposition to execute their PR taskings.
22	Time	To establish the PR intelligence architecture and network.
23	Time	To establish the PR communications architecture and network.
24	Time	For staff to readjust the ISR plan significantly.
25	Time	To issue the updated plan and to execute the updated plan.
26	Percent	Of significant adjustment the ISR plan required.

SECTION IV – ART 2.4: SUPPORT TO TARGETING AND INFORMATION SUPERIORITY

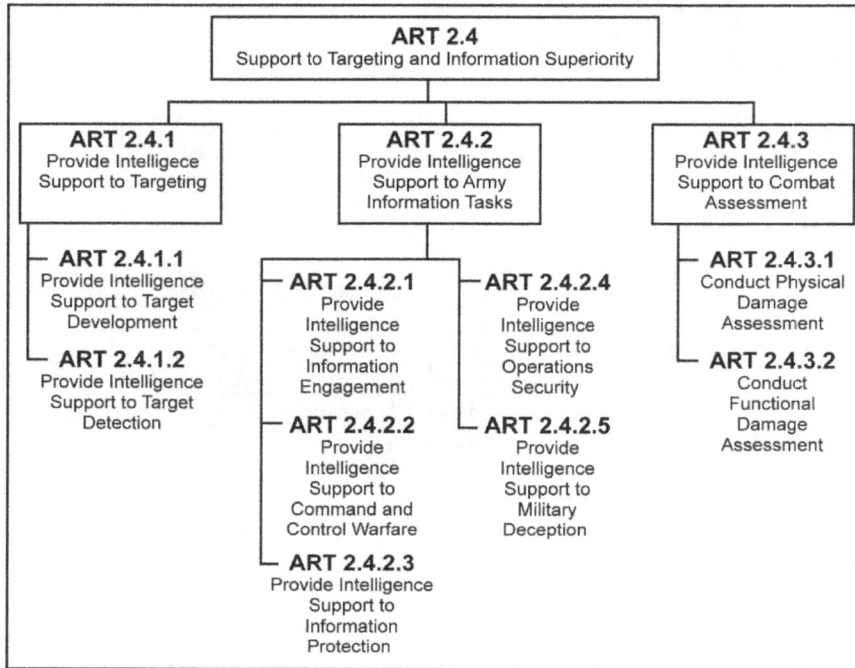

ART 2.4
Support to Targeting and Information Superiority

ART 2.4.1
Provide Intelligece Support to Targeting

ART 2.4.2
Provide Intelligence Support to Army Information Tasks

ART 2.4.3
Provide Intelligence Support to Combat Assessment

ART 2.4.1.1
Provide Intelligence Support to Target Development

ART 2.4.1.2
Provide Intelligence Support to Target Detection

ART 2.4.2.1
Provide Intelligence Support to Information Engagement

ART 2.4.2.2
Provide Intelligence Support to Command and Control Warfare

ART 2.4.2.3
Provide Intelligence Support to Information Protection

ART 2.4.2.4
Provide Intelligence Support to Operations Security

ART 2.4.2.5
Provide Intelligence Support to Military Deception

ART 2.4.3.1
Conduct Physical Damage Assessment

ART 2.4.3.2
Conduct Functional Damage Assessment

2-62. Intelligence support to targeting and information superiority is the task of providing the commander information and intelligence support for targeting through lethal and nonlethal actions. It includes intelligence support to the planning and execution of direct and indirect fires and the Army information tasks of information engagement, command and control warfare, information protection, OPSEC, and military deception, as well as assessing the effects of those operations. (FM 2-0) (USAICoE)

Note: This task branch supports both direct (ART 1.4) and indirect (ART 3.3) delivery of fires.

This task is also linked with ART 5.1.4.3 (Provide Combat Assessment).

No.	Scale	Measure
01	Yes/No	Targets for lethal and nonlethal action were identified, nominated, and prioritized by the entire staff.
02	Yes/No	Unit determined whether lethal, nonlethal, or a combination of lethal and nonlethal actions will achieve the best effect or the desired outcome.
03	Yes/No	Unit linked lethal and nonlethal targets to specific intelligence, surveillance, and reconnaissance (ISR) assets and included targets in the ISR plan.
04	Yes/No	Units identified ISR assets that can be retasked by the operations officer to acquire new lethal or nonlethal targets per the commander's targeting priorities.
05	Percent	Of information requirements identified for lethal and nonlethal effects.
06	Percent	Of information requirements identified for combat assessment and battle damage assessment.

No.	Scale	Measure
07	Percent	Of combat assessment and battle damage assessments that identified if targets achieved the desired effects or require reattack.
08	Percent	Of targeting databases developed, maintained, and updated.

ART 2.4.1 PROVIDE INTELLIGENCE SUPPORT TO TARGETING

2-63. The intelligence officer (supported by the entire staff) provides the fire support coordinator, information engagement officer, and electronic warfare officer, and other staff officers with information and intelligence for targeting the threat's forces and systems with direct and indirect lethal and nonlethal fires. The information and intelligence includes identification of threat capabilities and limitations. The targeting process uses the decide, detect, deliver, and assess methodology. The intelligence officer supports targeting by providing accurate, current intelligence and information to the staff and ensures the ISR plan supports the finalized targeting plan. (FM 2-0) (USAICoE)

Note: This task branch supports ART 3.2 (Detect and Locate Surface Targets) and is a byproduct of decide, detect, deliver, and assess.

No.	Scale	Measure
01	Yes/No	The intelligence officer identified threat command and control nodes, to include all aspects.
02	Yes/No	The intelligence officer identified threat communication systems, to include all aspects.
03	Yes/No	The intelligence officer identified threat computer systems, to include all aspects.
04	Yes/No	The fire support coordinator, information engagement officer, and electronic warfare officer received information and intelligence support for targeting of the threat's forces through lethal and nonlethal fires.
05	Yes/No	Risks to targeting cultural, historic sites, religious centers, medical facilities, natural resources, hazard areas (such as nuclear power plants, chemical facilities, and oil refineries) were assessed and included in target nomination criteria.
06	Time	To detect all aspects of the threat command and control nodes.
07	Time	To detect all aspects of the threat communications systems.
08	Time	To detect all aspects of the threat computer systems.
09	Time	To provide the fire support coordinator, information engagement officer, and electronic warfare officer, information and intelligence support for targeting of the threat's forces through lethal and nonlethal fires, to include updates.
10	Percent	Of threat command and control nodes vulnerable to electronic attack and electronic support.
11	Percent	Of threat computer systems vulnerable to computer network attack and computer network exploit.
12	Percent	Of threat command and control nodes disrupted and degraded.
13	Percent	Of threat command and control nodes monitored.
14	Percent	Of threat computer systems compromised.
15	Percent	Of threat computer systems monitored.

ART 2.4.1.1 PROVIDE INTELLIGENCE SUPPORT TO TARGET DEVELOPMENT

2-64. The systematic analysis of threat forces and operations to determine high-value targets (people, organizations, or military units the threat commander requires for successful completion of the mission), high-payoff targets (people, organizations, or military units whose loss to the enemy contributes significantly to the success of the friendly course of action), and systems and system components for potential attack through maneuver, fires, electronic means, or information engagement or operations. (FM 2-0) (USAICoE)

No.	Scale	Measure
01	Yes/No	Unit developed target critical components on high-value targets.
02	Yes/No	Unit refined target critical components as high-payoff targets.
03	Yes/No	Critical components passed to targeting cell in sufficient time to engage targets.
04	Yes/No	Unit identified, nominated, and prioritized targets for lethal or nonlethal action.
05	Yes/No	Staff integrated information engagement requirements into target development.
06	Yes/No	Staff identified target areas of interest associated with each lethal or nonlethal action.
07	Time	To identify, nominate, and prioritize targets for lethal or nonlethal action.
08	Time	To develop target list and perform target system analysis, critical components, vulnerability assessment, and target validation based on commander's guidance.
09	Time	To identify target areas of interest for each lethal or nonlethal action.
10	Percent	Of targeted system vulnerabilities identified correctly.
11	Number	Of correctly identified critical components in targeted systems.
12	Number	Of correctly identified targets for lethal and nonlethal action.

ART 2.4.1.2 PROVIDE INTELLIGENCE SUPPORT TO TARGET DETECTION

2-65. The intelligence officer establishes procedures for dissemination of targeting information. The targeting team develops the sensor and attack guidance matrix to determine the sensors required to detect and locate targets. The intelligence officer places these requirements into the intelligence, surveillance, and reconnaissance (ISR) synchronization tools for later incorporation into the ISR plan. (FM 2-0) (USAICoE)

No.	Scale	Measure
01	Yes/No	The intelligence officer linked targets to specific sensors for near real-time targeting and included targets in the ISR synchronization plan.
02	Yes/No	The intelligence officer included targets and sensors in the sensor and attack guidance matrix.
03	Yes/No	The intelligence officer disseminated targeting information to appropriate systems for lethal and nonlethal action.
04	Time	For sensor to pass targeting data to appropriate systems for lethal and nonlethal action.
05	Percent	Of targets linked to sensor and appropriate systems for lethal and nonlethal action.

ART 2.4.2 PROVIDE INTELLIGENCE SUPPORT TO ARMY INFORMATION TASKS

2-66. Intelligence support to information tasks is the task of providing the commander information and intelligence support for targeting through nonlethal actions. It includes intelligence support to the planning, execution of the Army information tasks (information engagement, command and control warfare, information protection, operations security, and military deception), as well as assessing the effects of those operations. Key activities reflected in this task include communications, planning, synchronization, and integration of intelligence into operation plans and orders. (FM 3-0) (USACAC)

Note: This task branch only addresses those intelligence tasks that support the conduct of Army information tasks. The actual conduct of information engagement, information protection operations, and activities related to command and control warfare are tasks addressed elsewhere in the FM 7-15:

- ART 3.4 (Integrate Command and Control Warfare) addresses the conduct of electronic command and control warfare operations.
- ART 6.3 (Conduct Information Protection) addresses the conduct of passive measures that protect information and systems.

- ART 6.11.3.1 (Conduct Counterintelligence Operations) addresses the conduct of that particular task which also relates to ART 6.3 (Conduct Information Protection).
- ART 5.3.1 (Integrate Information Engagement Capabilities) is the other activity related to Army information tasks.

No.	Scale	Measure
01	Yes/No	Intelligence support to targeting database maintained.
02	Yes/No	Information operations (IO) target identified, prioritized and nominated.
03	Yes/No	Intelligence support identified for each element of IO involved in the operation and integrated into the intelligence, surveillance, and reconnaissance plan.
04	Yes/No	Battle damage assessment performed on IO targets and target systems.
05	Yes/No	Combat assessment performed on IO targets.
06	Time	To determine support required for IO.
07	Time	To determine specific information requirements for IO.
08	Time	To determine the effects on the IO targets engaged.
09	Time	To provide combat assessments in support of IO.
10	Percent	Of IO requirements answered.

ART 2.4.2.1 PROVIDE INTELLIGENCE SUPPORT TO INFORMATION ENGAGEMENT

2-67. Military intelligence organizations operating outside U.S. territories support activities related to information engagement under some circumstances. (FM 2-0) (USAICoE)

No.	Scale	Measure
01	Yes/No	Unit identified intelligence support required for public affairs through intelligence preparation of the battlefield.
02	Yes/No	Unit provided intelligence and information to identify pertinent civil considerations that affect information engagement.
03	Yes/No	Unit provided intelligence support to identify populace perceptions, sentiments, attitudes, mores, allegiances, alliances, and behaviors.
04	Yes/No	Unit provided intelligence support to identify nongovernmental and international organizations in the operational environment.
05	Yes/No	Unit provided identification support of resources and capabilities of nongovernmental and international organizations in the operational environment.
06	Yes/No	Unit identified all relevant government agencies, organizations, or departments that affect public affairs activities.
07	Yes/No	Unit provided intelligence support to identify adversary misinformation, disinformation, and propaganda capabilities.
08	Yes/No	Unit provided intelligence support to identify the location, biases, and agenda of national media representatives in the operational environment.
09	Yes/No	Unit provided intelligence support to identify the location, biases, and agenda of international media representatives in the operational environment.
10	Yes/No	Unit provided intelligence support to identify trends reflected by the national and international media.
11	Yes/No	Unit provided intelligence and information to public affairs per all applicable regulations, policies, and laws.

ART 2.4.2.1.1 Provide Intelligence Support to Public Affairs

2-68. This task entails military intelligence organizations collecting and providing information and intelligence products concerning civil considerations in the area of operations to support public affairs activities. (FM 2-0) (USAICoE)

No.	Scale	Measure
01	Yes/No	Intelligence support required for public affairs identified through intelligence preparation of the battlefield.
02	Yes/No	Unit provided intelligence information to identify pertinent civil considerations that affect public affairs activities and operations.
03	Yes/No	Unit provided intelligence support to identify populace perceptions, sentiments, attitudes, mores, allegiances, alliances, and behaviors.
04	Yes/No	Unit provided intelligence support to identify nongovernmental and international organizations in the operational environment.
05	Yes/No	Unit identified all relevant government agencies, organizations, or departments that affect public affairs activities and operations.
06	Yes/No	Unit provided intelligence support to identify trends reflected by the national and international media.
07	Yes/No	Unit provided intelligence support to identify the location, biases, and agenda of national media representatives in the operational environment.
08	Yes/No	Unit provided intelligence support to identify the location, biases, and agenda of international media representatives in the operational environment.
09	Yes/No	Unit provided intelligence support to identify adversary misinformation, disinformation, and propaganda capabilities.
10	Yes/No	Unit provided intelligence and information to public affairs per all applicable regulations, policies, and laws.
11	Time	To identify pertinent civil considerations which affect public affairs activities and operations.
12	Time	To identify populace perceptions, sentiments, attitudes, mores, allegiances, alliances, and behaviors.
13	Time	To identify nongovernmental and international organizations in the operational environment.
14	Time	To identify all relevant government agencies, organizations, or departments that affect public affairs activities and operations.
15	Time	To identify trends reflected by the national and international media.
16	Time	To identify the location, biases, and agenda of national media representatives in the operational environment.
17	Time	To identify the location, biases, and agenda of international media representatives in the operational environment.
18	Time	To identify adversary or enemy misinformation, disinformation, and propaganda capabilities.
19	Percent	Of accurate information and intelligence provided to public affairs for civil considerations that affect public affairs activities and operations.

ART 2.4.2.1.2 Provide Intelligence Support to Psychological Operations

2-69. Psychological operations require information and intelligence to support analysis of foreign target audiences and their environment to include the following factors: political, military, economic, social, social, information, infrastructure, physical environment and time. Continuous and timely intelligence is required to assess target audience behavioral trends. Information and intelligence focus on target audience motivation and behavior; indicators of success or lack of success (measure of effectiveness); and the target audience's reaction to friendly, hostile, and neutral force actions. (FM 2-0) (USAICoE)

No.	Scale	Measure
01	Yes/No	Unit identified intelligence support required for psychological operations through intelligence preparation of the battlefield and support to targeting.
02	Yes/No	Unit responded to psychological operations specific information requirements concerning the civil considerations.
03	Yes/No	Unit completed combat assessment on psychological operations targets.
04	Time	To determine the effects on the psychological operations targets engaged.

No.	Scale	Measure
05	Time	To provide combat assessments in support of psychological operations.
06	Percent	Of psychological operations information requests to which unit responded.

ART 2.4.2.2 PROVIDE INTELLIGENCE SUPPORT TO COMMAND AND CONTROL WARFARE

2-70. Military intelligence organizations provide information to identify threat decisionmaking and command and control nodes, processes, and means in order of criticality. Intelligence also helps identify threat systems, activities, and procedures that may be vulnerable to command and control warfare. Additionally, intelligence plays a key role in evaluating and assessing the effectiveness of command and control warfare. (FM 2-0 (USAICoE)

> *Note:* This task supports electronic attack which employs jamming, electromagnetic energy, or directed energy against personnel, facilities, or equipment. This task identifies critical threat information systems and command and control nodes.

No.	Scale	Measure
01	Yes/No	Unit identified threat command and control nodes.
02	Yes/No	Unit identified threat communications systems.
03	Yes/No	Unit identified threat computer systems.
04	Yes/No	Intelligence support required for electronic attack identified through intelligence preparation of the battlefield and support to targeting.
05	Yes/No	Unit identified the threat's assets, processes, patterns, and means vulnerable to electronic attack.
06	Yes/No	Unit provided intelligence support to locate targets for electronic attack.
07	Yes/No	Unit provided intelligence support to determine if desired effects were achieved.
08	Yes/No	Unit provided information regarding target capabilities and vulnerabilities.
09	Yes/No	Unit provided information regarding which enemy systems are available against which to perform electronic attack.
10	Yes/No	Unit completed combat assessment for electronic attack.
11	Time	To determine support required for electronic attack.
12	Time	To determine specific information requirements for electronic attack.
13	Time	To provide combat assessments in support of electronic attack.
14	Percent	Of threat command and control nodes vulnerable to electronic attack.
15	Percent	Of threat command and control nodes vulnerable to electronic exploitation.
16	Percent	Of threat computer systems vulnerable to computer network attack.
17	Percent	Of threat command and control nodes disrupted and degraded.
18	Percent	Of threat computer systems compromised.
19	Percent	Of personnel, facilities, or equipment degraded, denied, disrupted, or destroyed by electronic attack.

ART 2.4.2.3 PROVIDE INTELLIGENCE SUPPORT TO INFORMATION PROTECTION

2-71. The military intelligence organizations provide information to identify threat command and control warfare capabilities and activities and tactics, techniques, and procedures. Intelligence provides information relating to computer network defense, physical security, operations security, counterdeception, and counterpropaganda. (FM 2-0) (USAICoE)

No.	Scale	Measure
01	Yes/No	Intelligence support required for information protection identified through intelligence preparation of the battlefield and support to targeting.
02	Yes/No	Unit identified threat offensive information capabilities.
03	Yes/No	Unit identified threat offensive information operations (IO) tactics, techniques, and procedures.
04	Yes/No	Unit identified friendly assets and forces that the threat could exploited.
05	Yes/No	Unit completed combat assessment on threat offensive IO.
06	Time	That threat offensive IO disrupts, degrades, or exploits friendly information systems.
07	Percent	Of known threat sensor coverage in area of operations.
08	Percent	Of attempted threat penetration of friendly information systems and facilities that were successfully defeated.
09	Percent	Of imitative electronic deception attempted and successful against our forces.
10	Percent	Of threat IO capabilities not covered by operations security measures.
11	Percent	Of procedures for information protection changed or modified based on the results of combat assessment.
12	Number	Of threat offensive IO attempts that disrupt, degrade, or exploit friendly information systems and facilities.
13	Number	Of imitative electronic deception operations against our forces.

ART 2.4.2.4 PROVIDE INTELLIGENCE SUPPORT TO OPERATIONS SECURITY

2-72. This task identifies capabilities and limitations of the threat's intelligence system including adversary intelligence objectives and means, procedures, and facilities to collect, process, and analyze information. This task supports the identification of indicators that adversary intelligence capabilities and systems might detect that could be interpreted or pieced together to obtain essential elements of friendly information in time to use against friendly forces. (FM 2-0) (USAICoE)

No.	Scale	Measure
01	Yes/No	Intelligence support required for operations security identified through intelligence preparation of the battlefield.
02	Yes/No	Unit identified capabilities and limitations of adversary, threat intelligence, and security services.
03	Yes/No	Unit identified operations security compromises.
04	Time	To input the information operations annex of the operation order.
05	Time	To identify potential compromises of essential elements of friendly information in area of operations (AO).
06	Number	Of adversary capabilities in the AO identified.
07	Number	Of successful adversary attempts to obtain information concerning friendly information systems.
08	Number	Of encrypted communications in AO.
09	Number	Of operations security measures selected tied to vulnerability analysis.
10	Number	Of operations security vulnerabilities tied to specific adversary capabilities by planners.

ART 2.4.2.5 PROVIDE INTELLIGENCE SUPPORT TO MILITARY DECEPTION

2-73. This task identifies capabilities and limitations of the threat's intelligence collection capabilities, systems, and means and identifies threat biases and perceptions. (FM 2-0) (USAICoE)

No.	Scale	Measure
01	Yes/No	Unit identified profiles of key threat leaders.
02	Yes/No	Unit outlined the threat decisionmaking processes, patterns, and biases.

No.	Scale	Measure
03	Yes/No	Unit identified the threat's decisionmaking dissemination processes, patterns, and means.
04	Yes/No	Unit identified the threat perceptions of the military situation in the area of operations.
05	Yes/No	Unit identified capabilities and limitations of threat intelligence and security services.
06	Yes/No	Unit completed combat assessment for deception operations.
07	Time	To determine support required for deception operations.
08	Time	To determine specific information requirements for deception.
09	Time	To provide combat assessments in support of deception.
10	Percent	Of threat forces and targets who were deceived.

ART 2.4.3 PROVIDE INTELLIGENCE SUPPORT TO COMBAT ASSESSMENT

2-74. Intelligence supports the assess activity of the operations process and targeting processes. The commander uses combat assessment to determine if targeting actions have met the attack guidance and if reattack is necessary to perform essential fires tasks and achieve the commander's intent for fires. The staff determines how combat assessment relates to specific targets by completing battle damage, physical damage, functional damage, and target system assessments. (FM 2-0) (USAICoE)

Note: This task branch supports ART 5.1.4 (Assess Tactical Situations and Operations), ART 5.1.4.3 (Provide Combat Assessment), and ART 5.1.4.3.1 (Perform Battle Damage Assessment). It is also associated with decide, detect, deliver, and assess.

No.	Scale	Measure
01	Yes/No	Unit identified intelligence, surveillance, and reconnaissance assets that can acquire information on target effects.
02	Yes/No	Unit provided an initial assessment of attacks.
03	Yes/No	Unit provided a full assessment of attacks.
04	Time	To identify and submit information requirements for lethal and nonlethal effects.
05	Time	To identify and submit information requirements for combat assessment and battle damage assessment.
06	Time	To perform combat assessment to identify if targets achieved the desired effects or require reattack.
07	Time	To make initial assessment of attacks after engagement.
08	Time	To provide initial assessment of attack effects to force commander.
09	Time	To complete full assessment of attack effects after engagement.
10	Time	To provide full assessment of attacks to force commander.
11	Time	To provide reattack recommendations.
12	Percent	Of high-payoff targets correctly assessed to meet attack guidance.

ART 2.4.3.1 CONDUCT PHYSICAL DAMAGE ASSESSMENT

2-75. Conduct physical damage assessment is a staff task that estimates the extent of physical damage to a target based on observed or interpreted damage. It is a post attack target analysis that is a coordinated effort among all units and the entire staff. (FM 2-0) (USAICoE)

No.	Scale	Measure
01	Yes/No	Unit identified intelligence, surveillance, and reconnaissance assets that can acquire information for physical damage assessment.
02	Yes/No	Unit provided an initial assessment of physical damage.
03	Yes/No	Unit provided a full assessment of attacks.

No.	Scale	Measure
04	Yes/No	Unit provided reattack recommendations.
05	Time	To identify and submit information requirements for physical damage assessment.
06	Time	To make initial assessment of physical damage after engagement.
07	Time	To provide initial assessment of physical damage to force commander.
08	Time	To complete full assessment of physical damage.
09	Time	To provide full assessment of attacks to force commander.
10	Percent	Of physical damage achieved within the commander's intent for fires.
11	Percent	Of targets correctly assessed for the commander's intent for fires.
12	Percent	Of targets unnecessarily reattacked.

ART 2.4.3.2 CONDUCT FUNCTIONAL DAMAGE ASSESSMENT

2-76. The staff conducts the functional damage assessment for the threat's remaining functional or operational capabilities. The assessment focuses on measurable effects. It estimates the threat's ability to reorganize or find alternative means to continue operations. The targeting working group and staff integrate analysis with external sources to determine if the commander's intent for fires has been met. (FM 2-0) (USAICoE)

No.	Scale	Measure
01	Yes/No	Unit identified intelligence, surveillance, and reconnaissance assets that can acquire information for functional damage assessment.
02	Yes/No	Unit provided an initial assessment of functional damage.
03	Yes/No	Unit provided a full assessment of attacks.
04	Yes/No	Unit provided reattack recommendation.
05	Time	To identify and submit information requirements for functional damage assessment.
06	Time	To make initial assessment of functional damage after engagement.
07	Time	To provide initial assessment of functional damage to force commander.
08	Time	To complete full assessment of functional damage.
09	Time	To provide full assessment of attacks to force commander.
10	Percent	Of functional damage achieved within the commander's intent for fires.
11	Percent	Of targets correctly assessed for the commander's intent for fires.
12	Percent	Of targets unnecessarily reattacked.

Chapter 3

ART 3.0: The Fires Warfighting Function

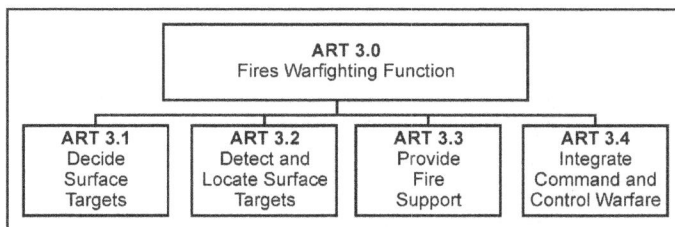

```
                    ┌─────────────────────────────┐
                    │          ART 3.0            │
                    │  Fires Warfighting Function │
                    └─────────────────────────────┘
         ┌──────────────┬──────────────┬──────────────┐
   ┌──────────┐   ┌──────────────┐ ┌──────────┐ ┌──────────────┐
   │ ART 3.1  │   │   ART 3.2    │ │ ART 3.3  │ │   ART 3.4    │
   │ Decide   │   │ Detect and   │ │ Provide  │ │ Integrate    │
   │ Surface  │   │Locate Surface│ │ Fire     │ │ Command and  │
   │ Targets  │   │   Targets    │ │ Support  │ │Control Warfare│
   └──────────┘   └──────────────┘ └──────────┘ └──────────────┘
```

The *fires warfighting function* is the related tasks and systems that provide collective and coordinated Army indirect fires, joint fires, and command and control warfare, including nonlethal fires, through the targeting process. It includes tasks associated with integrating and synchronizing the effects of these types of fires and command and control warfare—including nonlethal fires—with the effects of other warfighting functions. These are integrated into the concept of operations during planning and adjusted based on the targeting guidance. Fires normally contribute to the overall effect of maneuver but commanders may use them separately for the decisive operation and shaping operations. (FM 3-0) (USACAC)

Note: ART 5.1.4.3 (Provide Combat Assessment) addresses combat assessment (Battle Damage Assessment, Munitions Effects Assessments and Reattack Recommendation).

ART 6.1 (Employ Air and Missile Defense) addresses acquiring and attacking aerial targets.

SECTION I – ART 3.1: DECIDE SURFACE TARGETS

3-1. Analyze the situation relative to the mission, objectives, and capabilities. Identify and nominate specific vulnerabilities, high-value targets, and high-payoff targets that if influenced, degraded, delayed, disrupted, disabled, or destroyed will accomplish the commander's intent. (FM 6-20) (USAFAS)

No.	Scale	Measure
01	Yes/No	Targets selected allowed for accomplishing the unit mission and commander's intent.
02	Yes/No	Selected targets reviewed for compliance with rules of engagement.
03	Yes/No	Unit integrated nonlethal fires into the targeting process.
04	Time	To select and decide on attacking a high-payoff target once inside the execution cycle.
05	Time	To create a target nomination list.
06	Time	To complete target prioritization.
07	Time	To determine moving target intercept points.
08	Time	To issue prohibited target guidance.
09	Time	To pass commander's guidance to targeting agencies.
10	Percent	Of high-payoff targets discovered resulting in a reprioritized target list.
11	Percent	Of potential targets subjected to systematic analysis.
12	Percent	Of potential targets analyzed within an established time.

No.	Scale	Measure
13	Percent	Of selected targets that completed duplication checks.
14	Percent	Of selected high-payoff targets accurately identified.

SECTION II – ART 3.2: DETECT AND LOCATE SURFACE TARGETS

3-2. Perceive an object of possible military interest without confirming it by recognition (detect). Determine the placement of a target on the battlefield (locate). Target location can be expressed, for example, as a six-digit grid coordinate. (FM 6-20) (USAFAS)

Note: Contributions made by the intelligence warfighting function toward this task can be found in ART 2.4.1 (Provide Intelligence Support to Targeting).

No.	Scale	Measure
01	Time	To locate targets during surveillance and reconnaissance of defined target area of interest.
02	Percent	Of potential targets detected to targeting accuracy during surveillance and reconnaissance.
03	Percent	Of target locations verified before next targeting cycle.
04	Percent	Of designated high-payoff targets that have correct location data.
05	Number	Of high-payoff targets detected and located to targeting accuracy in the area of operations.

SECTION III – ART 3.3: PROVIDE FIRE SUPPORT

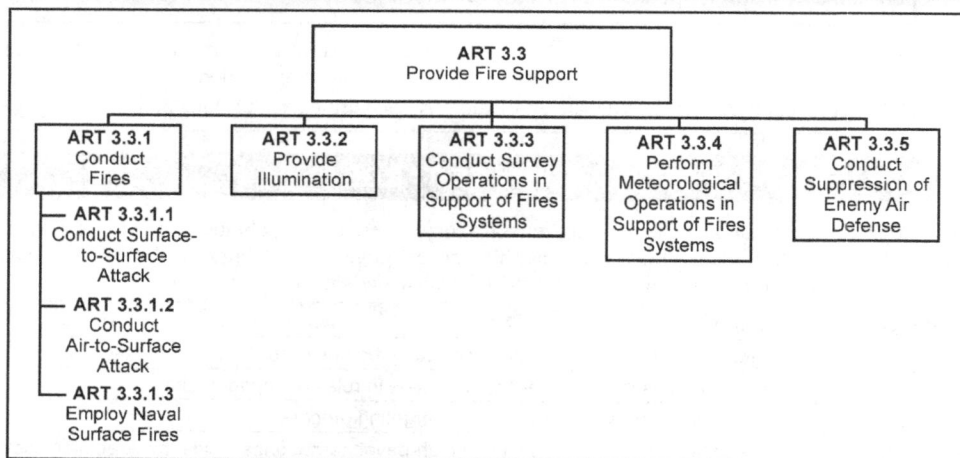

3-3. Provide collective and coordinated use of Army indirect fires, joint fires, and command and control warfare, including nonlethal fires, through the targeting process to support operations against surface targets. (FM 6-20) (USAFAS)

No.	Scale	Measure
01	Yes/No	Fires achieved the commander's fires guidance.
02	Yes/No	Commander directed that contingency plans be developed for the employment of lethal fires.

No.	Scale	Measure
03	Yes/No	Commander directed that contingency plans be developed for the employment of nonlethal fires.
04	Time	For a given fires weapon system to respond.
05	Time	To reattack target after battle damage assessment.
06	Percent	Of target attacks that achieve desired results.
07	Percent	Of targets selected for engagement that were engaged.
08	Percent	Of friendly areas of operations within the range of supporting fires systems.
09	Percent	Of fires missions synchronized with the maneuver of friendly units.
10	Percent	Of targets engaged that result in collateral damage.
11	Percent	Of targets engaged that result in friendly or neutral casualties.
12	Number	Of targets engaged that result in collateral damage.
13	Number	Of courses of action denied to an enemy or friendly force due to friendly fires efforts as determined from interrogations and after action reviews.
14	Number	Of targets engaged that result in friendly or neutral casualties.

ART 3.3.1 CONDUCT FIRES

3-4. Use fires weapon systems against troops, fortifications, materiel, or facilities. (FM 6-20) (USAFAS)

No.	Scale	Measure
01	Yes/No	Lethal and nonlethal fires achieved desired result.
02	Yes/No	Lethal and nonlethal fires followed rules of engagement.
03	Time	To get ordnance on target after initiating task.
04	Time	To complete attack after identifying target.
05	Time	To provide adjustment data after observing first rounds.
06	Time	To select targets for attack.
07	Time	To plan and coordinate naval surface fires attack.
08	Time	To prepare to conduct a surface-to-surface attack.
09	Percent	Of total fires missions requested by units executed.
10	Percent	Of the probability of a hit when selecting the correct munitions under existing conditions.
11	Percent	Of the probability of catastrophic, mobility, or firepower kill when selecting the correct munitions under existing conditions.
12	Percent	Of missions executed within a specified time.
13	Percent	Of missions flown and fired achieve desired target damage.
14	Percent	Of preplanned targets successfully attacked during operations.
15	Percent	Of desired results achieved by expected conclusion of a phase of the operation.
16	Percent	Of time-on-target missions accomplished on time.
17	Percent	Of enemy high-payoff target delivery systems engaged and destroyed by friendly forces.
18	Percent	Of friendly fires systems available to deliver ordnance.
19	Percent	Of enemy forces destroyed, delayed, disrupted, or degraded per FM 3-90 definitions.
20	Percent	Of total target list successfully engaged.
21	Percent	Of unplanned enemy targets successfully engaged.
22	Percent	Of unit basic load of ordnance available for use in lethal attack.
23	Percent	Of enemy high-payoff targets requiring more than one type of weapon system to ensure successful attack.

No.	Scale	Measure
24	Percent	Of lethal fires effort diverted by higher commanders to support their targeting priorities.
25	Number	Of fires systems available.

ART 3.3.1.1 CONDUCT SURFACE-TO-SURFACE ATTACK

3-5. Use ground-based, indirect-fire weapon systems to destroy, suppress, or neutralize enemy equipment (including aircraft on the ground), materiel, personnel, fortifications, and facilities. (FM 6-20) (USAFAS)

No.	Scale	Measure
01	Yes/No	Surface-to-surface attack achieved desired result.
02	Yes/No	Surface-to-surface attack followed rules of engagement.
03	Time	To get ordnance on target after initiating task.
04	Time	To complete attack after identifying target.
05	Time	To provide adjustment data after observing first rounds.
06	Time	To plan and coordinate surface-to-surface attack.
07	Time	To prepare for surface-to-surface attack.
08	Percent	Of total fires missions requested by units accomplished.
09	Percent	Of probability of a hit when selecting the correct munitions under sufficient conditions.
10	Percent	Of probability of catastrophic, mobility, or firepower kill given a hit when selecting correct munitions under sufficient conditions.
11	Percent	Of missions accomplished within a specified time.
12	Percent	Of missions fired that accomplished desired target damage.
13	Percent	Of preplanned targets successfully attacked during operation.
14	Percent	Of time-on-target missions accomplished on time.
15	Percent	Of enemy high-payoff target delivery systems engaged and destroyed by friendly forces.
16	Percent	Of friendly fires systems available to deliver ordnance.
17	Percent	Of enemy forces destroyed, delayed, disrupted, or degraded.
18	Percent	Of total target list successfully engaged.
19	Percent	Of immediate enemy targets successfully engaged.
20	Percent	Of unit basic load of ordnance available for use in lethal attack.
21	Percent	Of selected targets for which accurate coordinates are available.
22	Percent	Of ground based, indirect fire missions that result in collateral damage.
23	Percent	Of ground based, indirect fire missions that result in friendly or neutral casualties.
24	Number	Of fires systems available.
25	Number	Of ground based, indirect fire missions that result in collateral damage.
26	Number	Of ground based, indirect fire missions that result in friendly or neutral casualties.

ART 3.3.1.2 CONDUCT AIR-TO-SURFACE ATTACK

3-6. Use fixed- and rotary-wing aircraft-mounted weapon systems to destroy, suppress, or neutralize equipment (including aircraft on the ground), materiel, personnel, fortifications, and facilities. ART 3.3.1.2 includes the integration of fires from aerial platforms with other fires systems. (FM 3-09.32) (TRADOC-ALSA)

No.	Scale	Measure
01	Yes/No	Air-to-surface attacks allowed the commander to accomplish the mission within identified time and resource constraints.
02	Yes/No	Air-to-surface attacks followed rules of engagement.

No.	Scale	Measure
03	Time	To process air-to-surface attack requests through fires (fixed-wing) or maneuver (rotary-wing) channels as appropriate.
04	Time	For aerial systems to respond to mission request.
05	Time	For aircraft to identify target.
06	Time	To get ordnance on target after initiating air-to-surface attack request.
07	Percent	Of requested missions accomplished.
08	Percent	Of available aerial weapon systems (fixed- and rotary-wing).
09	Percent	Of missions where the ordnance carried by aerial weapon systems is appropriate for the target attacked.
10	Percent	Of missions requested directed to appropriate agency.
11	Percent	Of targets engaged by friendly aerial weapon systems (fixed- and rotary-wing).
12	Percent	Of friendly force operations delayed, disrupted, or modified due to lack of requested air support.
13	Percent	Of targets attacked achieve desired results.
14	Percent	Of maneuver forces having aerial weapon systems available.
15	Percent	Of air-to-surface attack missions that result in collateral damage.
16	Percent	Of aerial weapon systems having functioning identification, friend or foe systems.
17	Percent	Of air-to-surface missions cancelled because of weather restrictions.
18	Percent	Of air-to-surface attack missions that result in friendly or neutral casualties.
19	Number	Of aerial weapon systems available to support the commander.
20	Number	Of air-to-surface attack missions that result in collateral damage.
21	Number	Of air-to-surface attack missions that result in friendly or neutral casualties.

ART 3.3.1.2.1 Request Air-to-Surface Attack

3-7.   Request employment of Army special operations forces; Marine Corps, Navy, and Air Force aircraft; and other systems to deliver rocket, cannon, missile fires, and bombs on surface targets. (FM 3-09.32) (TRADOC-ALSA)

No.	Scale	Measure
01	Yes/No	Unit responded promptly to requests for an air-to-surface attack to provide support at the appropriate time.
02	Time	To process air support request.
03	Time	To get ordnance on target after initiating air request.
04	Percent	Of missions requested by Army commanders accomplished.
05	Percent	Of availability of supporting air agencies for maneuver forces.
06	Percent	Of missions requested directed to appropriate agency.

ART 3.3.1.2.2 Employ Close Air Support

3-8.   Employ aircraft in preplanned and immediate close air support (CAS) missions and joint air attack team operations to destroy, delay, disrupt, or suppress targets to support land operations. CAS requires positive identification, friend or foe and Army procedural or Army positive control of aircraft. (FM 6-20) (USAFAS)

No.	Scale	Measure
01	Yes/No	CAS destroyed, delayed, disrupted, or suppressed targets to support land operations.
02	Yes/No	CAS followed rules of engagement.
03	Time	To process CAS requests through fires channels.
04	Time	For CAS aircraft to respond to mission request.

No.	Scale	Measure
05	Time	For CAS aircraft to identify target.
06	Time	To get ordnance on target after initiating CAS request.
07	Percent	Of missions requested by Army commands accomplished.
08	Percent	Of CAS sorties allocated or distributed to the supported unit that were employed.
09	Percent	Of missions requested directed to appropriate agency.
10	Percent	Of requested targets engaged with friendly air support.
11	Percent	Of friendly force operations delayed, disrupted, or modified due to lack of requested CAS.
12	Percent	Of targets attacked achieving desired results.
13	Percent	Of CAS missions that result in collateral damage.
14	Percent	Of CAS missions cancelled by weather.
15	Percent	Of CAS missions flown with functioning identification, friend or foe systems and Army positive control of aircraft.
16	Percent	Of instances where CAS missions result in friendly or neutral casualties.
17	Number	Of CAS missions that result in collateral damage.
18	Number	Of instances where CAS missions result in friendly or neutral casualties.

ART 3.3.1.2.3 Employ Air Interdiction

3-9.   Employ aircraft to destroy, disable, disrupt, or delay enemy military potential before it can be used effectively against friendly ground forces. Employ aircraft far enough from friendly forces that detailed integration of each air mission with friendly ground maneuver is not required. (FM 6-20) (USAFAS)

No.	Scale	Measure
01	Yes/No	Air interdiction destroyed, delayed, disrupted, or suppressed targets to support land operations.
02	Yes/No	Air interdiction followed rules of engagement.
03	Time	To process air interdiction requests through fires channels.
04	Time	For air interdiction aircraft to respond to mission request.
05	Time	For air interdiction aircraft to identify target.
06	Time	To get ordnance on target after initiating air interdiction request.
07	Percent	Of missions requested by Army commands accomplished.
08	Percent	Of mission requests directed to appropriate agency.
09	Percent	Of air interdiction enemy systems and targets engaged with friendly air support.
10	Percent	Of friendly force operations delayed, disrupted, or modified due to lack of requested air interdiction.
11	Percent	Of targets attacked achieving desired results.
12	Percent	Of air interdiction missions accomplished without incidents of fratricide.
13	Percent	Of air interdiction missions cancelled by weather.
14	Percent	Of air interdiction missions flown with functioning identification, friend or foe systems and Army positive control of aircraft.
15	Percent	Of air interdiction missions that result in collateral damage.
16	Percent	Of air interdiction missions that result in friendly or neutral casualties.
17	Number	Of air interdiction missions that result in collateral damage.
18	Number	Of air interdiction missions that result in friendly or neutral casualties.

ART 3.3.1.3 EMPLOY NAVAL SURFACE FIRES

3-10. Employ fires provided by naval surface gun, missile, and electronic warfare systems to support units tasked with achieving the commander's intent. (FM 3-09.32) (TRADOC-ALSA)

No.	Scale	Measure
01	Yes/No	Naval surface fires achieved desired result.
02	Yes/No	Naval surface fires platform were within range.
03	Yes/No	Naval surface fires followed rules of engagement.
04	Time	To get ordnance on target after initiating task.
05	Time	To provide adjustment data after observing first rounds.
06	Time	To accomplish targeting for fires.
07	Time	To plan and coordinate for naval surface fires.
08	Percent	Of total fires missions requested by units executed.
09	Percent	Of the probability of a hit when selecting the correct munitions under sufficient conditions.
10	Percent	Of the probability of catastrophic, mobility, or firepower kill when selecting the correct munitions under existing conditions.
11	Percent	Of high-priority missions accomplished within a specified time.
12	Percent	Of missions fired accomplishing desired target damage.
13	Percent	Of preplanned targets successfully attacked during operations.
14	Percent	Of desired results achieved by the expected conclusion of a given phase of the operation.
15	Percent	Of naval cannon and missile systems available to deliver ordnance.
16	Percent	Of enemy forces destroyed, delayed, disrupted, or degraded.
17	Percent	Of total target list successfully engaged.
18	Percent	Of unplanned enemy targets successfully engaged.
19	Percent	Of the ship's basic load of ordnance available for use in lethal attack.
20	Percent	Of naval surface fire missions that result in collateral damage.
21	Percent	Of naval surface fire missions that result in friendly or neutral casualties.
22	Number	Of naval surface fires systems available.
23	Number	Of naval surface fire missions that result in collateral damage.
24	Number	Of naval surface fire missions that result in friendly or neutral casualties.

ART 3.3.2 PROVIDE ILLUMINATION

3-11. Provide friendly forces scheduled or on-call illumination fires to support night operations or harass the enemy. (FM 6-20) (USAFAS)

No.	Scale	Measure
01	Yes/No	Illumination support achieved desired result.
02	Yes/No	Weather allowed the illumination mission.
03	Time	To get illumination rounds on target after initiating task.
04	Time	To provide adjustment data after observing first rounds.
05	Time	To plan and coordinate for illumination.
06	Time	To prepare for firing unit for illumination mission.
07	Percent	Of total illumination missions requested by units accomplished.
08	Percent	Of friendly fires systems available to deliver illumination.
09	Percent	Of unplanned illumination missions accomplished.

No.	Scale	Measure
10	Percent	Of unit basic load of illumination rounds available.
11	Number	Of fires systems available to accomplish illumination missions.

ART 3.3.3 CONDUCT SURVEY OPERATIONS IN SUPPORT OF FIRES SYSTEMS

3-12. Use mechanical or electronic systems to determine dimensional relationships—such as locations, horizontal distances, elevations, directions, and angles—on the earth's surface. (FM 6-2) (USAFAS)

No.	Scale	Measure
01	Yes/No	Unit completed survey by time specified in order.
02	Yes/No	Survey order detailed priorities and accuracies for primary, alternate, and supplementary positions for firing units and target-locating element.
03	Time	To plan survey operation to include traverse, triangulation, and three-point resection.
04	Time	To prepare for survey operation.
05	Time	To execute survey operation.
06	Time	To enter a new survey control point into the database.
07	Time	To update survey control point in the database.
08	Time	From requesting information to providing desired survey information to firing units.
09	Percent	Of accuracy of survey operation.
10	Percent	Of accuracy of survey control available.
11	Percent	Of positioning and azimuth determining systems operational.
12	Number	Of positioning and azimuth determining systems available.

ART 3.3.4 PERFORM METEOROLOGICAL OPERATIONS IN SUPPORT OF FIRES SYSTEMS

3-13. This task includes using meteorological measuring instruments to determine necessary adjustments to individual weapon firing tables to increase the chances for first round hits, conserve ammunition, achieve surprise, and reduce chances for fratricide. Field artillery meteorological operations involve the determination of current atmospheric conditions. Atmospheric conditions along the trajectory of a projectile or rocket directly affect its accuracy and may cause the projectile or rocket to miss the desired point of impact. (FM 3-09.15) (USAFAS)

No.	Scale	Measure
01	Yes/No	Unit completed meteorological operations by time specified in order.
02	Yes/No	Meteorological order detailed frequency of meteorological observations.
03	Yes/No	Meteorological data obtained from Internet sources when organic meteorological measuring was not available.
04	Time	To prepare for meteorological operations.
05	Time	To execute meteorological operations.
06	Time	From requesting information to providing desired meteorological information to units firing.
07	Percent	Of meteorological measurement systems operational.
08	Number	Of meteorological measurement systems operational.

ART 3.3.5 CONDUCT SUPPRESSION OF ENEMY AIR DEFENSES

3-14. Neutralize, destroy, or temporarily degrade surface-based enemy tactical air defenses by destructive and disruptive means. Lethal suppression of enemy air defenses (SEAD) seeks the destruction of surface-based enemy tactical air defenses, such as target systems or operating personnel, by destructive means. Examples of destructive SEAD capabilities are bombs, air- and surface-to-surface missiles, air-scatterable mines, and field artillery. Nondestructive SEAD seeks to temporarily deny, degrade, deceive, delay, or neutralize surface-based enemy tactical air defense systems by disruptive means to increase aircraft survivability. (FM 6-20) (USAFAS)

Note: Disruptive means include the offensive and defensive activities discussed in ART 3.4 (Integrate Command and Control Warfare).

No.	Scale	Measure
01	Yes/No	Friendly aerial platforms accomplished mission without unacceptable losses to enemy air defense systems.
02	Time	To plan for the suppression of enemy air defense system.
03	Time	To respond to new requirements to suppress enemy air defense systems.
04	Time	To complete execution of all phases of the plan to suppress enemy air defenses.
05	Time	To prepare weapon systems and obtain munitions used in the suppression of enemy air defense systems.
06	Percent	Of available combat power dedicated toward SEAD.
07	Percent	Of enemy air defense systems destroyed.
08	Percent	Of enemy air defense systems temporarily neutralized by nondestructive means.
09	Percent	Of friendly air sorties attacked by enemy air defense.
10	Percent	Of enemy air defense that required reattack.
11	Percent	Of friendly air losses due to enemy air defense.
12	Percent	Probability of hitting the targeted enemy air defense system.
13	Percent	Probability of killing the targeted enemy air defense system, given a hit.
14	Percent	Of friendly suppression of enemy air defense system missions that accomplished their destruction or suppression mission.
15	Percent	Of SEAD missions that result in collateral damage.
16	Percent	Of SEAD missions that result in friendly or neutral casualties.
17	Number	And types of weapon systems and munitions used for SEAD.
18	Number	And types of enemy air defense systems that are permanently or temporarily suppressed.
19	Number	Of SEAD missions that result in collateral damage.
20	Number	Of SEAD missions that result in friendly or neutral casualties.

SECTION IV – ART 3.4: INTEGRATE COMMAND AND CONTROL WARFARE

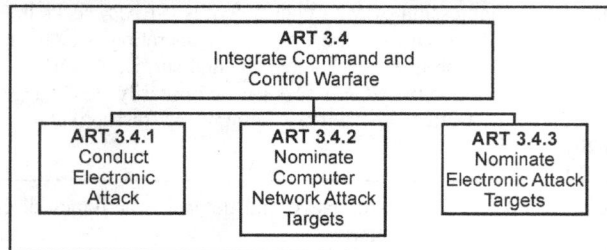

```
                    ART 3.4
              Integrate Command and
                 Control Warfare

   ART 3.4.1         ART 3.4.2         ART 3.4.3
   Conduct           Nominate          Nominate
   Electronic        Computer        Electronic Attack
   Attack          Network Attack       Targets
                     Targets
```

3-15. *Command and control warfare* is the integrated use of physical attack, electronic warfare, and computer network operations, supported by intelligence, to degrade, destroy, and exploit an enemy's or adversary's command and control system or to deny information to it. Command and control warfare combines lethal and nonlethal capabilities to attack the enemy's command and control capability. Command and control warfare degrades or destroys the enemy's information and its ability to use that information. In addition, command and control warfare attacks the enemy's ability to collect information about friendly forces. (FM 3-0) (USACAC)

ART 3.4.1 CONDUCT ELECTRONIC ATTACK

3-16. Employ electromagnetic or directed energy to attack personnel, facilities, or materiel to degrade, neutralize, or destroy enemy combat capability. Includes actions taken to prevent or reduce the enemy's effective use of the electromagnetic spectrum, such as jamming and antiradiation missiles, misinformation, intrusion, and meaconing. Electronic warfare supports attack by aviation through suppression of enemy air defenses operations. This task supports ART 3.0 command and control warfare through the use of offensive electronic attack to deceive, disrupt, deny, degrade or destroy the enemy's command and control capabilities. (FM 3-13) (USACAC)

> *Note:* Electronic attack includes both offensive and defensive activities to include countermeasures. Offensive electronic attack activities are generally conducted at the initiative of friendly forces. Examples include jamming an adversary's radar or command and control systems; using antiradiation missiles to suppress an adversary's air defenses; using electronic deception techniques to confuse an adversary's intelligence, surveillance, and reconnaissance systems; and using directed energy weapons to disable an adversary's equipment or capability. Defensive electronic attack activities use the electromagnetic system to protect personnel, facilities, capabilities, and equipment. Examples include self-protection and force protection measures such as use of expendables (flares and active decoys), jammers, towed decoys, directed energy infrared countermeasure systems, and counter radio-controlled improvised explosive device systems.

No.	Scale	Measure
01	Yes/No	Unit achieved electronic attack objective.
02	Yes/No	Electronic attack followed rules of engagement.
03	Time	To initiate electronic attack.
04	Time	To create a frequency deconfliction plan.
05	Time	To plan electronic attack.
06	Time	To prepare to conduct electronic attack including moving systems into place and completing rehearsals and precombat inspections.
07	Percent	Of operation plans and orders that integrate electronic attack with lethal fires.

No.	Scale	Measure
08	Percent	Of electronic attacks that achieve desired results on enemy.
09	Percent	Of tasked electronic attacks conducted.
10	Percent	Of reduction in enemy communications emissions after electronic attack.
11	Percent	Of enemy force degradation due to electronic attack.
12	Percent	Of available electronic attack systems that are mission capable.
13	Number	Of available electronic attack systems.

ART 3.4.2 NOMINATE COMPUTER NETWORK ATTACK TARGETS

3-17. Nominate targets to disrupt, deny, degrade, or destroy information in computers, computer networks, or the computers and networks themselves. This is a corps-level task. (FM 3-13) (USACAC)

No.	Scale	Measure
01	Yes/No	Unit identified computer network attack targets.
02	Yes/No	Unit nominated computer network attack targets following rules of engagement.
03	Time	To submit computer network attack targets to operational echelons.
04	Time	To receive results of computer network attack.
05	Percent	Of enemy information system not engaged by computer network attack that is targeted for attack by lethal systems.

ART 3.4.3 NOMINATE ELECTRONIC ATTACK TARGETS

3-18. Nominate targets operating within the electromagnetic spectrum to be destroyed, deceived, degraded, or neutralized. (FM 3-13) (USACAC)

No.	Scale	Measure
01	Yes/No	Unit identified electronic attack targets.
02	Yes/No	Unit nominated electronic attack targets following rules of engagement.
03	Time	To submit electronic attack targets to operational echelons.
04	Time	To receive results of electronic attack.
05	Percent	Of enemy information system not engaged by electronic attack that is targeted for attack by lethal systems.

This page intentionally left blank.

Chapter 4

ART 4.0: The Sustainment Warfighting Function

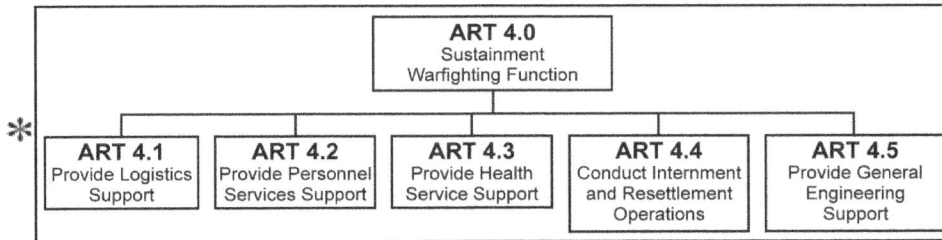

```
                          ┌────────────────────────┐
                          │        ART 4.0          │
                          │      Sustainment        │
                          │  Warfighting Function   │
                          └────────────────────────┘
   ┌──────────┬──────────────┬──────────────┬──────────────┬──────────────┐
┌────────┐ ┌────────────┐ ┌────────────┐ ┌────────────┐ ┌────────────┐
│ ART 4.1│ │  ART 4.2   │ │  ART 4.3   │ │  ART 4.4   │ │  ART 4.5   │
│Provide │ │Provide     │ │Provide     │ │Conduct     │ │Provide     │
│Logistics│ │Personnel   │ │Health      │ │Internment  │ │General     │
│Support │ │Services     │ │Service     │ │and         │ │Engineering │
│        │ │Support      │ │Support     │ │Resettlement│ │Support     │
│        │ │             │ │            │ │Operations  │ │            │
└────────┘ └────────────┘ └────────────┘ └────────────┘ └────────────┘
```

The *sustainment warfighting function* is the related tasks and systems that provide support and services to ensure freedom of action, extend operational reach, and prolong endurance. The endurance of Army forces is primarily a function of their sustainment. Sustainment determines the depth and duration of Army operations. It is essential to retaining and exploiting the initiative. Sustainment is the provision of the logistics, personnel services, and health service support necessary to maintain operations until mission accomplishment. Internment, resettlement, and detainee operations fall under the sustainment warfighting function and include elements of all three major subfunctions. (FM 3-0) (USACAC)

Note: This task and many of its subordinate tasks encompass environmental considerations.

+ Logistics is the science of planning, preparing, executing, and assessing the movement and maintenance of forces. In its broadest sense, logistics includes the design, development, acquisition, fielding, and maintenance of equipment and systems. Logistics integrates strategic, operational, and tactical support efforts within the joint operations area and schedules the mobilization and deployment of forces and materiel. (FM 4-0) (CASCOM)

SECTION I – ART 4.1: PROVIDE LOGISTICS SUPPORT

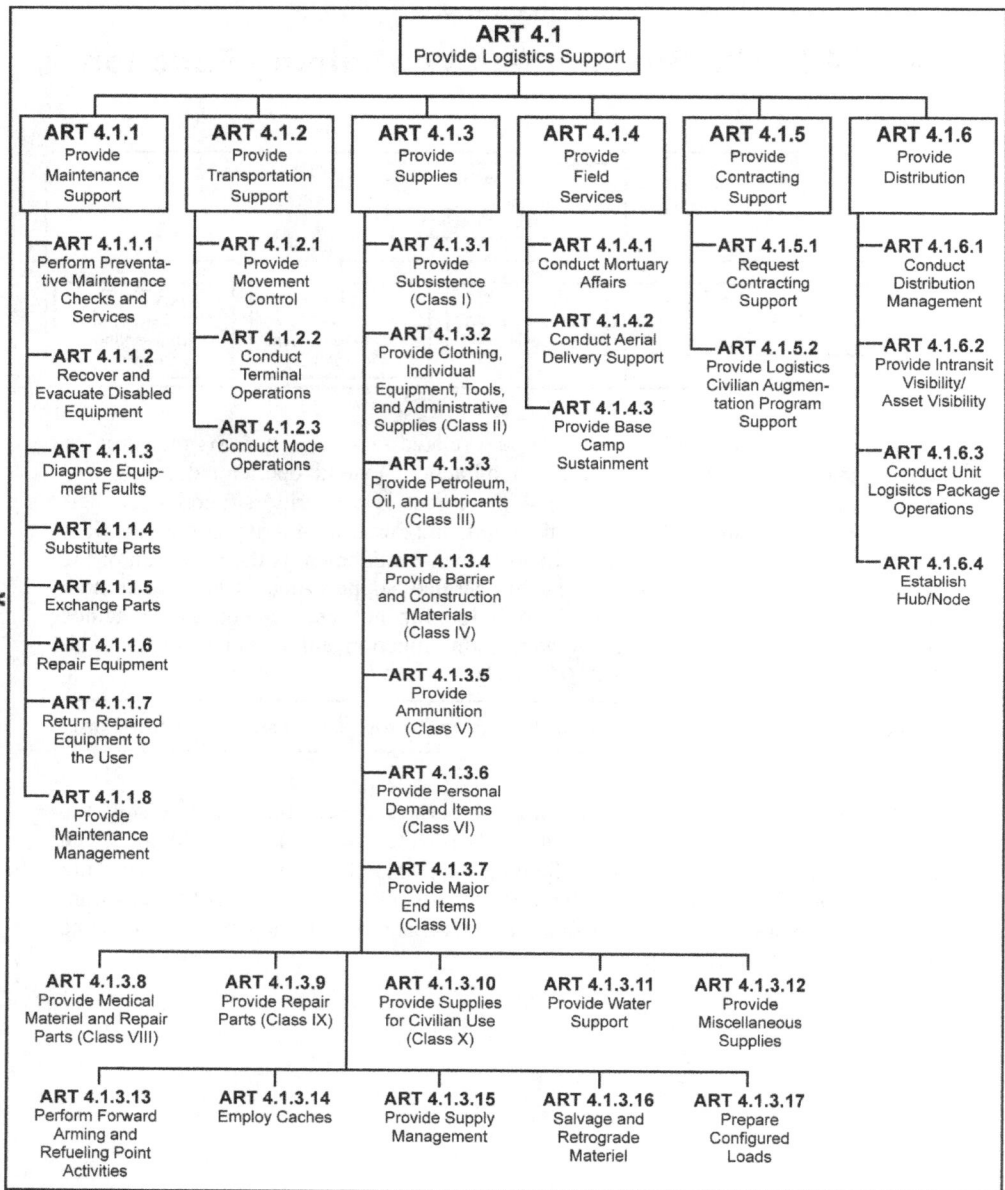

4-1 Logistics is the science of planning, preparing, executing, and assessing the movement and maintenance of forces. In its broadest sense, logistics includes the design, development, acquisition, fielding, and maintenance of equipment and systems. Logistics integrates strategic, operational, and tactical support efforts within the joint operations area and schedules the mobilization and deployment of forces and materiel. (FM 4-0) (CASCOM)

ART 4.1.1 PROVIDE MAINTENANCE SUPPORT

4-2.   Repair and maintain the availability of weapon systems and equipment. (FM 4-30.3) (CASCOM)

No.	Scale	Measure
01	Yes/No	Weapon systems and equipment supported the unit to accomplish its mission.
02	Time	Of average equipment downtime based on when equipment work order was submitted to maintenance organization.
03	Time	Of turnaround to repair priority combat equipment.
04	Time	To refine salvage and recovery plans after receipt of warning order.
05	Time	To diagnose malfunctioning equipment.
06	Time	To complete diagnosis, troubleshoot, and determine repair parts requirements for faulty equipment.
07	Time	To refine substitution policy after receipt of warning order.
08	Time	To diagnose, determine subsystem availability, and perform substitution to return major combat equipment to service.
09	Time	To refine direct exchange policy after receipt of warning order.
10	Time	To diagnose, determine direct exchange availability, and perform the exchange to return major combat equipment to service.
11	Time	Of average downtime for deadlined equipment using the direct exchange pipeline.
12	Time	To obtain replacement parts.
13	Time	To repair and return equipment.
14	Time	To refine the concept and policies for equipment repair, maintenance, and evacuation, and to establish maintenance facilities after receipt of warning order.
15	Percent	Of preventative maintenance checks and services (PMCS) tasks performed.
16	Percent	Of PMCS tasks deferred.
17	Percent	Of equipment operationally ready at any given time.
18	Percent	Of friendly damaged equipment recovered.
19	Percent	Of enemy abandoned equipment recovered.
20	Percent	Of damaged and abandoned materiel left on the battlefield due to failure to promptly report.
21	Percent	Of malfunctioning equipment deadlined with misdiagnosis.
22	Percent	Of equipment unavailable due to substitution or cannibalization.
23	Percent	Of substitutions completed by using salvage and disposal materiel.
24	Percent	Of substitutions completed by using authorized cannibalization within unit.
25	Percent	Of deadlined equipment returned to service through direct exchange.
26	Percent	Of direct exchanges unsuccessful due to faulty record keeping.
27	Percent	Of average equipment downtime.
28	Percent	Of time-phased force and deployment data maintenance units deployed and operational.
29	Percent	Of equipment deadlined for maintenance.
30	Percent	Of equipment deadlined for supply.
31	Percent	Of equipment failures successfully repaired.
32	Percent	Of maintenance capability provided by host nation.
33	Percent	Of maintenance facilities and sites secured from attack.

No.	Scale	Measure
34	Percent	Of operations degraded, delayed, or modified due to faulty maintenance management.
35	Percent	Of operations that address collection, classification, and disposition of enemy materiel.
36	Percent	Of captured enemy materiel collected, classified, and disposed of per instructions from appropriate materiel management center.
37	Number	Of PMCS tasks deferred.
38	Number	Of PMCS tasks performed.
39	Number	Of recovery operations that are performed successfully and to the time standard based on the applicable technical manuals.
40	Number	Of instances when unit operations are degraded, delayed, or modified due to lack of skilled mechanics and diagnostic equipment.

ART 4.1.1.1 PERFORM PREVENTATIVE MAINTENANCE CHECKS AND SERVICES

4-3. Perform preventative maintenance checks and services (PMCS) to identify potential problems quickly. ART 4.1.1.1 includes quick turnaround by component replacement, minor repairs, and performance of scheduled services at the operator, crew, company, and battalion and task force levels. ART 4.1.1.1 includes the performance of daily checks and scheduled services. (FM 4-30.3) (CASCOM)

No.	Scale	Measure
01	Yes/No	Unit performed PMCS per standards established in technical manuals for its equipment and with the applicable environmental considerations.
02	Time	Of average equipment downtime.
03	Time	Of turnaround for repair of priority combat equipment.
04	Percent	Of PMCS tasks performed.
05	Percent	Of PMCS tasks deferred.
06	Percent	Of equipment operationally ready.
07	Number	Of PMCS tasks.
08	Number	Of PMCS tasks deferred.
09	Number	Of PMCS tasks performed.

ART 4.1.1.2 RECOVER AND EVACUATE DISABLED EQUIPMENT

4-4. Obtain damaged, discarded, condemned, or abandoned multinational or enemy materiel (ground, aircraft, and marine). ART 4.1.1.2 includes the tactics, techniques, and procedures for recovering and evacuating disabled equipment. (FM 4-30.31) (CASCOM)

No.	Scale	Measure
01	Yes/No	Recovery and evacuation of disabled equipment contributed to the maintenance of unit combat power.
02	Time	To refine salvage and recovery plans after receipt of warning order.
03	Percent	Of friendly damaged equipment recovered.
04	Percent	Of damaged and abandoned materiel left on the battlefield due to failure to timely report.
05	Percent	Of enemy abandoned equipment recovered.
06	Number	Of instances when unit operations are degraded, delayed, or modified due to ongoing recovery operations.

ART 4.1.1.3 DIAGNOSE EQUIPMENT FAULTS

4-5. Identify malfunctions with on-board sensors, external test equipment, and visual inspections. ART 4.1.1.3 includes fault isolation or troubleshooting, battle damage or accident assessment, and differentiating between parts needing repair or replacement and serviceable parts or equipment. (FM 4-30.3) (CASCOM)

No.	Scale	Measure
01	Yes/No	Unit correctly identified equipment malfunctions and assessed equipment battle damage.
02	Time	To diagnose malfunctioning equipment.
03	Time	To complete diagnosis, troubleshoot, and determine repair parts requirements for faulty equipment.
04	Percent	Of equipment faults correctly diagnosed.
05	Percent	Of malfunctioning equipment deadlined with misdiagnosis.
06	Number	Of instances when unit operations are degraded, delayed, or modified due to lack of skilled mechanics and diagnostic equipment.

ART 4.1.1.4 SUBSTITUTE PARTS

4-6. Remove serviceable parts, components, and assemblies from unserviceable, economically repairable equipment, or from materiel authorized for disposal. Immediately reuse it in restoring a like item to a combat-operable or serviceable condition. ART 4.1.1.4 includes controlled exchange of parts and cannibalization. (FM 4-30.3) (CASCOM)

No.	Scale	Measure
01	Yes/No	Unit substitution policy increased the unit's available combat power.
02	Yes/No	Procedures for the substitution of parts were implemented with the applicable environmental considerations.
03	Time	To refine substitution policy after receipt of warning order.
04	Time	To diagnose and determine subsystem availability and to perform substitution to return major combat equipment to service.
05	Percent	Of equipment unavailable due to substitution and cannibalization.
06	Percent	Of substitutions completed using salvage and disposal materiel.
07	Percent	Of substitutions completed using authorized cannibalization within unit.

ART 4.1.1.5 EXCHANGE PARTS

4-7. Issue serviceable materiel in direct exchange for unserviceable materiel on an item-for-item basis. (FM 4-30.3) (CASCOM)

No.	Scale	Measure
01	Yes/No	Unit exchange policy did not degrade unit readiness.
02	Yes/No	Exchange procedures had applied environmental considerations.
03	Time	To refine direct exchange policy after receipt of warning order.
04	Time	To diagnose and determine direct exchange availability and to perform the exchange to return major combat equipment to service.
05	Time	Of average downtime for deadlined equipment using the direct exchange pipeline.
06	Percent	Of deadlined equipment returned to service through direct exchange.
07	Percent	Of direct exchanges unsuccessful due to faulty record keeping.

ART 4.1.1.6 REPAIR EQUIPMENT

4-8. Restore an item to serviceable condition by correcting a specific failure or unserviceable condition. These items include but are not limited to tactical and combat vehicles, aircraft, marine equipment, and information systems. ART 4.1.1.6 includes testing and checking equipment; adjusting, aligning, and repairing components and assemblies; repairing and modifying defective end items; replacing components and assemblies; removing and replacing piece parts; and marking and painting equipment. ART 4.1.1.6 also includes systems calibration, installation of modification work orders, and preventative replacement of parts before they can fail. (FM 4-30.3) (CASCOM)

No.	Scale	Measure
01	Yes/No	The time necessary to repair unit equipment did not take more than five percent longer than the sum of time necessary to recover the equipment to a maintenance site; diagnose equipment faults; order, obtain, and install replacement parts; and test and check repaired equipment.
02	Yes/No	Repairs were completed with environmental considerations—to include petroleum, oils, and lubricants products, chemical-agent-resistant coating, or other types of paint—and force protection.
03	Time	To obtain replacement parts.
04	Time	To repair equipment.
05	Time	Of average equipment downtime.
06	Time	Of turnaround for repair and return of critical combat equipment.
07	Time	To determine reason why equipment is malfunctioning.
08	Percent	Of equipment faults diagnosed correctly.
09	Percent	Of average equipment downtime.
10	Percent	Of time-phased force and deployment data maintenance units deployed and operational.
11	Percent	Of equipment deadlined for maintenance.
12	Percent	Of equipment deadlined for supply.
13	Percent	Of equipment failures successfully repaired.
14	Percent	Of damaged equipment salvaged.
15	Percent	Of maintenance capability provided by host nation.
16	Percent	Of unit maintenance capability diverted to other requirements.

ART 4.1.1.7 RETURN REPAIRED EQUIPMENT TO THE USER

4-9. Repair and return equipment to the battle or supply system. ART 4.1.1.7 includes providing operationally ready items to supply stream or float and repaired equipment to units. (FM 4-30.3) (CASCOM)

No.	Scale	Measure
01	Yes/No	Repaired equipment was returned to the user in a timely manner based on the situation.
02	Time	To obtain replacement parts.
03	Time	To repair and return equipment.
04	Time	Of average equipment downtime.
05	Time	Of turnaround for repair and return of critical combat equipment.
06	Percent	Of average equipment downtime.
07	Percent	Of time-phased force and deployment data maintenance units deployed and operational.
08	Percent	Of equipment deadlined for maintenance.
09	Percent	Of equipment deadlined for supply.
10	Percent	Of equipment failures successfully repaired.
11	Percent	Of damaged equipment salvaged.
12	Percent	Of maintenance capability provided by host nation.

ART 4.1.1.8 PROVIDE MAINTENANCE MANAGEMENT

4-10. Plan, coordinate, and synchronize maintenance operations to ensure maintenance operations are conducted efficiently with maintaining accurate maintenance records. (FM 4-30.3) (CASCOM)

No.	Scale	Measure
01	Yes/No	Procedures to provide maintenance management were successful in maintaining unit combat power and integrating environmental considerations.
02	Time	To obtain replacement parts.
03	Time	To repair equipment.
04	Time	To refine the concept and policies for equipment repair, maintenance, and evacuation and to establish maintenance facilities after receipt of warning order.
05	Time	Of average equipment downtime.
06	Time	Of turnaround for repair and return of critical combat equipment.
07	Percent	Of average equipment downtime.
08	Percent	Of time-phased force and deployment data maintenance units deployed and operational.
09	Percent	Of equipment deadlined for maintenance.
10	Percent	Of equipment deadlined for supply.
11	Percent	Of equipment failures successfully repaired.
12	Percent	Of damaged equipment salvaged.
13	Percent	Of maintenance capability provided by host-nation and contractor support.
14	Percent	Of sustainment area maintenance facilities secure.
15	Percent	Of operations degraded, delayed, or modified due to faulty maintenance management.
16	Percent	Of operation orders that address the collection, classification, and disposition of enemy materiel.
17	Percent	Of captured enemy materiel collected, classified, and disposed of per instructions from appropriate materiel management center.

ART 4.1.2 PROVIDE TRANSPORTATION SUPPORT

4-11. Move and transfer equipment, materiel, or personnel by towing, self-propulsion, or any means of carrier such as railways, highways, waterways, oceans, joint logistics over-the-shore, and airways. ART 4.1.2 includes technical operations and moving, evacuating, and transferring cargo, personnel, and equipment between transportation modes. This task includes environmental considerations imbedded due to issues of petroleum, oils, and lubricants and potential hazardous materials or other items to be transported. (FM 55-1) (CASCOM)

No.	Scale	Measure
01	Yes/No	Lack of transportation support did not delay, degrade, cause the modification of, or cancel unit operations.
02	Yes/No	Commander directed that contingency plans were developed for providing movement and transportation in the area of operations (AO).
03	Time	To refine transportation plan for AO after receipt of warning order.
04	Time	To establish a comprehensive movement plan after AO is assigned.
05	Time	For movement control battalion or team to begin operations after AO is assigned.
06	Time	That operations were delayed due to late arrivals of personnel and equipment.
07	Percent	Of difference between transportation plan and actual requirements in AO.
08	Percent	Of planned movement services support attained in AO.
09	Percent	Of allowable cabin load filled for AO lift sorties (not including staging and backload).
10	Percent	Of airfields with materials handling equipment.

No.	Scale	Measure
11	Percent	Of airlift sorties (not including staging and backhaul) flying at 90-percent allowable cabin load or better.
12	Percent	Of landing zones with materials handling equipment.
13	Percent	Of ports with materials handling equipment.
14	Percent	Of useable berths within a port facility.
15	Percent	Of scheduled transport movements completed on schedule.
16	Percent	Of fires delayed or canceled due to ammunition shortfall.
17	Percent	Of supplies moved to correct destination.
18	Percent	Of supplies lost or destroyed en route to destination in AO.
19	Number	Of passengers transported in AO per day in support of operations.
20	Number	Of ton-miles of supplies and equipment transported per day in AO.

ART 4.1.2.1 PROVIDE MOVEMENT CONTROL

4-12. Plan, route, schedule, control, coordinate, and provide in-transit visibility of personnel, units, equipment, and supplies moving via all modes of transportation (less pipeline) over air and surface lines of communications. (FM 4-01.30) (CASCOM)

No.	Scale	Measure
01	Yes/No	Procedures established to provide movement control in the area of operations (AO) did not delay, degrade, cause the modification of, or cancel unit operations.
02	Yes/No	Environmental considerations planning and procedures were present and being followed.
03	Yes/No	Commander directed that contingency plans be developed for providing movement and transportation in the AO.
04	Yes/No	Unit used transportation assets to backhaul waste for disposal.
05	Time	To refine movement plan for AO after receipt of warning order.
06	Time	To establish comprehensive movement control plan after AO is assigned.
07	Time	For AO movement control battalion or team to begin operation after AO is assigned.
08	Time	That operations are delayed due to late arrivals of personnel and equipment.
09	Time	To identify location of personnel and equipment in transit.
10	Percent	Of difference between movement plan requirements and actual requirements in AO.
11	Percent	Of planned movement control support attained in AO.
12	Percent	Of scheduled transport movements completed on schedule.
13	Percent	Of accurate position reports.
14	Number	Of passengers transported per day in AO in support of operations.
15	Number	Of ton-miles of supplies and equipment transported per day in AO.

ART 4.1.2.1.1 Provide Highway and Main Supply Route Regulation

4-13. Plan, route, schedule, and deconflict the use of highways and main supply routes (MSRs) to facilitate movement control. Highway regulation synchronizes with unit movement and maneuver. (FM 4-01.30) (CASCOM)

No.	Scale	Measure
01	Yes/No	Procedures established to provide highway and MSR regulation in the area of operations (AO) did not delay, degrade, cause the modification of, or cancel unit operations.
02	Time	To refine the highway and MSR regulation plan for AO after receipt of warning order.
03	Time	For movement control battalion or team to begin operation after AO is assigned.

No.	Scale	Measure
04	Time	To establish, publish, and distribute a comprehensive highway and MSR regulation plan after AO is assigned.
05	Time	Delay in highway and MSR movements due to late arrivals of personnel or equipment.
06	Time	To identify location of personnel and equipment in transit.
07	Percent	Of difference between highway and MSR regulation plan requirements and actual requirements in AO.
08	Percent	Of planned highway and MSR regulation support attained in AO.
09	Percent	Of scheduled highway and MSR movements that must be deconflicted.
10	Percent	Of accurate position reports.
11	Number	Of highway regulation teams available in the AO.

ART 4.1.2.1.2 Regulate Movement

4-14. Identify critical points where restrictions could slow down or stop movement. Critical points are facilities, terminals, ports, railheads, and cargo transfer points that, if congested, will limit the efficiency and effectiveness of the entire transportation network. (FM 4-01.30) (CASCOM)

No.	Scale	Measure
01	Yes/No	Procedures to regulate movement in the area of operations (AO) did not delay, degrade, cause the modification of, or cancel unit operations.
02	Time	To refine movement program for AO after receipt of warning order.
03	Time	To establish comprehensive port clearance plan after AO is assigned.
04	Time	For AO movement control battalion or team to begin operation after AO is assigned.
05	Time	Of delay in operations due to critical chokepoint clearance.
06	Time	To identify location of personnel and equipment in transit.
07	Percent	Of difference between movement program requirements and actual requirements in AO.
08	Percent	Of planned port clearance (air and sea) realized in AO.
09	Percent	Of scheduled transport movements completed on schedule.
10	Percent	Of accurate position reports.
11	Number	Of critical points identified.

ART 4.1.2.1.3 Conduct Support to Movement

4-15. Control movement with a traffic circulation plan that addresses the use of main supply route (names, direction of travel, size, and weight restrictions), checkpoints, rest and refuel areas, traffic control points, highway regulation points, and mobile patrols. It reflects any restrictive route features, such as direction of travel or size and weight restrictions, and critical points. ART 4.1.2.1.3 includes providing route signs and installing radio frequency interrogators at critical points to capture radio frequency tag data and report the data to in-transit visibility servers. (FM 4-01.30) (CASCOM)

No.	Scale	Measure
01	Yes/No	Mobility support operations enhanced the capabilities of units in the area of operations (AO) to accomplish missions.
02	Time	To refine traffic circulation plan for AO after receipt of warning order.
03	Time	To establish checkpoints, rest areas, refuel points, traffic control points, and highway regulation points to support the operation.
04	Time	To identify the location of personnel and equipment in transit.
05	Time	To produce and post route control signs.
06	Percent	Of difference between traffic circulation plan requirements and actual requirements in AO.

No.	Scale	Measure
07	Percent	Of routes classified (weight and size restrictions) in AO.
08	Percent	Of secured routes produced in AO.
09	Percent	Of scheduled transport movements completed on schedule.
10	Percent	Of accurate position reports.
11	Number	Of radio frequency interrogators used in AO.

ART 4.1.2.2 CONDUCT TERMINAL OPERATIONS

4-16. Provide an area or facility at which cargo, unit equipment, or personnel are loaded, unloaded, or handled in transit or transferred and reconfigured to another mode of transportation. ART 4.1.2.2 includes the preparation and reading of radio frequency tags to capture in-transit visibility and document inbound and outbound cargo movements. (FM 55-60) (CASCOM)

No.	Scale	Measure
01	Yes/No	Terminal operations in the area of operations did not delay, disrupt, cause the modification of, or eliminate unit courses of action.
02	Yes/No	Environmental considerations planning and procedures were present and being followed.
03	Time	To refine theater distribution plan after receipt of warning order.
04	Time	To establish comprehensive terminal operation plan after theater operation order execution date.
05	Time	To begin movement control operations.
06	Percent	Of difference between the theater distribution plan requirements and actual requirements.
07	Percent	Of planned terminal operations completed in area of operations.
08	Percent	Of scheduled transport movements completed on schedule.
09	Percent	Of terminal capacity utilized in theater per day.
10	Percent	Of required terminal capacity provided by host nation.
11	Percent	Of in-transit visibility transactions expected and actual.
12	Number	Of passengers processed per day through terminals in theater.
13	Number	Of terminals operating in theater at activation.
14	Number	Of tons of supplies and equipment handled per day in theater terminals.
15	Number	Of terminals with active in-transit visibility radio frequency identification (RFID) capabilities installed.
16	Number	Of in-transit visibility RFID transactions transmitted.

ART 4.1.2.2.1 Perform Arrival and Departure Airfield Control Group Activities

4-17. Coordinate, control, outload, and offload units and their equipment by air for deployment and redeployment. ART 4.1.2.2.1 requires marshaling of transported units, airfield reception, and outloading procedures as well as receiving and disposing of forces at the offload airfield. This task is the responsibility of the transported unit, its parent organization or installation, the arrival/departure airfield control group (A/DACG), and the Air Force tanker airlift control element mission support team. ART 4.1.2.2.1 includes the movement of sustainment cargo through the airfield and coordination for the transfer of the cargo to another mode of transportation. (FM 55-1) (CASCOM)

No.	Scale	Measure
01	Yes/No	A/DACG activities did not delay the unit's movement through the aerial port of embarkation or aerial port of debarkation beyond the time to prepare equipment and personnel for movement or to reconfigure equipment for operations after movement.
02	Time	To establish A/DACG program for area of operations (AO) after receipt of warning order.

No.	Scale	Measure
03	Time	To activate A/DACG on receipt of requirement in the AO.
04	Time	To provide requested materials handling equipment and logistic support to A/DACG.
05	Percent	Of difference between outload and offload requirements planned and the actual requirements in the AO.
06	Percent	Of planned arrivals and departures done on schedule in the AO.
07	Percent	Of scheduled passengers, unit equipment, and supplies moved on schedule.
08	Number	Of tons of supplies and equipment handled by A/DACG per day.
09	Number	Of passengers per day processed through A/DACG.

ART 4.1.2.2.2 Conduct Trailer, Container, and Flatrack Transfer Operations

4-18. Establish and operate a trailer, container, and flatrack transfer point for exchanging transportation platforms for line haul trucks operating over adjoining segments of a line haul route or in support of a distribution hub. ART 4.1.2.2.2 includes controlling and reporting equipment engaged in transfer operations. (FM 55-30) (CASCOM)

No.	Scale	Measure
01	Yes/No	Trailer, container, and flatrack transfer operations did not delay, restrict, cause the modification of, or cancel unit operations and mission.
02	Time	To refine theater distribution plan after receipt of warning order.
03	Time	To establish operational transfer points.
04	Percent	Of planned capability utilized.
05	Percent	Of in-transit visibility transactions expected and actual.
06	Percent	Of documentation errors corrected (digital and hard copy).
07	Number	Of trailers, containers, and flatracks processed through transfer point.
08	Number	Of trailers, containers, and flatracks stranded at transfer point.
09	Number	Of tons of supplies and materiel transferred at transfer point per day.
10	Number	Of unserviceable en route radio frequency identification (RFID) tag readers and interrogators effectively repaired by support element.
11	Number	Of transfer points with active in-transit visibility RFID capability installed.
12	Number	Of maintenance and repair actions completed at transfer points.

ART 4.1.2.2.3 Conduct Rail Transfer Operations

4-19. Coordinate, control, upload, and offload equipment by rail for deployment and redeployment. ART 4.1.2.2.3 requires marshaling of equipment, railhead reception, outloading procedures, and receiving and disposing of equipment at the offload railhead. ART 4.1.2.2.3 includes using rail for moving sustainment supplies and materiel, and the operation and maintenance of railway facilities, locomotives, and rolling stock. (FM 4-01.41) (CASCOM)

No.	Scale	Measure
01	Yes/No	A delay in moving or handling of personnel, supplies, and equipment by rail degraded, delayed, or modified unit operations.
02	Time	To refine theater distribution plan for area of operations (AO) after receipt of warning order.
03	Time	To establish comprehensive rail movement plan after AO is assigned.
04	Percent	Of difference between the theater distribution plan requirements and actual requirements in AO.
05	Percent	Of planned rail movements performed in AO.
06	Percent	Of scheduled rail transport movements completed on schedule.

No.	Scale	Measure
07	Percent	Of surface cargo in AO transported by railroad.
08	Percent	Of passengers stranded in transit for more than one day.
09	Percent	Of personnel, supplies, and equipment in AO that arrive at their destination on schedule.
10	Percent	Of operation degraded, delayed, or modified due to delays in moving or handling of personnel, supplies, and equipment.
11	Percent	Of rail transport capacity in AO used per day.
12	Percent	Of required rail transport provided by the host nation.
13	Percent	Of available rail network double tracked.
14	Percent	Of rolling stock operational.
15	Number	Of tons per day of supplies and equipment loaded at railheads.
16	Number	Of tons per day of supplies and equipment unloaded at railheads.
17	Number	Of passengers moved by rail transport per day.
18	Number	Of flatcars for oversized cargo.
19	Number	Of miles of rail per day repaired or upgraded.

ART 4.1.2.2.4 Conduct Marine Terminal Operations

4-20. Provide for loading, offloading, and in-transit handling of cargo and personnel at a seaport of embarkation and debarkation. ART 4.1.2.2.4 includes the transfer of cargo, equipment, and personnel from a marine terminal to another mode of ground or air transportation to facilitate onward movement and sustainment operations. Marine terminals are classified as fixed port facilities, unimproved port facilities, or bare beach port facilities. (FM 55-60) (CASCOM)

No.	Scale	Measure
01	Yes/No	A delay in moving or handling of personnel, supplies, and equipment degraded, delayed, or modified unit operations.
02	Time	To refine marine terminal service program for area of operations (AO) after receipt of warning order.
03	Time	To establish comprehensive marine terminal services plan after AO is assigned.
04	Time	For terminals to begin operation after AO is assigned.
05	Percent	Of difference between planned marine terminal service plan and actual requirements in AO.
06	Percent	Of planned marine terminal services support realized in AO.
07	Percent	Of scheduled transport movements completed on schedule.
08	Percent	Of marine terminal capacity in AO used per day.
09	Percent	Of required marine terminal capacity provided by host nation.
10	Percent	Of operations degraded, delayed, or modified due to delays in loading, unloading, or handling of personnel, supplies, and equipment at a marine terminal.
11	Percent	Of lost submerged items recovered or salvaged.
12	Number	Of tons per day of supplies and equipment handled at marine terminals.
13	Number	Of tons per day of supplies and equipment held in storage at marine terminals.
14	Number	Of tons per day of supplies and equipment handled in intermodal operations in AO on average.
15	Number	Of passengers per day vectored through marine terminals in AO.
16	Number	Of marine terminals operating in AO at activation.

ART 4.1.2.3 CONDUCT MODE OPERATIONS

4-21. Physically move supplies, unit equipment, individuals, and materiel on a transportation conveyance by a transportation mode or by unit means (individuals or unit organic means). ART 4.1.2.3 includes the relocation of ammunition supply and transfer points, supply support activities, and tactical joint logistics over-the-shore in support of the area of operations (AO) and mode operations. Operations include all modes of transportation and the use of military, contracted, and commercial transportation assets. (FM 55-1) (CASCOM)

No.	Scale	Measure
01	Yes/No	Delay in mode operations degraded, delayed, or modified unit operations.
02	Yes/No	Environmental considerations planning and procedures were present and being followed.
03	Yes/No	Unit facilitated internal travel of key leaders.
04	Time	To refine theater distribution plan for AO after receipt of warning order.
05	Time	To establish comprehensive movement support plan after AO is assigned.
06	Time	For movement control battalion or team to begin operation after AO is assigned.
07	Percent	Of difference between the theater distribution plan requirements and actual requirements in AO.
08	Percent	Of planned movement services support completed in AO.
09	Percent	Of scheduled transport movements completed on schedule.
10	Percent	Of operations degraded, delayed, or modified due to delays in moving or evacuating personnel, supplies, and equipment.
11	Percent	Of personnel, supplies, and equipment in AO that arrive on schedule.
12	Percent	Of passengers stranded in transit for more than one day.
13	Percent	Of movement capacity in AO utilized per day.
14	Percent	Of required transport services provided by host nation.
15	Number	Of tons per day of supplies and equipment moved by transport means in AO.
16	Number	Of tons per day of supplies and equipment in AO moved by organic units.
17	Number	Of passengers per day transported in AO.

ART 4.1.2.3.1 Move by Surface

4-22. Transport cargo, equipment, and personnel by waterways, railroads, highways, or other means, such as organic transportation. ART 4.1.2.3.1 includes the performance of logistic convoys. (FM 55-30) (CASCOM)

No.	Scale	Measure
01	Yes/No	Delay in surface transportation did not degrade, delay, or modify unit operations.
02	Time	To refine theater distribution plan for area of operations (AO) after receipt of warning order.
03	Time	To establish comprehensive surface movement plan after AO is assigned.
04	Time	For AO movement control battalion or team to begin operation after AO is assigned.
05	Percent	Of difference between planned theater distribution requirements and actual requirements in AO.
06	Percent	Of planned surface transport support completed in AO.
07	Percent	Of scheduled surface transport movements completed on schedule.
08	Percent	Of operations degraded, delayed, or modified due to delays in moving or evacuating personnel, supplies, and equipment.
09	Percent	Of surface cargo in AO transported by waterways.
10	Percent	Of surface cargo in AO transported by railroad.

No.	Scale	Measure
11	Percent	Of bulk fuel in AO transported by pipeline.
12	Percent	Of surface cargo in AO transported by wheeled or tracked vehicles.
13	Percent	Of personnel, supplies, and equipment in AO that arrive at their destination on schedule.
14	Percent	Of passengers stranded in transit for more than one day.
15	Percent	Of surface transport capacity in AO utilized per day.
16	Percent	Of required surface transport provided by host nation.
17	Number	Of tons per day of supplies and equipment in AO moved by surface transport.
18	Number	Of tons per day of supplies and equipment in AO moved by organic units.
19	Number	Of passengers per day transported by surface means in AO.

ART 4.1.2.3.2 Move by Air

4-23. Transport personnel, cargo, and equipment by aircraft. These assets include military, contracted, and commercial as well as strategic and theater fixed-wing airlift. ART 4.1.2.3.2 also includes the use of rotary-wing and Army operational support fixed-wing airlift as transportation platforms to move personnel, equipment, and sustainment supplies. (FM 55-1) (CASCOM)

No.	Scale	Measure
01	Yes/No	Delay in air transportation did not degrade, delay, or modify unit operations.
02	Time	To refine theater distribution plan for area of operations (AO) after receipt of warning order.
03	Time	To establish comprehensive air movement plan after AO is assigned.
04	Time	For AO movement control battalion or team to begin operation after AO is assigned.
05	Percent	Of difference between planned theater distribution requirements and actual requirements in AO.
06	Percent	Of planned air transport support completed in AO.
07	Percent	Of scheduled air transport movements completed on schedule.
08	Percent	Of operations degraded, delayed, or modified due to delays in moving or evacuating personnel, supplies, and equipment.
09	Percent	Of personnel, supplies, and equipment in AO that arrive at destination on schedule.
10	Percent	Of cargo in AO moved by air transport.
11	Percent	Of passengers stranded in transit for more than one day.
12	Percent	Of air transport capacity in AO utilized per day.
13	Percent	Of required air transport provided by the host nation.
14	Number	Of tons per day of supplies and equipment moved by air transport in AO.
15	Number	Of tons per day of supplies and equipment in AO moved by organic aviation units.
16	Number	Of passengers per day moved by air transport in AO.

ART 4.1.2.3.3 Move by Water

4-24. Provide for the movement of unit personnel, equipment, and sustainment cargo through and between Army water terminals. Water transport operations are conducted at established ocean and river ports, beach sites, and inland waterways. They are an integral part of inland waterway and shore-to-shore operations. Tasks include offshore ship discharge, inland waterway, and shore-to-shore operations for logistic purposes including logistics over-the-shore. (FM 55-50) (CASCOM)

Note: Amphibious operations as a task are addressed in ART 1.2.1.1.3 (Conduct an Amphibious Assault).

No.	Scale	Measure
01	Yes/No	Delay in water transportation did not degrade, delay, or modify unit operations.
02	Time	To refine theater distribution plan for area of operations (AO) after receipt of warning order.
03	Time	To establish comprehensive water transport movement plan after AO is assigned.
04	Percent	Of difference between planned theater distribution requirements and actual requirements in AO.
05	Percent	Of planned water transport movements performed in AO.
06	Percent	Of scheduled water transport movements completed on schedule.
07	Percent	Of surface cargo in AO transported via waterways.
08	Percent	Of passengers stranded in transit for more than one day.
09	Percent	Of personnel, supplies, and equipment in AO that arrive on schedule.
10	Percent	Of operation degraded, delayed, or modified due to delays in moving or handling of personnel, supplies, and equipment.
11	Percent	Of water transport capacity in AO used per day.
12	Percent	Of required water transport provided by host nation.
13	Number	Of tons per day of supplies and equipment moved by water transport.
14	Number	Of tons per day of supplies and equipment moved by organic units.
15	Number	Of passengers per day moved by water transport.

ART 4.1.3 PROVIDE SUPPLIES

4-25. Provide all classes of supply necessary to equip, maintain, and operate military units. This task also includes the commander responsibilities in coordination, percentages of supplies delivered, and amount of supplies issued and received. (FM 10-1) (CASCOM)

No.	Scale	Measure
01	Yes/No	Unit had supplies to conduct its mission.
02	Yes/No	Unit did not have to wait for supplies before it could conduct its mission.
03	Yes/No	Commander coordinated with higher, adjacent, supported, and supporting units to maintain all classes of supplies necessary to equip, maintain, and operate assigned, attached, and supported units in the area of operations (AO).
04	Time	Required to develop or update plans to establish support operations after receipt of warning order.
05	Time	To develop concept of support sustainment requirements after receipt of warning order.
06	Time	To reach time-phased operating and safety levels of supply in AO.
07	Percent	Of supply stockpiled in AO to support operations.
08	Percent	Of sustainment supply in AO supported by available facilities.
09	Percent	Of difference between planned and actual demand by type of supply in AO.
10	Percent	Of planned supply support completed in AO.
11	Percent	Of supplies available in AO compared to requirements by type of supply.
12	Percent	Of replenishment stocks in AO delivered on time.
13	Percent	Of shortfalls in supply that have acceptable alternatives.
14	Percent	Of required supplies delivered in AO.
15	Percent	Of operations degraded, delayed, or modified due to delays in moving or evacuating personnel, supplies, and equipment.
16	Percent	Of requisitions in AO filled.
17	Percent	Of required delivery date in AO met.
18	Percent	Of requisitions in AO filled from shelf stocks.

No.	Scale	Measure
19	Percent	Of critical replenishment stocks in AO that experienced late delivery.
20	Number	Of tons per day of supplies and equipment in AO delivered to operating forces by class of supply.

ART 4.1.3.1 PROVIDE SUBSISTENCE (CLASS I)

4-26. Provide food in bulk or prepackaged rations and packaged water. This task also includes the provision of health and comfort packages, such as toothbrushes, toothpaste, soap, disposable razors, and other personnel care items after the first 30 days of deployment and until Army and Air Force Exchange Service (AAFES) tactical field exchanges are operational. (FM 10-23) (CASCOM)

Note: The task of inspecting class I and class I sources is contained in ART 6.8.2 (Provide Veterinary Services).

No.	Scale	Measure
01	Yes/No	Unit had class I supplies to conduct its mission.
02	Yes/No	Unit did not have to wait for class I supplies before it could conduct its mission.
03	Yes/No	Unit ensured adequate health, food provisions, and security for belligerents.
04	Time	To develop or update plans to establish support operations after receipt of warning order.
05	Time	To develop concept of support sustainment requirements after receipt of warning order.
06	Time	To reach time-phased operating and safety levels of supply in the area of operations (AO).
07	Time	Of class I supply stockpiled in AO to support operations.
08	Time	Of sustainment supply in AO supported by available facilities.
09	Percent	Of difference between planned and actual demand by supply line in AO.
10	Percent	Of planned class I supply support performed in AO.
11	Percent	Of class I supplies available in AO compared to requirements.
12	Percent	Of replenishment stocks delivered on time in AO.
13	Percent	Of shortfalls in class I supply in AO that have acceptable alternatives.
14	Percent	Of required class I supplies in AO delivered.
15	Percent	Of operations degraded, delayed, or modified due to delays in moving class I supplies.
16	Percent	Of class I supply requisitions filled in AO.
17	Percent	Of required delivery date for class I supplies in AO performed.
18	Percent	Of critical replenishment stocks in AO that experienced late delivery.
19	Percent	Of class I supplies provided by host nation.
20	Percent	Of units ordering health and comfort packs after 90 and 180 days of deployment.
21	Number	Of tons per day of class I supply in AO delivered to operating forces.
22	Number	Of containers received in theater with only one menu item.
23	Number	Of containers received in theater with multiple lunch, dinner, or breakfast items.
24	Number	Of units receiving health and comfort packs after AAFES exchange or tactical field exchange has been established locally.
25	Number	Of days of supply of class I maintained by subsistence prime vendor.
26	Number	Of days of supply of class I maintained by contractor other than subsistence prime vendor.
27	Number	Of days of supply of class I maintained by subsistence platoons.

ART 4.1.3.2 Provide Clothing, Individual Equipment, Tools, and Administrative Supplies (Class II)

4-27. Provide clothing, individual equipment, tentage, organizational tool sets and kits, hand tools, geospatial products (maps), and administrative and housekeeping supplies and equipment. (FM 10-27) (CASCOM)

No.	Scale	Measure
01	Yes/No	Unit had class II supplies to conduct its mission.
02	Yes/No	Unit did not have to wait for class II supplies before it could conduct its mission.
03	Time	To refine supply support program for the area of operations (AO) after receipt of warning order.
04	Time	To develop concept of support sustainment requirements after receipt of warning order.
05	Time	To reach time-phased operating and safety levels of supply in AO.
06	Time	Of class II supply stockpiled in AO to support operations.
07	Time	Of sustainment supply in AO supported by available facilities.
08	Percent	Of difference between planned and actual demand by supply line in AO.
09	Percent	Of planned class II supply support completed in AO.
10	Percent	Of class II supplies available in AO compared to requirements.
11	Percent	Of replenishment stocks delivered on time in AO.
12	Percent	Of shortfalls in class II supply in AO that have acceptable alternatives.
13	Percent	Of required class II supplies in AO delivered.
14	Percent	Of operations degraded, delayed, or modified due to delays in moving class II supplies.
15	Percent	Of class II supply requisitions filled in AO.
16	Percent	Of required delivery date for class II supplies in AO completed.
17	Percent	Of critical replenishment stocks in AO that delivered late.
18	Percent	Of class II supplies provided by host nation.
19	Number	Of tons per day of class II supplies in AO delivered to operating forces.

ART 4.1.3.3 Provide Petroleum, Oils, and Lubricants (Class III)

4-28. Supply bulk fuel and packaged petroleum products. This task and its subordinate tasks will always include environmental considerations. (FM 10-67) (CASCOM)

No.	Scale	Measure
01	Yes/No	Unit had sufficient class III supplies to conduct its mission.
02	Yes/No	Unit did not have to wait for class III supplies before it could conduct its mission.
03	Yes/No	Spill planning and procedures related to environmental considerations were present and being followed.
04	Time	To develop replenishment concept after receipt of warning order.
05	Time	Of operational delay due to fuel shortages.
06	Time	Of supply of required fuel in place to support campaign.
07	Percent	And type of daily class III provided by host nation.
08	Percent	And type of required fuel delivered to theater.
09	Percent	Of fuel delivery capacity available in theater.
10	Percent	Of attempted deliveries destroyed by enemy action.
11	Percent	Of destroyed fuel deliveries anticipated and compensated for.
12	Percent	Of refueling capability available in theater at times and places needed.
13	Percent	Of total refueling assets available to support operational forces in theater.
14	Percent	Of time phased force and deployment data bulk fuel units deployed and operating.

No.	Scale	Measure
15	Number	Of available gallons of fuel lost to spills per day.
16	Number	Of gallons and types of fuel delivered to theater.

ART 4.1.3.3.1 Provide Bulk Fuel

4-29. Provide bulk fuels to units using tankers, rail tank cars, hose lines, bulk transporters, commercial pipeline and inland petroleum distribution system pipeline. ART 4.1.3.3.1 includes the completion of refueling on the move. (FM 10-67) (CASCOM)

No.	Scale	Measure
01	Yes/No	Unit had the necessary bulk class III supplies to conduct its mission.
02	Yes/No	Unit did not have to wait for bulk class III supplies before it could conduct its mission.
03	Yes/No	Spill planning and procedures related to environmental considerations were present and being followed.
04	Yes/No	Unit converted commercial grade fuel to military grade by injecting additives.
05	Time	To develop replenishment concept after receipt of warning order.
06	Time	Of operational delay due to fuel shortages.
07	Time	Of supply of required fuel in place to support campaign.
08	Percent	And type of daily class III bulk fuel provided by host nation.
09	Percent	And type of required bulk fuel delivered to theater.
10	Percent	Of bulk fuel deliveries completed compared to forecasted requirements.
11	Percent	Of bulk fuel delivery capacity available in theater.
12	Percent	Of attempted deliveries destroyed by enemy action.
13	Percent	Of destroyed bulk fuel deliveries anticipated and compensated for.
14	Percent	Of bulk refueling capability available in theater at times and places needed.
15	Percent	Of total bulk refueling assets available to support operational forces in theater.
16	Percent	Of time phased force and deployment data bulk fuel units deployed and operating.
17	Percent	Of available bulk fuel lost to spills.
18	Number	Of gallons per day of bulk fuel lost to spills.
19	Number	Of gallons and types of bulk fuel delivered to theater.

ART 4.1.3.3.2 Provide Packaged Petroleum, Oils, and Lubricants Products

4-30. Provide packaged products—including lubricants, greases, hydraulic fluids, compressed gasses, and specialty items—that are stored, transported, and issued in containers with a capacity of 55 gallons or less. (FM 10-67) (CASCOM)

No.	Scale	Measure
01	Yes/No	Unit had class III supplies to conduct its mission.
02	Yes/No	Unit did not have to wait for packaged class III supplies before it could conduct its mission.
03	Yes/No	Spill planning and procedures related to environmental considerations were present and being followed.
04	Time	To refine the supply support program for an area of operations after receipt of warning order.
05	Time	Of operational delay due to fuel shortages.
06	Time	Of supply of required packaged petroleum products in place to support operations.
07	Percent	And type of daily class III packaged petroleum products provided by host nation.
08	Percent	And type of required packaged petroleum products delivered to theater.

No.	Scale	Measure
09	Percent	Of packaged petroleum products deliveries completed compared to forecasted requirements.
10	Percent	Of attempted deliveries destroyed by enemy action.
11	Number	Of gallons per day and type of packaged petroleum products delivered to theater.

ART 4.1.3.3.3 Provide Petroleum Quality Assurance and Quality Surveillance

4-31. Quality assurance is performed by the government in determining requirements and specifications for petroleum products and related services. Quality surveillance includes measures used to determine and maintain the quality of government-owned petroleum products to ensure that such products are suitable for their intended use. Quality surveillance ensures that products meet quality standards after acceptance from the contractor as well as after transfer between government agencies or issue to users. (FM 10-67-2) (CASCOM)

No.	Scale	Measure
01	Yes/No	Class III supplies did not cause damage to the equipment when used per normal safeguards.
02	Yes/No	Unit did not have to delay operations to purge equipment fuel and lubrication systems of contamination.
03	Yes/No	Spill planning and procedures related to environmental considerations were present and being followed.
04	Time	To develop a program that ensures military petroleum products are procured under federal or military specification.
05	Percent	Of fuel meeting Department of Defense 4140.25-M inspection specifications.
06	Percent	Of fuel receiving tested per applicable regulations and standards.
07	Percent	Of laboratories provided and maintained for testing fuels and lubricants per applicable regulations.
08	Percent	Of agencies maintaining a quality surveillance program as prescribed in applicable regulations.
09	Percent	Of fuel certified by the Army Petroleum Center.
10	Percent	Of fuel products in excess of 10,000 gallons per DA PAM 710-2-1.
11	Percent	Of fuel products meeting standards established by AR 200-1 and the Air Pollution Abatement Program.
12	Percent	Of fuel tested per American Society for Testing and Materials or federal test methods.

ART 4.1.3.3.4 Conduct Aerial Refueling

4-32. Use Army special operations forces aviation capabilities to receive aerial refueling to extend the range of rotary-wing insertion and exfiltration platforms. (FM 3-05.60) (USAJFKSWCS)

No.	Scale	Measure
01	Yes/No	Unit had aerial refueling capabilities to conduct its mission.
02	Yes/No	Aerial refueling assets required to support mission accomplishment were where they were supposed to be when and with the quantities specified in the operation order.
03	Yes/No	Spill planning and procedures related to environmental considerations were present and being followed.
04	Time	To provide aerial fuel resupply concept after receipt of warning order.
05	Time	To conduct aerial refueling operations.
06	Percent	Of daily aerial refueling requirements provided by organic support assets.
07	Percent	Of required bulk fuel delivered to units.
08	Percent	Of bulk fuel delivery capacity available.

No.	Scale	Measure
09	Percent	Of required aerial refueling support assets available for the operation.
10	Percent	Of bulk aerial refueling capability available in theater at times and places needed.
11	Percent	Of total bulk refueling assets available to support operational forces in theater.
12	Number	Of gallons or pounds of fuel available in the area of operations.
13	Number	Of days of required fuel in place to support operation.
14	Number	Of total flying hours for each aircraft.
15	Number	Of kilometers between support locations and supported forces.
16	Number	Of aviation platforms requiring fuel.
17	Number	Of gallons per day of bulk aviation fuel lost to spills.

ART 4.1.3.3.5 Provide Retail Fuel

4-33. Provide retail fuels to individual systems from tankers, rail tank cars, hose lines, or bulk transporters. (FM 10-67) (CASCOM)

No.	Scale	Measure
01	Yes/No	Unit had the necessary bulk class III supplies to conduct its mission.
02	Yes/No	Spill planning and procedures related to environmental considerations were present and being followed.
03	Time	That the supply of required fuel was in place to support campaign.
04	Percent	And type of daily class III retail fuel requirements provided by host nation.
05	Percent	Of retail fuel deliveries completed compared to forecasted requirements.
06	Percent	Of available retail fuel lost to spills.
07	Number	Of gallons per day of retail fuel lost to spills.
08	Number	Of gallons and types of retail fuel delivered to users in the area of operations.

ART 4.1.3.4 PROVIDE BARRIER AND CONSTRUCTION MATERIALS (CLASS IV)

4-34. Provide construction materials including installed equipment and all fortification and barrier materials. (FM 10-27) (CASCOM)

No.	Scale	Measure
01	Yes/No	Unit had class IV supplies to conduct its mission.
02	Yes/No	Unit did not have to wait for class IV supplies before it could conduct its mission.
03	Time	Required to develop or update plans to establish support operations after receipt of warning order.
04	Time	To develop concept of support sustainment requirements after receipt of warning order.
05	Time	To reach time-phased operating and safety levels of supply in the area of operations (AO).
06	Percent	Of difference between projected engineer construction material requirements and actual requirements in AO.
07	Percent	Of planned class IV supply support completed in AO.
08	Percent	Of class IV supplies available in AO compared to requirements.
09	Percent	Of replenishment stocks delivered on time in AO.
10	Percent	Of shortfalls in class IV supply in AO that have acceptable alternatives.
11	Percent	Of required class IV supplies in AO delivered.
12	Percent	Of operations degraded, delayed, or modified due to delays in providing class IV supplies to the right locations in the right quantities.
13	Percent	Of class IV supply requisitions filled in AO.

No.	Scale	Measure
14	Percent	Of required delivery date for class IV supplies performed in AO.
15	Percent	Of critical replenishment stocks in AO that experienced late delivery.
16	Percent	Of class IV supplies provided by host nation.
17	Number	Of days of class IV supply stockpiled in AO to support campaign.
18	Number	Of days of sustainment supply in AO supported by available facilities.
19	Number	Of sawmills operating in the AO per environmental considerations.
20	Number	Of rock crushing facilities or quarries operating in the AO per environmental considerations.
21	Number	And types of class IV supply in tons per day delivered to forces in the AO.

ART 4.1.3.5 PROVIDE AMMUNITION (CLASS V)

4-35. Provide the right type and quantity of ammunition to the force. (FM 4-30.1) (CASCOM)

No.	Scale	Measure
01	Yes/No	Unit had class V supplies to conduct its mission.
02	Yes/No	Unit did not have to wait for class V supplies before it could conduct its mission.
03	Time	To determine suitable munitions available in theater after receipt of warning order.
04	Time	To develop replenishment concept after receipt of warning order.
05	Time	To deliver replenishment stocks after required delivery date.
06	Percent	Of required lift available.
07	Percent	Of minimum safety level of build-up stocks maintained at staging areas.
08	Percent	Of munitions at zero balance.
09	Percent	Of munitions lines below required supply rates.
10	Percent	Of time-phased force and deployment list (TPFDL) ammunition units deployed and operational.
11	Percent	Capacity of TPFDL ammunition units deployed and operational.
12	Percent	Of unit missions delayed due to shortfall in major equipment items.
13	Percent	Of high-payoff targets attacked with preferred munitions.

ART 4.1.3.5.1 Provide Munitions

4-36. Supply munitions—small arms ammunition, grenades, mines, rockets, missiles, tank and field artillery rounds, and nonlethal ammunition—to the force. (FM 4-30.1) (CASCOM)

No.	Scale	Measure
01	Yes/No	Unit had class V supplies to conduct its mission.
02	Yes/No	Unit did not have to wait for class V supplies before it could conduct its mission.
03	Time	After receipt of warning order to determine if suitable munitions are available in the area of operations.
04	Time	To deliver replenishment stocks after required delivery date (RDD).
05	Percent	Of minimum safety level of build-up stocks maintained at staging areas.
06	Percent	Of required reception and onward movement support available.
07	Percent	Of missions delayed due to shortfall of munitions.
08	Percent	Of fire missions delayed or not completed due to munitions shortfall.
09	Percent	Of high-payoff targets requiring reattack because preferred munitions not available.
10	Percent	Of replenishment stocks delivered before RDD.
11	Number	Of days of supply supported by available facilities.

ART 4.1.3.5.2 Provide Separate Loading Munitions

4-37. Supply munitions items, such as separate loading field artillery rounds, that have distinct components. (FM 4-30.1) (CASCOM)

No.	Scale	Measure
01	Yes/No	Unit had class V supplies to conduct its mission.
02	Yes/No	Unit did not have to wait for class V supplies before it could conduct its mission.
03	Time	After receipt of warning order to determine if suitable munitions are available in the area of operations.
04	Time	To deliver replenishment stocks of separate loading ammunition after required delivery date (RDD).
05	Percent	Of minimum safety level of build-up stocks maintained at staging areas.
06	Percent	Of required reception and onward movement support available.
07	Percent	Of fire missions delayed or not completed due to separate loading munitions shortfall.
08	Percent	Of high-payoff targets that require reattack because separate loading munitions not available.
09	Percent	Of replenishment stocks delivered before RDD.
10	Number	Of days of supply supported by available facilities.

ART 4.1.3.5.3 Provide Pyrotechnic and Specialty Items

4-38. Supply munitions items such as explosive bolts, ejection cartridges, and demolition charges and other specialty items that fall within the nonlethal category. (FM 4-30.1) (CASCOM)

No.	Scale	Measure
01	Yes/No	Unit had pyrotechnic and specialty items to conduct its mission.
02	Yes/No	Unit did not have to wait for pyrotechnic and specialty items before it could conduct its mission.
03	Time	After receipt of warning order to determine if suitable munitions are available in the area of operations.
04	Time	To deliver replenishment stocks of pyrotechnic or specialty items ammunition after required delivery date (RDD).
05	Percent	Of minimum safety level of build-up stocks maintained at staging areas.
06	Percent	Of required reception and onward movement support available.
07	Percent	Of missions delayed due to shortfall of pyrotechnic and specialty items.
08	Percent	Of fire missions delayed or not completed due to pyrotechnic and specialty items munitions shortfall.
09	Percent	Of high-priority targets that require reattack because pyrotechnic and specialty items munitions not available.
10	Percent	Of replenishment stocks delivered before RDD.
11	Number	Of days of supply supported by available facilities.

ART 4.1.3.6 PROVIDE PERSONAL DEMAND ITEMS (CLASS VI)

4-39. Coordinate and provide personal demand items such as health and hygiene products and nonmilitary sales items. (FM 10-1) (CASCOM)

No.	Scale	Measure
01	Yes/No	Unit had the necessary class VI supplies to conduct its mission.
02	Yes/No	Unit did not have to wait for class VI supplies before it could conduct its mission.
03	Time	Required to develop or update plans to establish support operations after receipt of warning order.

No.	Scale	Measure
04	Time	To develop concept of support sustainment requirements after receipt of warning order.
05	Time	To reach time-phased operating and safety levels of supply in area of operations (AO).
06	Percent	Of difference between planned and actual demand by supply line in AO.
07	Percent	Of class VI supplies available in AO compared to requirements.
08	Percent	Of replenishment stocks delivered on time in AO.
09	Percent	Of shortfalls in class VI supply in AO that have acceptable alternatives.
10	Percent	Of required class VI supplies in AO delivered.
11	Percent	Of planned class VI supply support completed in AO.
12	Percent	Of operations degraded, delayed, or modified due to delays in moving class VI supplies.
13	Number	Of days of class VI supply stockpiled in AO to support campaign.
14	Number	Of days of sustainment supply in AO supported by available facilities.
15	Number	Of tons per day of class VI supply in AO delivered to operating forces.

ART 4.1.3.7 PROVIDE MAJOR END ITEMS (CLASS VII)

4-40. Provide major end items such as launchers, tanks, mobile machine shops, and vehicles. (FM 10-1) (CASCOM)

No.	Scale	Measure
01	Yes/No	Unit had class VII supplies to conduct its mission.
02	Yes/No	Unit did not have to wait for class VII supplies before it could conduct its mission.
03	Time	To reach time-phased operating and safety levels of supply in the area of operations (AO).
04	Time	Required to develop or update plans to establish support operations after receipt of warning order.
05	Time	To develop concept of support sustainment requirements after receipt of warning order.
06	Percent	Of difference between planned and actual demand by supply line in AO.
07	Percent	Of planned class VII supply support performed in AO.
08	Percent	Of class VII supplies available in AO compared to requirements.
09	Percent	Of replenishment stocks delivered on time in AO.
10	Percent	Of shortfalls in class VII supply in AO that have acceptable alternatives.
11	Percent	Of required class VII supplies in AO delivered.
12	Percent	Of class VII supply requisitions filled in AO.
13	Percent	Of operations degraded, delayed, or modified due to delays in moving class VII supplies.
14	Percent	Of required delivery date for class VII supplies in AO completed.
15	Percent	Of critical replenishment stocks in AO delivered late.
16	Percent	Of class VII supplies provided by host nation.
17	Percent	Of class VII supplies enemy actions attempted to destroy.
18	Number	Of days of class VII supply stockpiled in AO to support operations.
19	Number	Of days of sustainment supply in AO supported by available facilities.
20	Number	Of tons per day of class VII supply in AO delivered to operating forces.

ART 4.1.3.8 PROVIDE MEDICAL MATERIEL AND REPAIR PARTS (CLASS VIII)

4-41. Provide class VIII medical materiel to include medical supplies, equipment, and medical peculiar repair parts. (FM 4-02.1) (USAMEDDC&S)

Note: ART 4.3.4 (Provide Medical Logistics) addresses the other aspects of medical logistics.

No.	Scale	Measure
01	Yes/No	Unit had class VIII supplies to conduct its mission.
02	Yes/No	Unit did not have to wait for class VIII supplies before it could conduct its mission.
03	Yes/No	Unit had class VIII medical peculiar repair parts to conduct its mission.
04	Yes/No	Unit did not have to wait for class VIII medical peculiar parts before it could conduct its mission.
05	Time	Required to develop or update plans to establish support operations after receipt of warning order.
06	Time	To transship class VIII supplies and medical equipment upon receipt of warning order.
07	Time	To provide emergency shipment of class VIII supplies in the area of operations (AO).
08	Time	To refine medical equipment maintenance and repair support program after receipt of warning order.
09	Time	To receive medical equipment peculiar repair parts after requisition.
10	Time	An average piece of medical equipment is not mission capable (awaiting parts).
11	Time	To requisition, procure, and provide critical medical equipment peculiar repair parts.
12	Percent	Of difference between planned and actual demand by supply line in AO.
13	Percent	Of planned class VIII supply support completed in AO.
14	Percent	Of class VIII supplies that require replenishment per day.
15	Percent	Of shortfalls in class VIII supply in AO that have acceptable alternatives.
16	Percent	Of required class VIII supplies in AO delivered.
17	Percent	Of operations degraded, delayed, or modified due to delays in moving class VIII supplies.
18	Percent	Of class VIII supply requisitions filled in AO.
19	Percent	Of required delivery date for class VIII supplies in AO performed.
20	Percent	Of critical replenishment stocks in AO delivered late.
21	Percent	Of class VIII supplies (meeting regulatory requirements) provided by host nation.
22	Percent	Of average medical equipment down.
23	Percent	Of time phased force and deployment list medical logistics units deployed and operational.
24	Percent	Of medical equipment deadlined for supply.
25	Percent	Of class VIII medical equipment peculiar repair parts requirements provided by the host nation.
26	Number	Of instances when medical capability is unavailable due to shortage or lack of class VIII supplies or equipment.
27	Number	Of tons per day of class VIII supplies (and medical peculiar repair parts) in AO delivered to operating forces.

ART 4.1.3.9 PROVIDE REPAIR PARTS (CLASS IX)

4-42. Provide any part, subassembly, assembly, or component required for installation in the maintenance of an end item, subassembly, or component. (FM 10-1) (CASCOM)

No.	Scale	Measure
01	Yes/No	Unit had class IX supplies to conduct its mission.
02	Yes/No	Unit did not have to wait for class IX supplies before it could conduct its mission.
03	Time	To refine supply support program after receipt of warning order.
04	Time	To receive repair parts after requisition.
05	Time	That an average piece of equipment is not mission capable (awaiting parts).

No.	Scale	Measure
06	Time	To requisition, procure, and provide critical repair parts.
07	Percent	Of average equipment downtime.
08	Percent	Of time-phased force and deployment data maintenance units deployed and operational.
09	Percent	Of equipment deadlined for supply.
10	Percent	Of transportation units deployed and operational.
11	Percent	Of class IX requirements provided by host nation.
12	Number	Of tons per day of class IX supplies in the area of operations delivered to operating forces.

ART 4.1.3.10 PROVIDE SUPPLIES FOR CIVILIAN USE (CLASS X)

4-43. Provide materials to support nonmilitary programs, such as agriculture and economic development. (FM 10-1) (CASCOM)

No.	Scale	Measure
01	Yes/No	Unit had class X supplies to conduct its mission.
02	Yes/No	Unit did not have to wait for class X supplies before it could conduct its mission.
03	Yes/No	U.S. and host-nation laws and regulations allowed civilians to use supplies provided.
04	Time	To establish liaison with appropriate host-nation civilian government officials in the area of operations (AO) after receipt of mission.
05	Time	To coordinate host-nation support agreements after AO is assigned.
06	Percent	Of sustainment supplies in AO procured from host-nation sources.
07	Percent	Of logistic effort in AO provided by host nation.
08	Number	Of facilities used by U.S. units in AO provided by host nation.
09	Number	Of host-nation support agreements in effect in AO.
10	Number	Of U.S. military units that have host-nation liaison officers assigned in AO.
11	Number	Of tons per day of class X supplies in AO delivered for civilian use.

ART 4.1.3.11 PROVIDE WATER SUPPORT

4-44. ART 4.1.3.11 includes purification, distribution, storage, and quality surveillance of water. This task incorporates environmental considerations. (FM 10-52) (CASCOM)

Note: ART 4.1.7 (Provide General Engineering Support) addresses construction, repairing, maintenance, and operations of permanent and semipermanent water facilities, such as the drilling of water wells.

No.	Scale	Measure
01	Yes/No	Unit had potable and nonpotable water supplies to conduct its mission.
02	Yes/No	Unit did not have to wait for potable and nonpotable water before it could conduct its mission.
03	Time	Required to develop or update plans to establish support operations after receipt of warning order.
04	Time	To develop concept of support sustainment requirements after receipt of warning order.
05	Time	To reach time-phased operating and safety levels of supply in the area of operations (AO).
06	Percent	Of difference between planned and actual demand by supply line in AO.
07	Percent	Of planned potable water support completed in AO.
08	Percent	Of potable water generation equipment available in AO compared to requirements.
09	Percent	Of required potable water generated in AO.

No.	Scale	Measure
10	Percent	Of shortfalls in potable water generation and distribution equipment in AO that have acceptable alternatives.
11	Percent	Of operations degraded, delayed, or modified due to delays in generating and distributing potable water.
12	Percent	Of potable water (bottled) provided by host nation.
13	Number	And types of potable water generation equipment stockpiled in AO to support operations.
14	Number	Of days of sustainment supply in AO supported by available facilities.
15	Number	Of gallons per day of potable water in AO delivered to operating forces.
16	Number	Of gallons or liters of potable water packaged daily at each bottling plant.

ART 4.1.3.11.1 Purify Water

4-45. Provide and operate water purification equipment and facilities to provide potable water for consumption and use where purified water is required. (FM 10-52) (CASCOM)

No.	Scale	Measure
01	Yes/No	Unit had potable water supplies to conduct its mission.
02	Yes/No	Unit did not have to wait for potable water before it could conduct its mission.
03	Yes/No	Spill planning and procedures related to environmental considerations were present and being followed.
04	Time	To refine field services program for the area of operations (AO) after receipt of warning order.
05	Percent	Of difference between planned and actual demand by supply line in AO.
06	Percent	Of planned capacity of water purification support performed in AO.
07	Percent	Of operational water purification facilities and equipment in AO.
08	Percent	Of required gallons of water provided per day in AO.
09	Percent	Of water distribution system operational on average in AO.
10	Percent	Of total potable water production capacity used on average in AO.
11	Number	Of water purification teams in AO.
12	Number	Of water sources available in AO.
13	Number	Of days of water supply on hand in AO.
14	Number	Of gallons of potable water required per person per day in AO.
15	Number	Of gallons of potable water provided per person per day in AO.

ART 4.1.3.11.2 Provide Packaged Water

4-46. Provide potable water that is packaged, stored, transported, and issued in containers with a capacity .5 to 10 liters. (FM 10-52) (CASCOM)

No.	Scale	Measure
01	Yes/No	Unit had purified and packaged enough potable water to conduct its mission.
02	Yes/No	Unit had potable packaged water to conduct its mission.
03	Yes/No	Unit did not have to wait for package potable water before it could conduct its mission.
04	Time	To refine supply support program for the area of operations (AO) after receipt of warning order.
05	Time	Of operational delay due to potable packaged water shortages.
06	Percent	Of supply of required packaged potable water in place of support operations.
07	Percent	Of daily potable packaged water provided by host nation.
08	Percent	Of potable packaged water delivered to theater.

No.	Scale	Measure
09	Percent	Of potable packaged water deliveries completed compared to forecasted requirements.
10	Number	Of gallons or liters per day of potable packaged water delivered to theater.

ART 4.1.3.12 PROVIDE MISCELLANEOUS SUPPLIES

4-47. Provide miscellaneous supplies and captured materials. (FM 10-1) (CASCOM)

No.	Scale	Measure
01	Yes/No	Unit had miscellaneous supplies to conduct its mission.
02	Yes/No	Unit did not have to wait for miscellaneous supplies before it could conduct its mission.
03	Time	Required to develop or update plans to establish support operations after receipt of warning order.
04	Time	To develop concept of support for miscellaneous supply requirements after receipt of warning order.
05	Time	To reach time-phased operating and safety levels of supply in the area of operations (AO).
06	Time	To certify captured supplies as being safe substitutes for U.S supplies.
07	Percent	Of difference between planned and actual demand by supply line in AO.
08	Percent	Of planned supply support for miscellaneous supplies completed in AO.
09	Percent	Of miscellaneous supplies available in AO compared to requirements.
10	Percent	Of miscellaneous replenishment stocks delivered on time in AO.
11	Percent	Of shortfalls in miscellaneous supplies that have acceptable alternatives.
12	Percent	Of daily supply requirements that can be supported by using captured supplies.
13	Percent	Of required miscellaneous supplies in AO delivered.
14	Percent	Of operations degraded, delayed, or modified due to delays in moving miscellaneous supplies.
15	Percent	Of miscellaneous supply requisitions filled in AO.
16	Percent	Of required delivery date for miscellaneous supplies in AO realized.
17	Percent	Of critical replenishment stocks in AO that experienced late delivery.
18	Percent	Of miscellaneous supplies provided by host nation.
19	Number	Of days of miscellaneous supplies stockpiled in AO to support operations.
20	Number	Of days of miscellaneous supplies in AO supported by available facilities.
21	Number	Of tons of miscellaneous supplies per day delivered to operating forces.

ART 4.1.3.13 PERFORM FORWARD ARMING AND REFUELING POINT ACTIVITIES

4-48. Establish a temporary facility organized, equipped, and deployed to provide fuel and ammunition to employ helicopter units. A forward arming and refueling point (FARP) is located closer to the area of operations than the aviation unit's sustainment area. Implementing a forward arming and refueling point includes environmental considerations. (FM 3-04.104) (USAAWC)

No.	Scale	Measure
01	Yes/No	FARP site contained supplies to allow the aviation element to continue its mission.
02	Yes/No	FARP site was operational when and where necessary to support the aviation element and had applied environmental considerations.
03	Time	To provide forward aerial fuel and ammunition concept of support after receipt of warning order.
04	Time	To complete bulk refuel and rearm operations at the forward arming and refueling point, once aircraft arrived.
05	Percent	Of bulk fuel and ammunition distribution capacity available.
06	Percent	Of supplies destroyed by enemy action.

No.	Scale	Measure
07	Percent	Of communications resources and support personnel available to support mission.
08	Percent	Of materials handling equipment available to support each mission and site.
09	Number	And types of aviation platforms requiring resupply.
10	Number	And types of supplies to support operation.
11	Number	Of gallons of bulk fuel delivered to forward arming and refueling point.
12	Number	And types of ordnance delivered to forward arming and refueling point.
13	Number	Of gallons or pounds of bulk fuel provided by organic assets.
14	Number	And types of ordnance provided by organic assets.
15	Number	Of total flying hours for each aircraft and number of miles between support location.
16	Number	Of kilometers between support location.
17	Number	And types of platforms providing security.
18	Number	Of kilometers that the location of supported forces is offset from the location of supporting forces.
19	Number	And types of supplies destroyed by enemy action.
20	Number	Of casualties occurring during the conduct of the forward arming and refueling point operations resulting from enemy action and accidents.

ART 4.1.3.14 EMPLOY CACHES

4-49. Hide supply stocks in isolated locations by procedures such as burial, concealment, or submersion to support the operations of a designated force. ART 4.1.3.14 can routinely occur during the conduct of retrograde, evasion and recovery, and special operations. (FM 7-85) (USAIS)

No.	Scale	Measure
01	Yes/No	Caches allowed the unit to continue its mission.
02	Yes/No	Unit operations were not delayed by an inability to find the cache due to inaccurate reporting of the cache's location.
03	Yes/No	Unit establishing the cache used cover and concealment to hide the location of the cache.
04	Time	To develop caches concept of support and determine specific quantities and types of supplies to place in caches after receipt of warning order.
05	Time	To deliver replenishment stocks before required delivery date.
06	Percent	Of required lift by mode available to emplace caches.
07	Percent	Of supplies destroyed by enemy forces.
08	Percent	Of supplies destroyed by environmental factors, to include damage from animals.
09	Percent	Of caches whose locations were accurately reported.
10	Number	Of caches established.
11	Number	And types of vehicles supported from each cache location.
12	Number	And types of enemy forces operating in the area of operations.

ART 4.1.3.15 PROVIDE SUPPLY MANAGEMENT

4-50. Provide management for all classes of supplies and materiel. ART 4.1.3.15 includes procedures established for requesting, procuring, and issuing supplies while maintaining accountability and security of supply stocks. (FM 4-0) (CASCOM)

No.	Scale	Measure
01	Yes/No	Unit operations were not delayed by an inability to perform supply management.
02	Yes/No	Supply management enhanced the unit's ability to accomplish its mission.
03	Time	To develop concept of support sustainment requirements after receipt of warning order.

No.	Scale	Measure
04	Time	Required to develop or update plans to establish support operations after receipt of warning order.
05	Time	To reach time-phased operating and safety levels of supply in area of operations (AO).
06	Percent	Of difference between planned and actual demand by supply line in AO.
07	Percent	Of planned supply support delivered in AO.
08	Percent	Of supplies available in AO compared to requirements.
09	Percent	Of replenishment stocks in AO delivered on time.
10	Percent	Of shortfalls in supply that have acceptable alternatives.
11	Percent	Of required supplies delivered in AO.
12	Percent	Of operations degraded, delayed, or modified due to delays in moving or evacuating personnel, supplies, and equipment.
13	Percent	Of requisitions in AO filled.
14	Percent	Of required delivery date in AO met.
15	Percent	Of requisitions in AO filled from shelf stocks.
16	Percent	Of critical replenishment stocks in AO that experienced late delivery.
17	Number	Of days of supplies stockpiled in AO to support operations.
18	Number	Of days of sustainment supplies in AO supported by available facilities.
19	Number	Of tons of supplies and equipment per day delivered to operating forces.

ART 4.1.3.15.1 Request Supplies

4-51. Submit a supply request to a supporting supply element. ART 4.1.3.15.1 includes company-sized and smaller units requesting all types of supplies from the organization such as the unit supply section, battalion support platoon, forward support company, or headquarters and distribution company responsible for sustaining them. ART 4.1.3.15.1 also includes determining requirements and on-hand or due-in stocks, preparing requisitions, and determining the source and location of supply items. (FM 10-1) (CASCOM)

No.	Scale	Measure
01	Yes/No	Requesting supplies did not delay the conduct of unit operations.
02	Time	To reach time-phased operating and safety levels of supply in area of operations (AO).
03	Time	To process requisition in AO for replacement supplies and equipment.
04	Percent	Of difference between planned and actual demand by supply line in AO.
05	Percent	Of planned supply support delivered in AO.
06	Percent	Of supplies available in AO compared to requirements.
07	Percent	Of replenishment stocks in AO delivered on time.
08	Percent	Of shortfalls in supplies that have acceptable alternatives.
09	Percent	Of required supplies delivered in AO.
10	Percent	Of operations degraded, delayed, or modified due to delays in processing requisitions for needed supplies and equipment.
11	Percent	Of requisitions in AO filled.
12	Percent	Of required delivery date in AO met.
13	Percent	Of requisitions in AO filled from shelf stocks.
14	Percent	Of critical replenishment stocks in AO that experienced late delivery.
15	Percent	Of supply requisitions returned due to errors in requisition.
16	Number	Of days of supplies stockpiled in AO to support operations.
17	Number	Of days of sustainment supplies in AO supported by available facilities.
18	Number	Of requisitions processed per day in AO.
19	Number	Of tons of supplies and equipment in AO delivered per day to operating forces.

ART 4.1.3.15.2 Receive Supplies

4-52. Replenish stocks to maintain required levels of supply. ART 4.1.3.15.2 includes maintaining unit basic loads or supply stocks required for a specific mission. It also includes determining the type and quantity of shipment, and performing quality assurance. (FM 10-1) (CASCOM)

No.	Scale	Measure
01	Yes/No	Receipt of supplies did not delay the conduct of unit operations.
02	Time	Required to develop or update plans to establish support operations after receipt of warning order.
03	Time	To develop concept of support sustainment requirements after receipt of warning order.
04	Time	To reach time-phased operating and safety levels of supply in the area of operations (AO).
05	Time	To process requisition in AO for replacement supplies and equipment.
06	Percent	Of difference between planned and actual demand by supply line in AO.
07	Percent	Of planned supply support delivered in AO.
08	Percent	Of supplies available in AO compared to requirements.
09	Percent	Of replenishment stocks in AO delivered on time.
10	Percent	Of shortfalls in supply that have acceptable alternatives.
11	Percent	Of required supplies delivered in AO.
12	Percent	Of operations degraded, delayed, or modified due to errors in processing requisitions in AO for needed supplies and equipment.
13	Percent	Of requisitions in AO filled.
14	Percent	Of required delivery date in AO met.
15	Percent	Of critical replenishment stocks in AO delivered late.
16	Percent	Of supplies received in AO that are in error due to faulty requisitioning.
17	Percent	Of errors in requisitioning of supplies in AO found through quality assurance audit.
18	Number	Of days of sustainment supply in AO supported by available facilities.
19	Number	Of requisitions processed per day in AO.
20	Number	Of tons of supplies and equipment delivered per day to operating forces.

ART 4.1.3.15.3 Procure Supplies

4-53. Obtain resources that may be available through such sources as local purchase, support agreements developed at command echelons, and foraging. ART 4.1.3.15.3 includes determining requirements for commercial sources and processing captured supplies. (FM 10-1) (CASCOM)

No.	Scale	Measure
01	Yes/No	Procedures to procure supplies did not negatively impact the unit's ability to accomplish its mission.
02	Yes/No	Procedures to procure supplies had implemented all applicable environmental considerations.
03	Time	Required to develop or update plans to establish support operations after receipt of warning order.
04	Time	To develop concept of support sustainment requirements after receipt of warning order.
05	Time	To process requisition in area of operations (AO) for replacement supplies and equipment from procurement sources.
06	Time	To reach time-phased operating and safety levels of supply in the AO.
07	Time	To certify captured supplies as being safe substitutes for U.S. supplies.
08	Percent	Of difference between planned and actual demand by supply line in AO.
09	Percent	Of planned supply support completed in AO.

No.	Scale	Measure
10	Percent	Of supplies procured in AO compared to requirements.
11	Percent	Of replenishment stocks in AO delivered on time.
12	Percent	Of shortfalls in supply that have acceptable alternatives.
13	Percent	Of required procurements completed in AO.
14	Percent	Of operations degraded, delayed, or modified in AO due to procurement shortfall.
15	Percent	Of requisitions in AO filled by procurement from inter-Service support agreements.
16	Percent	Of required delivery date in AO met for supply procurement items.
17	Percent	Of procurement requirements in AO provided by host nation.
18	Percent	Of critical replenishment stocks from procurement in AO that experienced late delivery.
19	Percent	Of supplies received through procurement agencies in AO that are in error due to faulty requisitioning.
20	Percent	Of procured items obtained from captured enemy supplies and equipment.
21	Number	Of requisitions for procurement items processed per day in AO.
22	Number	Of tons of supplies and equipment procured per day for operating forces.

ART 4.1.3.15.4 Issue Supplies

4-54. Provide supplies to using activities and units. ART 4.1.3.15.4 includes determining procedures of issue that maintain supply accountability, establish or operate transfer and distribution points, and reissue supplies. (FM 10-1) (CASCOM)

No.	Scale	Measure
01	Yes/No	Procedures to issue supplies did not negatively impact the unit's ability to accomplish its mission.
02	Time	Required to develop or update plans to establish support operations after receipt of warning order.
03	Time	To develop supply distribution system after receipt of warning order.
04	Percent	Of difference between planned and actual demand by supply line in the area of operations (AO).
05	Percent	Of planned supply distribution capability delivered in AO.
06	Percent	Of supply distribution points established in AO compared to requirements.
07	Percent	Of supplies and equipment in AO issued via supply point distribution.
08	Percent	Of supplies and equipment in AO issued via unit distribution.
09	Percent	Of operations degraded, delayed, or modified in AO due to time delays caused by failure to distribute supplies and equipment to the operating forces.
10	Percent	Of supply distribution capability in AO provided by host nation.
11	Number	Of major supply distribution points established in AO.

ART 4.1.3.16 SALVAGE AND RETROGRADE MATERIEL

4-55. Save, rescue, or retrograde condemned, discarded, or abandoned property and materiel, or operational stocks and supplies not consumed for reuse, refabrication, or scrapping. ART 4.1.3.16 includes receiving materiel at collection points, classifying materiel, and disposing of the materiel. This task includes environmental considerations. (FM 10-1) (CASCOM)

No.	Scale	Measure
01	Yes/No	Procedures to salvage and retrograde materiel integrated environmental considerations and did not negatively impact on the unit's ability to accomplish its mission.
02	Yes/No	Unit moved retrograde containers, flat racks, and containerized roll-in and roll-off platforms to distribution point.

No.	Scale	Measure
03	Yes/No	Unit tracked the flow of retrograde materiel in the distribution system.
04	Yes/No	Unit synchronized retrograde support operations and established return priority of shipping containers, aerial delivery platforms, and flatracks.
05	Time	Required to develop or update plans to establish support operations after receipt of warning order.
06	Time	To develop concept of salvage and restoration after receipt of warning order.
07	Percent	Of difference between planned and actual demand by supply line in the area of operations (AO).
08	Percent	Of planned salvage capability performed in AO.
09	Percent	Of salvage collection points established in AO compared to requirements.
10	Percent	Of salvaged supplies and equipment in AO reusable as is.
11	Percent	Of salvaged supplies and equipment in AO reusable after refabrication.
12	Percent	Of salvaged supplies and equipment in AO converted to scrap metal.
13	Percent	Of salvaged supplies and equipment in AO discarded as unusable.
14	Percent	Of salvaged property in AO converted to use by the operating forces.
15	Percent	Of salvage capability in AO provided by host nation.
16	Number	Of salvage collection points established in AO.

ART 4.1.3.17 PREPARE CONFIGURED LOADS

4-56. Configure a load at a supply activity for a user. (FMI 4-93.41) (CASCOM)

No.	Scale	Measure
01	Yes/No	Configured loads supported the unit in accomplishing its mission.
02	Yes/No	Procedures to prepare configured loads did not negatively impact on the supported unit's ability to accomplish its mission.
03	Time	Required to develop or update plans to establish support operations after receipt of warning order.
04	Time	Of longevity of each type of projected mission in the area of operations (AO).
05	Percent	Of difference between planned and actual demand by supply line in AO.
06	Percent	Of host-nation support available in AO.
07	Number	Of personnel in AO requiring support.
08	Number	And type of vehicles requiring support in AO.
09	Number	Of composite items within a single request for each type of unit in AO.
10	Number	And type of weapon systems and other equipment in each supported unit that require resupply.
11	Number	Of rounds of ammunition each weapon system in the supported unit consumes per mission.
12	Number	And types of transport used to move supplies.
13	Number	Of days of supply for all classes or line numbers of supply on hand.

ART 4.1.4 PROVIDE FIELD SERVICES

4-57. Environmental considerations embedded in providing field services support are many and varied and will have an effect on every subordinate task. (FM 4-0) (CASCOM)

No.	Scale	Measure
01	Yes/No	Procedures to provide field services did not negatively impact on supported units' ability to perform missions.
02	Time	Required to develop or update plans to establish field services operations after receipt of warning order.
03	Time	To establish field services operations for military, Army civilians, qualifying contractors, and other personnel in the area of operations.

ART 4.1.4.1 CONDUCT MORTUARY AFFAIRS

4-58. Provide for the care and disposition of deceased personnel. ART 4.1.4.1 includes search and recovery, collection, decontamination (if necessary), evacuation, establishment of tentative identification, and temporary burial. Mortuary affairs support also includes the inventory, safeguard, and evacuation of personal effects of deceased personnel. (FM 4-20.64) (CASCOM)

No.	Scale	Measure
01	Yes/No	Mortuary affairs requirements were balanced against mission requirements and unit morale.
02	Yes/No	The remains of every U.S. Service member who dies in the area of operations (AO) was accounted for.
03	Yes/No	The remains of every U.S. Service member who dies in the AO was provided mortuary services.
04	Yes/No	Appropriate measures were taken to preserve remains in accordance with current doctrine and policy.
05	Yes/No	Nonmortuary affairs units had a unit search and recovery team trained as a minimum on a semiannual basis.
06	Yes/No	Nonmortuary affairs units had sufficient mortuary affairs supplies on hand to conduct unit level mortuary affairs recovery mission.
07	Time	To refine mortuary affairs plan for AO after receipt of warning order.
08	Time	To coordinate mortuary affairs policy and procedures with the joint mortuary affairs office.
09	Time	Of delay in evacuation of remains, on average.
10	Time	To coordinate transportation of remains to continental United States, on average.
11	Time	To establish temporary interment facilities.
12	Time	Of delay in identification, care, and evacuation or disposition of remains due to lack of adequate mortuary affairs units.
13	Percent	Of difference between projected sustainment field service requirements and actual requirements in AO.
14	Percent	Of mortuary affairs operations, to include mortuary affairs collection points, personal effects depot, theater mortuary evacuation points, and temporary interment sites established in AO compared to requirements.
15	Percent	Of records of deceased and missing personnel in AO maintained accurately.
16	Percent	Of records of personal effects in AO maintained accurately.
17	Percent	Of personal effects of deceased and missing personnel in AO processed accurately.
18	Percent	Of personal effects of deceased and mission personnel in AO properly accounted for and safeguarded.
19	Number	Of remains processed within a given time.

ART 4.1.4.2 PROVIDE AERIAL DELIVERY SUPPORT

4-59. Using fixed-wing, rotorcraft and hybrid aircraft, provide supplies and equipment by airland, airdrop, freedrop and slingload operations and methods. ART 4.1.4.2 includes the provision of aerial delivery equipment and systems, consisting of parachute packing, air item maintenance, external slingload, and rigging supplies and equipment. (FM 4-20.41) (CASCOM)

No.	Scale	Measure
01	Yes/No	Procedures to provide aerial delivery did not negatively affect supported unit's ability to perform its missions.
02	Yes/No	Theater logistics planners had planned for the use of intratheater aerial delivery support.
03	Time	To refine field services program for the area of operations (AO) after receipt of warning order.
04	Time	From conduct of Joint Precision Airdrop System (JPADS) airdrop mission to receipt of retrograde JPADS systems at parachute rigger unit.
05	Percent	Of difference between projected field service requirements and actual requirements in AO.
06	Percent	Of planned capacity of aerial delivery support completed in AO.
07	Percent	Of operations degraded, delayed, or modified due to lack of aerial delivery support and equipment.
08	Percent	Of equipment delivered undamaged.
09	Percent	Of personnel who received injuries during aerial delivery operations.
10	Percent	Of aerial deliveries on time and target.
11	Percent	Of aerial equipment recovered.
12	Percent	Of supplies not received due to enemy action.
13	Percent	Of supplies and equipment not delivered due to enemy action.
14	Number	Of aerial delivery support operations denied, degraded, delayed, or modified due to lack of aerial delivery support and equipment.
15	Number	Of tonnage delivered via intratheater airland operations (fixed wing).
16	Number	Of tonnage delivered via intratheater rotorcraft operations.
17	Number	Of tonnage delivered via intratheater airdrop operations.
18	Number	Of personnel who received injuries during aerial delivery operations.
19	Number	Of precision airdrop mission by weight category.
20	Number	Of low cost air drop system missions.
21	Number	Of low cost low altitude air drop missions.
22	Number	Of freedrop missions.

ART 4.1.4.3 PROVIDE BASE CAMP SUSTAINMENT

4-60. Provide base camp sustainment facilities and services to Soldiers and other authorized personnel conducting full spectrum operations. Provision of this support to authorized individuals and units occurs regardless of their physical location within or external to a base, facility, installation, camp, or station. ART 4.1.4.3 includes the provision of clothing and textile repair support, hygiene services (shower, laundry, and latrine support), nutrition support, and general purpose shelters and systems. Environmental considerations for a base camp are significant and similar to those for an installation. (FM 4-20.07) (CASCOM)

Note: ART 4.1.3.11.1 (Purify Water) addresses water purification support.

ART 4.1.7 (Provide General Engineering Support) addresses construction, repairing, maintenance, and operations of permanent and semipermanent water facilities, such as the drilling of water wells.

ART 4.1.7.5 (Provide Facilities Engineer Support) addresses waste management; the acquisition, management, and disposal of real estate; firefighting support; and the construction, management, and maintenance of bases and installations.

No.	Scale	Measure
01	Yes/No	Procedures to provide base camp sustainment did not negatively affect supported unit's ability to perform its missions.
02	Yes/No	Environmental considerations planning and procedures were present and being followed.
03	Time	To refine base camp sustainment program for area of operations (AO) after receipt of warning order.
04	Percent	Of difference between projected base camp sustainment requirements and actual requirements in AO.
05	Percent	Of planned base camp sustainment capacity reached in AO.
06	Percent	Of operations degraded, delayed, or modified due to lack of adequate base camp sustainment.
07	Percent	Of base camp sustainment requirements that can be performed by host nation, third nation, or contractors.
08	Number	And types of base camp sustainment facilities available in AO.

ART 4.1.4.3.1 Provide Clothing and Textile Repair Support

4-61. Provide clean, serviceable clothing; restore clothing or light textiles; and provide clothing exchange. ART 4.1.4.3.1 includes collecting repairable clothing and textiles. (FM 10-16) (CASCOM)

No.	Scale	Measure
01	Yes/No	Procedures to provide clothing and textile repair services did not negatively affect supported unit's ability to perform its missions.
02	Time	To refine field services program for the area of operations (AO) after receipt of warning order.
03	Time	To coordinate clothing and textile repair operations in AO.
04	Time	To repair clothing and textiles in AO, on average.
05	Percent	Of difference between projected sustainment field service requirements and actual requirements in AO.
06	Percent	Of planned capacity of field service clothing and textile repair performed in AO.
07	Percent	Of required production rate of clothing and textile repair reached in AO.
08	Percent	Of required production rate of clothing and textile repair that can be provided by host-nation or contract sources.
09	Number	Of clothing and textile repair units in AO.
10	Number	Of items per month of clothing and textile repaired in AO facilities.
11	Number	Of tons per month of clothing and textiles out of service for renovation or repair.
12	Number	Of items by type identified as non-reparable due to exceeding authorized repair times.
13	Number	Of tons per month of clothing and textiles out of service for renovation or repair.

ART 4.1.4.3.2 Provide Hygiene Support

4-62. Provide hygiene services—showers, laundries, and latrines. ART 4.1.4.3.2 includes obtaining the fresh water and cleaning materials necessary to provide these services. It also includes preparing the shelters and drainage necessary to perform these services in addition to operating shower and delousing units, laundering clothing, and re-impregnating clothing. This task includes environmental considerations involving health service and engineering support. (FM 42-414) (CASCOM)

Note: ART 4.1.3.6 (Provide Personal Demand Items [Class VI]) addresses the provision of health and comfort packages containing personnel care items, such as disposable razors and sanitary napkins, necessary for personnel hygiene when Army and Air Force Exchange Service tactical field exchanges are not operational.

ART 4.1.3.11.1 (Purify Water) addresses water purification support.

ART 4.1.7 (Provide General Engineering Support) addresses construction, repairing, maintenance, and operations of permanent and semipermanent water facilities, such as the drilling of water wells.

ART 4.1.7.5.1 (Provide Waste Management) addresses waste management, to include wastewater collection and treatment and refuse collection and disposal.

No.	Scale	Measure
01	Yes/No	Procedures to provide hygiene support did not negatively affect supported unit's ability to perform its missions.
02	Yes/No	Procedures to provide hygiene support did not favor sustainment units over ground maneuver units.
03	Yes/No	Environmental considerations planning and procedures were present and being followed.
04	Time	To refine hygiene support program for the area of operations (AO) after receipt of warning order.
05	Time	To coordinate hygiene support operations with medical authorities in AO.
06	Time	To establish hygiene (shower, laundry, and latrine) facilities for personnel in AO.
07	Percent	Of difference between projected hygiene support requirements and actual requirements in AO.
08	Percent	Of planned hygiene support capacity produced in AO.
09	Percent	Of required production rate of potable water delivered in AO.
10	Percent	Of personal daily water requirements provided in AO.
11	Percent	Of required hygiene support (shower, laundry, and latrine) equipment and materials available in AO.
12	Percent	Of available water sources in AO cleared by medical authorities for use in hygiene support.
13	Percent	Of required hygiene support (shower, laundry, and latrine) requirements that can be provided by host nation, third nation, or contractors.
14	Percent	Of force provider and shower laundry and clothing repair teams having shower water reuse capability.
15	Number	And capabilities of laundry and shower units and latrine providers available in AO.
16	Number	Of units not receiving laundry support every 7 days.
17	Number	Of units not receiving shower support every 7 days.
18	Number	Of gallons of shower water being used and recycled for reuse at each shower site daily.

ART 4.1.4.3.3 Provide Nutrition Support

4-63. Provide nutrition support to Soldiers at all echelons. ART 4.1.4.3.3 includes obtaining, preparing, and serving prepared rations; performing in-field kitchen sanitation; operating field kitchens, installation dining facilities, and hospital dining facilities; and preparing products for shipment. This task also includes providing bakery products and nutritional care. (FM 10-23) (CASCOM)

No.	Scale	Measure
01	Yes/No	Procedures to provide nutritional support (food preparation, serving, field kitchen sanitation, and accounting for rations) did not negatively affect supported unit's ability to perform its missions.
02	Yes/No	Environmental considerations planning and procedures were present and being followed.
03	Time	To refine nutrition support program for the area of operations (AO) after receipt of warning order.
04	Percent	Of difference between projected nutrition support requirements and actual requirements in AO.
05	Percent	Of planned capacity for nutrition support reached in AO.
06	Percent	Of personnel in AO receiving at least one hot meal per day.
07	Percent	Of meals served to non-Department of Defense personnel.
08	Percent	Of personnel in AO receiving three meals per day.
09	Percent	Of field kitchens temporarily closed due to sanitation violations.
10	Percent	Of nutritional support requirements that can be performed by host-nation, third-nation, or contractor support personnel.
11	Number	Of hot meals served in AO in a given time.
12	Number	Of personnel in AO requiring nutritional support.
13	Number	Of days of supply of meals, ready to eat available in AO.
14	Number	Of field kitchens available in AO.
15	Number	Of available water sources and platforms in AO.
16	Number	Of nutritional support (food service) personnel in AO.
17	Number	Of units still consuming meals ready to eat as the sole source of nutrition after 21 days.
18	Number	Of units or personnel supported by table of organization and equipment cooks and equipment.
19	Number	Of units or personnel supported by other than table of organization and equipment cooks and equipment.

ART 4.1.4.3.4 Provide General Purpose Shelters and Systems

4-64. Provide shelters, heaters, and environmental control units to provide shelter from the elements for Soldiers and units. ART 4.1.4.3.4 includes providing environmental control units, lightweight or quick-erect shelters, and environmentally safe, lightweight heaters. (FM 4-20.07) (CASCOM)

No.	Scale	Measure
01	Yes/No	Procedures to provide general purpose shelters and systems did not negatively affect supported unit's ability to perform missions.
02	Yes/No	Environmental considerations planning and procedures were present and being followed.
03	Time	To refine the area of operations (AO) program for general purpose shelters and system after receipt of warning order.
04	Percent	Of difference between projected shelter and systems requirements and actual requirements in AO.
05	Percent	Of planned shelter and system capacity completed in AO.
06	Percent	Of operations degraded, delayed, or modified due to lack of applicable general purpose shelters and systems in AO.
07	Percent	Of general purpose shelters and systems erected undamaged.
08	Percent	Of unit Soldiers billeted in other than unit tents.
09	Percent	Of general purpose shelters and systems requirements that can be performed by host-nation, third-nation, or contractor support personnel.

No.	Scale	Measure
10	Number	Of instances of mission delays or failures due to lack of general purpose shelters and systems.
11	Number	And types of general purpose shelters and environmental control units available in AO.

ART 4.1.5 PROVIDE CONTRACTING SUPPORT

4-65. Provide contracting support to obtain or provide supplies, services, and construction in support of operations. Contracting provides a responsive alternative procedure to increasing support force capability to perform a mission and support all phases of an operation. (FM 100-10-2) (CASCOM)

No.	Scale	Measure
01	Yes/No	Contracting support services assisted the supported unit in accomplishing its mission.
02	Yes/No	Environmental considerations planning and procedures were present and being followed.
03	Time	To develop contracting plans and policies.
04	Time	To establish contract office in the area of operations (AO).
05	Time	To prepare and forward contracting reports.
06	Time	To coordinate legal and financial management aspects of contracts.
07	Time	To coordinate inspection and quality control of contracted services.
08	Time	To coordinate delivery of contracted service and supplies.
09	Time	To provide contract status updates to the commander and the principal assistant responsible for contracting.
10	Time	To cross level contracting assets to meet changes in contracting requirements.
11	Time	To monitor contractor performance.
12	Time	To provide technical advice and assistance to staffs, subordinate units, and contractor's representatives.
13	Time	To establish working relationship with contractors and vendors in AO.
14	Percent	Of available time spent monitoring contract administration.
15	Percent	Of contracts that must be coordinated with other agencies, such as staff judge advocate and host nation.
16	Percent	Of contracts executed in time to meet commander's intent.
17	Percent	Of contracted services and supplies provided on time and to standard.
18	Number	Of contracts issued within a given time.
19	Number	Of contracting officers located in the AO and the size of their individual warrants.

ART 4.1.5.1 REQUEST CONTRACTING SUPPORT

4-66. The role of the supporting contract office is to procure urgently needed supplies, services, and minor construction in support of contingency operations. Activities must understand that all requirements must be obtained by first using existing military supply channels. Only items and services that cannot be met by normal supply channels or host-nation support or items and services that are critical to mission accomplishment that cannot be obtained in a timely manner will be procured through contracting. Activities requesting contracting support must obtain proper funding prior to submitting request to the contracting officer. Timely and proper planning of support requirements by the requiring activity determines the adequacy of the support provided. (FM 100-10-2) (CASCOM)

No.	Scale	Measure
01	Yes/No	Unit determined need for contracting support.
02	Yes/No	Unit developed acquisition plan and conducted market research.
03	Yes/No	Unit developed independent government cost estimate.

No.	Scale	Measure
04	Yes/No	If a supply item is required, unit described need in terms of capability rather than name brand.
05	Yes/No	If service contract, unit developed performance work statement and nominated a contracting officer representative for training.
06	Time	To submit justification and authorizations, as required.
07	Time	To obtain proper funding by submitting a DA Form 3953 (Purchase Request and Commitment).
08	Time	To submit all completed documents to contracting for execution.

ART 4.1.5.2 PROVIDE LOGISTICS CIVILIAN AUGMENTATION PROGRAM SUPPORT

4-67. Logistics civilian augmentation program (LOGCAP) provides contingency planning to deployed forces and provides logistics and minor engineer and construction services. (FMI 4-93.41) (CASCOM)

No.	Scale	Measure
01	Yes/No	LOGCAP planners had prepared country or regional plans to provide baseline for support of contingency operations.
02	Yes/No	Army Service component command determined a need for LOGCAP support to augment or replace component force capability requirement.
03	Yes/No	Headquarters, Department of the Army approved LOGCAP use in specific operations.
04	Yes/No	Unit provided notice to proceed to contractor.
05	Time	To prepare statement of work.
06	Time	To fund task orders.
07	Time	To deploy contractor and Army team LOGCAP management team.
08	Time	To definitize the cost of the task order with the contractor.
09	Time	To modify statements of work as requiring units needs change.
10	Time	To provide contract status updates to the procurement contracting officer responsible for advising the commander.
11	Time	To monitor contractor performance.
12	Time	To provide technical advice and assistance to staffs, tactical customer units, and contracting officer's technical representatives.
13	Time	To close out task order at end of period of performance.

ART 4.1.6 PROVIDE DISTRIBUTION

4-68. Distribution is the operational process of synchronizing all elements of the logistics system to deliver the "right things" to the "right place" at the "right time" in the "right mode" to support the commander. (FM 100-10-1) (CASCOM)

No.	Scale	Measure
01	Yes/No	Unit developed the area of operations (AO) distribution plan.
02	Yes/No	Unit synchronized the movement of supplies throughout the AOs.
03	Yes/No	Unit maintained visibility of equipment and supplies moving throughout the AO using information systems.
04	Yes/No	Unit established comprehensive surface movement plans.

ART 4.1.6.1 CONDUCT DISTRIBUTION MANAGEMENT

4-69. Distribution management is the function of synchronizing and coordinating a complex of networks and functional components to achieve responsive, customized solutions to Soldier requirements. (FM 100-10-1) (CASCOM)

No.	Scale	Measure
01	Yes/No	Inability of the distribution system to get the right supplies to the right unit at the right time did not delay, degrade, or prevent unit operations.
02	Yes/No	Unit developed the area of operations (AO) distribution plan.
03	Yes/No	Unit collected, analyzed, and monitored in-transit visibility distribution data using information systems to control the distribution flow.
04	Yes/No	Unit maintained asset visibility of major end items inside the AO using information systems.
05	Yes/No	Unit tracked the flow of retrograde classes of supply in the distribution system.
06	Yes/No	Unit synchronized retrograde support operations, establishing return priority of shipping containers, aerial delivery platforms, and flatracks.
07	Yes/No	Unit optimized the distribution network (information systems, transportation, supply, and storage) to maximize throughput, maintaining a continuous flow of resources.
08	Yes/No	Unit maintained visibility of distribution assets (ground assets, delivery platforms, containers, flatracks, and aerial delivery platforms) using information systems.
09	Yes/No	Unit identified and resolved distribution bottlenecks in AO.
10	Yes/No	Unit reallocated and redirected classes of supply to meet urgent or critical needs based on sustainment data.
11	Yes/No	Unit established priorities for off-loading transportation assets in the AO.
12	Yes/No	Unit managed distribution of controlled classes of supply.
13	Time	To set up transportation modes within theater.
14	Time	To redirect inbound shipments.
15	Time	To repackage and ship redirected supplies and equipment.
16	Percent	Of required items of supply transiting the distribution pipeline.
17	Percent	Of visibility and control maintained over the distribution pipeline within and external to the AO.
18	Percent	Of unit operations delayed, degraded, or modified due to lack of any or all classes of supply.
19	Percent	Of flexibility to provide resources from host-nation or other agencies.
20	Percent	Of retrograde of containers, flat racks, and containerized roll-in and roll-off platforms to distribution point.
21	Percent	Of tie-down straps, cargo nets, pallets returned with flat racks, containerized roll-in and roll-off platforms, and trailers.
22	Percent	Able to maintain in-transit visibility of distribution pipeline and assets flowing through pipeline.
23	Percent	Of sustainment shipments redirected to meet current or near term operational requirements.
24	Percent	Of shipments requiring repackaging of loads.

ART 4.1.6.2 PROVIDE IN-TRANSIT VISIBILITY/ASSET VISIBILITY

4-70. Provide in-transit visibility by continuously updating the location of units, equipment, personnel, and supplies as they travel throughout the transportation and distribution system. Provide commanders with critical information and allow for shipment diversion based on changing battlefield priorities. (FM 55-1) (CASCOM)

No.	Scale	Measure
01	Yes/No	Unit maintained in-transit visibility over personnel, equipment, and supplies moving throughout the area of operations (AO).
02	Yes/No	Unit maintained visibility of distribution assets (ground assets, delivery platforms, containers, flatracks, and aerial delivery platforms) using information systems.

No.	Scale	Measure
03	Yes/No	Unit tracked the flow of major end items inside the AO, using information systems.
04	Time	To refine plan for AO to identify en route locations for radio frequency identification (RFID) tag readers and interrogators.
05	Time	For satellite transponder location information relating to convoy movements to be reported to the appropriate regional in-transit visibility server.
06	Time	To establish fixed or mobile RFID tag readers and interrogators at highway and rail arrival gates, barge arrival gates, and airfields per in-transit visibility plan.
07	Time	To establish the RFID system and RFID readers and interrogators set to collect tag data.
08	Time	For the theater support command to assign support element responsibility to manage collection activities at designated interrogation locations.
09	Time	To pass unit cargo movement data to the global transportation network via Transportation Coordinator's Automated Information for Movement System II.
10	Time	To scan smart cards for all deploying Soldiers at designated integration locations and to pass them to the global transportation network.
11	Time	To establish procedures to remove and properly dispose of RFID tags and satellite transponders.
12	Time	To identify location and status of critical assets.
13	Time	To track location, track status of movement, and identity of units, personnel, and supplies.
14	Percent	Of en route location RFID tag readers and interrogators effectively repaired by support element.
15	Percent	Of RFID tagged unit equipment, vehicles, 463L pallets, and containers that are read and interrogated and the data automatically reported to the appropriate regional in-transit visibility server.
16	Percent	Of RFID tags and smart card data that are complete and accurate.
17	Number	Of designated support elements that scan military shipping label and smart cards at designated locations.

ART 4.1.6.3 CONDUCT UNIT LOGISTICS PACKAGE OPERATIONS

4-71. A logistics package (LOGPAC) is a grouping of multiple classes of supply and supply vehicles under the control of a single convoy commander. Daily LOGPACs contain a standardized allocation of supplies. Special LOGPACs can also be dispatched as needed. Implementing a unit logistics package includes environmental considerations. (FM 4-90.7) (CASCOM)

No.	Scale	Measure
01	Yes/No	Unit LOGPAC contained supplies to allow the unit to continue its mission.
02	Yes/No	Unit operations were not delayed by LOGPAC failure to arrive at the resupply site at the specified time.
03	Yes/No	Unit LOGPAC site was operational when and where necessary to support the receiving unit and had applied environmental considerations.
04	Time	Required to develop or update plans to establish support operations after receipt of warning order.
05	Time	To establish comprehensive surface movement plan on receipt of the warning order.
06	Time	For transport vehicles to upload supplies and equipment for supported units.
07	Times	For transport vehicles to travel to supported units, resupply the units, and return to point of origin.
08	Percent	Of difference between planned and actual demand by supply line in the area of operations (AO).
09	Percent	Of projected surface transport available in AO.

No.	Scale	Measure
10	Percent	Of operations degraded, delayed, or modified due to delays in moving or evacuating personnel, supplies, and equipment.
11	Percent	Of surface cargo in AO transported by wheeled or tracked vehicles.
12	Percent	Of roads available to transport supplies and equipment.
13	Percent	Of supplies lost to enemy action.
14	Percent	Of supplies transported by host nation.
15	Percent	Of personnel, supplies, and equipment in AO that arrive at destination on schedule.
16	Number	Of passengers transported per day by surface means in AO.
17	Number	Of kilometers between the brigade support area and field trains, and between the field trains and supported company.
18	Number	Of tons per day of supplies and equipment moved by surface transport in AO.
19	Number	Of tons per day of supplies and equipment in AO moved by organic units.

ART 4.1.6.4 ESTABLISH A HUB/NODE

4-72. Establish centralized distribution points where cargo is originated, processed for onward movement, or terminated. (FM 4-01.30) (CASCOM)

No.	Scale	Measure
01	Yes/No	Delay in node operations degraded, delayed, or modified unit operations.
02	Yes/No	Environmental considerations planning and procedures were present and being followed.
03	Time	To refine theater distribution plan for the area of operations (AO) after receipt of warning order.
04	Time	To establish comprehensive movement support plan after AO is assigned.
05	Time	For AO movement control battalion or team to begin operation after AO is assigned.
06	Percent	Of difference between the theater distribution plan requirements and actual requirements in AO.
07	Percent	Of planned movement services support completed in AO.
08	Percent	Of scheduled transport movements completed on schedule.
09	Percent	Of operations degraded, delayed, or modified due to delays in moving or evacuating personnel, supplies, and equipment.
10	Percent	Of personnel, supplies, and equipment in AO that arrived on schedule.
11	Percent	Of passengers stranded in transit for more than one day.
12	Percent	Of movement capacity in AO utilized per day.
13	Percent	Of required transport services provided by host nation.
14	Number	Of tons per day of supplies and equipment moved by transport means in AO.
15	Number	Of tons per day of supplies and equipment in AO moved by organic units.
16	Number	Of passengers per day transported in AO.

ART 4.1.6.4.1 Conduct Aerial Port of Debarkation Operations

4-73. Receive and offload cargo, equipment, and personnel from arriving flights. Process cargo, equipment, and personnel and begin transport to final destination within area of operations (AO). (FM 4-01.30) (CASCOM)

No.	Scale	Measure
01	Yes/No	Delay in airfield clearance operations degraded, delayed, or modified unit operations.
02	Yes/No	Environmental considerations planning and procedures were present and being followed.

No.	Scale	Measure
03	Time	To refine theater distribution plan for aerial port of debarkation operations after receipt of warning order.
04	Time	To establish aerial port of debarkation operations support plan after AO is assigned.
05	Time	For aerial port if debarkation movement control team to begin operation after AO is assigned.
06	Percent	Of difference between the theater distribution plan requirements and actual requirements in AO.
07	Percent	Of planned movement services support conducted by aerial port of debarkation.
08	Percent	Of scheduled transport movements completed on schedule.
09	Percent	Of operations degraded, delayed, or modified due to delays in moving or evacuating personnel, supplies, and equipment.
10	Percent	Of personnel, supplies, and equipment in aerial port of debarkation that arrive on schedule.
11	Percent	Of passengers stranded in transit for more than one day.
12	Number	Of tons per day of supplies and equipment arrived in aerial port of debarkation.
13	Number	Of passengers per day transported from aerial port of debarkation.

ART 4.1.6.4.2 Conduct Seaport of Debarkation Operations

4-74. Receive and offload cargo and equipment from arriving vessels. Process cargo and equipment and begin transport to final destination within the area of operations (AO). (FM 4-01.30) (CASCOM)

No.	Scale	Measure
01	Yes/No	Delay in seaport clearance operations degraded, delayed, or modified unit operations.
02	Yes/No	Environmental considerations planning and procedures were present and being followed.
03	Time	To refine theater distribution plan for seaport of debarkation operations after receipt of warning order.
04	Time	To establish seaport of debarkation operations support plan after AO is assigned.
05	Time	For seaport of debarkation movement control team to begin operation after AO is assigned.
06	Percent	Of difference between the theater distribution plan requirements and actual requirements in AO.
07	Percent	Of planned movement services support conducted by seaport of debarkation.
08	Percent	Of scheduled transport movements completed on schedule.
09	Percent	Of operations degraded, delayed, or modified due to delays in moving or evacuating personnel, supplies, and equipment.
10	Percent	Of supplies and equipment in seaport of debarkation that arrive on schedule.
11	Number	Of tons per day of supplies and equipment moved by transport means from seaport of debarkation.
12	Number	Of tons per day of supplies and equipment arrived in seaport of debarkation.

ART 4.1.6.4.3 Conduct Hub Operations

4-75. Sort and distribute inbound cargo from wholesale supply sources (airlifted, sealifted, and ground transportable) from within the theater. (FM 4-01.30) (CASCOM)

No.	Scale	Measure
01	Yes/No	Delay in hub operations degraded, delayed, or modified unit operations.
02	Yes/No	Environmental considerations planning and procedures were present and being followed.
03	Time	To refine theater distribution plan for hub operations after receipt of warning order.

No.	Scale	Measure
04	Time	To establish distribution hub plan after the area of operations (AO) is assigned.
05	Time	For distribution hub to begin operation after AO is assigned.
06	Percent	Of difference between the theater distribution hub planned requirements and actual requirements in AO.
07	Percent	Of planned movement services support conducted by distribution hub.
08	Percent	Of scheduled transport movements completed on schedule.
09	Percent	Of operations degraded, delayed, or modified due to delays in moving or evacuating personnel, supplies, and equipment.
10	Percent	Of supplies and equipment that arrive on schedule.
11	Number	Of tons per day of supplies and equipment moved by transport means from distribution hub.
12	Number	Of tons per day of supplies and equipment arrived in distribution hub.

*ART 4.1.7 PROVIDE GENERAL ENGINEER SUPPORT

4-76. ART 4.1.7 has been modified and moved to ART 4.5.1.

SECTION II – ART 4.2: PROVIDE PERSONNEL SERVICES SUPPORT

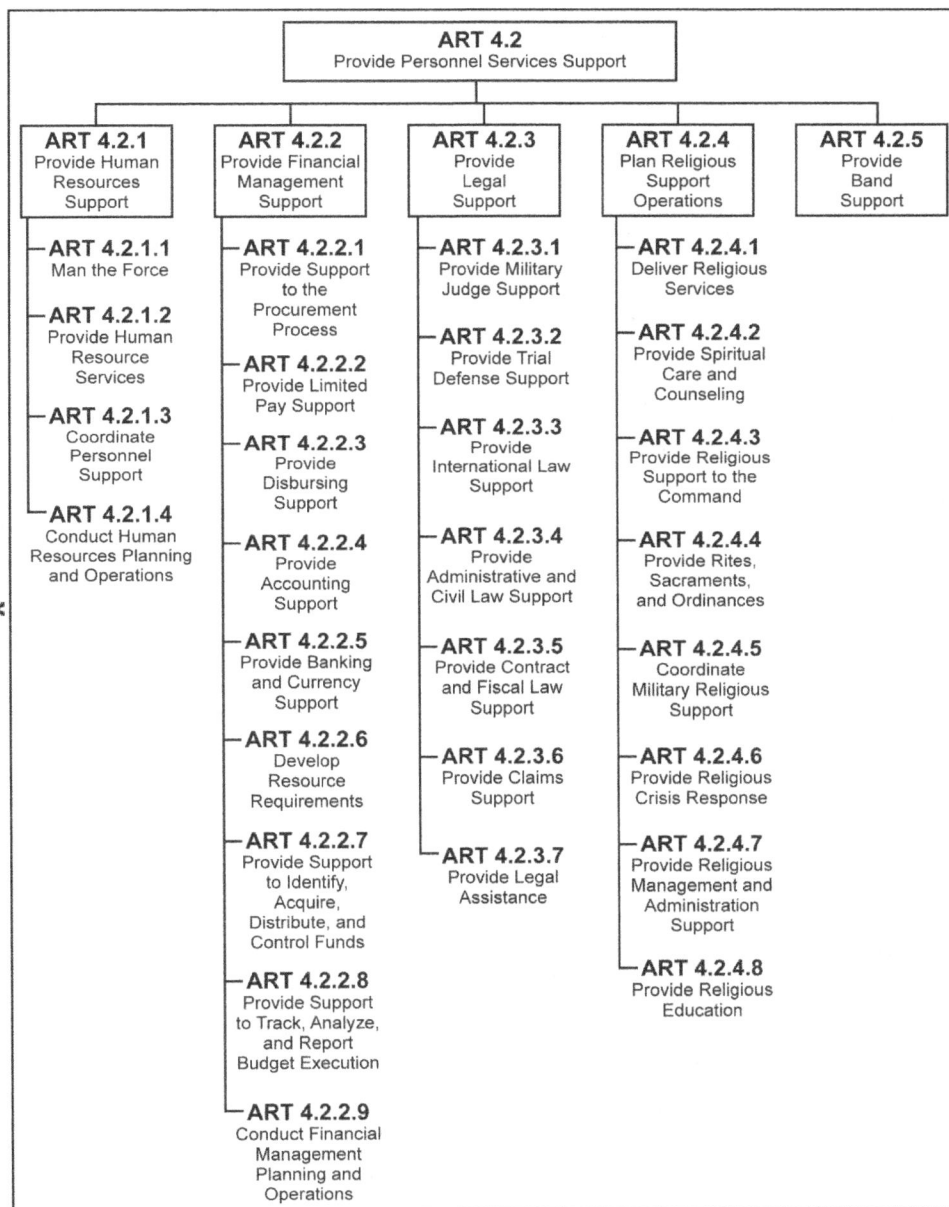

ART 4.2
Provide Personnel Services Support

ART 4.2.1
Provide Human Resources Support

- **ART 4.2.1.1**
  Man the Force
- **ART 4.2.1.2**
  Provide Human Resource Services
- **ART 4.2.1.3**
  Coordinate Personnel Support
- **ART 4.2.1.4**
  Conduct Human Resources Planning and Operations

ART 4.2.2
Provide Financial Management Support

- **ART 4.2.2.1**
  Provide Support to the Procurement Process
- **ART 4.2.2.2**
  Provide Limited Pay Support
- **ART 4.2.2.3**
  Provide Disbursing Support
- **ART 4.2.2.4**
  Provide Accounting Support
- **ART 4.2.2.5**
  Provide Banking and Currency Support
- **ART 4.2.2.6**
  Develop Resource Requirements
- **ART 4.2.2.7**
  Provide Support to Identify, Acquire, Distribute, and Control Funds
- **ART 4.2.2.8**
  Provide Support to Track, Analyze, and Report Budget Execution
- **ART 4.2.2.9**
  Conduct Financial Management Planning and Operations

ART 4.2.3
Provide Legal Support

- **ART 4.2.3.1**
  Provide Military Judge Support
- **ART 4.2.3.2**
  Provide Trial Defense Support
- **ART 4.2.3.3**
  Provide International Law Support
- **ART 4.2.3.4**
  Provide Administrative and Civil Law Support
- **ART 4.2.3.5**
  Provide Contract and Fiscal Law Support
- **ART 4.2.3.6**
  Provide Claims Support
- **ART 4.2.3.7**
  Provide Legal Assistance

ART 4.2.4
Plan Religious Support Operations

- **ART 4.2.4.1**
  Deliver Religious Services
- **ART 4.2.4.2**
  Provide Spiritual Care and Counseling
- **ART 4.2.4.3**
  Provide Religious Support to the Command
- **ART 4.2.4.4**
  Provide Rites, Sacraments, and Ordinances
- **ART 4.2.4.5**
  Coordinate Military Religious Support
- **ART 4.2.4.6**
  Provide Religious Crisis Response
- **ART 4.2.4.7**
  Provide Religious Management and Administration Support
- **ART 4.2.4.8**
  Provide Religious Education

ART 4.2.5
Provide Band Support

*

4-77. Personnel services are those sustainment functions related to Soldiers' welfare, readiness, and quality of life. Personnel services complement logistics by planning for and coordinating efforts that provide and sustain personnel. (FM 3-0) (USACAC)

*ART 4.2.1 PROVIDE HUMAN RESOURCES SUPPORT

4-78. Perform activities and tasks to sustain human resources (HR) functions of manning the force; HR services; personnel support; and HR planning and operations in support of deployed forces. HR support maximizes operational effectiveness and facilitates support to Soldiers, their families, DOD civilians, and contractors authorized to accompany the force. (FM 1-0) (USAAGS)

No.	Scale	Measure
01	Yes/No	HR support and procedures assisted the supported unit to accomplish its mission.
02	Yes/No	Developed a comprehensive plan to provide HR support.
03	Yes/No	Unit identified adequate resources and deployed the resources as part of the early entry element.
04	Yes/No	Unit identified location and support requirements for HR operations.
05	Yes/No	Adequate personnel information management was available.
06	Yes/No	Unit maintained personnel accountability and tracking of personnel entering or departing the organization or theater.
07	Time	To complete required coordination.
08	Time	To establish postal operations.
09	Time	To establish morale, welfare, and recreation programs and services.
10	Time	To establish HR communications nodes
11	Time	To deploy casualty liaison teams, personnel accountability teams and Theater Gateway Personnel Accountability Team.
12	Time	To perform personnel accounting functions associated with theater gateways and other inter/intra theater ports.
13	Time	To provide/receive HR support on request.
14	Time	To produce strength reports that accurately reflects the strength of the unit.
15	Percent	Of personnel meeting personnel readiness requirements.
16	Percent	Of HR resources in place and operational.
17	Percent	Of casualty reports processed in accordance with established timeframes.

*ART 4.2.1.1 MAN THE FORCE

4-79. Manning combines anticipation, movement, and skillful positioning of personnel so that the commander has the personnel required to accomplish the mission. ART 4.2.1.1 involves personnel readiness management, personnel accountability, strength reporting, retention, and management of personnel information. (FM 1-0) (USAAGS)

No.	Scale	Measure
01	Yes/No	The supported unit had sufficient personnel to accomplish its mission.
02	Yes/No	Unit maintained accountability of personnel transiting the organization or theater.
03	Time	To establish HR organizations as required.
04	Time	To access HR systems enablers and accurately post changes to the personnel database of record.
05	Time	To coordinate transportation and life support of transiting personnel.
06	Percent	Of unit and nonunit personnel scheduled to deploy or redeploy per scheduled dates or times.
07	Percent	Of unit personnel requirements met by Deployment-Day.
08	Percent	Of individuals, teams, platoons, and companies resourced for operations.
09	Percent	Of transiting personnel processed on a daily basis or in accordance with the distribution plan.

No.	Scale	Measure
10	Percent	Of reports submitted in accordance with established time lines.
11	Number	Of operations degraded, delayed, or modified due to personnel shortages.

*ART 4.2.1.1.1 Perform Personnel Readiness Management

4-80. Distribute Soldiers and Army civilians to subordinate commands based on documented manpower requirements, authorizations, and predictive analysis in support of the commander's plans and priorities. (FM 1-0) (USAAGS)

No.	Scale	Measure
01	Yes/No	Soldiers/individuals distributed per commander's priorities and documented manpower authorizations.
02	Yes/No	Strength management numbers were accurately maintained.
02	Yes/No	Accurate individual personnel readiness data was available in a timely manner to make personnel readiness decisions.
03	Time	On average for managing unit/individual readiness.
04	Time	Delay in providing replacements due to operational priorities.
05	Time	Delay in providing replacements due to transportation shortfalls.
06	Percent	Of reception, replacement, rest and recuperations, redeployment, and return to duty personnel record transactions completed correctly for individuals transiting the organization/theater.
07	Percent	Of military personnel files that have incorrect data entered (based on sample survey)
08	Number	Of operations degraded, delayed, or modified due to personnel shortages.
09	Number	Of replacement personnel provided by the national provider.

*ART 4.2.1.1.2 Conduct Personnel Accounting

4-81. Personnel accounting is the by-name recording of specific data on individuals' as they arrival and departure from units, duty status changes, change in location, MOS or specialty codes, grade changes, and so on. (FM 1-0) (USAAGS)

No.	Scale	Measure
01	Yes/No	Personnel accounting enhanced the unit's ability to accomplish its mission.
02	Yes/No	All transiting personnel are tracked and accountability maintained.
03	Time	To perform personnel accounting of transiting individuals/units into or out of the organization or theater.
04	Time	To integrate transiting Soldiers/individuals personnel accounting data into the theater database.
05	Percent	Of personnel data transactions completed.

*ART 4.2.1.1.3 Conduct Strength Reporting

4-82. Strength reporting is the numerical end product of the personnel accountability process, it is based on fill versus authorizations, and drives Army readiness and personnel readiness management. Strength reporting reflects the combat power of a unit and is used to monitor unit strength, prioritize replacements, execute strength distribution, and make tactical and human resources (HR) support decisions. (FM 1-0) (USAAGS)

No.	Scale	Measure
01	Yes/No	Strength reporting enhanced the unit's ability to accomplish its mission.
02	Yes/No	Strength reports for subordinate organizations are consolidated and reported to higher.
03	Yes/No	Strength reports have been reconciled to match database of record.

No.	Scale	Measure
04	Time	To consolidate subordinate personnel summaries.
05	Time	To integrate transiting Soldiers and other personnel accounting data into the theater database.
06	Percent	Of personnel data transactions completed meeting established submission guidelines.
07	Percent	Of reports submitted on a timely basis.

*ART 4.2.1.1.4 Provide Personnel Information Management

4-83. Collecting, processing, storing, displaying, and disseminating of relevant human resources (HR) information about units and personnel. This HR information includes Soldiers, attached joint, international and multinational military personnel and civilians (Department of Defense, interagency, and contractor employees) authorized to accompany the force. (FM 1-0) (USAAGS)

No.	Scale	Measure
01	Yes/No	The conduct of reception, replacement, rest and recuperation, redeployment, and return to duty (R5) operations supported unit mission accomplishment.
02	Time	To perform personnel accounting of transiting individuals or units in theater.
03	Time	To deploy and operate theater human resources teams such as theater gateway, R5, and casualty liaison teams.
04	Time	To integrate transiting personnel accounting data into the theater database.
05	Time	To coordinate transportation requirements.
06	Time	To coordinate life support for transiting personnel.
07	Percent	Of individuals or units processed daily.

*ART 4.2.1.1.5 Conduct Retention Operations

4-84. Retention improves readiness, aligns forces, and maintains Army end strength. Employ the four phases of the Army Career Counseling System to increase retention and reduce unit level attrition by advising leaders and developing and counseling Soldiers. (FM 1-0) (USAAGS)

No.	Scale	Measure
01	Yes/No	Retention supported unit mission accomplishment.
02	Time	To provide retention information (on average).
03	Time	To provide retention services (on average).
04	Time	To retain personnel to fill current positions (on average).
05	Percent	Of reenlistment actions processed correctly.
06	Percent	Of eligible personnel retained.
07	Percent	Of reenlistments occurring during a designated time.
08	Number	Of retention actions processed correctly.
09	Number	Of military personnel supported (given in an average).

*ART 4.2.1.2 PROVIDE HUMAN RESOURCES SERVICES

4-85. Human resources (HR) services are functions which directly impact a Soldier's status, assignment, qualifications, financial status, career progression, and quality of life which allows the Army leadership to effectively manage the force. HR services include the functions of essential personnel services (EPS), Postal, and Casualty operations. HR services include essential personnel services, casualty operations, and postal operations. (FM 1-0) (USAAGS)

No.	Scale	Measure
01	Yes/No	The unit's ability to accomplish its mission was enhanced because of the quality or quantity of personnel service support provided.
02	Time	To process an individual action.
03	Time	To coordinate or collect essential personnel services information.
04	Percent	Of total actions processed in specified time.
05	Percent	Of actions processed incorrectly.
06	Percent	Of actions returned for additional information.
07	Percent	Of HR services capabilities in place and operational after area of operations is assigned.
08	Percent	Of casualty and postal reports submitted in a specified time.

*ART 4.2.1.2.1 Conduct Casualty Operations

4-86. Collect, record, process, verify, and report casualty information from unit level to Department of the Army. (FM 1-0) (USAAGS)

No.	Scale	Measure
01	Yes/No	Families of personnel in the area of operations who become casualties were accurately notified in a timely and compassionate manner.
02	Time	To establish casualty liaison team at medical facilities and other required locations.
03	Time	To accurately record and report casualty information.
04	Time	To obtain evacuation reports from medical facilities.
05	Time	Of average delay in reporting and processing casualties reports.
06	Time	To provide casualty information to commanders.
07	Time	To appoint Summary Court Martial Officer and AR 15-6 Investigating officer.
08	Time	To prepare Next of Kin letters and process personnel actions.
09	Time	To complete a line of duty investigation.
09	Percent	Of total number of casualties not reported to Department of the Army within 12 hours of incident.

*ART 4.2.1.2.2 Perform Essential Personnel Services

4-87. Essential personnel services include customer service, awards and decorations, evaluation reports, promotions and reductions, transfers and discharges, , leaves and passes, military pay, personnel action request and other S-1 support (officer procurement, line-of-duty investigations, AR 15-6 investigations, suspension of favorable actions/Bars to reenlistment, citizenship/naturalization, congressional inquiries, identification card and tags). (FM 1-0) (USAAGS)

No.	Scale	Measure
01	Yes/No	Personnel actions let Soldiers know their contributions were valued by that organization.
02	Yes/No	Every Soldier, Army civilian or authorized contractor in the area of operations has required identification documents.
03	Yes/No	Unit recognition program fairly, equitably, and accurately recognized contributions made by unit or individual.
04	Yes/No	Unit personnel promotions and reductions occurred on a timely, fair and equitable basis.
05	Time	To process action (on average).
06	Time	To verify eligibility for ID documents (on average).
07	Time	To process award or decoration (on average)
08	Time	To process evaluation reports (on average).

No.	Scale	Measure
09	Time	To process promotion or reduction (on average).
10	Time	To process leave or pass (on average)
11	Time	To initiate and process line of duty investigations (on average).
12	Percent	Of number of actions returned for incompleteness.
13	Percent	Of personnel actions processed correctly.
14	Percent	Of line of duty investigations completed in a timeline prescribed by Army regulation.
15	Percent	Of pay inquiries successfully completed.
16	Percent	Of evaluation reports submitted after timelines prescribed by Army regulation.
17	Percent	Of accuracy in the preparation of ID documents.
18	Number	Of promotions per grade in a given period.
19	Number	Of identification documents processed and issued.
20	Number	Of pay inquiries resolved during a specific period.

*ART 4.2.1.2.3 Conduct Postal Operations

4-88. Postal operations provide a network to process mail and provide postal services. Processing mail involves receiving, separating, sorting, dispatching, and redirecting ordinary and accountable mail; completing international mail exchange; handling casualty and enemy prisoner of war mail; and screening for contaminated or suspicious mail. Postal services involve selling stamps; cashing and selling money orders; providing registered (including classified up to secret), insured, and certified mail services; and processing postal claims and inquiries. (FM 1-0) (USAAGS)

No.	Scale	Measure
01	Yes/No	Military Mail Terminal was established and effective.
02	Yes/No	Adequate personnel and equipment was available to support postal operations.
03	Yes/No	Unit members can send and receive mail.
04	Yes/No	Postal operations established in a timely manner that enhances force morale.
05	Time	To process and distribute mail, after receipt.
06	Time	For mail to transit from CONUS to overseas addressee (on average).
07	Percent	Of required military mail terminals and post offices established within planned timelines.
08	Percent	Of routes that have alternative routing sites.
09	Percent	Of routes that have daily delivery.
10	Percent	Of processed mail undeliverable.
11	Number	Of tons of backlogged mail, by class per day.

*ART 4.2.1.3 COORDINATE PERSONNEL SUPPORT

4-89. Personnel support activities encompass those functions and activities which contribute to unit readiness by promoting fitness, building morale and cohesion, enhancing quality of life, and by providing recreational, social, and other support services for Soldiers, Department of Defense civilians, and other personnel who deploy with the force. Personnel support encompasses the following functions: morale, welfare, and recreation (MWR), command interest programs, community support activities and band operations. (FM 1-0) (USAAGS)

Note: ART 4.2.5 Provide Band Support provides additional support to ART 4.2.1.3.

No.	Scale	Measure
01	Yes/No	Command interest programs supported unit readiness and morale activities.
02	Yes/No	All deployed unit members have access to human resource and community activity programs.
03	Time	To routinely establish command interest programs.
04	Percent	Of planned command interest resource programs in place and operational.
05	Number	Of personnel who have access to command interest programs, MWR and community activity programs.

*ART 4.2.1.3.1 Conduct Command Interest Programs

4-90. Army human resource programs are critical to sustain individual and unit readiness. These include the equal opportunity program, sexual harassment program, substance abuse prevention program, and weight control program. (FM 1-0) (USAAGS)

No.	Scale	Measure
01	Yes/No	All unit members felt they are valued members of the unit.
02	Yes/No	Community interest programs are fully established and enhance individual and unit readiness.
03	Time	Necessary to implement community interest programs or resolve an individual case.
04	Percent	Of planned community interest programs in place and operational.
05	Percent	Of community interest program cases successfully closed or completed.
06	Number	Of command interest programs required.
07	Number	Of personnel required to support community interest programs.
08	Number	Of cases successfully closed or completed.

*ART 4.2.1.3.2 Provide Morale, Welfare, and Recreation and Community Support Activities

4-91. Provide Soldiers, Army civilians, and other authorized personnel with recreational and fitness activities, goods, and services. The morale, welfare, and recreation network provides unit recreation and sports programs and rest areas for brigade-sized and larger units. Community support programs include the American Red Cross and family support. (FM 1-0) (USAAGS)

No.	Scale	Measure
01	Yes/No	Unit personnel and other authorized individuals had safe means to release some of the stress imposed on them as a result of their participation in full spectrum operations.
02	Yes/No	Quality of morale, welfare, and recreation program met unit and individual needs.
03	Time	To coordinate for the establishment of adequate recreation or fitness facilities in AO.
04	Percent	Of deployed units that have access to American Red Cross programs.
05	Percent	Of personnel with access to adequate recreational or fitness facilities.
06	Percent	Of deployed personnel who have access to rest and recuperation facilities.
07	Percent	Of units that have active family and community support programs.
08	Number	Of hours per day allotted to personal leisure, recreational, and fitness activities.

*ART 4.2.1.4 CONDUCT HUMAN RESOURCES PLANNING AND OPERATIONS

4-92. Performs and coordinates functions and activities needed to conduct and sustain human resources (HR) support operations. ART 4.2.1.4 includes planning and mission preparations, staff coordination, and establishment and operations of HR data nodes. (FM 1-0) (USAAGS)

No.	Scale	Measure
01	Yes/No	Human resource support was included as part of the planning process
02	Yes/No	Human resource support was coordinated and supports operations
03	Yes/No	HR connectivity to data and voice communications nodes was established.
04	Time	To establish HR connectivity to data and voice nodes.

*ART 4.2.1.4.1 Perform Human Resources Planning

4-93. Human resources (HR) planning support the commander's mission requirements. HR planners conducts mission and planning analysis, creates possible courses of action, analyze and compare courses of action, recommends a solution, and produce an operation plan or order annex. (FM 1-0) (USAAGS)

No.	Scale	Measure
01	Yes/No	Human resource support was integrated into the staff process.
02	Yes/No	Human resource support was integrated into the command budget cycle.
03	Yes/No	Human resource element conducted mission analysis.
04	Yes/No	Human resource element produced a human resource support plan to support mission.
05	Time	To coordinate and plan human resource support for mission.
06	Time	To integrate written plan into mission operation order or fragmentary order.
07	Time	Needed to plan for resources.
08	Number	Of days required to determine requirements.
09	Number	Of days required to determine support requirements.
10	Yes/No	Operational communications nodes provided the ability to maintain the common operational picture (COP).
11	Percent	Of coordinating operations occurring during a designated time.
12	Number	Of coordinating actions processed correctly.

*ART 4.2.1.4.2 Operate Human Resources Command and Control Nodes

4-94. Establish, operate, and maintain connectivity to human resources (HR) data and voice communications nodes for HR operations. HR command and control nodes include those required for all HR operations, across commands and echelons, and to higher and lower elements. (FM 1-0) (USAAGS)

No.	Scale	Measure
01	Yes/No	Established connectivity to data communications nodes and procedures supported unit mission accomplishment.
02	Yes/No	Adequate equipment and personnel were available to establish required connectivity to data and voice communications nodes.
03	Yes/No	Operational communications nodes provided the ability to maintain common operational picture.
04	Time	To establish required connectivity to communications nodes for HR operations.
05	Time	To coordinate and establish connectivity to communications and sustainment operations.
06	Number	Of HR nodes established in a specified time.
07	Number	Of critical HR data and voice communications nodes required for operations.

ART 4.2.2 PROVIDE FINANCIAL MANAGEMENT SUPPORT

4-95. Provide financial management support to commanders. Financial management is composed of two mutually supporting core functions: finance operations and resource management operations. Finance operations include developing policy and guidance; providing advice to commanders; disbursing support to the procurement process, banking and currency; accounting; and providing limited pay support. Resource management operations include providing advice to commanders; maintaining accounting records; establishing a management control process; developing resource requirements; identifying, acquiring, distributing, and controlling funds; and tracking, analyzing, and reporting budget execution. (FM 1-06) (USAFMS)

No.	Scale	Measure
01	Yes/No	Finance and resource management support enhanced the supported units' ability to accomplish its mission.
02	Yes/No	Operations were funded and reimbursed properly per policy guidance.
03	Yes/No	Unit developed a comprehensive plan to provide financial management support.
04	Yes/No	Access to requisite financial management systems was available.
05	Time	To refine finance support plan after receipt of warning order.
06	Time	To refine the resource management support plan after receipt of warning order.
07	Time	To establish financial management procurement support policies and guidance.
08	Time	To establish pay support policies and guidance.
09	Time	To establish disbursing operations.
10	Time	To establish accounting support policies and guidance.
11	Time	To establish banking and currency support.
12	Time	To identify, acquire, distribute, and control funding sources.
13	Time	To develop resource requirements.
14	Time	To track, analyze, and report budget execution.
15	Time	To establish finance operations management internal control process.
16	Percent	Of disbursing transactions, accounting actions, and pay actions processed in an accurate and timely manner.
17	Percent	Of contracts paid per the Prompt Payment Act per month.
18	Number	Of accounts maintained per month.
19	Number	Of accounts audited per month.
20	Cost	In dollars per month held in local depository accounts.
21	Cost	In dollars per month for contracted services in the area of operations.
22	Cost	In dollars per month for foreign national labor in the area of operations.
23	Cost	In dollars per month transacted in foreign currency.
24	Cost	In dollars per month disbursed in Department of the Treasury checks.
25	Cost	In dollars per month obligated.

ART 4.2.2.1 PROVIDE SUPPORT TO THE PROCUREMENT PROCESS

4-96. Providing support to the procurement process encompasses support to the logistics system and contingency contracting efforts. This support includes two areas: contracting support and commercial vendor services support. (FM 1-06) (USAFMS)

No.	Scale	Measure
01	Yes/No	Support to the procurement process enhanced the supported units' ability to accomplish their mission.
02	Yes/No	Accuracy of payments to vendors was within acceptable limits.
03	Yes/No	Payments occurred within established timelines.
04	Time	To refine finance support plan after receipt of warning order.
05	Time	To coordinate with legal and contracting.
06	Number	Of contracts paid per month.
07	Number	Of contracts paid per month by currency type and method of payment.
08	Cost	In dollars per month of contracts paid by currency type and method of payment.

ART 4.2.2.2 PROVIDE LIMITED PAY SUPPORT

4-97. Provide limited U.S. and non-U.S. pay support. Limited U.S. pay support ensures personnel from all Services are receiving financial support. This support includes making casual payments; cashing checks; processing travel pay; converting local currency; receiving manual savings deposit program payments; and support to noncombatant evacuation operations. Non-U.S. pay support includes payments to enemy prisoners of war, civilian internees, host-nation employees, and day laborers. (FM 1-06) (USAFMS)

No.	Scale	Measure
01	Yes/No	Limited pay support enhanced the supported units' ability to accomplish their mission.
02	Time	To refine finance support plan after receipt of warning order.
03	Time	To process transactions by type on average.
04	Percent	Of transactions accurately paid.
05	Percent	Of transactions audited per month.
06	Number	Of transactions performed per month.
07	Cost	In dollars per month for foreign national labor in the area of operations.

ART 4.2.2.3 PROVIDE DISBURSING SUPPORT

4-98. Provide disbursing support. Make payments on prepared and certified vouchers; receive collections; receive and control all currencies and precious metals; cash negotiable instruments; train and clear paying agents; support the rewards program; make claims and solatium payments; administer the stored value card program; support paper check conversions; fund financial management units; determine the need for currency (U.S. and foreign); and make foreign currency conversions. (FM 1-06) (USAFMS)

No.	Scale	Measure
01	Yes/No	Disbursement support enhanced the supported unit's ability to accomplish its mission.
02	Time	To refine finance support plan after receipt of warning order.
03	Time	To train, fund, and clear paying agents.
04	Time	To determine amount of foreign currency needed to support operation.
05	Percent	Of transactions by type audited per month.
06	Percent	Of disbursing transactions without errors.
07	Number	Of transactions performed per month by type.
08	Number	Of paying agents trained, funded, and cleared per month.
09	Number	Of currency conversions transacted per month.
10	Cost	In dollars per month of foreign currency on hand.

ART 4.2.2.4 PROVIDE ACCOUNTING SUPPORT

4-99. Maintain appropriated and nonappropriated funds accounting records and report the status of funds disbursed or collected. (FM 1-06) (USAFMS)

No.	Scale	Measure
01	Yes/No	Accounting support enhanced the supported unit's ability to accomplish its mission.
02	Time	To refine finance support plan after receipt of warning order.
03	Time	To establish, coordinate, and maintain nonappropriated fund accounting after establishing the area of operations.
04	Time	To coordinate and maintain appropriated fund accounting.
05	Percent	Of audit samples without accounting errors.
06	Number	Of reports reconciled per month.
07	Number	Of accounts maintained per month.
08	Number	Of accounts audited per month.

ART 4.2.2.5 PROVIDE BANKING AND CURRENCY SUPPORT

4-100. Provide banking and currency support to include supplying U.S. currency, foreign currencies, Treasury checks, and precious metals. Coordinate with host-nation banking industry to establish local depository accounts and banking procedures. (FM 1-06) (USAFMS)

No.	Scale	Measure
01	Yes/No	Banking and currency support enhanced the supported unit's ability to accomplish its mission.
02	Time	To refine the finance support plan after receipt of warning order.
03	Time	To establish and maintain central funding support after establishing the area of operations.
04	Time	To establish required electronic funds transfer accounts.
05	Time	To coordinate or establish host-nation banking.
06	Number	Of local depository accounts established.
07	Number	Of transactions per month by type of foreign currency.
08	Number	Of U.S. currency transactions per month by payment method.
09	Cost	In dollars per month transacted in foreign currency by type.
10	Cost	In dollars per month disbursed in U.S. currency by payment method.

ART 4.2.2.6 DEVELOP RESOURCE REQUIREMENTS

4-101. Provide advice to commanders on development of resource requirements. Determine and validate costs for mission support; provide accurate and detailed determination of costs; determine what resources are available; determine when resources are needed throughout the fiscal years; make resources available at the time and amount needed; develop budgets; address fiscal issues in the area of operations; complete manpower assessments; and determine phasing and supporting schedules. (FM 1-06) (USAFMS)

No.	Scale	Measure
01	Yes/No	Developed resource requirements enhanced the supported unit's ability to accomplish its mission.
02	Time	To refine the resource management plan after receipt of warning order.
03	Time	To identify resource requirements.
04	Time	To determine and validate mission costs.
05	Time	To estimate cost of future operations.
06	Percent	Of operations funded and reimbursed properly per policy guidance.

ART 4.2.2.7 PROVIDE SUPPORT TO IDENTIFY, ACQUIRE, DISTRIBUTE, AND CONTROL FUNDS

4-102. Provide advice to commanders on resource management implications; identify correct funds to support commander's requirements; acquire funds to support those requirements; and distribute in accordance with commander's priorities. Establish controls to monitor and track expenditure of funds. Participate in the planning, programming, budgeting, and execution process. (FM 1-06) (USAFMS)

No.	Scale	Measure
01	Yes/No	Identification, acquisition, distribution, and control of funds enhanced the supported unit's ability to accomplish its mission.
02	Yes/No	Operations were funded or reimbursed properly per policy guidance.
03	Yes/No	Identification, acquisition, distribution, and control of funds complied with fiscal law.
04	Time	To refine the resource management plan after receipt of warning order.
05	Time	To provide guidance to commands on funding procedures for operations after establishment of the area of operations.

No.	Scale	Measure
06	Time	To establish management internal control process.
07	Time	To complete required legal and contracting coordination.
08	Number	Of distribution documents and resource allocation documents provided per month.

ART 4.2.2.8 PROVIDE SUPPORT TO TRACK, ANALYZE, AND REPORT BUDGET EXECUTION

4-103. Track and analyze budget execution to provide commanders with reports and recommendations that facilitate decisionmaking. Determine phasing and supporting schedules; establish cost-capturing procedures; establish reporting procedures; and develop measures of effectiveness. (FM 1-06) (USAFMS)

No.	Scale	Measure
01	Yes/No	Tracking, analyzing and reporting budget execution support enhanced the supported unit's ability to accomplish its mission.
02	Yes/No	Unit had access to requisite financial management systems.
03	Time	To establish cost capturing procedures.
04	Time	To establish reporting procedures.
05	Number	Of cost reports provided per month.
06	Number	Of obligating documents per month.

ART 4.2.2.9 CONDUCT FINANCIAL MANAGEMENT PLANNING AND OPERATIONS

4-104. Performs and coordinates functions and activities needed to conduct and sustain financial management support operations. ART 4.2.2.9 includes planning and mission preparations, staff coordination, establishment and operations of Financial Management data nodes. (FM 1-06) (USAFMS)

No.	Scale	Measure
01	Yes/No	Financial management support was included as part of the planning process.
02	Yes/No	Financial management support was coordinated and supports operations.
03	Yes/No	Finance management element produced financial management plan to support mission.
04	Yes/No	Financial management connectivity to data and voice communications nodes was established.
05	Time	To coordinate and plan financial management support for mission.
06	Time	Needed to plan and coordinate for funding and special programs.
07	Time	To establish financial management connectivity to data and voice nodes.
08	Number	Of units in place and operational to support mission.

ART 4.2.3 PROVIDE LEGAL SUPPORT

4-105. Provide operational law support in all legal disciplines (including military justice, administrative and civil law, international and operational law, contract and fiscal law, claims, and legal assistance) in support of the command, control, and sustainment of operations. (FM 27-100) (TJAGLCS)

Note: ART 5.5.1.2.1 (Provide Law and Order) and ART 5.5.1.3 (Provide Military Justice Support) are included in ART 5.0 (The Command and Control Warfighting Function).

No.	Scale	Measure
01	Yes/No	Legal support services enhanced the supported unit's ability to accomplish its mission.
02	Time	To refine the legal services program for the area of operations (AO) after receipt of warning order.
03	Time	Between commander's requests for and receipt of legal advice or support.
04	Time	To prepare legal estimates.

No.	Scale	Measure
05	Time	Between requests for briefings on rules of engagement or law of war and actual presentation of the briefing.
06	Time	To review existing interagency or multinational agreements.
07	Percent	Of issues correctly identified, analyzed, and resolved to support command and control missions.
08	Percent	Of legal opinions that reflect an accurate view of law.
09	Percent	Of legal opinions that answer the client's questions clearly and concisely.
10	Percent	Of legal opinions in a form that is useful to the client.
11	Percent	Of opinions formatted in compliance with regulatory requirements.
12	Percent	Of opinions that are reviewed by a supervisor before release.
13	Percent	Of deployments requiring augmentation of legal personnel.
14	Percent	Of judge advocates and support personnel with working knowledge of current automated Army information systems.
15	Percent	Of judge advocates with access to automated Army information systems.
16	Percent	Of judge advocates and support personnel with access to Legal Automation Army-Wide System.
17	Percent	Of operationally ready vehicles dedicated to legal support.
18	Percent	Of core legal disciplines provided in support of unit.
19	Percent	Of operational cells with a judge advocate detailed.
20	Percent	Of missions where judge advocate participates in mission analysis.
21	Percent	Of targets reviewed by a judge advocate.
22	Percent	Of entities requiring legal liaison having a designated judge advocate liaison.
23	Percent	Of crisis management team meetings attended by a judge advocate.
24	Percent	Of units or Soldiers that receive legal briefings on rules of engagement or law of war, status-of-forces agreements, and host-nation law before deployment.
25	Number	Of judge advocates required to provide support in more than one core legal discipline.
26	Number	Of judge advocates required to provide support both in AO and at home station.
27	Number	Of vehicles dedicated for legal support.

ART 4.2.3.1 PROVIDE MILITARY JUDGE SUPPORT

4-106. Preside over courts-martial, supervise military judges, promulgate rules of court, and supervise the military magistrate program to include the review of pretrial confinement, confinement pending the outcome of foreign criminal charges, and the issuance of search, seizure, or apprehension authorizations. (FM 27-100) (TJAGLCS)

No.	Scale	Measure
01	Yes/No	Military judges supported the accomplishment of the supported unit's mission.
02	Time	Between referral of a case for trial by courts-martial and detailing of the military judge.
03	Time	Between referral of a case for trial by courts-martial and the arraignment.
04	Time	From pretrial confinement to military magistrate review.
05	Time	Between completion of the record of trial and the military judge's authentication.
06	Percent	Of trials in which the military judge leads "bridging the gap" mentoring sessions.
07	Percent	Of pretrial confinement cases overruled by the military judge.
08	Percent	Of search and seizure authorization later suppressed by the military judge.
09	Percent	Of the unit covered by military judge support.
10	Number	Of counsel having a copy of the rules of court.

ART 4.2.3.2 PROVIDE TRIAL DEFENSE SUPPORT

4-107. Provide personal legal advice to Soldiers related to criminal allegations; represent Soldiers in courts-martial and adverse administrative proceedings. (FM 27-100) (TJAGLCS)

No.	Scale	Measure
01	Yes/No	Accused Soldiers' legal rights were protected.
02	Time	Between a Soldier's request for and the scheduling of an appointment for legal advice.
03	Time	Between the scheduling of an appointment and the actual appointment date.
04	Percent	Of Soldiers electing to retain their detailed or individually requested military counsel.
05	Percent	Of Soldiers appearing before administrative boards represented by military counsel.
06	Percent	Of legal issues correctly identified and analyzed.
07	Percent	Of legal opinions that answer clients' questions clearly and concisely.
08	Percent	Of Soldiers receiving advice on adverse administrative actions from legal assistance instead of trial defense services.
09	Percent	Of units covered by trial defense service support.

ART 4.2.3.3 PROVIDE INTERNATIONAL LAW SUPPORT

4-108. Implement the Department of Defense law of war program. Assist with international legal issues relating to U.S. forces overseas. Advise concerning the legal basis for conducting operations and the use of force; advise concerning the legal status of forces; monitor foreign trials and confinement of Army personnel and their family members; perform legal liaison with the International Committee of the Red Cross and host-nation legal authorities; and advise concerning legal issues in intelligence operations, security assistance, counterdrug operations, stability operations, and civil assistance activities. (FM 27-100) (TJAGLCS)

No.	Scale	Measure
01	Yes/No	International law support services enhanced the supported unit's ability to accomplish its mission.
02	Time	Between discovery of possible law of war violations and report to higher headquarters.
03	Time	Between foreign confinement of Army personnel and notification to the U.S. legal liaison.
04	Time	Between reporting of a potential law of war violation and the decision whether to investigate.
05	Time	Between capture of an individual and determination of status under Article 5, Geneva Convention of 1949.
06	Percent	Of required international agreements on hand.
07	Percent	Of commanders or Soldiers who receive legal briefings on rules of engagement or law of war, status-of-forces agreement, and host-nation law before deployment.
08	Percent	Of targets reviewed by a judge advocate.
09	Percent	Of law of war allegations that are reported.
10	Percent	Of foreign trials and confinement of Army personnel and family members that comply with status-of-forces agreement requirements.
11	Percent	Of international law issues correctly identified, analyzed, and resolved.
12	Percent	Of legal opinions that answer the client's questions clearly and concisely.
13	Percent	Of legal opinions that are reviewed by a supervisor before release.
14	Percent	Of entities requiring legal liaison having a designated judge advocate liaison.
15	Percent	Of orders, plans, and policies reviewed for compliance with international legal obligations.
16	Percent	Of foreign trials observed by a qualified U.S. trial observer.
17	Number	Of U.S. law of war violations.

ART 4.2.3.4 PROVIDE ADMINISTRATIVE AND CIVIL LAW SUPPORT

4-109.   Advise commanders and litigate on behalf of the Army. Provide legal advice and representation for the command. The practice of civil law includes environmental law, military installations law, regulatory law, intellectual property law, and cases within the U.S. magistrate program and felony prosecution program, as applicable. (FM 27-100) (TJAGLCS)

No.	Scale	Measure
01	Yes/No	Administrative law support services enhanced the supported unit's ability to accomplish its mission.
02	Yes/No	Civil law support services enhanced the supported unit's ability to accomplish its mission.
03	Time	Between a request for legal review and completion of the review.
04	Time	Between requests for briefings on environmental law and actual presentation of the briefings.
05	Time	To review environmental orders.
06	Time	To review environmental consent agreements and settlements with federal, state, and local officials.
07	Percent	Of financial disclosure forms completed and filed on time.
08	Percent	Of request for opinions that are received before the legally significant decisions.
09	Percent	Of financial liability investigations of property loss that are not legally sufficient at the second review.
10	Percent	Of conscientious objection issues identified before deployment.
11	Percent	Of family care plan failure issues identified before deployment.
12	Percent	Of personnel requiring ethics training who receive the training.
13	Percent	Of family advocacy case review committee meetings attended by a judge advocate.
14	Percent	Of issues correctly identified, analyzed, and resolved.
15	Percent	Of legal opinions that answer the client's questions clearly and concisely.
16	Percent	Of legal opinions in a form that is useful to the client.
17	Percent	Of opinions formatted in compliance with regulatory requirements.
18	Percent	Of legal opinions that are reviewed by a supervisor before release.
19	Percent	Of plans reviewed to ensure environmental laws are followed.
20	Percent	Of environmental baseline studies conducted within a given timeframe.
21	Percent	Of units coordinating with the staff judge advocate on environmental enforcement activities.
22	Number	Of litigation cases arising from employee grievances, discrimination complaints, and unfair labor practices.

ART 4.2.3.5 PROVIDE CONTRACT AND FISCAL LAW SUPPORT

4-110.   Provide legal advice and assistance to procurement officials during all phases of the contracting process, overseeing an effective procurement fraud abatement program; and providing legal advice to the commander concerning battlefield acquisition, contingency contracting, use of the Logistics Civil Augmentation Program, acquisition and cross-servicing agreements, the commercial activities program, and overseas real estate and construction. Provide legal advice on the proper use and expenditure of funds, interagency agreements for logistics support, security assistance, and support to nonfederal agencies and organizations. (FM 27-100) (TJAGLCS)

No.	Scale	Measure
01	Yes/No	Contract law support services enhanced the supported unit's ability to accomplish its mission
02	Yes/No	Fiscal law support services enhanced the supported unit's ability to accomplish its mission

No.	Scale	Measure
03	Time	Between requests for procurement legal advice and actual opinion rendered.
04	Time	To review international acquisition agreements and contingency contracting matters.
05	Time	To draft legal opinions on foreign military sales cases.
06	Time	To provide legal opinions on proper use and expenditure of funds.
07	Time	To review contract for legal sufficiency.
08	Percent	Of issues correctly identified, analyzed, and resolved.
09	Percent	Of legal opinions that answer the client's questions clearly and concisely.
10	Percent	Of legal opinions in a form that is useful to the client.
11	Percent	Of opinions formatted in compliance with regulatory requirements.
12	Percent	Of legal opinions that are reviewed by a supervisor before release.
13	Percent	Of civil law judge advocates with immediate access to fiscal law codes and regulations.
14	Percent	Of contracts reviewed by a judge advocate.
15	Number	Of contracts reviewed by a judge advocate.

ART 4.2.3.6 PROVIDE CLAIMS SUPPORT

4-111. Investigate, process, adjudicate, and settle claims on behalf of and against the United States per statute, regulation, Department of Defense directives, and international or interagency agreements. Categories of claims include claims for property damage of Soldiers and employees arising incident to service, torts alleged against Army civilians or military personnel acting within the scope of employment, and claims by the United States against individuals who injure Army personnel or damage Army property. (FM 27-100) (TJAGLCS)

No.	Scale	Measure
01	Yes/No	Claims services enhanced the supported unit's ability to accomplish its mission.
02	Time	To adjudicate a small claim.
03	Time	To adjudicate a large claim.
04	Time	Between a claimant's request for forms and actual receipt of forms.
05	Time	To investigate personal property claims.
06	Time	To investigate medical malpractice claims.
07	Time	To investigate federal tort claims.
08	Time	Between identification of and approval for payment of ex gratia claims.
09	Time	Between requests for briefings on claims procedures and actual presentations.
10	Time	Between entry into a foreign area and obtaining translation service and local legal advice.
11	Percent	Of claims received with all substantiation included.
12	Percent	Of claims investigated and paid in the area of operations.
13	Percent	Of claims offices missing equipment necessary to investigate claims (for example, digital cameras).
14	Percent	Of personnel in claims office proficient in the use of all equipment necessary to investigate claims (for example, digital cameras).
15	Percent	Of claims received for reconsideration.
16	Percent	Of claims adjudicated consistent with law, regulation, and U.S. Army claims services policies.
17	Percent	Of units that have appointed unit claims officers.
18	Percent	Of base camps with documentation of preexisting conditions.
19	Percent	Of filed claims paid or transferred before redeployment.
20	Percent	Of large personal property claims that are inspected by claims personnel.

No.	Scale	Measure
21	Number	Of judge advocates in staff judge advocate offices on orders as claims officers.
22	Number	Of claims briefings given before deployment.
23	Cost	Of claims resolved in area of operations within a given time.

ART 4.2.3.7 PROVIDE LEGAL ASSISTANCE

4-112. Provide personal civil legal services to Soldiers, their family members, and other eligible personnel. Provide support to combat readiness exercises, premobilization legal preparation, Soldier readiness program processing, demobilization briefings, and noncombatant evacuation operations. Provide federal and state income tax assistance, ministerial and notary services, legal counseling, legal correspondence, negotiation, legal document preparation and filing, limited in-court representation, legal referrals, and mediation. Operate preventive law programs. (FM 27-100) (TJAGLCS)

No.	Scale	Measure
01	Yes/No	Legal assistance support services enhanced the supported unit's ability to accomplish its mission.
02	Time	Between a request for an appointment and the actual appointment.
03	Time	Between submitting a completed will worksheet and the client's review of the will.
04	Time	Between the client's review of the draft will and the final will signing.
05	Time	Between requests for briefings and actual presentations.
06	Time	Before deployment that Soldiers receive Soldier readiness program packets.
07	Time	To provide client with notary services.
08	Percent	Of clients whose problems are resolved in one visit.
09	Percent	Of Soldiers who use the tax assistance program versus commercial tax preparation services.
10	Percent	Of attorneys trained to provide trust and estate planning.
11	Percent	Of nonattorneys in staff judge advocate offices authorized to perform notary duties.
12	Percent	Of attorneys authorized to represent clients in civilian court.
13	Percent	Of legal assistance personnel trained on drafting library will programs.
14	Percent	Of units having income tax assistance available.
15	Percent	Of wills prepared to include trust and estate planning.
16	Percent	Of legal assistance services that are provided to family members and retirees.
17	Percent	Of client issues correctly identified and resolved.
18	Percent	Of documents written in simple format so that clients can readily understand them.

ART 4.2.4 PLAN RELIGIOUS SUPPORT OPERATIONS

4-113. Religious support operations undergird and fortify the Warrior Ethos, especially in operations overseas. The comprehensive integration of religious support operations is the means by which the free exercise of religion for Soldiers and their families occurs in the future force. Religious support operations provide for the spiritual, ethical, and moral needs of Soldiers, family members, and authorized Army civilians (to include contractors deploying with the force) at all levels. Religious support operations also support command and staff's conduct of information engagement operations by advising commanders and staff on religious aspects of the local environment. (FM 1-05) (USACHCS)

No.	Scale	Measure
01	Yes/No	Religious support operations supported the unit's ability to accomplish its mission.
02	Yes/No	Unit developed a comprehensive religious support plan for the operational environment addressing both core capabilities of religious leader and principal religious advisor.
03	Yes/No	Unit planned for faith group coverage to include general Protestant, Roman Catholic, Orthodox, Jewish, Buddhist, Islamic, and others.

No.	Scale	Measure
04	Yes/No	Unit developed training program for lay leaders to perform worship services.
05	Yes/No	Unit assessed the spiritual readiness of Soldiers and units to include the moral and ethical climate.
06	Yes/No	Unit planned for religious support for multinational forces.
07	Yes/No	Unit developed standing operating procedures for religious support to the caregiver and mass casualties.
08	Number	Of unit ministry teams in the area of operations.

ART 4.2.4.1 DELIVER RELIGIOUS SERVICES

4-114. Provide or perform collective and denominational religious worship services and religious coverage in the operational environment. Deliver or provide for memorial ceremonies, memorial services, and funerals. (FM 1-05) (USACHCS)

No.	Scale	Measure
01	Yes/No	Religious services met the needs of the supported unit's personnel.
02	Yes/No	Unit provided for faith group services to include general Protestant, Roman Catholic, Orthodox, Jewish, Buddhist, Islamic, and others.
03	Yes/No	Lay leaders were identified to perform worship services, as needed.
04	Yes/No	Unit identified location for services.
05	Yes/No	Unit identified and executed force protection plan for services.
06	Yes/No	Unit identified and planned transportation requirements to enable unit ministry teams to get to identified locations.
07	Yes/No	Unit planned for seasonal religious celebrations.
08	Number	Of hours per week spent delivering worship services.
09	Number	Of memorial ceremonies or services and funerals completed.

ART 4.2.4.2 PROVIDE SPIRITUAL CARE AND COUNSELING

4-115. Provide spiritual care and counseling to Soldiers, family members, and authorized Army civilians (to include contractors deploying with the force) with spiritual comfort, moral support, encouragement. (FM 1-05) (USACHCS)

No.	Scale	Measure
01	Yes/No	Unit ministry team completed the military decisionmaking process to identify religious care and counseling needs for the unit.
02	Yes/No	Religious care and counseling supported the mission of the unit's personnel.
03	Time	Between the unit ministry team receiving a request for counseling and the counseling.
04	Time	For individual to be seen by a chaplain.
05	Time	To move in the operational environment to provide religious care and counseling.
06	Time	To develop resources and supporting agencies to refer individual for additional care.
07	Percent	Of Soldiers seen who require follow-up counseling.
08	Number	Of Soldiers seen who require referral services.
09	Number	Of hours per week spent providing religious care and counseling.
10	Number	Of hours per week planning and analyzing needs and trends for spiritual care and counseling that impact mission readiness.

ART 4.2.4.3 PROVIDE RELIGIOUS SUPPORT TO THE COMMAND

4-116. Advise the commander on issues of religion, ethics, and morale (as affected by religion), including the religious needs of all personnel for whom the commander is responsible. (FM 1-05) (USACHCS)

No.	Scale	Measure
01	Yes/No	Unit performed unit analysis to determine the current religious, moral, and ethical climate within the unit and the area of operations.
02	Yes/No	Unit performed mission analysis to determine the impact on the religious, moral, and ethical climate within the unit and the area of operations.
03	Yes/No	Unit performed a religious area analysis to determine the impact of religion on the unit's mission.
04	Yes/No	Unit completed a religious intelligence preparation of the battlefield to determine trigger points that would affect the mission.
05	Yes/No	Unit participated in the information engagement working group as a sitting member.
06	Time	To prepare a religious area analysis.
07	Time	To complete unit analysis.
08	Time	To prepare a religious intelligence preparation of the battlefield.
09	Time	To advise the commander on enemy prisoners of war, civilian detainees, and refugees.
10	Time	To advise the commander on issues concerning subordinate unit ministry teams.

ART 4.2.4.4 PROVIDE RITES, SACRAMENTS, AND ORDINANCES

4-117. Provide for sacraments, rites, and ordinances per the tenets of the denomination or faith group. Army chaplains meet all faith group and denominational nonworship religious support requirements. Chaplains support the religious diversity to guarantee the Constitutional rights of Soldiers. (FM 1-05) (USACHCS)

Note: Rites, sacraments, and ordnances include marriages, burials, baptisms, confirmations, and blessings.

No.	Scale	Measure
01	Yes/No	Unit identified faith group requirements for sacraments, rites, and ordnances.
02	Yes/No	Unit developed service matrix for sacramental requirements.
03	Yes/No	Unit developed lay leader coverage plan to provide sacraments, rites, and ordnances.

ART 4.2.4.5 COORDINATE MILITARY RELIGIOUS SUPPORT

4-118. The unit ministry team plans, coordinates, and resources precise religious support per the factors of mission, enemy, terrain and weather, troops and support available, time available, civil considerations. Unit ministry teams respond to crises across the spectrum of conflict. (FM 1-05) (USACHCS)

No.	Scale	Measure
01	Yes/No	Unit ministry team was integrated into staff planning and mission.
02	Yes/No	Unit ministry team was integrated into the command budget cycle.
03	Yes/No	Unit ministry team completed mission analysis.
04	Yes/No	Unit ministry team produced religious support plan to support mission.
05	Time	To plan for resources.
06	Time	To write religious support plan.
07	Time	To collaborate with staff elements.

ART 4.2.4.6 PROVIDE RELIGIOUS CRISIS RESPONSE

4-119. The unit ministry team assists the command through prevention, intervention, mitigation, and normalization of crisis events. It integrates all crisis-helping agencies to support the needs of the combatant commander. Unit ministry team responds to crises operating across the spectrum of conflict from homeland security to humanitarian and civic assistance. (FM 1-05) (USACHCS)

No.	Scale	Measure
01	Yes/No	Unit developed plan and resources for crisis intervention.
02	Yes/No	Unit identified symptoms of combat trauma.
03	Yes/No	Unit developed pastoral self-care resources for trauma.
04	Yes/No	Unit completed training to harden Soldiers spiritually for deployment.
05	Time	To perform critical stress defusing.
06	Time	To perform critical stress debriefings.
07	Time	To refer individuals for follow-up care.
08	Percent	Of Soldiers with symptoms of combat trauma.

ART 4.2.4.7 PROVIDE RELIGIOUS MANAGEMENT AND ADMINISTRATION SUPPORT

4-120.   Manage and administer chaplaincy personnel, facilities, equipment, materiel, funds, and logistics. (FM 1-05) (USACHCS)

No.	Scale	Measure
01	Yes/No	Unit ministry team understood chaplain life cycle.
02	Yes/No	Unit ministry team updated table of organization and equipment and table of distribution and allowances requirements.
03	Yes/No	Unit ministry team developed quarterly training guidance for the command training guidance.
04	Yes/No	Unit ministry team maintained hand receipts for all facilities and equipment.
05	Yes/No	Unit ministry team planned for resupply and distribution of essential ecclesiastical supplies.
06	Time	To spend on personnel management.
07	Time	To spend on internal management and administrative activities.

ART 4.2.4.8 PROVIDE RELIGIOUS EDUCATION

4-121.   The unit ministry team provides, performs, and integrates religious education and faith sustaining activities to meet the military religious support needs of Soldiers and their families. The unit ministry team enables religious education and spiritual formation through classes, studies, groups, meetings, retreats, and discussion groups and by providing religious educational material and curriculum. (FM 1-05) (USACHCS)

No.	Scale	Measure
01	Yes/No	Unit ministry team performed the military decisionmaking process to determine religious educational needs and faith specific requirements.
02	Yes/No	Religious educational programs supported the mission readiness of the unit.
03	Yes/No	Religious educational materials were appropriate for the faith specific and educational program.
04	Yes/No	Unit determined needs requirements for director of religious education.
05	Time	To supervise the director of religious education.
06	Time	To assess the effectiveness of the religious educational program on mission readiness.
07	Time	To plan and identify the resource requirements to support religious educational program.
08	Number	Of personnel who attend religious educational programs.
09	Number	Of hours per week holding and supervising religious educational programs.

ART 4.2.5 PROVIDE BAND SUPPORT

4-122.   Provide music for all operations to instill in our Soldiers the will to fight and win, foster the support of our citizens, and promote our national interests at home and abroad. (FM 1-0) (USAAGS)

No.	Scale	Measure
01	Yes/No	Band support contributed to mission accomplishment.
02	Time	To rehearse the music required for the mission.
03	Time	To rehearse drill and ceremony required for the mission.
04	Time	To coordinate the performance of a mission.
05	Time	To arrange logistic and administrative support for the band.
06	Time	To obtain recommendations and legal advice from the staff judge advocate.
07	Percent	Of authorized personnel required to perform the specific mission.
08	Percent	Of authorized musical equipment on hand and serviceable.

*SECTION III – ART 4.3: PROVIDE HEALTH SERVICE SUPPORT

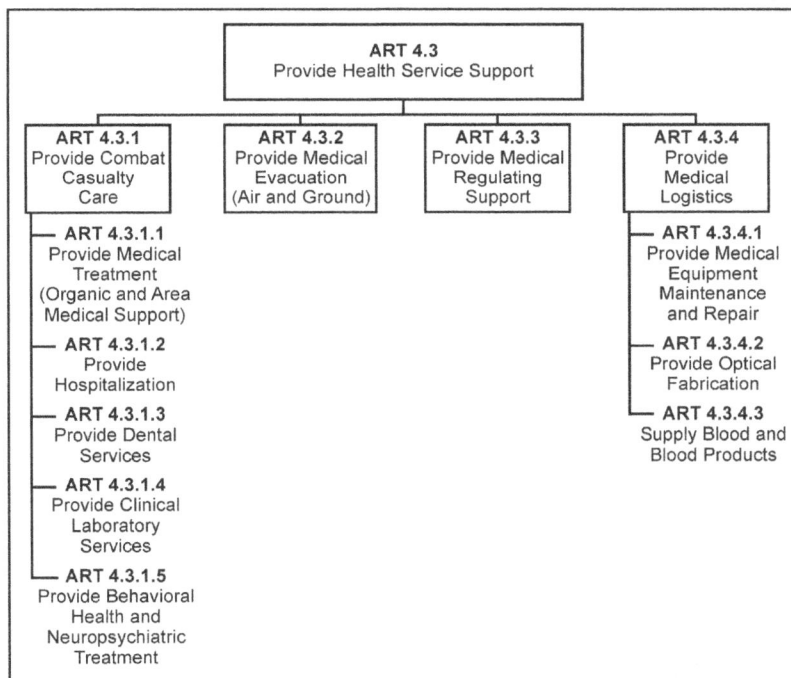

```
                          ART 4.3
                  Provide Health Service Support

   ART 4.3.1          ART 4.3.2          ART 4.3.3          ART 4.3.4
 Provide Combat    Provide Medical    Provide Medical       Provide
   Casualty          Evacuation         Regulating          Medical
    Care          (Air and Ground)       Support          Logistics

  ART 4.3.1.1                                             ART 4.3.4.1
Provide Medical                                         Provide Medical
  Treatment                                               Equipment
(Organic and Area                                        Maintenance
Medical Support)                                          and Repair

  ART 4.3.1.2                                             ART 4.3.4.2
   Provide                                              Provide Optical
Hospitalization                                          Fabrication

  ART 4.3.1.3                                             ART 4.3.4.3
Provide Dental                                        Supply Blood and
  Services                                             Blood Products

  ART 4.3.1.4
Provide Clinical
  Laboratory
   Services

  ART 4.3.1.5
Provide Behavioral
  Health and
Neuropsychiatric
  Treatment
```

4-123. The Army Health System is a component of the Military Health System that is responsible for operational management of the health service support (HSS) and force health protection (FHP) missions. The Army Health System includes all mission support services performed, provided, and arranged by the Army Medical Department to support HSS and FHP mission requirements for the Army. The HSS mission promotes, improves, conserves, or restores the mental and physical wellbeing of Soldiers and, as directed other personnel and is comprised of three elements: casualty care, medical evacuation, and medical logistics: casualty care encompasses the treatment aspects of organic and area medical support, hospitalization (to include treatment of chemical, biological, radiological, and nuclear patients), dental treatment, behavioral health/neuropsychiatric treatment, and clinical laboratory services; medical evacuation (to include en route care and medical regulating); and medical logistics(to include blood and blood products). (FM 4-02.2) (USAMEDDC&S)

Note: Health service support is closely related to force health protection. Health service support includes the requirement to ensure adequate health, safety, quality assurance, food provisions, and sanitation for belligerents.

*ART 4.3.1 PROVIDE COMBAT CASUALTY CARE

4-124. Casualty care encompasses a number of Army Medical Department functions. It groups organic and area medical treatment, hospitalization, dental services, clinical laboratory services, and behavioral health and neuropsychiatry. (FM 4-02) (USAMEDDC&S)

Note: The preventive aspects of dentistry and combat and operational stress control are addressed under ART 6.8 (Provide Force Health Protection).

No.	Scale	Measure
01	Yes/No	Comprehensive casualty care provided in AO conserved the fighting strength.
02	Yes/No	Combat medics were proficient in clinical skills.
03	Percent	Of units supported with organic health service support resources.
04	Percent	Of units supported requiring medical support on an area basis.
05	Percent	Of difference between planned hospital resources and actual requirements in the AO.
06	Percent	Of supported forces requiring behavioral health and neuropsychiatric treatment.
07	Percent	Of supported forces requiring dental treatment.
08	Percent	Of patients requiring clinical laboratory support.
09	Percent	Of patients requiring treatment for injuries related to chemical, biological, radiological, and nuclear munitions.

*ART 4.3.1.1 PROVIDE MEDICAL TREATMENT (ORGANIC AND AREA MEDICAL SUPPORT)

4-125. Provide medical treatment (organic and area support) for all units in the AO. Examine and stabilize patients. Evaluate wounded and disease and nonbattle injuries. Examine the general medical status to determine treatment and medical evacuation precedence. (FM 4-02) (USAMEDDC&S)

No.	Scale	Measure
01	Yes/No	Area medical support and treatment met the unit's health services needs.
02	Yes/No	Organic medical support and treatment met unit's health service needs.
03	Time	To refine medical treatment (organic and area medical support) program for AO after receipt of warning order.
04	Time	To publish estimates of medical sustainment and anticipated resupply.
05	Time	To expand medical treatment facilities to full capacity and full functionality.
06	Time	To deploy required additional medical specialists after AO is assigned.
07	Percent	Of difference between planned medical treatment (organic and area medical support) requirements and actual requirements in AO.
08	Percent	Of planned capacity of medical treatment (organic and area medical support) support performed in AO.
09	Percent	Of patient encounters recorded in individual health and/or electronic medical record.
10	Percent	Of personnel treated who are from other Services.
11	Percent	Of personnel treated who are from multinational or host-nation forces.
12	Percent	Of total of medically treated persons that are enemy prisoners of war and detained or retained personnel.
13	Percent	Of total casualties treated per day in AO who are noncombatants.

No.	Scale	Measure
14	Percent	Of total casualties per time treated in AO who returned to duty.
15	Percent	Of total casualties per time treated in AO who are evacuated per the stated theater evacuation policy.
16	Percent	Of personnel in AO who have access to optometry services.
17	Number	Of patients treated by a forward surgical team.
18	Number	Of patients treated for traumatic brain injury.
19	Number	If physical therapists available to provide Role 1 and 2 care.

*ART 4.3.1.2 PROVIDE HOSPITALIZATION

4-126. Hospitalization resources are medical treatment facilities capable of providing inpatient care and services. Hospitalization continues the medical care provided at Roles 1 and 2 of the Army Health System. It also provides a far forward surgical capability that provides essential care in theater, outpatient services, and ancillary support (pharmacy, clinical laboratory, radiology services, and nutrition care). Within theater, the hospitalization capability includes returning those patients to duty within the limits of the theater evacuation policy. This conserves the fighting strength by returning trained manpower to the tactical commander. It also provides stabilizing care to facilitate the evacuation of those patients who will not recover from their injuries or illnesses within the stated theater evacuation policy to facilities capable of providing required care. Theater hospitals may be augmented with hospital augmentation teams to provide specific specialty care. (FM 4-02.10) (USAMEDDC&S)

No.	Scale	Measure
01	Yes/No	Number of beds and services available in the AO was equal to or greater than the peak demand for these services.
02	Time	To refine hospital operations program for AO after receipt of warning order.
03	Time	To reach full functionality after activation of AO.
04	Percent	Of difference between planned hospitalization requirements and actual requirements in AO.
05	Percent	Of difference between hospitalization capacity realized and planned hospitalized capacity.
06	Percent	Of casualties per day in AO that require hospitalization.
07	Percent	Of hospital beds in AO utilized per month.
08	Percent	Of hospitalized patients who require further medical care outside the AO or in CONUS.
09	Percent	Of patients returning to duty from the hospital within the theater evacuation policy.
10	Percent	Of patients requiring radiology services.
11	Percent	Of patients requiring routine pharmacy support.
12	Percent	Of patients requiring specialized formulary pharmacy support.
13	Percent	Of patients requiring surgical care.
14	Percent	Of patients requiring inpatient medical care.
15	Percent	Of patients requiring renal hemodialysis (when augmented).
16	Percent	Of patients requiring support from special care team (when augmented to provide humanitarian assistance, disaster relief, or support to other stability or civil support operations).
17	Percent	Of hospital minimal care patients exceeding organic minimal care capabilities (minimal care capability augmentation required).
18	Percent	Of patients requiring head and neck surgical support (when augmented).
19	Percent	Of patients requiring infectious disease investigative and consultation services (when augmented).

No.	Scale	Measure
20	Percent	Of patients requiring enhanced anatomic pathology, chemistry, and microbiology support (when augmented).
21	Percent	Of difference between actual nutrition care support requirements in AO and planned nutrition care support.
22	Number	Of outpatient visits completed in a month.
23	Number	Of casualties per day in AO who require hospitalization.
24	Number	Of patients requiring special diets.
25	Number	Of supported units or personnel provided training in the Army health promotion program.
26	Number	Of cases requiring telemedicine support (when augmented).
27	Percent	Of totally of medically treated persons that are enemy prisoners of war and detained and retained personnel.
28	Percent	Of patients treated for mild traumatic brain injury.
29	Percent	Of patient encounters recorded in individual health and/or electronic medical record.
30	Percent	Of patients treated by a forward surgical team

*ART 4.3.1.3 PROVIDE DENTAL SERVICES

4-127. Prevent and treat dental disease and injury. ART 4.3.1.3 includes providing operational dental care, which consists of emergency dental care and essential dental care, and comprehensive care that normally is performed only in fixed facilities in CONUS or in at least a Role 3 facility. (FM 4-02.19) (USAMEDDC&S)

No.	Scale	Measure
01	Yes/No	Nonavailability of unit personnel because of dental problems did not degrade, delay, or disrupt unit operations.
02	Time	To refine dental service support program for AO after receipt of warning order.
03	Time	To establish comprehensive dental plan after AO is assigned.
04	Time	In advance required to schedule routine dental appointments in AO.
05	Percent	Of difference between planned dental service support requirements and actual requirements in AO.
06	Percent	Of planned dental support performed in AO.
07	Percent	Of personnel in AO rated as class I or class II dental.
08	Percent	Of personnel in AO rated as class III or class IV dental.
09	Percent	Of personnel in AO with no dental rating.
10	Percent	Of units with effective dental health care programs in AO.
11	Percent	Of dental capacity in use per day in AO.
12	Percent	Of dental patients requiring evacuation to role 3 dental care facilities.
13	Percent	Of dental patients requiring evacuation out of theater.
14	Percent	Of dental patients requiring oral or maxillofacial surgery.
15	Number	Of emergency dental cases per quarter in AO.
16	Percent	Of enemy prisoners of war/detainees requiring dental treatment.

*ART 4.3.1.4 PROVIDE CLINICAL LABORATORY SERVICES

4-128. Perform clinical laboratory diagnostic procedures in support of medical treatment activities. (FM 4-02.10) (USAMEDDC&S)

No.	Scale	Measure
01	Yes/No	Unit operations were not degraded, delayed, or disrupted nor was the health of unit personnel endangered by the nonavailability of clinical diagnostic laboratory services.
02	Time	To refine clinical diagnostic laboratory service plans for AO after receipt of warning order.
03	Time	To establish required clinical diagnostic laboratory services plan after AO is assigned.
04	Time	Of turnaround for clinical diagnostic laboratory testing results.
05	Percent	Of difference between planned clinical diagnostic laboratory requirements and actual requirements of the AO.
06	Percent	Of planned clinical diagnostic laboratory support performed in the AO.
07	Percent	Of required laboratory functionality in place and fully functional at activation of the AO.
08	Percent	Of laboratory capacity in use per day in AO.
09	Number	Of clinical laboratory procedures performed in AO per month.

*ART 4.3.1.5 PROVIDE BEHAVIORAL HEALTH AND NEUROPSYCHIATRIC TREATMENT

4-129. Provide medical treatment for behavioral health and neuropsychiatric medical conditions. (FM 4-02.51) (USAMEDDC&S)

No.	Scale	Measure
01	Yes/No	Absence of command personnel due to behavioral health and neuropsychiatric treatment in a medical treatment facility did not degrade, delay, or disrupt unit operations.
02	Time	To refine behavioral health and neuropsychiatric treatment program for AO after receipt of warning order.
03	Time	To establish comprehensive behavioral health and neuropsychiatric treatment plan after AO is assigned.
04	Percent	Of difference between planned behavioral health and neuropsychiatric treatment requirements and actual requirements in the AO.
05	Percent	Of psychiatric support completed in the AO versus planned psychiatric support.
06	Percent	Of required psychiatrists at activation of AO.
07	Percent	Of other required behavioral health and neuropsychiatric treatment personnel at activation of AO.
08	Percent	Of personnel in AO requiring behavioral health and neuropsychiatric treatment per quarter.
09	Percent	Of behavioral health and neuropsychiatric patients returned to duty in the AO.
10	Number	Of behavioral health and neuropsychiatric patients requiring medical evacuation from the AO.
11	Number	Of available occupational therapists.

*ART 4.3.2 PROVIDE MEDICAL EVACUATION (AIR AND GROUND)

4-130. Provide direct and area air and ground medical evacuation (MEDEVAC) support to; evacuate sick, injured, or wounded personnel (U.S. joint, interagency, intergovernmental and multinational forces; enemy prisoners of war; detained or retained personnel; and when authorized civilian personnel) from the point of injury or wounding, casualty collection points, battalion aid stations, ambulance exchange points, or any other designated points where casualties have been located to be evacuated to a higher role medical treatment facility (MTF) in the area of operations (AO). Provide medical care en route while transporting patients to, between, and from treatment facilities in the AO. Provide transport of patients from MTF to ports of debarkation for evacuation out of theater. Support personnel recovery operations. Provide emergency pickup, transport, and delivery of medical equipment, medications, blood products, class VIII supplies, medical personnel and military working dogs to, between, and from MTFs in the AO as required. (FM 4-02.2) (USAMEDDC&S)

Note: All pickup, transport, and delivery support missions must be conducted in a timely and efficient manner to conserve the fighting force and prevent the loss, damage, or spoilage of medical equipment, medications, blood products, and class VIII supplies.

No.	Scale	Measure
01	Yes/No	Unit medically evacuated and provided medical care en route for wounded, sick, and injured personnel to, between, and from MTFs without their medical condition declining due to the mode of evacuation or the time required to evacuate.
02	Yes/No	Unit provided medical care en route for wounded, sick, and injured personnel to, between, and from MTFs without their medical condition declining due to the time required to evacuate.
03	Yes/No	Unit executed a property exchange (for litters, blankets, and litter straps and patient movement items to prevent degraded, delayed, or disrupted medical evacuation operations.
04	Time	To refine response times for MEDEVAC, emergency pickup, transport, and delivery missions in the AO after receipt of a warning order.
05	Percent	Of difference between planned and actual requirements for medical evacuations, emergency pickup, transport, and delivery missions in the AO.
06	Percent	Of planned support capacity of MEDEVAC, emergency pickup, transport, and delivery missions in the AO.
07	Percent	Of wounded, sick, and injured personnel requiring patient movement items.
08	Number	Of wounded, sick, and injured personnel per day in AO evacuated from battlefield by evacuation precedence: urgent, urgent-surgery, priority, routine, and convenience.
09	Number	Of wounded, sick, and injured personnel per day evacuated from the AO.
10	Number	Of wounded, sick, and injured personnel evacuated to MTFs by nonstandard evacuation platforms in the AO.
11	Number	Of patients hospitalized in the AO that exceeds the theater evacuation policy.
12	Number	Of wounded, sick, and injured military working dogs requiring evacuation.

*ART 4.3.3 PROVIDE MEDICAL REGULATING SUPPORT

4-131. Medical regulating entails identifying the patients awaiting evacuation, locating the available hospital beds, and coordinating the transportation means for movement. The formal medical regulating system begins at Role 3 hospitals. (FM 4-02.2) (USAMEDDC&S)

No.	Scale	Measure
01	Yes/No	Unit regulated the evacuation of wounded, sick, and injured personnel to appropriate medical treatment facilities.
02	Yes/No	Unit required patient movement items available when and where needed.
03	Yes/No	Lack of patient movement items degraded, delayed, or disrupted medical treatments.
04	Percent	Of patients requiring medical regulating in the AO.
05	Percent	Of patients requiring medical regulating out of the AO.
06	Percent	Of patients requiring patient movement items.
07	Number	Of medical regulating requests coordinated through the joint patient movement requirements center, theater patient movement requirements center, or global patient movement requirements center.
08	Number	Of patients hospitalized in AO that exceeds the theater evacuation policy.
09	Number	Of neuropsychiatric patients requiring evacuation.

*ART 4.3.4 PROVIDE MEDICAL LOGISTICS

4-132. Provide class VIII medical materiel, medical equipment maintenance and repair, production of medical gases, medical contracting support, health facilities planning and management, and blood management for all U.S. forces. When serving as the area of operations (AO) single integrated medical logistics manager/theater lead agent for medical materiel, supply of medical materiel will be extended to other Services. (FM 4-02.1) (USAMEDDC&S)

Note: This task is related to ART 4.1.3.8 (Provide Medical Materiel and Repair Parts [Class VIII]).

No.	Scale	Measure
01	Yes/No	Medical logistics and blood management in the AO did not degrade, delay, or disrupt unit operations and medical treatment of wounded, injured, and sick Soldiers.
02	Yes/No	Distribution system got the right supplies to the right unit at the right time.
03	Yes/No	Unit maintained in-transit visibility of distribution systems and assets flowing through the system.
04	Time	To refine medical logistics operations for AO after receipt of warning order.
05	Time	To transship class VIII supplies and medical equipment after AO is assigned.
06	Time	To provide emergency shipment of class VIII materiel within AO.
07	Time	To transship class VIII supplies and medical equipment on AO activation.
08	Time	To set up medical supply transportation modes within theater.
09	Percent	Of planned medical logistics capacity completed in AO.
10	Percent	Of difference between planned medical logistics operations requirements and actual requirements in AO.
11	Percent	Of planned Class VIII materiel support delivered in AO.
12	Percent	Of Class VIII supplies that require replenishment per month.
13	Percent	Of required items of supply transiting the distribution system.
14	Percent	Of unit operations delayed, degraded, or modified due to lack of medical supplies, equipment, repair parts, or blood.
15	Percent	Of necessary or required Class VIII resources (meeting regulatory requirements) obtained from host nation or other agencies.
16	Percent	Of blood products in the system required to be disposed of in accordance with environmental considerations.
17	Number	Of instances that medical capability unavailable due to shortage or lack of Class VIII supplies.

*ART 4.3.4.1 PROVIDE MEDICAL EQUIPMENT MAINTENANCE AND REPAIR

4-133. Provide medical equipment maintenance and repair of deployed medical equipment. (FM 4-02.1) (USAMEDDC&S)

No.	Scale	Measure
01	Yes/No	Nonavailability of medical equipment did not degrade, delay, or disrupt unit operations and medical treatment of wounded, injured, and sick Soldiers.
02	Time	To provide emergency repair of medical equipment in AO.
03	Time	To refine medical equipment maintenance and repair operations program for AO after receipt of warning order.
04	Number	Of medical equipment (each type) with remote prognosis or diagnostics capability in the AO.

No.	Scale	Measure
05	Number	Of difference between planned medical equipment maintenance and repair operations requirements and actual requirements in AO.Of repairs completed using remote prognostics or diagnostics in AO.
06	Number	Of medical equipment (each type) in AO with original equipment manufacturer training required.
07	Number	Of medical equipment that requires repair per month.
08	Number	Of instances when medical capability was unavailable due to inability to repair medical equipment in AO.
09	Number	Of incidents involving medical equipment suspected of malfunctioning and causing further injury or death of a Soldier in the AO.
10	Number	Of incidents involving medical equipment that actually caused further injury or death of a Soldier in the AO.
11	Number	Of repairs with remote prognostic or diagnostic capability which could not be remedied in the AO.

*ART 4.3.4.2 PROVIDE OPTICAL FABRICATION

4-134. Provide manufacturing of single and multivision lens, protective mask inserts, and eyewear repair. (FM 4-02.1) (USAMEDDC&S)

No.	Scale	Measure
01	Yes/No	Nonavailability of single and multivision lens and eyewear for unit personnel did not degrade, delay, or disrupt unit operations.
02	Time	To refine optical fabrication and repair operations program for AO after receipt of warning order.
03	Time	To transship optometry equipment after AO is assigned.
04	Time	Required in advance to schedule optometry appointment in AO.
05	Percent	Of difference between planned optical fabrication and repair requirements and actual requirements in AO.
06	Percent	Of planned optometry support performed in AO.
07	Percent	Of personnel in AO requiring optometry facilities.
08	Number	Of optical devices in AO per quarter.

*ART 4.3.4.3 SUPPLY BLOOD AND BLOOD PRODUCTS

4-135. Supply whole blood and blood products, such as packed red blood cells, with a varying of blood groups and types. (FM 4-02.1) (USAMEDDC&S)

No.	Scale	Measure
01	Yes/No	Nonavailability of blood and blood products did not degrade, delay, or disrupt medical treatment of wounded, injured, and sick Soldiers.
02	Time	To refine blood management program for AO after receipt of warning order.
03	Time	To establish system for collection, storage, and distribution of blood products in AO.
04	Time	To initially coordinate blood requirements and distribution of blood and blood products to support medical treatment facilities in AO.
05	Percent	Of difference between planned blood management requirements and actual requirements in AO.
06	Percent	Of planned blood and blood products support attained in AO.
07	Percent	Of personnel in AO requiring blood or blood products per quarter.
08	Percent	Of required blood and blood products on hand in AO.

No.	Scale	Measure
09	Percent	Of daily blood reports submitted on time to joint blood program office.
10	Percent	Of blood products in AO required to be disposed of in accordance with applicable environmental regulations.
11	Number	Of units of required blood products per initial admission maintained in AO.
12	Time	To establish for collection, storage, and distribution of blood products in the AO.
13	Time	To initially coordinate blood requirements and distribution of blood and blood products with medical treatment facilities in the AO and with the Joint Blood Program Office.
14	Percent	Of planned blood management capacity produced in the AO.
15	Percent	Of required blood products on hand.
16	Percent	Of blood products in the system that must be disposed.
17	Percent	Of daily blood reports submitted to the Joint Blood Program office within the prescribed time limit.
18	Percent	Of blood products in the system required to be disposed of in accordance with environmental considerations.

+ SECTION IV – ART 4.4: CONDUCT INTERNMENT/RESETTLEMENT OPERATIONS

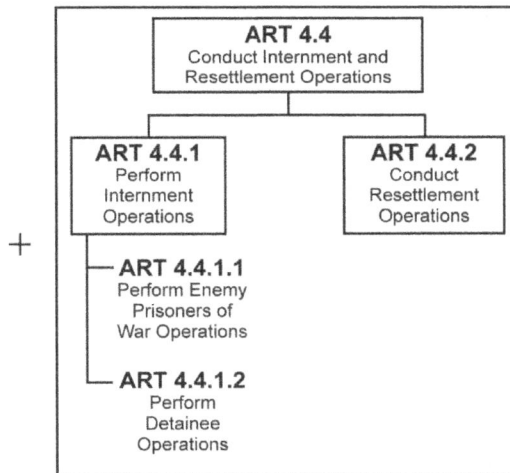

```
                        ART 4.4
                  Conduct Internment and
                  Resettlement Operations

         ART 4.4.1                    ART 4.4.2
          Perform                      Conduct
         Internment                  Resettlement
         Operations                   Operations

    — ART 4.4.1.1
      Perform Enemy
      Prisoners of
      War Operations

    — ART 4.4.1.2
      Perform
      Detainee
      Operations
```

4-136. Internment and resettlement (I/R) operations include the two major categories of internment operations and resettlement operations. These categories are further refined focusing on specific types of detainees and U.S. military prisoners while discriminating between civilian internees included as part of internment and those dislocated civilians (DCs) that may be retained as part of resettlement operations. Internment operations focus on all types of detainees and U.S. military prisoners. Although a part of internment operations, confinement refers to U.S military prisoners rather than internment (U.S. military prisoners are covered under ART 5.8.2). Resettlement operations focus on DCs, those civilians that are not part of the population of detainees. Military police conduct internment/resettlement operations to shelter, sustain, guard, protect, and account for populations (enemy prisoners of war (EPWs), civilian internees, DCs, or U.S. military prisoners) as a result of military or civil conflict, of natural or man-made disaster, or to facilitate criminal prosecution. Internment involves detaining a population or group that poses some level of threat to military operations. Resettlement involves quartering members of a population or group for their protection. These operations inherently control the movement and activities of their specific population for imperative reasons of security, safety, or intelligence gathering. (FM 3-39.40) (USAMPS)

> *Note:* ART 4.1.7.5.4 (Construct, Manage, and Maintain Bases and Installations) addresses the construction, management, and maintenance of bases and installations to include facilities such as those required for the internment of detainees.

ART 4.4.1 PERFORM INTERNMENT OPERATIONS

4-137. Activities performed by units when they are responsible for interning detainees, to include combatants, retained persons, and civilian internees. It ensures the safety and humane treatment of the incarcerated individuals, the maintenance of law and order within the facilities, as well as the safety of the guards and the surrounding civilian and military facilities and installations. All captured or detained personnel, regardless of status, shall be treated humanely and in accordance with the Detainee Treatment Act of 2005 and DODD 2310.1E, and no person in the custody or under the control of Department of Defense, regardless of nationality or physical location, shall be subject to torture or cruel, inhumane, or degrading treatment or punishment, in accordance with and as defined in U.S. law. (FM 3-39.40) (USAMPS)

No.	Scale	Measure
01	Yes/No	Internment activities did not prevent or seriously interfere with accomplishment of the unit's primary tactical mission.
02	Yes/No	Operations of U.S. forces performing internment activities observed international laws of war, U.S. laws and regulations, rules of engagement, and status-of-forces agreements.
03	Yes/No	Operations of U.S. forces performing internment activities observed local customs, mores, and taboos where possible.
04	Yes/No	Detainees received adequate amounts of appropriate food, water, clothing, housing, and medical care.
05	Yes/No	Order established and maintained in detention facilities.
06	Yes/No	Environmental regulations, laws, and considerations were taken into account during planning and present in procedures being followed.
07	Time	To refine plans for the collection and internment of detainees.
08	Time	To move detainees to their respective internment facilities from point of capture and detainee collection point.
09	Time	To construct camps to house detainees.
10	Time	To develop and enforce internment control measures, such as lists of controlled supplies and contraband.
11	Time	To forward intelligence information collected to unit intelligence staff.
12	Time	From interrogation to internment of civilian internees.
13	Time	To assign International Serial Number.
14	Percent	Of interned detainees requiring medical treatment provided in the AO.
15	Percent	Of interned detainees requiring medical treatment requiring evacuation out of the AO.
16	Percent	Of friendly force personnel in AO involved in maintaining internment facilities.
17	Percent	Of detained civilians released after interrogation.
18	Percent	Of unit sustainment requirements provided by detainees.
19	Percent	Of unit sustainment capabilities needed to support internment operations.
20	Number	Of civilian internees who can be resettled within a given time.
21	Number	Of civilian internees.

No.	Scale	Measure
22	Quantity	By type of supply needed to support internment operations.
23	Number	Of detainees interned.
24	Number	Of facilities to intern civilian internees established.
25	Number	Of internment facilities established.
26	Number	Of military working dogs needed for patrol or the detection of narcotics and explosives.
27	Ratio	Of guards to detainees or civilian internees.

ART 4.4.1.1 PERFORM ENEMY PRISONERS OF WAR OPERATIONS

4-138. Provide safe and humane treatment for enemy prisoners of war (EPW). This task includes the collection, screening, processing, transfer, internment, safeguarding, and release of EPW. (FM 3-39.40) (USAMPS)

No.	Scale	Measure
01	Yes/No	Conduct of internment activities did not prevent or seriously interfere with accomplishment of the unit's primary tactical mission.
02	Yes/No	Operations of U.S. forces conducting internment activities observed international laws of war, U.S. laws and regulations, rules of engagement, and U.S.-host-nation agreements.
03	Yes/No	Operations of U.S. forces conducting internment activities observed local customs, mores, and taboos where possible.
04	Yes/No	EPWs received necessary food, water, pay, clothing, housing, and medical care.
05	Yes/No	Order established and maintained in EPW facilities.
06	Time	To construct facilities to intern EPWs.
07	Time	To develop and enforce internment control measures, such as lists of controlled supplies and contraband.
08	Time	To conduct repatriation operations at the end of hostilities.
09	Percent	Of EPWs requiring medical treatment in the area of operations (AO).
10	Percent	Of EPWs requiring medical treatment requiring evacuation out of the AO.
11	Percent	Of unit sustainment capabilities needed to support internment operations.
12	Percent	Of unit sustainment capabilities needed to support internment operations.
13	Number	Of EPWs.
14	Number	Of facilities established to intern EPWs.
15	Number	Of military working dogs needed for patrol or the detection of narcotics and explosives.
16	Number	Of dollars paid to EPWs for work performed.
17	Number	Of retained persons used to support EPW operations.
18	Ratio	Of guards to EPWs.

ART 4.4.1.2 PERFORM DETAINEE OPERATIONS

4-139. Provide safe and humane treatment for civilian internees and combatant's not classified as enemy prisoners of war (EPWs) in accordance with the Geneva Conventions. This task includes the collection, screening, processing, transfer, internment, safeguarding, and release of EPWs. (FM 3-39.40) (USAMPS)

No.	Scale	Measure
01	Yes/No	Conduct of internment activities did not prevent or seriously interfere with accomplishment of the unit's primary tactical mission.
02	Yes/No	Operations of U.S. forces conducting internment activities observed international laws of war, U.S. laws and regulations, rules of engagement, and U.S.-host-nation agreements.
03	Yes/No	Operations of U.S. forces conducting internment activities observed local customs, mores, and taboos where possible.
04	Yes/No	Civilian internee's members of armed groups received necessary food, water, pay, clothing, housing, and medical care.
05	Yes/No	Order established and maintained in internment facilities.
06	Time	To construct facilities to intern detainees.
07	Time	To develop and enforce internment control measures, such as lists of controlled supplies and contraband.
08	Time	To conduct repatriation operations at the end of hostilities.
09	Percent	Of detainees requiring medical treatment in the area of operations (AO).
10	Percent	Of detainees requiring medical treatment requiring evacuation out of the AO.
11	Percent	Of unit sustainment capabilities needed to support internment operations.
12	Number	Of detainees.
13	Number	Of facilities established to intern detainees.
14	Number	Of military working dogs needed for patrol or the detection of narcotics and explosives.
15	Ratio	Of guards to detainees.

ART 4.4.2 CONDUCT RESETTLEMENT OPERATIONS

4-140. Provide support for resettlement of dislocated civilians to include their safety and security. This task includes controlling the movement of civilians, providing relief to human suffering, protecting civilians from combat operations or other threats, and establishing resettlement facilities in support of civil affairs operations. Establishing a facility requires collection, screening, processing, evacuation, housing, safeguarding, and releasing displaced civilians. ART 4.4.2 occurs in temporary and long-term facilities and points. (FM 3-39.40) (USAMPS)

No.	Scale	Measure
01	Yes/No	Resettlement activities did not prevent or seriously interfere with accomplishment of the unit's primary tactical mission.
02	Yes/No	Operations of U.S. forces performing resettlement activities observed international laws of war, U.S. laws and regulations, rules of engagement, and status-of-forces agreements.
03	Yes/No	Operations of U.S. forces performing resettlement activities observed local customs, mores, and taboos where possible.
04	Yes/No	Dislocated civilians received necessary food, water, pay, clothing, housing, and medical care.
05	Yes/No	Unit supervised incarceration process and transfer to prison facilities for dislocated civilians guilty of criminal activity.
06	Yes/No	Environmental considerations planning and procedures were present and being followed.
07	Yes/No	Unit determined the reliability of local markets to meet needs.
08	Yes/No	Unit provided emergency food, water, sanitation, shelter, and medicine.

No.	Scale	Measure
09	Yes/No	Unit coordinated with other donors and humanitarian agencies.
10	Yes/No	Unit established registration and health screening mechanisms.
11	Yes/No	Unit assessed prevalence for human immunodeficiency virus (HIV) and acquired immunodeficiency syndrome (AIDS).
12	Time	To refine plans for the movement, security and support of resettlement of dislocated civilians.
13	Time	To establish dislocated civilian collection points.
14	Time	To develop and enforce resettlement and population control measures, such as curfew, travel permits, and lists of controlled supplies and contraband.
15	Time	To forward intelligence information collected to unit intelligence staff.
16	Time	To move dislocated civilians to their respective resettlement facilities from their initial collection points.
17	Percent	Of dislocated civilians who received medical treatment.
18	Percent	Of friendly force personnel in area of operations involved in maintaining dislocated civilian resettlement facilities.
19	Percent	Of unit sustainment capabilities needed to support resettlement operations.
20	Number	And types of supplies needed to support resettlement operations.
21	Number	Of dislocated civilians.
22	Number	Of resettlement facilities and camps established.
23	Ratio	Of guards to dislocated civilians.

*SECTION V – ART 4.5: PROVIDE GENERAL ENGINEERING SUPPORT

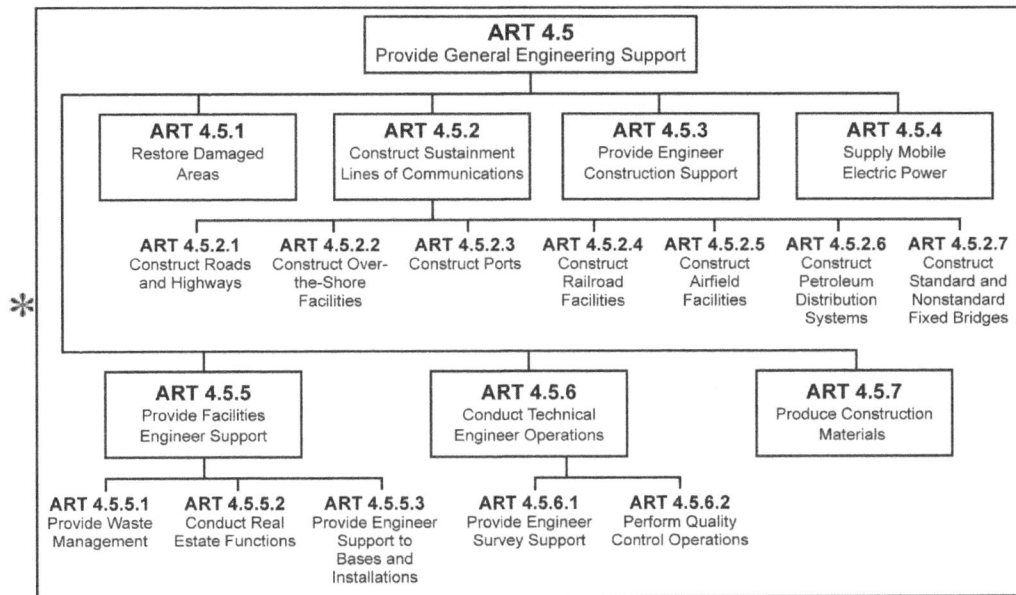

4-141. General engineering modifies, maintains, and protects the physical environment, including infrastructure, facilities, airfields, lines of communication and bases, protection of natural and cultural resources, terrain modification and repair, firefighting and aircraft crash rescue services, and selected explosive hazard activities. (FM 3-34.400) (USAES)

No.	Scale	Measure
01	Yes/No	Quantity or quality of general engineering support in the area of operations (AO) did not degrade or delay unit operations.
02	Yes/No	Environmental considerations planning and procedures were present and being followed.
03	Time	To assess and repair airfields for aviation operations throughout the AO.
04	Time	To construct and restore damaged utilities in AO.
05	Time	To refine general engineering support program for AO after receipt of warning order.
06	Time	To identify and marshal forces, equipment, and materials to construct or repair facilities in the AO.
07	Time	To construct or repair damaged lines of communications in AO to include aerial ports of debarkation and seaports of debarkation.
08	Time	Between arrival of building supplies and equipment and construction of sustainment facilities.
09	Time	To complete construction of sustaining base facilities in the AO.
10	Time	To begin building facilities (after final project approval and authorization).
11	Time	To have those bases identified in operation plan or operations order.
12	Time	To validate asset requests.
13	Time	To have assets at requesting location.
14	Percent	Of preventative maintenance activities completed based on the recommended activities from each systems owner's manual or generally accepted maintenance standards such as ASHRAE (American Society of Heating, Refrigerating, and Air Conditioning Engineers).
15	Percent	Of difference between planned general engineering support requirements and actual requirements in AO.
16	Percent	Of planned general engineering capability performed in AO.
17	Percent	Of supplies protected from the elements by weatherproof covers.
18	Percent	Of maintenance facilities protected from the elements.
19	Percent	Of overall cargo and equipment deliveries accommodated by sustaining base.
20	Percent	Of required installation throughput capacity available at execution.
21	Percent	Of tasks correctly assigned (correct engineers, location, and time).
22	Percent	Of general engineering support provided by host nation.
23	Number	In square meters of temporary facilities emplaced or constructed per day.
24	Number	In square meters of permanent facilities emplaced or constructed per day.

*ART 4.5.1 RESTORE DAMAGED AREAS

4-142. Inspect and repair surface and underwater facilities or restore terrain damaged by combat (such as clear rubble and restore electrical power), natural disaster, environmental accidents, or other causes. (FM 3-34) (USAES)

No.	Scale	Measure
01	Yes/No	Restoration completed per the schedule the operation order specifies.
02	Yes/No	Commander of the unit restoring a damaged area planned operations, established priorities, and allocated assets.
03	Yes/No	Restoration was per the standards the operation order specified.
04	Yes/No	Environmental considerations planning and procedures were present and being followed.
05	Time	To locate and stockpile repair materials.

No.	Scale	Measure
06	Time	To refine area damage control plan for the area of operations (AO) after receipt of warning order.
07	Time	To establish communications between the unit restoring the damaged areas and the unit or organization controlling the AO where the area to be restored is located.
08	Time	To perform engineer estimate to identify and prioritize potential tasks and determine required specialized support from engineers, explosive ordnance disposal, and other units; bill for needed materials; locate routes; identify replacement facilities; identify available host-nation assets; and perform other tasks as appropriate.
09	Time	To construct an expedient or alternate facility or bypass while restoration or repair is being competed if required.
10	Time	To repair facilities damaged by combat or natural disaster.
11	Time	To restore damaged utilities in AO.
12	Percent	Of difference between planned area damage control requirements and actual requirements in AO.
13	Percent	Of planned general engineering capability performed in AO.
14	Percent	Of facilities damaged beyond repair.
15	Percent	Of operations in AO degraded, delayed, or modified due to combat or natural disaster damage.
16	Percent	Of general restoration support provided by host nation.
17	Percent	Of restoration effort completed.
18	Number	Of Soldiers and civilians impacted by combat or natural disaster in the AO.
19	Number	And type of facilities damaged by combat or natural disaster in the AO.

*ART 4.5.2 CONSTRUCT SUSTAINMENT LINES OF COMMUNICATIONS

4-143. Construct and maintain land, water, and air routes that connect an operating military force with one or more bases of operations and along which supplies and reinforcements move. Sustainment lines of communications include main and alternate supply routes. (FM 3-34.400) (USAES)

No.	Scale	Measure
01	Yes/No	An inability to use lines of communications located in the area of operations (AO) did not degrade or delay unit operations.
02	Yes/No	Unit developed detailed plans for project.
03	Yes/No	Unit inspected project for quality control and ensured that the project was completed on time.
04	Yes/No	Environmental considerations planning and procedures were present and being followed.
05	Time	To reconnoiter to determine how the local environment will affect construction and if existing facilities or natural resources are available near the construction or maintenance site. This includes determining terrain features and their effect on the project; environmental considerations; problems involved in traveling to and from work site; what will be needed to keep the job site drained before, during, and after construction; and soil type and effort required to allow vehicle traffic and construction.
06	Time	To coordinate additional personnel, equipment, and critical items.
07	Time	To review available information in construction directive, intelligence reports, and site investigation to develop an operation plan or order.
08	Time	To plan the project including construction estimate, construction directive, and quality control.
09	Time	To prepare construction estimate including preparing a project activity list and a construction sequencing network; preparing materials, equipment, and personnel estimates; determining activity work rates; and preparing critical path.

No.	Scale	Measure
10	Time	To prepare construction directive and issue it to construction unit. Directive states the exact assignment, project location, and start and completion times; specifies additional personnel, equipment, and materials available; prioritizes the entire project; and specifies type and frequency of construction reports, time needed for special procurement, and coordination instructions with user agency.
11	Time	To monitor construction and perform quality assurance inspections.
12	Time	To perform final inspection of finished construction and turn it over to the user.
13	Time	To construct and maintain sustainment lines of communications.
14	Time	That scheduled arrivals in AO are delayed due to interruption in lines of communication (on average).
15	Percent	Of difference between planned and actual requirements for lines of communications construction and maintenance requirements.
16	Percent	Of force becoming casualties due to enemy action or accidents during construction or repair.
17	Percent	Increase in the carrying capability of the lines of communications due to construction or maintenance of the lines of communications.
18	Percent	Of planned general engineering capability performed in AO.
19	Percent	Of personnel in AO required for building and maintaining lines of communications.
20	Percent	Of general engineering support provided by host nation.
21	Number	Of lines of communications requiring construction or maintenance in AO.
22	Number	Of instances of delays in scheduled arrivals due to interruption of lines of communications.

*ART 4.5.2.1 CONSTRUCT ROADS AND HIGHWAYS

4-144. Determine road network requirements (such as classify roads in the area of operations (AO) per location, trafficability, and degree of permanence, traffic-bearing capabilities, and improvements needed). Maintain and repair existing roads (such as inspect and supervise, stockpile materials, keep road surfaces in usable and safe condition, prompt repair, correct basic cause of surface failure, and perform maintenance inspections) and construct new roads (such as route reconnaissance, site selection, surveys, drainage, construction, paving, and soil stabilization). (FM 3-34.400) (USAES)

Note: For construction of combat roads and trails to support maneuver of tactical forces, see ART 1.6.2.1 (Construct and Maintain Combat Roads and Trails).

No.	Scale	Measure
01	Yes/No	An inability to construct or maintain highways and roads in the AO within the time the construction directive specifies did not degrade or delay unit operations.
02	Yes/No	Unit developed detailed plans for project.
03	Yes/No	Unit inspected project for quality control and ensured that the road or highway construction project was completed on time.
04	Yes/No	Environmental considerations planning and procedures were present and being followed.
05	Time	To reconnoiter to determine how the local environment will affect roads and highway construction and determine if existing facilities or natural resources are available near the construction or maintenance site. This includes determining terrain features and their effect on the project; environmental considerations; problems involved in traveling to and from work site; what will be needed to keep the job site drained before, during, and after construction; and soil type and effort required to allow vehicle traffic and construction.
06	Time	To review available information in construction directive, intelligence reports, and site investigation to develop an operation plan or order.

No.	Scale	Measure
07	Time	To plan the road or highway project. This includes construction estimate, construction directive, and quality control.
08	Time	To prepare a road or highway construction estimate. This includes preparing a project activity list and a construction sequencing network; determining material, equipment, or personnel estimates; determining work rates for activities; and preparing critical path.
09	Time	To prepare road or highway construction directive and issue it to construction unit. Directive states the exact assignment, project location, and start and completion times; specifies additional personnel, equipment, and materials available; prioritizes the entire project; and specifies type and frequency of construction reports, time needed for special procurement, and coordination instructions with user agency.
10	Time	To coordinate additional personnel, equipment, and critical items.
11	Time	To monitor construction and perform quality assurance inspections.
12	Time	To perform final inspection of finished road or highway and turn it over to the user.
13	Time	For building and maintaining roads and highways.
14	Time	That scheduled arrivals in AO are delayed on the average due to interruptions in roads and highways by combat actions or natural disasters.
15	Percent	Of difference between planned and actual requirements for road and highway construction and maintenance requirements.
16	Percent	Of force becoming casualties due to enemy action or accidents during road and highway construction or repair.
17	Percent	Increase in the carrying capability of a road or highway due to construction or maintenance.
18	Percent	Of planned road or highway construction or maintenance capability realized in AO.
19	Percent	Of personnel in AO required for building and maintaining roads and highways.
20	Percent	Of road and highway construction and repair capability provided by host nation.
21	Percent	Of existing roads and highways in AO improved.
22	Percent	Of roads and highways in AO that can be used in their current condition by unit combat and tactical vehicles.
23	Percent	Of unit operations degraded, delayed, or modified in AO due to road or highway impassability.
24	Percent	Of roads and highways in the AO damaged by enemy fire or natural disaster.
25	Percent	Of roads upgraded from gravel to asphalt.
26	Number	And types of quarries required to support this task.
27	Number	Of roads and highways in the AO requiring construction or maintenance.
28	Number	Of roads and highways constructed or improved in the AO.
29	Number	Of kilometers of roads and highways constructed or improved in the AO within a given time.
30	Number	Of instances of delays in scheduled arrivals due to interruption of roads and highways in the AO by combat actions or natural disaster.
31	Number	Of instances in which troop movement or sustaining operations were prevented due to road or highway impassability.
32	Number	Of road or highway maintenance inspections performed per month in the AO.

*ART 4.5.2.2 CONSTRUCT OVER-THE-SHORE FACILITIES

4-145. Provide construction, repair, and maintenance support to logistics over-the-shore operations. Construct piers and causeways. Prepare and stabilize beaches. Construct access and egress routes. Provide access to marshalling and storage areas and adjoining logistics over-the-shore sites, which may also need constructing. Provide road and rail links to existing lines of communications. Construct utility systems and petroleum, oils, and lubricants storage and distribution systems. (FM 5-480) (USAES)

No.	Scale	Measure
01	Yes/No	An inability to construct or maintain over-the-shore facilities within the time the construction directive specifies did not degrade or delay unit operations.
02	Yes/No	Unit developed detailed plans for project.
03	Yes/No	Unit inspected over-the-shore facility projects for quality control and ensured that the project was completed on time.
04	Yes/No	Environmental considerations planning and procedures were present and being followed.
05	Time	To reconnoiter to determine how the local environment will affect over-the-shore facilities construction and determine if existing facilities or natural resources are available near the construction or maintenance site. This includes determining terrain features and their effect on the project; environmental considerations; problems involved in traveling to and from work site; what will be needed to keep the job site drained before, during, and after construction; and soil type and effort required to allow vehicle traffic and construction.
06	Time	To review available information in construction directive, intelligence reports, and site investigation to develop an operation plan or order.
07	Time	To plan the over-the-shore facility project. This includes construction estimate, construction directive, and quality control.
08	Time	To prepare a construction estimate for the over-the-shore facility. This includes preparing a project activity list and a construction sequencing network; preparing material, equipment, or personnel estimates; determining work rates for activities; and preparing critical path.
09	Time	To prepare construction directive for the over-the-shore facility and issue it to the construction unit. Directive states the exact assignment, project location, and start and completion times; specifies additional personnel, equipment, and materials available; prioritizes the entire project; and specifies type and frequency of construction reports, time needed for special procurement, and coordination instructions with user agency.
10	Time	To coordinate additional personnel, equipment, and critical items.
11	Time	To monitor construction and perform quality assurance inspections.
12	Time	To perform final inspection of finished over-the-shore facility and turn it over to the user.
13	Time	For building and maintaining over-the-shore facilities.
14	Time	That scheduled arrivals in the area of operations (AO) are delayed (on average) due to interruptions in the construction and maintenance of over-the-shore facilities by combat actions or natural disasters.
15	Percent	Of difference between planned and actual requirements for the construction or maintenance of over-the-shore facilities in the AO.
16	Percent	Of force that becomes casualties due to enemy action or accidents during the construction or maintenance of over-the-shore facilities.
17	Percent	Of increase in the throughput capability of a surface port due to the construction or maintenance of over-the-shore facilities.
18	Percent	Of planned construction or maintenance capability for over-the-shore facilities completed in AO.
19	Percent	Of personnel in AO required for building and maintaining over-the-shore facilities.
20	Percent	Of over-the-shore facilities in the AO damaged by enemy fire or natural disaster.
21	Percent	Of over-the-shore facilities in AO that can be used in their current condition.
22	Percent	Of unit operations degraded, delayed, or modified in AO due to an inability to use existing over-the-shore facilities.
23	Percent	Of over-the-shore construction or repair capability provided by host nation.
24	Percent	Of existing over-the-shore facilities improved in AO.
25	Percent	Of existing logistics over-the-shore facilities connected to existing roads, pipelines, or railroads.
26	Percent	Of supplies lost or destroyed during logistics over-the-shore offload activities in the AO.

No.	Scale	Measure
27	Number	Of over-the-shore facilities in the AO requiring construction or maintenance.
28	Number	And types of over-the-shore facilities such as piers, causeways, and marshaling or storage sites constructed or improved in the AO.
29	Number	And types of over-the-shore facilities in the AO damaged by enemy fire or natural disaster.
30	Number	Of meters of piers, causeways, and beaches constructed, improved, or stabilized in the AO within a given time.
31	Number	Of instances of delays in scheduled arrivals due to the destruction or damage of over-the-shore facilities in the AO by combat actions or natural disaster.
32	Number	Of instances that troop movement or sustaining operations were prevented due to an inability to use over-the-shore facilities.
33	Number	Of port facility inspections performed per month in the AO.

*ART 4.5.2.3 CONSTRUCT PORTS

4-146. Construct and rehabilitate ship unloading and cargo handling facilities in the area of operations (AO). Repair and maintenance can include emergency repair, major repair, rehabilitation of breakwater structures, and expedients. (FM 5-480) (USAES)

No.	Scale	Measure
01	Yes/No	An inability to construct or maintain seaport facilities within the time the construction directive specifies did not degrade or delay unit operations.
02	Yes/No	Unit developed detailed plans for project.
03	Yes/No	Unit inspected seaport projects for quality control and ensured the project was completed on time.
04	Yes/No	Environmental considerations planning and procedures were present and being followed.
05	Time	To review available information in construction directive, intelligence reports, and site investigation to develop an operation plan or order.
06	Time	To plan the seaport construction project. This includes construction estimate, construction directive, and quality control.
07	Time	To reconnoiter to determine how the local environment will affect the construction and maintenance of seaports. This includes determining if existing facilities or natural resources are available near the work site; terrain features and their effects on the project; environmental considerations; problems involved in traveling to and from work site; what will be needed to keep the job site drained before, during, and after construction; and soil type and effort required to allow vehicle traffic and construction.
08	Time	To coordinate additional personnel, equipment, and critical items.
09	Time	To monitor construction and perform quality assurance inspections.
10	Time	To perform final inspection of finished seaport and turn it over to the user.
11	Time	For building and maintaining port facilities.
12	Time	To prepare a construction estimate for the seaport. This includes preparing a project activity list and a construction sequencing network; preparing material, equipment, and personnel estimates; determining work rates for activities; and preparing critical path.
13	Time	To prepare construction directive for the seaport and issue it to the construction unit. This directive states the exact assignment, project location, and start and completion times; specifies additional personnel, equipment, and materials available; prioritizes the entire project; and specifies type and frequency of construction reports, time needed for special procurement, and coordination instructions with user agency.
14	Percent	Of difference between planned and actual requirements for the construction or maintenance of port facilities in the AO.
15	Percent	Of force that become casualties due to enemy action or accidents during the construction or maintenance of port facilities.

No.	Scale	Measure
16	Percent	Of increase in the throughput capability of a surface port due to the construction or maintenance of port facilities.
17	Percent	Of planned construction or maintenance capability for port facilities completed in AO.
18	Percent	Of personnel in AO required for building and maintaining port facilities.
19	Percent	Of port construction or repair capability provided by host nation.
20	Percent	Of existing port facilities improved in AO.
21	Percent	Of port facilities in AO that can be used in their current condition.
22	Percent	Of unit operations degraded, delayed, or modified in AO due to an inability to use existing port facilities.
23	Percent	Of port facilities in the AO damaged by enemy fire or natural disaster.
24	Percent	Of existing logistic port facilities connected to existing roads, pipelines, or railroads.
25	Percent	Of supplies lost or destroyed during logistic offload activities in the AO.
26	Percent	Of underwater habitat destroyed.
27	Number	Of port facilities in the AO requiring construction or maintenance.
28	Number	And types of port facilities—such as piers, causeways, cranes, and marshaling or storage sites—constructed or improved in the AO.
29	Number	And types of port facilities in the AO damaged by enemy fire or natural disaster.
30	Number	Of meters of breakwater, piers, and causeways, constructed or improved in the AO within a given time.
31	Number	Of port facility inspections performed per month in the AO.
32	Number	Of instances of delays in scheduled arrivals due to the destruction or damage of port facilities in the AO by combat actions or natural disaster.
33	Number	Of instances in which troop movement or sustaining operations were prevented due to an inability to use port facilities.

*ART 4.5.2.4 CONSTRUCT RAILROAD FACILITIES

4-147. Provide construction, major rehabilitation, and major repair of railroads. ART 4.1.7.2.4 includes all design, new construction, and modification of existing railroads to meet military traffic needs. (FM 3-34.400) (USAES)

No.	Scale	Measure
01	Yes/No	An inability to construct or maintain railroad facilities within the time the construction directive specifies did not degrade or delay unit operations.
02	Yes/No	Unit developed detailed plans for project.
03	Yes/No	Unit inspected railroad projects for quality control and ensured the project was completed on time.
04	Yes/No	Environmental considerations planning and procedures were present and being followed.
05	Time	To reconnoiter to determine how the local environment will affect the construction and maintenance of railroad facilities. This includes determining if existing facilities or natural resources are available near the work site; terrain features and their effect on the project; environmental considerations; problems involved in traveling to and from work site; what will be needed to keep the job site drained before, during, and after construction; and soil type and effort required to allow vehicle traffic and construction.
06	Time	To review available information in construction directive, intelligence reports, and site investigation to develop an operation plan or order.
07	Time	To plan the railroad facility construction project—includes construction estimate, construction directive, and quality control.
08	Time	To prepare a construction estimate for railroad facilities. This includes preparing a project activity list and a construction sequencing network; preparing material, equipment, or personnel estimates; determining work rates for activities; and preparing critical path.

No.	Scale	Measure
09	Time	To prepare construction directive for the railroad facility and issue it to the construction unit. This directive states the exact assignment, project location, and start and completion times; specifies additional personnel, equipment, and materials available; prioritizes the entire project; and specifies type and frequency of construction reports, time needed for special procurement, and coordination instructions with user agency.
10	Time	To coordinate additional personnel, equipment, and critical items.
11	Time	To monitor construction and perform quality assurance inspections.
12	Time	To perform final inspection of finished railroad facilities and turn it over to the user.
13	Time	To construct or maintain railroad facilities.
14	Time	Of delay in scheduled arrivals in the area of operations (AO) (on average) due to interruptions in the construction or maintenance of railroad facilities by combat actions or natural disasters.
15	Percent	Of difference between planned and actual requirements for the construction or maintenance of railroad facilities in the AO.
16	Percent	Of force that become casualties due to enemy action or accidents during the construction or maintenance of railroad facilities.
17	Percent	Of increase in the throughput capability of a railroad port due to the construction or maintenance of railroad facilities.
18	Percent	Of planned construction or maintenance capability for railroad facilities completed in AO.
19	Percent	Of personnel in AO required for building and maintaining railroad facilities.
20	Percent	Of railroad construction or repair capability provided by host nation.
21	Percent	Of existing railroad facilities improved in AO.
22	Percent	Of railroad facilities in AO that can be used in current condition.
23	Percent	Of unit operations degraded, delayed, or modified in AO due to an inability to use existing railroad facilities.
24	Percent	Of railroad facilities in the AO damaged by enemy fire or natural disaster.
25	Percent	Of existing logistic facilities connected to existing railroads.
26	Number	Of railroad facilities in the AO requiring construction or maintenance.
27	Number	And types of railroad facilities constructed or improved in the AO.
28	Number	And types of railroad facilities in the AO damaged by enemy fire or natural disaster.
29	Number	Of kilometers of rail lines constructed or improved in the AO within a given time.
30	Number	Of instances of delays in scheduled arrivals due to the destruction or damage of railroad facilities in the AO by combat actions or natural disaster.
31	Number	Of instances in which troop movement or sustaining operations were prevented due to an inability to use rail facilities.
32	Number	Of tons per day of supplies transported.
33	Number	Of railroad facility inspections performed per month in the AO.

*ART 4.5.2.5 CONSTRUCT AIRFIELD FACILITIES

4-148. Provide for planning military airfields; new airfield and heliport construction, expansion, and rehabilitation; and maintenance and repair of airfields and heliports in the area of operations (AO). (FM 5-430-00-2) (USAES)

No.	Scale	Measure
01	Yes/No	An inability to construct or expand airfield facilities within the time the construction directive specifies did not degrade or delay unit operations.
02	Yes/No	Airfield facilities supported the aircraft for which they were designed.
03	Yes/No	Unit inspected airfield or heliport projects for quality control.

No.	Scale	Measure
04	Yes/No	Airfield or helipad project completed on time.
05	Yes/No	Environmental considerations planning and procedures were present and being followed.
06	Time	To prepare engineer construction estimate that determines the effort to meet the requirements, assign operational and construction responsibilities, and determine additional personnel and equipment requirements.
07	Time	To prepare construction directive for the airfield or heliport and issue it to the construction unit. This directive states the exact assignment, project location, and start and completion times; specifies additional personnel, equipment, and materials available; prioritizes the entire project; and specifies type and frequency of construction reports, time needed for special procurement, and coordination instructions with user agency.
08	Time	To reconnoiter and evaluate the site for suitability and conditions, integrate environmental considerations, identify construction problems and possible courses of action, and update or revise the engineer estimate.
09	Time	To coordinate for and receive engineer assets to perform task.
10	Time	To monitor construction and perform quality assurance inspections.
11	Time	To perform location survey to establish permanent benchmarks for vertical control and well-marked points for horizontal control.
12	Time	To perform construction layout survey.
13	Time	To perform earthwork estimation that calculates the earthwork volume or quantity, determines final grade balancing of cuts and fills, and determines most economical haul of materials.
14	Time	To design a storm-drainage system.
15	Time	To conduct clearing, grubbing, and stripping operations.
16	Time	To conduct subgrade and base-course operations.
17	Time	To stabilize soil and provide dust control if required.
18	Time	To install surface matting if required.
19	Time	To conduct airfield marking operations.
20	Time	To install airfield lighting.
21	Time	To mark all obstructions.
22	Time	To prepare and submit status, progress, or completion reports to higher headquarters.
23	Time	To establish job site security.
24	Time	That scheduled arrivals in AO are delayed on the average due to interruptions in the construction, expansion, or maintenance of airfield or heliport facilities by combat actions or natural disasters.
25	Percent	Of difference between planned and actual requirements for the construction or maintenance of airfields or heliports and their associated support facilities in the AO.
26	Percent	Of force becoming casualties due to enemy action or accidents during the construction or maintenance of airfields or heliports.
27	Percent	Of increase in the throughput capability of an airfield or heliport due to the construction or maintenance of aviation support facilities.
28	Percent	Of planned airfield or heliport construction or maintenance capability completed.
29	Percent	Of personnel in AO required for building and maintaining airfields, heliports, and their associated aviation support facilities.
30	Percent	Of airfield or helipad construction or repair capability provided by host nation.
31	Percent	Of existing airfields or heliports and their associated aviation support facilities improved in AO.
32	Percent	Of existing airfields or heliports and their associated aviation support facilities in AO that can be used in their current condition.
33	Percent	Of unit operations degraded, delayed, or modified in AO due to an inability to use existing airfields or helipads.

No.	Scale	Measure
34	Percent	Of airfield or heliport and aviation support facilities in the AO damaged by enemy fire or natural disaster.
35	Percent	Of existing logistic facilities with access to existing airfields or heliports.
36	Number	Of airfields constructed, expanded, or rehabilitated in the AO.
37	Number	Of heliports constructed or rehabilitated in AO.
38	Number	Of airfields or heliports and aviation support facilities in the AO requiring construction or maintenance.
39	Number	And types of airfields or heliports and associated aviation support facilities in the AO damaged by enemy fire or natural disaster.
40	Number	Of meters of airfield runway constructed, improved, or repaired in the AO within a given time.
41	Number	Of instances of delays in scheduled arrivals due to the destruction or damage of airfields and helipads in the AO by combat actions or natural disaster.
42	Number	Of instances in which troop movement or sustaining operations were prevented due to an inability to use airfields or helipads and associated aviation support facilities.
43	Number	Of tons per day of supplies transported by aviation platforms in the AO.
44	Number	Of passengers per day transported by aviation in the AO.
45	Number	Of inspections of aviation support infrastructure completed per month in the AO.

*ART 4.5.2.6 CONSTRUCT PETROLEUM DISTRIBUTION SYSTEMS

4-149.   Provide construction, major rehabilitation, and major repair of water and petroleum pipelines and tank farms. ART 4.1.7.2.6 includes all design, new construction, and modification of existing pipelines and tank farms to meet military traffic needs. This task will always include environmental considerations. (FM 5-482) (USAES)

No.	Scale	Measure
01	Yes/No	An inability for building and maintaining pipelines and tank farms within the time the construction directive specifies did not degrade or delay unit operations.
02	Yes/No	Size of storage tanks or tank farm was adequate for anticipated demand.
03	Yes/No	Unit constructed the system per plans and specifications.
04	Yes/No	The system was operational and leak proof.
05	Yes/No	Buried pipes were below frost line and deep enough that vehicle movement did not damage system.
06	Yes/No	Environmental regulations or considerations observed during construction or repair of petroleum distribution systems.
07	Yes/No	Environmental considerations planning and procedures were present and being followed.
08	Time	To respond to reportable tasks.
09	Time	To reconnoiter to evaluate the site for suitability and conditions, identify construction problems and possible courses of action, and update or revise the engineer estimate.
10	Time	To prepare engineer construction estimate that determines the effort needed to meet the requirements, assign operational and construction responsibilities, and determine additional personnel and equipment requirements.
11	Time	To prepare construction directive for the pipeline or tank farm and issue it to the construction unit. This directive states the exact assignment, project location, and start and completion times; specifies additional personnel, equipment, and materials available; prioritizes the entire project; and specifies type and frequency of construction reports, time needed for special procurement, and coordination instructions with user agency.
12	Time	To coordinate for and receive engineer assets to perform task.
13	Time	To monitor construction and perform quality assurance inspections.

No.	Scale	Measure
14	Time	To perform location survey to establish permanent benchmarks for vertical control and well-marked points for horizontal control.
15	Time	To perform construction layout survey.
16	Time	To perform earthwork estimation that calculates the earthwork volume or quantity, determines final grade balancing of cuts and fills, and determines most economical haul of materials.
17	Time	To excavate trenches per construction or repair plans.
18	Time	To lay pipe, make connections, install valves, and perform pressure tests.
19	Time	To conduct backfill and tamping operations.
20	Time	To construct pipeline supports and bracing for locations where the pipeline must be above ground.
21	Time	To construct pipeline suspension bridges for locations where the pipeline must be above ground.
22	Time	To install pipeline pumping stations.
23	Time	To ensure water distribution system functions properly.
24	Time	To install storage tanks or liquid storage facilities.
25	Time	To install underwater pipeline.
26	Percent	Of difference between planned and actual requirements for water and petroleum pipelines and tank farms in an area of operations (AO).
27	Percent	Of planned construction or repair program completed.
28	Number	Of pipelines constructed, expanded, or rehabilitated in the AO.
29	Number	Of tank farms constructed or rehabilitated in AO.
30	Number	Of kilometers of pipelines and tank farms in the AO required to support unit operations.
31	Number	Of pipelines and tank farms and associated support facilities in the AO damaged by enemy fire or natural disaster.
32	Number	Of meters of pipeline constructed, improved, or repaired in the AO within a given time.
33	Number	Of liters or metric tons of water or bulk petroleum products currently stored in tank farms in the AO.
34	Number	Of instances in which troop movement or sustaining operations were prevented due to lack of water or bulk petroleum products.
35	Number	Of liters or metric tons of supplies transported per day by pipelines in the AO.
36	Number	Of pipeline, tank, or pumping station inspections performed per month in the AO.
37	Number	Of casualties because of accidents during the construction, repair, or maintenance of pipelines and tank farms.
38	Number	Of incidents that result in the release of hazardous material because of accidents or spills resulting from combat actions.
39	Number	Of liters or metric tons of hazardous material released.
40	Number	Of water wells drilled in the AO.
41	Number	Of leaks per day.

*ART 4.5.2.7 CONSTRUCT BRIDGES

4-150.   Provide construction and repair of bridges. ART 4.5.2.7 includes all design, new construction, and modification of existing bridges to meet military traffic needs. (FM 3-34.343) (USAES)

No.	Scale	Measure
01	Yes/No	An inability for building and maintaining standard and nonstandard fixed bridges within the time the construction directive specifies did not degrade or delay unit operations.
02	Yes/No	Fixed bridges supported the traffic loads for which they were designed.
03	Yes/No	Unit inspected fixed bridge projects for quality control.

No.	Scale	Measure
04	Yes/No	Bridge construction projects completed on time.
05	Yes/No	Environmental considerations planning and procedures were present and being followed.
06	Time	To prepare engineer construction estimate that determines the effort needed to meet gap crossing requirements, assign operational and construction responsibilities, and determine additional personnel and equipment requirements.
07	Time	To reconnoiter to evaluate proposed bridge site for suitability and conditions, identify construction problems and possible courses of action, and update or revise the engineer estimate.
08	Time	To adapt standard fixed bridge construction designs to specific situation.
09	Time	To coordinate for and receive engineer assets to perform task.
10	Time	To prepare construction directive or operation order to construct or maintain a fixed bridge. This directive states exact assignment, project location, and start and completion times; specifies additional personnel, equipment, and materials available; prioritizes projects; and specifies type and frequency of construction reports, time needed for special procurement, and required coordination with user agency.
11	Time	To perform location survey to establish permanent benchmarks for vertical control and well-marked points for horizontal control.
12	Time	To monitor construction and perform quality assurance inspections.
13	Time	To perform construction layout survey.
14	Percent	Of difference between planned and actual requirements for fixed bridge construction or maintenance in the area of operations (AO).
15	Percent	Of force becoming casualties due to enemy actions or accidents while building or maintaining fixed bridges.
16	Percent	Of increase in the throughput capability of a line of communications or main supply route because of the replacement of tactical assault bridges with fixed bridges.
17	Percent	Of planned fixed bridge construction or maintenance capability completed in AO.
18	Percent	Of personnel in AO required for building and maintaining fixed bridges.
19	Percent	Of fixed bridge construction or repair capability provided by host nation.
20	Percent	Of existing fixed bridges repaired or improved in AO.
21	Percent	Of existing fixed bridges in AO that can be used in current condition.
22	Percent	Of unit operations degraded, delayed, or modified in AO due to an inability to use existing fixed bridges.
23	Percent	Of fixed bridges in the AO damaged by enemy fire or natural disaster.
24	Percent	Of tactical assault bridging in the AO replaced by fixed bridges.
25	Number	Of fixed bridge kits available for employment in the AO.
26	Number	Of fixed bridges constructed, improved, or rehabilitated in the AO.
27	Number	Of existing fixed bridges in the AO requiring maintenance or repair.
28	Number	And types of fixed bridges in the AO damaged by enemy fire or natural disaster.
29	Number	Of meters of gaps crossed by fixed bridges constructed, improved, or repaired in the AO within a given time.
30	Number	Of instances of delays in scheduled arrivals due to the destruction or damage of fixed bridges in the AO by combat actions or natural disaster.
31	Number	Of instances in which troop movement or sustaining operations were prevented due to an inability to use a fixed bridge.
32	Number	Of tons per day of bridge construction supplies required in the AO.
33	Number	Of inspections of fixed bridges performed per month in the AO.

*ART 4.5.3 PROVIDE ENGINEER CONSTRUCTION SUPPORT

4-151. Construct or renovate facilities. ART 4.1.7.3 includes construction of orderly, distribution, and storage facilities; construction or renovation of fixed facilities; construction, repair, maintenance, and operation of permanent and semipermanent water facilities, such as wells for water; and dismantling fortifications. (FM 3-34) (USAES)

No.	Scale	Measure
01	Yes/No	Engineer construction support provided the supported unit in accomplishing its mission in the area of operations (AO).
02	Yes/No	Environmental considerations planning and procedures were present and being followed.
03	Time	To refine general engineering service program for the AO after receipt of warning order.
04	Time	To prepare engineer construction estimate that determines the effort needed to meet the requirements, assign operational and construction responsibilities, and determine additional personnel and equipment requirements.
05	Time	To establish demobilization camps.
06	Time	To reconnoiter to evaluate the site for suitability and conditions, identify construction problems and possible courses of action, and update or revise the engineer estimate.
07	Time	To prepare construction directive for the construction or renovation of fixed facilities and issue it to the construction unit. This directive states the exact assignment, project location, and start and completion times; specifies additional personnel, equipment, and materials available; prioritizes the entire project; and specifies type and frequency of construction reports, time needed for special procurement, and coordination instructions with user agency.
08	Time	To coordinate for and receive engineer assets to perform task.
09	Time	To monitor construction and perform quality assurance inspections.
10	Time	To perform location survey to establish permanent benchmarks for vertical control and well-marked points for horizontal control.
11	Time	To perform construction layout survey.
12	Time	To develop concept of engineer construction support after receipt of warning order.
13	Time	Between arrival of building supplies and equipment and construction of sustainment facilities in the AO.
14	Time	To decommission demobilization camps.
15	Percent	Of difference between planned construction report requirements and actual requirements in the AO.
16	Percent	Of planned engineer construction support capability realized in the AO.
17	Percent	Of engineer construction support provided by host nation.
18	Percent	Of engineer construction projects damaged by combat action or natural disaster.
19	Percent	Of preventative maintenance activities completed based on the recommended activities from each systems owner's manual or generally accepted maintenance standards such as the American Society of Heating, Refrigerating, and Air Conditioning Engineers.
20	Percent	Of permanent facilities emplaced or constructed.
21	Number	Of water wells drilled in AO.
22	Number	Of storage facilities constructed in AO.
23	Number	Of pipelines constructed in AO.
24	Number	Of fixed facilities constructed or renovated in AO.
25	Number	Of square meters of temporary storage facilities emplaced or constructed per day.
26	Number	Of facilities holding enemy prisoners of war per current international conventions and standards.

*ART 4.5.4 SUPPLY MOBILE ELECTRIC POWER

4-152.   Supply electric power generation and distribution to military units through mobile generation and a tactical distribution grid system. ART 4.1.7.4 includes power production, power distribution, and power management. (FM 3-34.480) (USAES)

Note: ART 7.3.5 (Support Economic and Infrastructure Development) addresses providing electrical power to nonmilitary organizations.

No.	Scale	Measure
01	Yes/No	Mobile electric power met users' needs.
02	Yes/No	Electrical system constructed and power generation and regulation devices installed per operation order specifications and within the time stated in the directive.
03	Yes/No	Mobile electric power systems adhered to local and national electric code specifications.
04	Yes/No	Environmental considerations planning and procedures were present and being followed.
05	Time	To refine mobile electric power service program for the area of operations (AO) after receipt of warning order.
06	Time	To prepare engineer construction estimate that determines the effort needed to meet the requirements, assign operational and construction responsibilities, and determine additional personnel and equipment requirements.
07	Time	To reconnoiter to evaluate the site for suitability and conditions, identify construction problems and possible courses of action, and update or revise the engineer estimate.
08	Time	To prepare construction directive for a facility to house mobile electric power generators, power grid substations or transformers, and electric power lines and issue it to the construction unit. This directive states the exact assignment, project location, and start and completion times; specifies additional personnel, equipment, and materials available; prioritizes the entire project; and specifies type and frequency of construction reports, time needed for special procurement, and coordination instructions with user agency.
09	Time	To coordinate for and receive engineer assets to perform task.
10	Time	To monitor construction and perform quality assurance inspections.
11	Time	To perform location survey to establish permanent benchmarks for vertical control and well-marked points for horizontal control.
12	Time	To perform construction layout survey.
13	Time	To verify accuracy of construction plans and specifications to include ensuring the bill of materials includes all required materials to complete construction.
14	Time	To rough in the structure to accommodate electrical service.
15	Time	To install cable and conduit.
16	Time	To complete installation by connecting joints; grounding system at service entrance; connecting bonding circuit; attaching wire to switch terminal, ceiling and wall outlets, fixtures, and devices; and connecting service entrance cable and fusing or circuit breaker panels.
17	Time	To test and repair the system.
18	Percent	Of difference between planned mobile electric power requirements and actual requirements in the AO.
19	Percent	Of planned mobile electric power generation and distribution capability gained in the AO.
20	Percent	Of units in AO that require mobile generation power.
21	Percent	Of electrical power in AO generated by mobile generation units and distributed through a tactical grid.
22	Percent	Of electrical power in AO provided by existing power generation facilities and distributed through a commercial grid.
23	Percent	Of power generation systems operational.

No.	Scale	Measure
24	Percent	Of required kilowatt hours provided by mobile generation units.
25	Percent	Of power provided in the AO that meets voltage, frequency, and amperage standards.
26	Number	And types of mobile generation systems required meeting user requirements.
27	Number	Of kilometers of electric power lines that form the tactical grid in the AO.
28	Number	Of substations and transformers required by the tactical grid.

*ART 4.5.5 PROVIDE FACILITIES ENGINEER SUPPORT

4-153. Sustain military forces in the theater by providing waste management; acquiring, maintaining, and disposing real property; providing firefighting support; and constructing, managing, and maintaining bases and installations. ART 4.1.7.5 includes the design of facilities. (FM 3-34.400) (USAES)

No.	Scale	Measure
01	Yes/No	The provision of facilities engineering support did not cause the abandonment, modification, or delay in execution of the unit's chosen course of action.
02	Yes/No	Facility systems constructed to plan specifications within allotted time.
03	Yes/No	Facility engineering systems safeguarded the health of Soldiers and noncombatants in the area of operations (AO).
04	Yes/No	Environmental considerations planning and procedures were present and being followed.
05	Time	To refine facilities engineering support program in the AO.
06	Time	To reconnoiter to evaluate the site for suitability and conditions, identify construction problems and possible courses of action, and update or revise the engineer estimate.
07	Time	To prepare construction directive for facility engineering support facilities and issue it to the construction unit. This directive states the exact assignment, project location, and start and completion times; specifies additional personnel, equipment, and materials available; prioritizes the entire project; and specifies type and frequency of construction reports, time needed for special procurement, and coordination instructions with user agency.
08	Time	To coordinate for and receive engineer assets to perform facility engineer task.
09	Time	To monitor construction and perform quality assurance inspections.
10	Time	To verify accuracy of construction plans and specifications to include ensuring the bill of materials includes all required materials to complete construction.
11	Time	To design new construction requirements.
12	Time	To manage and administer facilities engineering program in the AO.
13	Time	To develop guidance for acquiring, managing, and disposing real estate in the AO.
14	Time	To complete environmental baseline surveys on real estate being considered for acquisition or use by U.S. forces.
15	Time	To inventory installed and personal property located on installations.
16	Time	To maintain facility engineering records.
17	Time	To complete legal and environmental reviews of real estate transactions in the AO.
18	Time	To develop a system for submitting real estate claims in the AO.
19	Percent	Of real estate required to conduct and support unit operations acquired.
20	Percent	Of required real estate and facilities provided by host nation.
21	Percent	Of difference between planned and actual requirements for facilities engineering in the AO.
22	Percent	Of planned facilities acquired or constructed in the AO.
23	Percent	Of required facilities provided by host nation.
24	Percent	Of existing facilities modernized in the AO.
25	Percent	Of existing facilities that can be used in current condition.
26	Percent	Of existing facilities damaged by combat actions or natural disaster.

No.	Scale	Measure
27	Percent	Of nonbattle injuries and disease in the AO attributable to inadequate facility engineer support.
28	Percent	Of each utility's (water, wastewater, power, and natural gas) reliability factor in each base camp.
29	Number	Of kilograms or liters and types of waste, refuse, and hazardous materials produced per day in the AO.
30	Number	And types of facilities constructed or acquired to support unit operations.

*ART 4.5.5.1 PROVIDE WASTE MANAGEMENT

4-154. Operate, maintain, or upgrade existing utilities. Construct, operate, and maintain new utilities systems for the purpose of waste management. ART 4.1.7.5.1 includes wastewater collection and treatment systems, refuse collection, and disposal. Special consideration is given to disposing hazardous waste. (FM 3-34.400) (USAES)

No.	Scale	Measure
01	Yes/No	The supported unit was not delayed, disrupted, or prevented from accomplishing its mission. Soldiers and civilians residing in the area of operations (AO) were not placed at risk of injury or disease because of the improper collection, treatment, and disposal of sewage, refuse, and hazardous waste.
02	Yes/No	Sewer system constructed to plan specifications within allotted time.
03	Yes/No	Waste management facilities safeguarded the health of Soldiers and noncombatants in the AO.
04	Yes/No	Sewage in the AO stabilized so that it did not overload the disposal media in lake, stream, or drain field.
05	Yes/No	Environmental considerations planning and procedures were present and being followed.
06	Yes/No	Unit considered or included the Overseas Environmental Baseline Guidance Document or final governing standards in construction.
07	Yes/No	Unit used transportation assets to backhaul waste for disposal.
08	Time	To refine waste management program after receipt of warning order.
09	Time	To prepare engineer construction estimate that determines the effort needed to meet the waste management requirements in the AO, assign operational and construction responsibilities, and determine additional personnel and equipment requirements.
10	Time	To reconnoiter to evaluate the site for suitability and conditions, identify construction problems and possible courses of action, and update or revise the engineer estimate.
11	Time	To prepare construction directive for a sewage or hazardous treatment facility and issue it to the construction unit. This directive states the exact assignment, project location, and start and completion times; specifies additional personnel, equipment, and materials available; prioritizes the entire project; and specifies type and frequency of construction reports, time needed for special procurement, and coordination instructions with user agency.
12	Time	To coordinate for and receive engineer assets to perform task.
13	Time	To monitor construction and perform quality assurance inspections.
14	Time	To perform location survey to establish permanent benchmarks for vertical control and well-marked points for horizontal control.
15	Time	To perform construction layout survey.
16	Time	To install sheeting and bracing on sewer trenches.
17	Time	To verify accuracy of construction plans and specifications to include ensuring the bill of materials includes all required materials to complete construction.
18	Time	To install or repair plumbing and sewage pipes in facilities.
19	Time	To install lavatories or sinks, water closet, and urinals.

No.	Scale	Measure
20	Percent	Of sewage or hazardous material produced per day in the AO that are disposed of in a manner that safeguards the health of Soldiers and noncombatants and the environment.
21	Percent	Of difference between planned and actual requirements for waste management in the AO.
22	Percent	Of planned waste management capabilities completed in AO.
23	Percent	Of required waste management capabilities provided by host nation.
24	Percent	Of existing waste management capabilities modernized in the AO.
25	Percent	Of existing waste management capabilities that can be used in their current condition.
26	Percent	Of existing waste management capabilities in AO damaged by combat actions or natural disaster.
27	Percent	Of nonbattle injuries and disease in the AO attributable to inadequate waste management.
28	Percent	Of waste reduced through recycling.
29	Number	Of kilograms and types of hazardous material produced per day in the AO.
30	Number	Of liters of sanitary sewage produced per day in the AO.
31	Number	Of liters of industrial sewage produced per day in the AO.
32	Number	Of liters of storm sewage produced by individual storms in the AO.
33	Number	Of liters of ground water that enters the sewage system per day.
34	Number	Of cesspools constructed in the AO.
35	Number	And capacity of septic tanks constructed in the AO.
36	Number	And capacity of sewage treatment plants constructed in the AO.
37	Number	Of kilograms per day of disinfectants added to chemically treat sewage in the AO.
38	Number	And capacity of sewage lagoons constructed in the AO.
39	Number	Of meters of sewer systems constructed per day in the AO.

*ART 4.5.5.2 CONDUCT REAL ESTATE FUNCTIONS

4-155. Furnish technical real estate guidance and perform additional real estate duties. (FM 3-34) (USAES)

No.	Scale	Measure
01	Yes/No	The acquisition, management, and disposition of real estate in the area of operations (AO) did not cause the abandonment, modification, or delay in execution of the unit's chosen course of action.
02	Yes/No	Environmental considerations planning and procedures were present and being followed.
03	Time	To acquire, manage, administer, and dispose of real estate in the AO.
04	Time	To develop guidance for the acquisition, management, and disposition of real estate in the AO.
05	Time	To complete environmental baseline surveys on real estate being considered for acquisition or use by U.S. forces.
06	Time	To inventory installed and personal property located on installations.
07	Time	To maintain real estate records.
08	Time	To complete legal and environmental reviews of real estate transactions in the AO.
09	Time	To develop a system for submitting real estate claims in the AO.
10	Percent	Of real estate required to conduct and support unit operations acquired.
11	Percent	Of required real estate and facilities provided by host nation.
12	Number	Of real estate teams operational in the AO.
13	Number	Of hectares of real estate acquired, managed, or disposed of in an AO within a given time.
14	Number	Of property claims submitted in the AO within a given time.
15	Cost	Of real estate restoration activities necessary to dispose of real estate in the AO.

No.	Scale	Measure
16	Cost	Of claims for damages to real estate in the AO.
17	Cost	Of rent for the use of real estate in the AO.

*ART 4.5.5.3 PROVIDE ENGINEER SUPPORT TO BASES AND INSTALLATIONS

4-156. Manage and maintain or upgrade existing facilities and utilities. Construct, manage, and maintain new facilities and utility systems for bases and installations. (FM 3-34.400) (USAES)

No.	Scale	Measure
01	Yes/No	The construction, management, and maintenance of bases and installations in the area of operations (AO) contributed toward unit mission accomplishment.
02	Yes/No	Construction requirements determined per existing doctrine and regulations.
03	Yes/No	Environmental considerations planning and procedures were present and being followed.
04	Time	To plan maintenance and repair of fixed facilities.
05	Time	To determine potential requirements for repairing damage resulting from combat actions and natural disasters.
06	Time	To plan the operation and maintenance or upgrade of existing utilities, such as electrical generating and distribution systems, waste water collection and treatment systems, and other special utilities systems including cooling and refrigeration, compressed air, and heating systems.
07	Time	To plan and perform fire prevention and protection programs in the AO.
08	Time	To plan refuse collection and disposal.
09	Percent	Of maintenance and repair of facilities that can be performed by unit self-help teams.
10	Percent	Of facilities in the AO meeting initial and temporary standards.
11	Number	Of incinerators and landfills operating in the AO.
12	Number	Of metric tons of refuse per day that are recycled in the AO.
13	Number	Of metric tons of hazardous waste per day disposed of per appropriate laws and regulations.
14	Number	And types of ports, bases, and installations in the AO.
15	Number	Of Soldiers supported by bases and installations in the AO.
16	Number	And types of engineer units used for building and maintaining bases and installations located in the AO.
17	Cost	Of base and installation construction, management, and maintenance.

*ART 4.5.6 CONDUCT TECHNICAL ENGINEER OPERATIONS

4-157. Provide technical support to engineering services in the area of operations. ART 4.5.6 includes quality assurance and control inspections, materials testing, surveying. Technical Engineering provides oversight to the regulatory construction, safety and environmental standards. Quality control for construction projects and facilities upgrades. Quality control includes planning, designing, and monitoring the construction process to achieve a desired end result. (FM 5-412) (USAES)

No.	Scale	Measure
01	Yes/No	Unit operations were not delayed by an inability to provide technical engineering support.
02	Yes/No	Technical engineering enhanced the supported unit's ability to accomplish its mission.
03	Time	To develop plans and designs in support of requirements after receipt of warning order.
04	Time	Required to develop or update plans to support operations after receipt of warning order.
05	Percent	Of planned technical engineering support delivered in the area of operations (AO).
06	Percent	OF requested technical engineering support delivered in the AO.
07	Percent	Of technical engineering support available in the AO compared to requirements.

No.	Scale	Measure
08	Percent	Of operations degraded, delayed, or modified due to delays in providing technical engineer support.
09	Percent	Of accuracy of survey operation.
10	Number	Of facilities inspected in the AO to support operations.
11	Number	Of projects requiring technical engineer support.
12	Number	Of technical engineer inspections in the AO to support operations.

*ART 4.5.6.1 PROVIDE ENGINEER SURVEY SUPPORT

4-158. Use mechanical or electronic systems to determine dimensional relationships—such as locations, horizontal distances, elevations, directions, and angles—on the earth's surface. ART 4.5.6.1 includes airfield surveys and obstacle evaluation assessments within airfield operational surfaces. (FM 3-34.400) (USAES)

No.	Scale	Measure
01	Yes/No	Unit completed survey by time specified in order.
02	Yes/No	Survey order detailed the priorities and accuracies required by the requesting unit.
03	Time	To plan survey operation to include traverse, triangulation, and three-point resection.
04	Time	To prepare for survey operation.
05	Time	To execute survey operation.
06	Time	To enter a new survey control point into the database.
07	Time	To update survey control point in the database.
08	Time	From requesting information to providing desired survey information to units.
09	Percent	Of accuracy of survey operation.
10	Percent	Of accuracy of survey control available.
11	Percent	Of positioning and azimuth determining systems operational.
12	Number	Of positioning and azimuth determining systems available.

*ART 4.5.6.2 PERFORM QUALITY CONTROL OPERATIONS

4-159. Quality control and quality assurance are performed by the government to determine that requirements and specifications are met. Quality control and quality assurance ensures products meet quality and safety standards, before acceptance from contractors, as well as transfer between government agencies or issued to users. Quality control is imperative to ensure that materials meet the critical construction tolerances/standards. (FM 5-42) (USAES)

Note: Quality control includes planning, designing and monitoring the construction process to achieve a desired end result.

No.	Scale	Measure
01	Yes/No	Unit inspected project for compliance with appropriate standards and completion in accordance with the established timeline.
02	Yes/No	Unit quality control and quality assurance program did not cause delay of project.
03	Time	To monitor construction and conduct quality assurance inspections.
04	Time	To perform final inspection of finished project and turn over to the user.
05	Time	To coordinate inspection and quality surveillance of contracted project specifications.
06	Time	To monitor contractor performance.
07	Time	To provide technical advice and assistance to staffs, subordinate units, and contracting officers' representatives.

No.	Scale	Measure
08	Percent	Of construction material analysis completed to ensure compliance to project specifications.
09	Percent	Of soils analysis completed to ensure compliance to construction project specifications.

*ART 4.5.7 PRODUCE CONSTRUCTION MATERIALS

4-160. Produce limited types of construction materials to support military operations. (FM 3-34.400) (USAES)

No.	Scale	Measure
01	Yes/No	Procedures to provide construction material production support did not negatively affect supported unit's ability to perform its missions.
02	Yes/No	Environmental considerations planning and procedures were present and being followed.
03	Time	To reconnoiter to evaluate the site for suitability and conditions, identify problems and possible courses of action.
04	Time	Required to set up quarry operations.
05	Time	Required to set up asphalt production operations.
06	Percent	Of required production rate of concrete delivered in the area of operations (AO).
07	Percent	Of required production rate of mineral product delivered in the AO.
08	Percent	Of required production rate of asphalt delivered in the AO.
09	Number	Of cubic yards of concrete produce per day in the AO.
10	Number	Of tons of asphalt produce per day in the AO.
11	Number	Of tons of mineral products produce per day in the AO.

This page intentionally left blank.

Chapter 5
ART 5.0: Conduct Command and Control

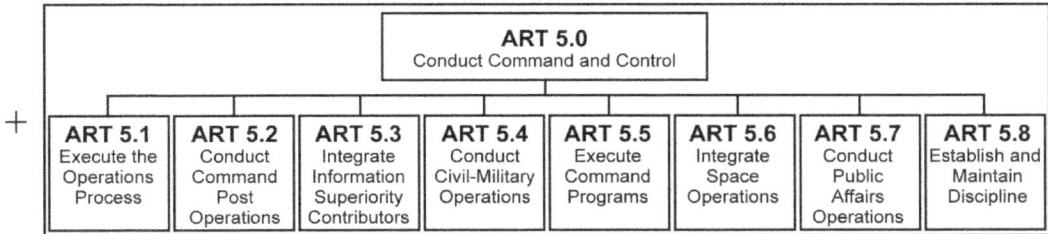

ART 5.0 Conduct Command and Control							
ART 5.1 Execute the Operations Process	**ART 5.2** Conduct Command Post Operations	**ART 5.3** Integrate Information Superiority Contributors	**ART 5.4** Conduct Civil-Military Operations	**ART 5.5** Execute Command Programs	**ART 5.6** Integrate Space Operations	**ART 5.7** Conduct Public Affairs Operations	**ART 5.8** Establish and Maintain Discipline

The *command and control warfighting function* is the related tasks and systems that support commanders in exercising authority and direction. It includes those tasks associated with acquiring friendly information, managing relevant information, and directing and leading subordinates. Through command and control, commanders integrate all warfighting functions to accomplish the mission. (FM 3-0) (USACAC)

+ Command and control is the exercise of authority and direction by a properly designated commander over assigned and attached forces in the accomplishment of a mission. Commanders perform command and control functions through a command and control system. (FM 6-0) (USACAC)

SECTION I – ART 5.1: EXECUTE THE OPERATIONS PROCESS

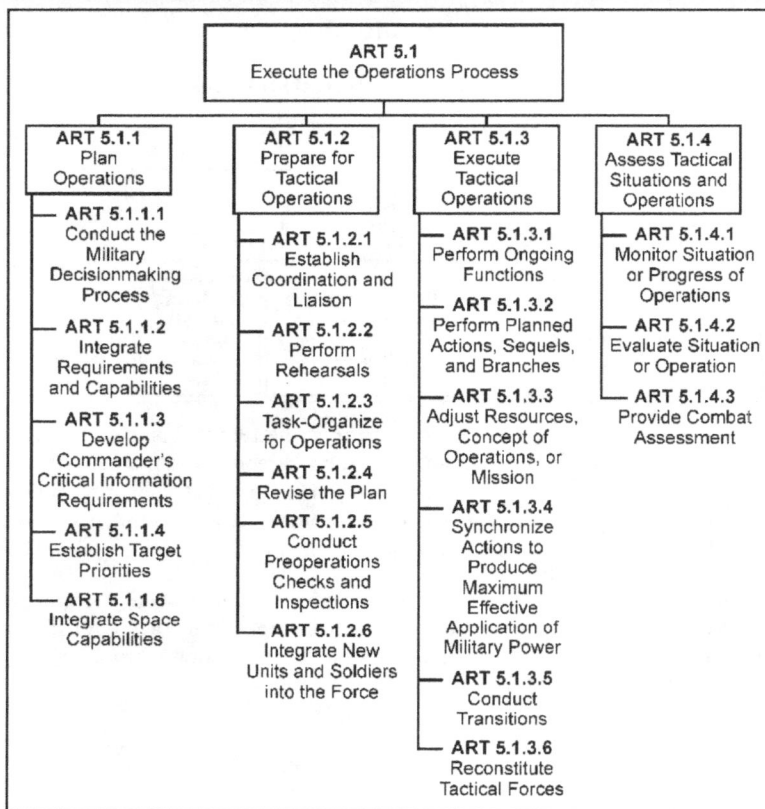

5-1. The operations process consists of the major command and control activities performed during operations: planning, preparing, executing, and continuously assessing the operation. Battle command drives the operations process. These activities may be sequential or simultaneous. They are usually not discrete; they overlap and recur as circumstances demand. Commanders use the operations process to help them decide when and where to make decisions, control operations, and provide command presence. (FM 3-0) (USACAC)

ART 5.1.1 PLAN OPERATIONS

5-2. Translate the commander's visualization into a specific course of action (COA) for preparation and execution. Produce orders and plans that communicate the selected COA, commander's intent, and decisions to subordinates, focusing on the expected results. (FM 5-0) (USACAC)

No.	Scale	Measure
01	Yes/No	Tactical planning produced a commander's decision, communicated an effective methodology, and facilitated mission accomplishment.
02	Yes/No	Subordinate unit staffs were able to understand the order produced by the military decisionmaking process (MDMP).
03	Yes/No	Time was available for plan rehearsal and refinement.

No.	Scale	Measure
04	Yes/No	Time management principles, such as maintaining established timeline, were used effectively.
05	Yes/No	Environmental considerations planning and procedures were present and being followed.
06	Time	Before execution to publish and deliver plan.
07	Time	To complete an iteration of troop leading procedures or the MDMP.
08	Percent	Of subordinate units that receive orders in time to plan, prepare, and execute.
09	Percent	Of time used to provide commander's intent.
10	Percent	Of available planning time allowed for subordinate planning and preparation.
11	Percent	Of subordinate commands clear about their immediate objectives.
12	Number	Of times a staff digresses to the earlier MDMP steps, such as going back to mission analysis issues while developing COAs.

ART 5.1.1.1 CONDUCT THE MILITARY DECISIONMAKING PROCESS

5-3. Employ the logic and techniques of a prescribed analytical process, the military decisionmaking process (MDMP), to determine a unit's restated mission. Develop courses of action (COAs) to accomplish the mission. Recommend the most effective COA. Prepare orders and plans to communicate the selected COA, commander's intent, and decisions to subordinates and coalition forces, focusing on the expected results. (FM 5-0) (USACAC)

Note: See FM 5-0 for a complete discussion of the MDMP.

No.	Scale	Measure
01	Yes/No	Prescribed analytic process produced effective decision and order to guide execution.
02	Yes/No	Milestone criteria for plan development met.
03	Yes/No	Environmental considerations planning and procedures were present and being followed.
04	Percent	Of completed planning documents passed to subordinates to allow parallel planning.
05	Percent	Of troop leading procedures or the MDMP completed correctly.

ART 5.1.1.1.1 Receive the Mission

5-4. To be given a mission by higher headquarters or deduce a need for a change in the current mission. This task involves preparing for mission analysis, to include collecting materials for analysis, receiving the commander's preliminary guidance, determining requirements and time available, and sending warning orders to subordinates. (FM 5-0) (USACAC)

No.	Scale	Measure
01	Yes/No	Commander and staff understood all of the specified and implied tasks contained within their mission.
02	Yes/No	Commander provided adequate initial guidance.
03	Time	After receipt of mission, to issued initial planning guidance.
04	Time	To alert staff of receipt of new mission.
05	Time	To issue warning order.

ART 5.1.1.1.2 Perform Mission Analysis

5-5. Analyze the received mission to define the tactical problem and begin to determine solutions through the identification of specified and implied tasks. It results in a restated mission, the commander's guidance, commander's intent, initial commander's critical information requirement (CCIR), planned use of available time, and a warning order. (FM 5-0) (USACAC)

No.	Scale	Measure
01	Yes/No	Unit developed mission analysis briefing for presentation to the commander.
02	Yes/No	Unit developed and approved restated mission, commander's guidance, commander's intent, CCIRs, use of available time, and warning order.
03	Yes/No	Mission statement included who, what, when, where, and why of the mission.
04	Yes/No	Unit performed time or distance analysis.
05	Yes/No	Unit developed assumptions to replace missing or unknown facts that are necessary for continued planning.
06	Yes/No	Commander issued planning guidance to staff and subordinate commands.
07	Yes/No	Staff developed and maintained running estimate pertaining to their area of expertise.
08	Time	To initiate preliminary movement.
09	Time	To update operational timeline.
10	Percent	Of critical information and running estimates reviewed before mission analysis.
11	Percent	Of major topics within the intelligence preparation of the battlefield for which assessments are completed.
12	Percent	Of assumptions that proved to be either invalid or unrealistic and significantly affected the operation.
13	Percent	Of constraints identified that affect the operation significantly.
14	Percent	Of specified tasks derived in mission analysis and carried into planning.
15	Percent	Of implied tasks derived in mission analysis and carried into planning.
16	Percent	Of essential tasks derived in mission analysis and carried into planning.
17	Percent	Of specified and implied tasks that are identified as essential tasks and included in the mission statement, commander's intent, or concept of operations paragraphs of the operation order.
18	Percent	Of forces identified as required to perform the essential tasks.
19	Percent	Completeness of commander's guidance (coverage of functional responsibilities).
20	Percent	Of planning time used to issue guidance.
21	Percent	Of rules of engagement (ROE) clearly understood.
22	Percent	Of subordinates accepting commander's intent without requests for clarification.
23	Number	Of amendments issued to planning guidance (due to requests for clarification).
24	Number	Of requests for clarification of planning guidance received from subordinate headquarters.
25	Number	Of misunderstood ROE.
26	Number	Of revisions to commander's intent.

ART 5.1.1.1.3 Develop Courses of Action

5-6. Design a course of action (COA) for analysis, evaluation, and selection as the one to accomplish the mission most effectively. It includes analyzing relative combat power, generating options, arraying initial forces, developing schemes of maneuver, assigning headquarters, and preparing COA statements and sketches. The commander has the option of directing a specific course of action because of time available, staff proficiency, or other reasons. (FM 5-0) (USACAC)

No.	Scale	Measure
01	Yes/No	Distinguishable and complete COAs developed in terms of feasibility, suitability, and acceptability for mission accomplishment if executed.
02	Time	To provide the commander with suitable, feasible, and acceptable COAs after receipt of operation order or warning order.

No.	Scale	Measure
03	Time	To prepare complete COA statements and sketches.
04	Percent	Of COAs that are complete.
05	Percent	Of nonselected COAs considered for military deception.
06	Percent	Of COAs suitable—solves the problem and is legal and ethical.
07	Percent	Of COAs feasible—fits within available resources.
08	Percent	Of COAs acceptable—worth the cost or risk.
09	Percent	Of COAs distinguishable—differs significantly from other solutions.
10	Percent	Of COAs presented to commander that were suitable, feasible, acceptable, and distinct from one another.
11	Number	Of COAs developed, as per commander's guidance.

ART 5.1.1.1.4 Analyze Courses of Action

5-7. Develop criteria for success and examine each course of action (COA) for its advantages and disadvantages with respect to those criteria. This task normally includes the technique of war gaming as described in FM 5-0. Visualize each COA objectively; focus intelligence preparation of the battlefield requirements; identify coordination requirements; anticipate critical operational events determine conditions and resources required for success; and assess suitability, feasibility, acceptability, and operational risk of the COA. (FM 5-0) (USACAC)

No.	Scale	Measure
01	Yes/No	Unit identified advantages and disadvantages of COAs and criteria of success for evaluation.
02	Yes/No	Unit reviewed and revised commander's critical information requirements, as necessary, during the war-gaming process.
03	Yes/No	Composite risk management plan developed for COA analysis.
04	Yes/No	Unit applied evaluation criteria (measures of effectiveness or measures of performance) to the war-gaming analysis.
05	Yes/No	Methods applied during war-gaming analysis included belt, box, or avenue in depth.
06	Yes/No	Synchronization matrix or sketch note worksheet used during war-gaming analysis.
07	Time	To complete COA analysis (war-gaming).
08	Percent	Of completeness of COAs (war-gaming).
09	Percent	Of conformance of analysis (war-gaming) to doctrine.
10	Percent	Of branches and sequels experienced identified in COAs.
11	Percent	Of capabilities ultimately required identified in COA analysis (war-gaming).
12	Percent	Of COAs analyzed against potential enemy COAs.
13	Number	Of limitations (ultimately identified during execution) identified during analysis.
14	Number	Of criteria of comparison and success identified during COA analysis (war-gaming).
15	Number	Of decision points and critical events identified and applied to commander's critical information requirement during war gaming.

ART 5.1.1.1.5 Compare Courses of Action

5-8. Evaluate courses of action (COAs) against each other and against criteria of success to determine the most effective one for mission accomplishment. Recommend it for selection. This comparison also considers risk, positioning for future operations, flexibility, and subordinate exercise of initiative. (FM 5-0) (USACAC)

No.	Scale	Measure
01	Yes/No	Selected comparison criteria allowed for definitive comparison of COAs.
02	Yes/No	Unit developed composite risk management plan used during COA comparison.
03	Percent	Of comparison criteria eliminated before comparison.
04	Percent	Of comparison criteria eventually used, defined, and weighted before comparison began.

ART 5.1.1.1.6 Approve a Course of Action

5-9. Decide and approve a course of action (COA) that is most advantageous to mission accomplishment and is within the higher commander's intent. Refine commander's intent and commander's critical information requirement to support selected COAs. Issue any additional guidance to subordinate commanders and staff. Prepare and issue warning order. (FM 5-0) (USACAC)

No.	Scale	Measure
01	Yes/No	COA brief developed and presented to commander.
02	Yes/No	Commander evaluated COAs, selected a COA, and modified or rejected all presented COAs.
03	Yes/No	Modified COA or new COA created a new war game to consider products deriving from that COA.
04	Yes/No	Revised commander's intent adequately addressed key tasks for force as whole, wider purpose. It is expressed in four to five sentences or bullets.
05	Yes/No	Commander decided level of risk to accomplish the mission and approved control measures.
06	Time	To issue warning orders.

ART 5.1.1.1.7 Produce a Plan or Order

5-10. Prepare a plan or order to implement the selected course of action per the commander's decision by turning it into a clear, concise concept of operations and required support. The plan includes annexes and overlays as necessary to implement the plan. The plan or order accurately conveys information that governs actions to be taken and is completed in the correct format. This includes the establishment of graphic control measures, including fire support coordination measures. (FM 5-0) (USACAC)

No.	Scale	Measure
01	Yes/No	Orders or plans accomplished the mission and commander's intent. They were completed with sufficient time for the force to complete required preparatory actions before execution and communicated effectively.
02	Yes/No	Commander's intent refined and adequately addressed key tasks for force as whole, wider purpose; it was expressed in four to five sentences.
03	Time	To issue warning orders, as required.
04	Time	Before execution to reissue commander's concept and intent.
05	Time	To prepare plans and orders (after deciding on mission concept and commander's intent).
06	Time	To obtain approval of plans and orders.
07	Time	To issue plan or order (after approved).
08	Percent	Of functional responsibilities covered in operation plan.
09	Percent	Of accurate information in plans and orders issued and disseminated to subordinate units.
10	Percent	Of accurate information in operation order or plan to meet established objectives.
11	Number	Of instances where the operation plan or order conflicts with standards under conduct effects of war and international convention.

ART 5.1.1.2 INTEGRATE REQUIREMENTS AND CAPABILITIES

5-11. Combine and coordinate capabilities of forces and warfighting functions in effective combinations to meet requirements for mission accomplishment established by planning. (FM 5-0) (USACAC)

No.	Scale	Measure
01	Yes/No	Combinations and timings of forces and warfighting functions contributed to mission accomplishment.
02	Yes/No	Unit integrated information superiority contributors to enhance rapid and accurate situational understanding that initiate or govern actions to accomplish tactical missions.
03	Yes/No	Unit leveraged information superiority contributors that support making more precise and timely decisions than the enemy does.
04	Yes/No	Unit integrated command and control warfare, operations security, military deception, and/or information engagement causing the enemy to make inappropriate, untimely, or irrelevant decisions.
05	Yes/No	Unit established measures of performance.
06	Yes/No	Unit planned transition operations.
07	Yes/No	Monitoring planned.
08	Time	Delay in initiating phase of operation.
09	Time	Before execution for force to execute matrix with sequence and timing of each subordinate task throughout the operation.
10	Time	To modify plans and actions due to operational contingencies.
11	Percent	Of assigned and supporting forces coordinated to synchronize operation in right place at right time.
12	Number	Of potential cross-boundary fratricides identified and eliminated by force headquarters.
13	Number	Of uncoordinated element or activity actions causing disruption or delay of U.S. or multinational plans and objectives.

ART 5.1.1.3 DEVELOP COMMANDER'S CRITICAL INFORMATION REQUIREMENTS

5-12. Analyze information requirements against a mission and commander's intent. Identify, prepare, and recommend for designation by the commander those information requirements critical to facilitating timely decisionmaking. (FM 3-0) (USACAC)

No.	Scale	Measure
01	Yes/No	Answering the developed commander's critical information requirements (CCIRs) provided the commander with the information needed to make decisions.
02	Time	Since last CCIRs update.
03	Time	To promulgate CCIRs for collection.
04	Time	To look at future CCIRs (depends on scale of operation and level of headquarters).
05	Percent	Of answered CCIRs.
06	Percent	Of CCIRs initiated by commander.
07	Percent	Of CCIRs initiated by staff.
08	Percent	Of information (pieces or types) commander needs to make decision listed as CCIRs.
09	Number	Of active CCIRs.

ART 5.1.1.4 ESTABLISH TARGET PRIORITIES

5-13. Identify and recommend for selection targets by categories in precedence to mission accomplishment or individual targets for attack by any warfighting functions during the operation. (FM 6-20) (USAFAS)

No.	Scale	Measure
01	Yes/No	Target priorities supported the accomplishment of the mission.
02	Time	To establish target priorities.
03	Percent	Of targets attacked in priority prescribed.
04	Percent	Of required priority targets correctly identified.
05	Percent	Of unknown targets that would have been priority if correctly identified.

ART 5.1.1.5 INTEGRATE SPACE CAPABILITIES

5-14. Plan, coordinate, and integrate space-based capabilities and products (national, military, and commercial) to support the command and control of tactical planning and execution (the military decisionmaking process and conduct of operations) to effectively meet requirements established for mission accomplishment. Coordinate across all staff elements to identify space-based asset support. Determine essential, specified, and implied tasks from a space perspective in mission analysis. Develop space input to intelligence preparation of the battlefield and other staff estimates Provide space-based support options, space-based products that support concept of operations development, and space-based products that support course of action analysis and statements in course of action development Develop space running estimate. Coordinate for additional operational space capabilities to support mission requirements space-based input to event templates, synchronization matrix, decision support template, and communications and sustainment estimates. (FM 3-14) (USASMDC)

No.	Scale	Measure
01	Yes/No	Unit analyzed higher headquarters orders from a space perspective.
02	Yes/No	Unit recommended space-specific specified, implied, and essential tasks.
03	Yes/No	Unit identified space assets available to support mission requirements.
04	Yes/No	Unit identified space and terrestrial environmental impacts on mission.
05	Yes/No	Unit integrated available Army space support team and space support element into planning.
06	Yes/No	Unit initiated coordination for required operational space capabilities.
07	Yes/No	Unit provided initial space input to intelligence preparation of the battlefield.
08	Yes/No	Unit identified space control and force protection measures required to maintain space superiority.
09	Yes/No	Unit coordinated with other staff sections to identify space assets available to assist.
10	Yes/No	Unit developed and maintained running space estimate.
11	Yes/No	Unit incorporated space estimate into mission analysis.
12	Yes/No	Unit provided space input to mission analysis.
13	Yes/No	Unit recommended space-specific priority intelligence requirement to the intelligence officer.
14	Yes/No	Unit provided space input to unit course of action development.
15	Yes/No	Unit finalized space input to the warning and operation orders.
16	Yes/No	Unit prepared space operations annex to operation plan and order.

ART 5.1.2 PREPARE FOR TACTICAL OPERATIONS

5-15. To initiate and perform activities by the unit before execution to improve its ability to conduct the operation. (FM 6-0) (USACAC)

No.	Scale	Measure
01	Yes/No	Preparatory activities prepared the unit to accomplish its mission.
02	Time	For subordinate forces to complete required preparations.
03	Time	For force to complete required staff, unit, and individual preparations.

No.	Scale	Measure
04	Time	Available to prepare before execution.
05	Percent	Of required preparations completed by execution time.
06	Percent	Of completed preparations that improved force capability to accomplish the mission.

ART 5.1.2.1 ESTABLISH COORDINATION AND LIAISON

5-16. Exchange information to inform, integrate, and deconflict actions by forces and warfighting functions during operations to reduce duplication, confusion, and problems. Liaison, included in coordination, involves maintaining physical contact and communication between elements of military forces to ensure mutual understanding and unity of purpose and action. (FM 6-0) (USACAC)

No.	Scale	Measure
01	Yes/No	Coordination of plans and synchronization of actions between organizations contributed to mission accomplishment.
02	Time	Between receiving alert order and establishing liaison structure.
03	Time	For liaison officers to communicate new orders or information to multinational elements of force.
04	Time	Since liaison officers attached to force headquarters last received situation update from their own units.
05	Time	For force headquarters liaison officers to contact force headquarters on behalf of unit to which they were sent.
06	Time	For force staff sections to contact liaison officer attached to force headquarters.
07	Time	For parent unit to contact its liaison officer.
08	Percent	Of adjacent units or agencies with liaison to force.
09	Percent	Of liaison personnel with required security clearances and identification credentials.
10	Percent	Of units or agencies with missing or late information.
11	Number	Of instances when friendly forces orders or taskings were significantly delayed.

ART 5.1.2.2 PERFORM REHEARSALS

5-17. Practice an action or operation before actual performance of that action. Rehearsals allow participants to become familiar with and translate the tactical plan into a visual impression that orients them to both their operational environment and other units during execution of the operation. They also imprint a mental picture of the sequence of key actions within the operation and provide a forum for subordinate leaders and units to coordinate. (FM 6-0) (USACAC)

No.	Scale	Measure
01	Yes/No	Unit placed rehearsals on the operational timeline as part of the operation order.
02	Yes/No	Rehearsal improved all participants' familiarity with the tactical plan, their roles within that plan, and coordination.
03	Yes/No	Unit specified type of rehearsal.
04	Yes/No	Unit specified technique of rehearsal.
05	Yes/No	Unit specified roles and responsibilities of participants.
06	Yes/No	Unit identified and prioritized events to be rehearsed.
07	Yes/No	Subordinates reviewed their mission, commander's intent, and concept of operations in relationship to time (such as by timelines or phases).
08	Yes/No	Subordinates discussed and resolved warfighting functions coordination issues.
09	Yes/No	Unit made major changes to the existing plan.
10	Time	To document and distribute results of the rehearsal.

No.	Scale	Measure
11	Percent	Of the operation's phases or objectives rehearsed.
12	Percent	Of developed branch (or contingency) plans reviewed.

ART 5.1.2.3 TASK-ORGANIZE FOR OPERATIONS

5-18. Transfer available resources and establish command and support relationships per the plan or order to carry out the operation. Organizing for operations includes moving the unit's location as required, exchanging standing operating procedures, establishing communications and liaison, and leading briefings and rehearsals. (FM 6-0) (USACAC)

No.	Scale	Measure
01	Yes/No	Units conducting operations properly transferred necessary resources per established command and supported relationships to accomplish their mission.
02	Yes/No	Units closed on new assembly areas before execution.
03	Time	For force to transition to or from tactical battle formation.
04	Time	From planned execution time force transitions to or from tactical battle formation.
05	Time	To move forces into locations to facilitate tactical commanders' plans for implementing subordinate plan.
06	Time	To complete force movement to new assembly areas (from which to maneuver).
07	Time	To assign subordinate forces to new tactical formations.
08	Time	Until subordinate units are prepared to send and receive data and do parallel planning with new headquarters.
09	Percent	Of force moved into position to facilitate tactical commander's plans.
10	Percent	Of required logistics in place on schedule.
11	Percent	Of allocated forces in place at operation execution.
12	Percent	Of required logistics stockpiled or positioned before operation D-day or H-hour.
13	Percent	Of designated forces massed in designated assembly area at specified time.
14	Percent	Of force postured physically to execute plan's tactical maneuver.

ART 5.1.2.4 REVISE THE PLAN

5-19. Adjust the plan based upon updated relevant information or further analysis of the plan, if necessary. (FM 6-0) (USACAC)

No.	Scale	Measure
01	Yes/No	Revision and refinements of plan contributed to accomplishing the mission.
02	Yes/No	Unit revised and refined the plan after receipt of updated relevant information, such as answered commander' critical information requests.
03	Time	To adjust plan after receipt of updated relevant information.
04	Time	To revise original plan after recognizing planning assumptions invalid or information updated.
05	Percent	Of plan adjusted with each revision.
06	Percent	Of accurate adjustments by execution.

ART 5.1.2.5 CONDUCT PREOPERATIONS CHECKS AND INSPECTIONS

5-20. Inspect and check unit and individual preparations for operations to ensure units, Soldiers, and systems are fully capable and ready to accomplish the mission. (FM 6-0) (USACAC)

No.	Scale	Measure
01	Yes/No	Inspections and corrections of deficiencies found unit, Soldiers, and systems improved their capabilities and readiness to accomplish the mission.
02	Yes/No	Environmental considerations planning and procedures were present and being followed.
03	Time	To perform checks and inspections.
04	Time	To correct deficiencies found during inspection.
05	Percent	Of units, individuals, and materiel not ready for operation.
06	Percent	Of deficiencies corrected before operation.

ART 5.1.2.6 INTEGRATE NEW UNITS AND SOLDIERS INTO THE FORCE

5-21. Assimilate new units and Soldiers into the force in a posture that allows them to contribute effectively to mission accomplishment during an operation. This includes receiving and introducing them to the force and the operational environment; orienting them on their place and role in the force and the operation; establishing command and control, sustainment, and communications within the force; and training them in the unit standing operating procedures and mission-essential task list for the operation. (FM 6-0) (USACAC)

No.	Scale	Measure
01	Yes/No	Integration of new units and Soldiers made them contribute more effectively to mission accomplishment.
02	Time	To receive individuals into new organizations.
03	Time	To introduce individuals to the operational environment and the operation.
04	Time	To receive new organizations into the force.
05	Time	To introduce new organizations to the operational environment and the operation.
06	Time	To train new organizations and individuals for the operation.
07	Percent	Of individuals successfully integrated into new organizations.
08	Percent	Of organizations successfully integrated into force.
09	Percent	Of new Soldiers within the unit who must be assimilated.
10	Number	Of new Soldiers within the unit who must be assimilated.

ART 5.1.3 EXECUTE TACTICAL OPERATIONS

5-22. Put a plan into action by applying military power to accomplish the mission. Use situational understanding to assess progress and make execution and adjustment decisions. (FM 6-0) (USACAC)

No.	Scale	Measure
01	Yes/No	Mission accomplished per higher commander's intent.
02	Yes/No	Mission accomplished per specified timeline.
03	Yes/No	Mission accomplishment did not detract from unit's capability of continuing or being assigned future missions and operations.
04	Yes/No	Mission accomplished without excessive expenditure of resources.
05	Yes/No	Actions synchronized using the rapid decisionmaking and synchronization process.
06	Time	To accomplish mission.
07	Percent	Of enemy actions successfully countered.
08	Number	Of opportunities for success exploited.
09	Ratio	Of friendly versus enemy casualties (Soldiers and weapon systems) to accomplish mission.

ART 5.1.3.1 Perform Ongoing Functions

5-23. Perform routine tasks during execution essential to satisfactory mission accomplishment. (FM 6-0) (USACAC)

No.	Scale	Measure
01	Yes/No	Unit performed routine management tasks as necessary to accomplish the mission without the need to take extraordinary efforts to ensure correct performance.
02	Percent	Of all functions performed adequately during operations.
03	Percent	Of individual functions performed adequately during operations.
04	Number	Of friendly missions affected by failure to perform one or more functions.

ART 5.1.3.1.1 Focus Assets on the Decisive Operation

5-24. Continuously survey all assets and ensure that they are in position and tasked to support the decisive operation or main effort for a phase of an operation or that they are moving to a position where they can provide that support. (FM 6-0) (USACAC)

No.	Scale	Measure
01	Yes/No	Unit allocation of combat power regularly supported the decisive operation necessary to accomplish the mission.
02	Percent	Of combat power assets directed to decisive operation.
03	Number	Of incidents of excessive combat power used to execute shaping operations.
04	Number	Of incidents of combat power resources improperly positioned to support decisive operation.

ART 5.1.3.1.2 Adjust Commander's Critical Information Requirements and Essential Elements of Friendly Information

5-25. Continuously update commander's critical information requirements (CCIRs) and essential elements of friendly information (EEFI) during operations with routine review by the commander and staff. Analyze information requirements against changing operational circumstances and designate information requirements that affect decisionmaking and the success of the mission and decisive operation. (FM 6-0) (USACAC)

No.	Scale	Measure
01	Yes/No	Unit continuously adjusted CCIRs and EEFI during operations with review by commander and staff.
02	Time	Of lag between changing information and update of CCIRs and EEFI.
03	Number	Of CCIRs and EEFI requirements not updated or changed.

ART 5.1.3.1.3 Adjust Graphic Control Measures

5-26. Change graphic control measures, including fire support coordination measures, any time there is significant movement of forces or special operations force activity in the area of operations. (FM 6-0) (USACAC)

No.	Scale	Measure
01	Yes/No	Adjustment of graphic control measures reflected changes in the factors of mission, enemy, terrain and weather, troops and support available, time available, civil considerations and was timely and effective.
02	Time	Between operations and update of graphic control measures.
03	Percent	Of graphic control measures requiring amendment to facilitate operations.

No.	Scale	Measure
04	Number	Of graphic control measures not updated.
05	Number	Of significant movement of forces compared to the number and timeliness of graphic changes.

ART 5.1.3.1.4 Manage Sustainment Force Positioning

5-27. Move or cause sustainment forces to be moved and positioned where they can contribute the maximum support to the operation. Determine where the sustainment forces are and where they must be to continuously provide effective support. Allow adequate planning and execution time to support mission requirements. (FM 6-0) (USACAC)

No.	Scale	Measure
01	Yes/No	Commander repositioned sustainment units and activities as necessary to respond to tactical operations without negative effects on friendly operations.
02	Time	To produce essential sustainment repositioning without negative effects on friendly operations.
03	Time	Between identification of need for repositioning and commencement of repositioning.
04	Percent	To which friendly operations affected by failures in sustainment positioning.
05	Number	Of friendly operations adversely affected by failure in positioning.

ART 5.1.3.1.5 Manage Use and Assignment of Terrain

5-28. Ensure that adequate space, including the use of routes, is available at the right time to support critical activities, especially the decisive operation in the area of operations. (FM 6-0) (USACAC)

No.	Scale	Measure
01	Yes/No	Adequate maneuver space and routes supported the decisive operation.
02	Time	Since last check of terrain management status.
03	Number	Of friendly unit operations affected by lack of terrain management.
04	Number	Of incidents where friendly units dispute assignment of terrain or routes.

ART 5.1.3.1.6 Maintain Synchronization

5-29. Take actions to preserve the arrangement of military actions in time, space, and purpose to produce maximum relative military power at a decisive place and time. (FM 6-0) (USACAC)

No.	Scale	Measure
01	Yes/No	Combat power was available at the decisive place and time to accomplish the mission.
02	Percent	Of assigned and supporting forces executed operation with right people, in right place, and at right time.
03	Percent	Of force or subordinate missions and support carried out as planned.
04	Percent	Of friendly forces actively contributing to conduct of operation.
05	Percent	Of support requirements met at time and as required.
06	Percent	Of policies and procedures for establishment and coordination of logistics, maintenance, transportation, and other support completed.
07	Number	Of missions accomplished without appropriate coordination.

ART 5.1.3.1.7 Control Tactical Airspace

5-30. Maximize the combat effectiveness of all tactical airspace users to include aerial systems in support of the operation. Prevent fratricide, enhance air defense operations, and permit greater flexibility of tactical operations. Prepare and implement the tactical airspace control plan and associated airspace control measures. (FM 3-52) (USACAC)

No.	Scale	Measure
01	Yes/No	The control of tactical airspace allowed airspace users to support tactical operations while preventing fratricide.
02	Yes/No	The control of tactical airspace enhanced the conduct of air operations.
03	Yes/No	The control of tactical airspace enhanced the conduct of air defense operations.
04	Time	To confirm identity of unidentified friendly target.
05	Time	Since last publication of airspace control orders.
06	Percent	Of friendly aircraft destroyed by friendly air defense.
07	Percent	Of friendly aircraft destroyed by friendly fire.
08	Percent	Of positive identification false negatives (friendly identified as enemy).
09	Percent	Of positive identification false positives (enemy identified as friendly).
10	Percent	Of fixed-wing sorties receive clearances needed to accomplish the mission.
11	Percent	Of rotary-wing sorties receive clearances needed to accomplish the mission.
12	Percent	Of air defense operations that did not interfere with other operations.
13	Number	Of attacks by enemy air forces and the time that they occur.
14	Number	Of air-to-air mishaps in the area of operations.

ART 5.1.3.1.8 Control a Tactical Insertion of Forces

5-31. Command and control an Army tactical insertion in support of campaign objectives. The insertion force may employ organic aviation assets or assets from other Services for transport to the objective area. (FM 3-52) (USACAC)

No.	Scale	Measure
01	Yes/No	Mission accomplished per higher commander's intent.
02	Yes/No	Mission accomplished per timeline.
03	Yes/No	Risk assessment conducted during planning phase.
04	Yes/No	Army personnel recovery included in plan.
05	Yes/No	Rehearsals conducted.
06	Yes/No	En route communications maintained between forcible entry force and controlling headquarters.
07	Yes/No	Liaison established with supporting movement element prior to H-hour.
08	Yes/No	Preassault fires executed prior to H-hour.
09	Percent	Of friendly casualties occurring during forcible entry operation.

ART 5.1.3.2 PERFORM PLANNED ACTIONS, SEQUELS, AND BRANCHES

5-32. Implement actions anticipated by and outlined in the plan of operations based upon evaluation of progress as within the criteria of success in the plan. (FM 6-0) (USACAC)

No.	Scale	Measure
01	Yes/No	Unit executed planned actions, sequels, and branches based upon the evaluation of progress.
02	Time	To refine planned branch or sequel for use by force.
03	Percent	Of planned branches and sequels meeting requirements of current operation.

ART 5.1.3.3 ADJUST RESOURCES, CONCEPT OF OPERATIONS, OR MISSION

5-33. Take actions to modify the operation (or major activities) to exploit opportunities or resolve progress problems as a result of evaluation of the progress of the operation against the criteria of success. (FM 6-0) (USACAC)

No.	Scale	Measure
01	Yes/No	Commander adjusted unit plan to exploit opportunities or resolve problems occurring during execution.
02	Time	To adjust original plan after assessing progress and recognizing threat or opportunity.
03	Number	Of adjustments effective in seizing opportunity or countering threat.

ART 5.1.3.4 SYNCHRONIZE ACTIONS TO PRODUCE MAXIMUM EFFECTIVE APPLICATION OF MILITARY POWER

5-34. Arrange military actions by forces and warfighting functions in time, space, and purpose to produce maximum relative military power at a decisive place and time. (FM 6-0) (USACAC)

No.	Scale	Measure
01	Yes/No	Commander synchronized unit actions to accomplish the mission.
02	Time	To resynchronize warfighting functions after execution or adjustment decision.
03	Percent	Of required resynchronization accomplished in time available.
04	Percent	Of assigned and supporting forces coordinated to synchronize operation in right place at right time.
05	Number	Of friendly actions.
06	Number	Of uncoordinated element or activity actions causing disruption or delay of U.S. or multinational plans and objectives.

ART 5.1.3.4.1 Coordinate Actions Within a Staff Section

5-35. Exchange information and arrange actions to inform, integrate, and deconflict actions within a staff section during operations to reduce duplication, confusion, and problems. (FM 6-0) (USACAC)

No.	Scale	Measure
01	Yes/No	Unit deconflicted actions within a staff section.
02	Time	Spent to coordinate within staff section.
03	Percent	Of actions coordinated within staff section before disseminating further.
04	Incidents	Of actions uncoordinated within staff section causing disruption or delay of operation.
05	Incidents	Of actions uncoordinated within staff section affecting others' actions.

ART 5.1.3.4.2 Synchronize Actions Among Staff Sections (Coordinating, Special, and Personal)

5-36. Coordinate actions among staff sections in arranging military actions in time, space, and purpose by warfighting functions to produce the maximum relative military power at a decisive place and time. This includes informing of, integrating, and deconflicting actions undertaken by or directed by staff sections during operations to reduce duplication, confusion, and problems among the staff sections. (FM 6-0) (USACAC)

Note: ART 5.1.3.4.2 includes the targeting process outlined in JP 3-60 and FM 6-20-10.

No.	Scale	Measure
01	Yes/No	Staff section cooperated in arranging military actions in time, space, and purpose.
02	Time	Spent to coordinate among staff sections.
03	Percent	Of actions coordinated among staff sections before disseminating further.

No.	Scale	Measure
04	Number	Of uncoordinated actions between staff sections causing disruption or delay of unit operations.
05	Number	Of actions uncoordinated among staff section affecting others' actions.

ART 5.1.3.4.3 Review Orders of Subordinate Organizations

5-37. Ensure that all relevant information and factors issued in subordinate orders contribute to vertical warfighting function synchronization. ART 5.1.3.4.3 informs the staff about whom they will have to coordinate. It reveals potential conflicts and problems among subordinate forces, higher headquarters, adjacent, and other units that might affect or be affected by the subordinate plan and the headquarters' plan. It resolves conflicts and problems among forces before they affect preparations or operations, or resolves the damage. (FM 6-0) (USACAC)

No.	Scale	Measure
01	Yes/No	Commander resolved conflicts and problems between subordinates before they affected preparations or operations.
02	Percent	Of subordinate organization orders reviewed.
03	Number	Of discrepancies between subordinates organizations' orders and unit order force order that could have affected operation.

ART 5.1.3.4.4 Synchronize Force Operations

5-38. Arrange military actions by subordinate forces and the force as a whole in time, space, and purpose to produce maximum relative military power at a decisive place and time. (FM 6-0) (USACAC)

No.	Scale	Measure
01	Yes/No	Synchronized force operations allowed the unit to accomplish its mission.
02	Percent	Of assigned and supporting forces participating in operation in right place at right time.
03	Number	Of uncoordinated element or activity actions causing disruption or delay of operation.

+ART 5.1.3.4.5 Advise the Command

5-38.1 Staffs contribute to achieving the commander's intent by fulfilling their functional responsibilities within the authority the commander delegates to them. Effective staffs and staff members provide commanders with timely relevant information and well-analyzed recommendations. Staff members inform and advise the commander and other staff members concerning matters pertaining to their respective and related functional responsibilities and assigned duties. (FM 6-0) (USACAC)

No.	Scale	Measure
01	Yes/No	Staff section provided quality and timely information in support of decision making processes.
02	Yes/No	Staff assisted the commander minimize unnecessary risks by assessing hazards within their respective and related functional responsibilities and duties.
03	Yes/No	Staff informed and advised the commander and other staff members concerning all matters in their individual functional responsibilities.
04	Yes/No	Staff informed and advised the commander and other staff members concerning capabilities, limitations, requirements, availability, and employment of resources.

No.	Scale	Measure
05	Yes/No	Staff informed and advised the commander and other staff members concerning capabilities, limitations, requirements, availability, and employment of supporting forces.
06	Yes/No	Staff informed and advised the commander and other staff members concerning directives and policy guidance from higher headquarters.
07	Yes/No	Staff functions are synchronized by the Chief of Staff or Executive Officer to provide a singular product that is not a series of individual staff estimates.
08	Yes/No	Staff coordinated with supporting units and higher headquarters for operational needs beyond unit capability.
09	Time	To evaluate, update, and disseminate new information with higher, lower, adjacent, supported and supporting organizations, which facilitate collaborative planning, rapid execution of commander's orders, and a shared situational understanding of the operational environment.

ART 5.1.3.5 CONDUCT TRANSITIONS

5-39. Transitions mark intervals between the ongoing operation and full execution of branches and sequels. Transitions often mark the change from one dominant type of operation, such as offense, to another such as stability. Commanders at all levels must possess the mental agility to rapidly transition from one type of operation to another. For example, at lower echelons, transitions occur when one formation passes through another, or when units must breach an obstacle belt. Links between phases and the requirement to transition between phases are critically important. Commanders establish clear conditions for how and when these transitions occur during planning. Although phases are distinguishable to friendly forces, the operational design conceals these distinctions from opponents through concurrent and complementary joint and Army actions. (FM 3-0) (USACAC)

No.	Scale	Measure
01	Yes/No	Unit conducted transition from one phase to another and maintained seamless continuity of operations.
02	Yes/No	Commander provided new graphic control measures.
03	Yes/No	Commander adopted new task organization, if required.
04	Yes/No	Commander issued new priorities.
05	Yes/No	Commander issued new rules of engagement.
06	Yes/No	Commander determined possible branches or sequels for likely next phase.

ART 5.1.3.6 RECONSTITUTE TACTICAL FORCES

5-40. Reconstitution is the extraordinary action taken by commanders to restore units to a desired level of combat effectiveness, commensurate with mission requirements and available resources. The major elements of reconstitution are reorganization, assessment, and regeneration. Reconstitution requires a decision by the commander having control of the required resources. (FM 100-9) (CASCOM)

This page intentionally left blank.

No.	Scale	Measure
01	Yes/No	Unit conducted reconstitution and maintained seamless continuity of operations.
02	Yes/No	Commander provided new graphic control measures.
03	Yes/No	Commander adopted new task organization, if required.

ART 5.1.3.6.1 Reorganize Units as Part of a Reconstitution Effort

5-41. Reorganization is action to shift resources within a degraded unit to increase its combat effectiveness. Commanders of all types of units at each echelon conduct reorganization. They reorganize before considering regeneration. Reorganization may be immediate or deliberate. (FM 100-9) (CASCOM)

No.	Scale	Measure
01	Yes/No	Unit reorganized and maintained seamless continuity of operations.
02	Yes/No	Commander provided new graphic control measures.
03	Yes/No	Commander adopted new task organization, if required.

ART 5.1.3.6.2 Regenerate Units and Organizations as Part of a Reconstitution Effort

5-42. Regeneration is the rebuilding of a unit, involving large-scale replacement of personnel, equipment, and supplies. This is a higher level of reorganization than the unit can do during normal reorganization without major personnel resources. (FM 100-9) (CASCOM)

No.	Scale	Measure
01	Yes/No	Unit regenerated and maintained seamless continuity of operations.
02	Yes/No	Commander provided new graphic control measures.
03	Yes/No	Commander adopted new task organization, if required.

ART 5.1.4 ASSESS TACTICAL SITUATIONS AND OPERATIONS

5-43. Continuously monitor the situation and progress of the operation. Evaluate the situation or operation against measures of performance or measures of effectiveness to make decisions and adjustments throughout the operations process (planning, preparing, executing, and assessing). (FM 6-0) (USACAC)

No.	Scale	Measure
01	Yes/No	Assessment supported decisionmaking and adjustments during the operations process.
02	Time	To compare situation against criteria of success.
03	Time	For commander to assess progress.
04	Percent	Of enemy actions or operations forecast.
05	Percent	Of time event of interest occurs without options available.
06	Percent	Of accuracy of commander's assessment of progress.

ART 5.1.4.1 MONITOR SITUATION OR PROGRESS OF OPERATIONS

5-44. Collect relevant information on the situation or operation to evaluate the actual situation or progress of the operation and support decisionmaking. (FM 6-0) (USACAC)

No.	Scale	Measure
01	Yes/No	Relevant information collected accurately reflected the situation or progress of the operation and supported decisionmaking.
02	Time	Of lag in currency of information on adjacent military forces or non-Department of Defense agencies.
03	Time	To obtain information on changes to tactical situation.
04	Time	Since update of force situation.

No.	Scale	Measure
05	Time	To access current situation.
06	Percent	Of time that commander learns of emerging tactical event from staff.
07	Number	Of incidents in which the commander was surprised (not briefed) by critical or emerging event.
08	Number	Of instances when commanders learn of emerging events from sources outside their staff.

ART 5.1.4.2 EVALUATE A SITUATION OR OPERATION

5-45. Analyze and compare the actual situation or progress of the operation against criteria of success. Highlight variances between the planned situation at that time in the operation and the current situation, and forecast the degree of mission accomplishment. (FM 6-0) (USACAC)

No.	Scale	Measure
01	Yes/No	Evaluation reflected reality of the degree of mission accomplishment and forecasted the degree of mission accomplishment.
02	Time	To evaluate progress or situation and determine type of decision.
03	Time	To complete evaluation of situation or progress.
04	Percent	Of accuracy of evaluation of situation or progress.
05	Percent	Of accurate friendly evaluations.
06	Number	Of opportunities or threats recognized.

ART 5.1.4.2.1 Develop Running Estimates

5-46. Use the running estimate procedure—consisting of significant facts, events, and conclusions based on analysis—as the staff element's means of assessing within their functional field. Running estimates consider both quantifiable and intangible aspects of military operations. They are as thorough as time permits and updated regularly as part of an ongoing process. They support assessing throughout the operations process. Running estimates also support the commander's visualization of the operation. (FM 6-0) (USACAC)

Note: The intelligence running estimate is the product of intelligence preparation of the battlefield. See ART 2.2.1 (Perform Intelligence Preparation of the Battlefield).

No.	Scale	Measure
01	Yes/No	Running estimates were accurate and supported the commander's visualization of the operation.
02	Time	Into future that planning branches have been developed.
03	Time	From receipt of information to complete or update running estimate.
04	Percent	Of decision points that have branches.
05	Percent	Of enemy actions or operations affected course of battle, but not forecast.
06	Percent	Of forecast branches appeared at execution.

ART 5.1.4.2.2 Evaluate Progress

5-47. Conclude whether the variances from the criteria of success of the current situation or progress of the operation are significant enough to affect accomplishing the mission or meeting the commander's intent and warrant adjustment. (FM 6-0) (USACAC)

No.	Scale	Measure
01	Yes/No	Command and control system effectively concluded when the force met the commander's intent or needed to adjust its course of action.
02	Time	For commander or staff to forecast degree of mission accomplishment based on variance.
03	Time	From receipt of information to complete evaluation of progress.
04	Percent	Of accurate evaluation of variances.
05	Percent	Of accurate forecast of progress and meaning of forecast.

ART 5.1.4.3 PROVIDE COMBAT ASSESSMENT

5-48. Determine the overall effectiveness of firepower employment during military operations. Firepower employed can be lethal and nonlethal in nature and delivered by direct and indirect fire systems. The objective of combat assessment is to identify recommendations for maneuver operations. Combat assessment is the overarching concept that consists of battle damage and munitions effects assessments while providing reattack recommendations. (FM 2-0) (USAIC&FH)

No.	Scale	Measure
01	Yes/No	The commander determined the overall damage inflicted on the enemy by those direct and indirect fires employed during the conduct of operations.
02	Time	To commence subsequent operations or restrike while awaiting combat assessment.
03	Time	To complete combat assessment.
04	Time	To provide full assessment of attacks to force commander.
05	Time	To provide initial combat assessment of attacks to force commander.
06	Time	To perform the combat assessment functions of battle damage assessment, mission assessment, and munitions effectiveness assessment based on commander's guidance or objectives.
07	Percent	Of high-payoff targets assessed as killed later assessed as being mission capable.
08	Percent	Difference between higher-level and force assessment of effectiveness.
09	Percent	Of targets with combat assessment data available.
10	Percent	Of targets unnecessarily reattacked.

ART 5.1.4.3.1 Perform Battle Damage Assessment

5-49. Provide a timely and accurate estimate of damage resulting from the application of military force, either lethal or nonlethal, against a target. Battle damage assessment provides commanders with a timely and accurate snapshot of their effect on the enemy. This helps commanders determine when or if their targeting effort is meeting their objectives. This task also includes determining if the application of military force results in the release of hazardous material. (FM 6-20) (USAFAS)

Note: This task is supported by the ART 2.4.3 (Provide Intelligence Support to Combat Assessment). It is also associated with the decide, detect, deliver, and assess process.

No.	Scale	Measure
01	Yes/No	The commander accurately determined the damage to the enemy from the overall effects of firepower employed during the conduct of military operations.
02	Time	To commence subsequent operations or reattack (following receipt of assessment).
03	Time	To complete full assessment of attacks after time on target.
04	Time	To make initial assessment of attacks after time on target.
05	Time	To provide full assessment of attacks to force commander.
06	Time	To provide initial battle damage assessment of attacks to force commander.

No.	Scale	Measure
07	Percent	Of high-payoff targets assessed as killed later assessed as being mission capable.
08	Percent	Of targets have battle damage assessment based on more than one type of system.
09	Percent	Of targets unnecessarily reattacked.
10	Percent	Of difference between higher-level and force assessment of effectiveness.

ART 5.1.4.3.2 Perform Munitions Effects Assessment

5-50. Assess the military force in terms of the lethal and nonlethal weapon systems and munitions effectiveness. This assessment is used as the basis of recommendation for changes to increase the effectiveness of methodology, tactics, weapon system, munitions, and weapon delivery parameters. Munitions effects assessment takes place concurrently and interactively with battle damage assessment. This assessment is primarily the responsibility of operations and fire support personnel, with input from the intelligence warfighting functions. (FM 6-20-10) (USAFAS)

No.	Scale	Measure
01	Yes/No	The commander accurately determined the overall effects of munitions and weapon systems employed against specific types of targets during the conduct of military operations.
02	Time	To commence subsequent operations or reattack (following receipt of munitions effects assessment).
03	Time	To complete munitions effects assessment after attack.
04	Time	To provide full assessment of attacks to force commander.
05	Time	To provide initial munitions effects assessment of attacks to force commander.
06	Percent	Of high-payoff targets assessed as killed later assessed as being mission capable.
07	Percent	Of high-payoff targets that require reattack.
08	Percent	Of high-payoff targets successfully attacked.
09	Percent	Difference between higher level and force assessment of munitions effectiveness.
10	Percent	Of targets unnecessarily reattacked.

ART 5.1.4.3.3 Provide Reattack Recommendations

5-51. Make recommendations to the commander, considering the level to which operational objectives have been met regarding reattack and other recommendations that address operational objectives relative to target, target critical elements, target systems, and enemy combat strengths. (FM 6-20-10) (USAFAS)

No.	Scale	Measure
01	Yes/No	Reattack recommendations were effective and efficient.
02	Time	To commence subsequent operations or execute other options.
03	Time	To provide full assessment of attacks to joint force commander.
04	Time	To assess reattack requirement (after attack).
05	Time	To provide initial reattack assessment of attacks to force commander.
06	Percent	Of high-payoff target attacks unsuccessful.
07	Percent	Of high-payoff targets assessed as killed later assessed as being mission capable.
08	Percent	Of high-payoff targets that require reattack.
09	Percent	Of high-payoff targets successfully attacked.
10	Percent	Of difference between higher level and force assessment of reattack requirement.
11	Percent	Of targets unnecessarily reattacked.

SECTION II – ART 5.2: CONDUCT COMMAND POST OPERATIONS

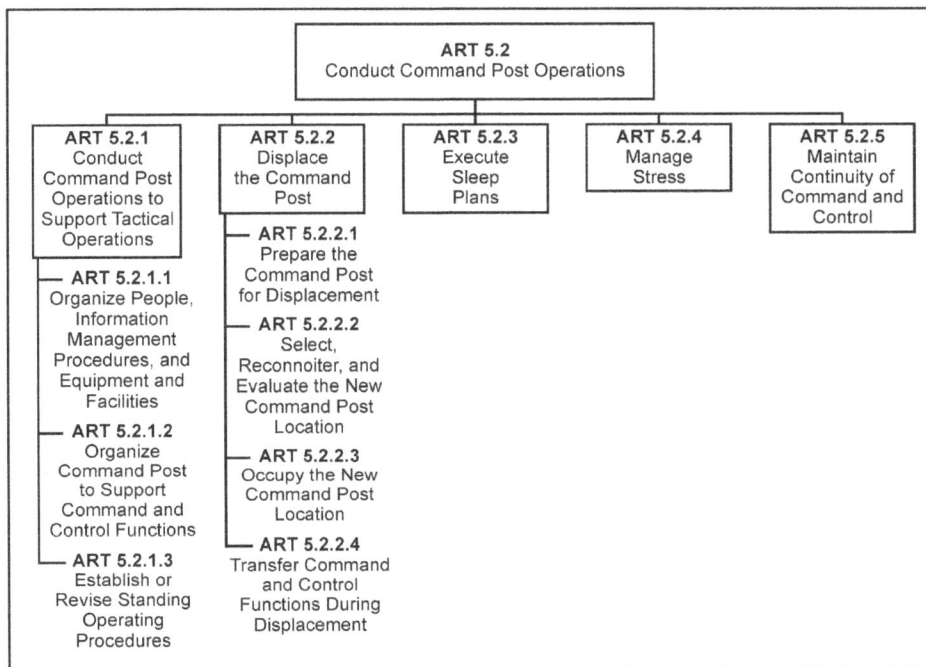

5-52. Organize, create or erect, operate, and move the principal facility used by the commander to exercise command and control of tactical operations. The command post performs the command and control warfighting functions discussed in succeeding subtasks. (FM 6-0) (USACAC)

Note: ART 6.5.3 (Establish Local Security) and ART 6.11.1 (Conduct Operations Security) address tasks inherently associated with conduct of command post operations. The measures of performance for those tasks are not included with the measures of performance of any of the tasks in this chapter since they are separate tasks.

No.	Scale	Measure
01	Yes/No	Command post supported commanders in exercising command and control of their units to accomplish the mission within the time and parameters specified by the higher commander and as stated in the operation order.
02	Time	To form command post from fully operational headquarters.
03	Time	To determine command post structure.
04	Time	To accomplish missions.
05	Percent	Of tactical actions or operations able to be executed.
06	Percent	Of missions accomplished.

ART 5.2.1 CONDUCT COMMAND POST OPERATIONS TO SUPPORT TACTICAL OPERATIONS

5-53. Organize, create or erect, and operate the command post in a manner that allows it to perform command and control (C2) functions effectively for a particular operation. (FM 6-0) (USACAC)

No.	Scale	Measure
01	Yes/No	Command post supported commanders in exercising C2 of their units to accomplish the mission within the time the operation order specifies.
02	Yes/No	Command post could communicate critical information with higher and subordinate headquarters, adjacent headquarters, and supported headquarters in near real time.
03	Time	To establish command post for operations.
04	Percent	Of C2 functions performed to standard.
05	Percent	Of C2 nodes that possess required communications capabilities.
06	Percent	Of essential C2, communications, and computer systems accessible from all subordinate locations.
07	Percent	Of required staff positions and materiel filled.
08	Number	Of times that subordinate commanders are unable to communicate with force headquarters.

ART 5.2.1.1 ORGANIZE PEOPLE, INFORMATION MANAGEMENT PROCEDURES, AND EQUIPMENT AND FACILITIES

5-54. Establish relationships among the personnel, information management, procedures, and equipment and facilities essential for using and communicating the common operational picture and execution information to achieve situational understanding and to direct the conduct (planning, preparing for, executing, and assessing) of operations. (FM 6-0) (USACAC)

No.	Scale	Measure
01	Yes/No	Organization and level of resources met command and control (C2) system requirements in supporting effective C2.
02	Time	To submit host-nation supportability requests.
03	Time	To obtain host-nation supportability approval.
04	Time	To form force staff (from activation order).
05	Time	To staff and equip C2 system.
06	Time	To establish and approve C2 system architecture.
07	Number	Of required C2 system resources identified.
08	Percent	Of required C2 system resources provided.

ART 5.2.1.2 ORGANIZE THE COMMAND POST TO SUPPORT COMMAND AND CONTROL FUNCTIONS

5-55. Arrange command post equipment and facilities in a manner that effectively supports the personnel performing command and control functions for a specific operation or situation. (FM 6-0) (USACAC)

No.	Scale	Measure
01	Yes/No	Command post organization facilitated performing command and control functions for a specific operation.
02	Time	After constitution of command post to establish "daily battle rhythm."
03	Percent	Of normal operations covered by procedures.

No.	Scale	Measure
04	Number	Of incidents of friendly forces orders or taskings significantly delayed because of unclear relationships within headquarters.
05	Number	Of incidents of friendly forces orders or taskings significantly delayed.

ART 5.2.1.3 ESTABLISH OR REVISE STANDING OPERATING PROCEDURES

5-56. Create or modify a set of instructions covering those tasks and functions that lend themselves to a definite or standing procedure without a loss of effectiveness; the standing operating procedures is effective unless ordered otherwise to meet altered conditions. (FM 6-0) (USACAC)

No.	Scale	Measure
01	Yes/No	Unit's standing operating procedures or its revisions facilitated mission accomplishment and warfighting functions integration.
02	Yes/No	Commands had policies and procedures for operation and maintenance of information systems.
03	Yes/No	Commands had restoration plans for critical failures.
04	Percent	Of procedures revised during operations.
05	Percent	Of total "down" time for operational command and control, communications, and computer systems attributed to failure to follow established policies and procedures.

ART 5.2.2 DISPLACE THE COMMAND POST

5-57. Move or transfer the principal facility used by the commander to exercise command and control of tactical operations of a specific operation. (FM 6-0) (USACAC)

No.	Scale	Measure
01	Yes/No	Command post displaced when necessary and resumes supporting the commander within the time the operation order specifies.
02	Time	To displace and establish command post in new location.
03	Time	To resume full operations in new location.
04	Number	Of incidents of degraded command and control effectiveness during displacement.
05	Number	Of incidents of operations affected negatively by displacement.

ART 5.2.2.1 PREPARE THE COMMAND POST FOR DISPLACEMENT

5-58. Execute activities before movement to facilitate the command post move. These activities include, but are not limited to, dismantling information systems and associated networks, dismantling the facilities and equipment, and packing elements of the command post per load plans. (FM 6-0) (USACAC)

No.	Scale	Measure
01	Yes/No	Command post prepared for displacement within the time either the unit standing operating procedures or operation order specifies.
02	Yes/No	Unit planned for continuous communications during command post displacement.
03	Time	Before displacement to publish plan.
04	Time	To dismantle information systems and networks, facilities, and equipment.
05	Time	To pack for displacement.
06	Percent	Of command post packed per loading plans or standing operating procedures.

ART 5.2.2.2 SELECT, RECONNOITER, AND EVALUATE THE NEW COMMAND POST LOCATION

5-59. Decide and confirm the suitability of the location to which the command post should displace, including the time of and for movement. (FM 6-0) (USACAC)

Note: ART 1.3 (Conduct Tactical Troop Movements) covers movement of command posts.

No.	Scale	Measure
01	Yes/No	Selection, reconnaissance, and evaluation of the new command post location resulted in a new location that effectively supports the exercise of command and control.
02	Time	To form reconnaissance party.
03	Time	For reconnaissance party to decide and confirm suitability of new location.
04	Time	For reconnaissance party to communicate findings to command post.
05	Percent	Of decisions of reconnaissance party valid for command post functions.

ART 5.2.2.3 OCCUPY THE NEW COMMAND POST LOCATION

5-60. Execute activities following a tactical or administrative movement to establish and conduct command post operations. (FM 6-0) (USACAC)

No.	Scale	Measure
01	Yes/No	Occupation of the new command post location allowed the command post to support the commander effectively in the exercise of command and control and within the time the unit standing operating procedures or operation order specifies.
02	Yes/No	Location of new command post met security requirements.
03	Time	To account for 100 percent of personnel and equipment after last arrival at final destination.
04	Time	To unpack equipment from transport.
05	Time	To set up facilities and information systems and networks for operation.
06	Time	To reestablish communications links after arrival.
07	Percent	Of command and control functions performed effectively after arrival.

ART 5.2.2.4 TRANSFER COMMAND AND CONTROL FUNCTIONS DURING DISPLACEMENT

5-61. Reassign command and control (C2) functions from the old location to the new location. Maintain C2 functions during the move without disrupting performance of C2 functions for the force. (FM 6-0) (USACAC)

No.	Scale	Measure
01	Yes/No	C2 functions transfer allowed the unit C2 system to maintain C2 functions without disruption.
02	Time	To transfer C2 to alternate command post.
03	Percent	Of effectiveness of alternate command post.
04	Percent	Of communications with subordinate headquarters during displacement.
05	Percent	Of effectiveness of C2 functions during displacement.
06	Percent	Of C2 functions not performed during displacement.

ART 5.2.3 EXECUTE SLEEP PLANS

5-62. Identify, counter, and minimize the degrading effects of night operations and sleep loss on units as they execute continuous operations. (FM 6-22.5) (USAMEDDC&S)

No.	Scale	Measure
01	Yes/No	Unit sleep plan provided adequate rest for effective performance over time.
02	Yes/No	Unit prepared sleep plan.
03	Yes/No	Unit complied with sleep plan during operations.
04	Percent	Of Soldiers obtaining sleep within the parameters of FM 6-22.5.
05	Number	Of Soldiers unable to perform due to lack of sleep.

ART 5.2.4 MANAGE STRESS

5-63. Identify, counter, and minimize the degrading effects of stress on units as they execute continuous operations. (FM 6-22.5) (USAMEDDC&S)

No.	Scale	Measure
01	Yes/No	Unit identified, countered, and minimized effects of stress on Soldiers to avoid degrading unit performance and mission accomplishment.
02	Number	Of misbehavior incidents due to effects of stress.
03	Number	Of decisions degraded by stress.

ART 5.2.5 MAINTAIN CONTINUITY OF COMMAND AND CONTROL

5-64. Ensure—through succession of command and the ability of the commander to exercise command and control (C2) continuously from any point in the area of operations—continuity in the exercise of the authority of command and continuity in the performance of functions, tasks, or duties necessary to accomplish a military operation or mission. ART 5.2.5 includes maintaining the functions and duties of the commander as well as the supporting functions and duties performed by the staff and others acting under the authority and direction of the commander. (FM 6-0) (USACAC)

No.	Scale	Measure
01	Yes/No	Unit maintained continuity of C2 throughout the mission.
02	Time	That commander was not available for command functions.
03	Time	Of communications loss with subordinates.
04	Time	To restore commander's understanding of current situation.
05	Time	To transfer command from one commander to another.
06	Number	Of times communications with the commander were lost.
07	Number	Of times the commander lost communications with subordinates.

SECTION III – ART 5.3: INTEGRATE INFORMATION SUPERIORITY CONTRIBUTORS

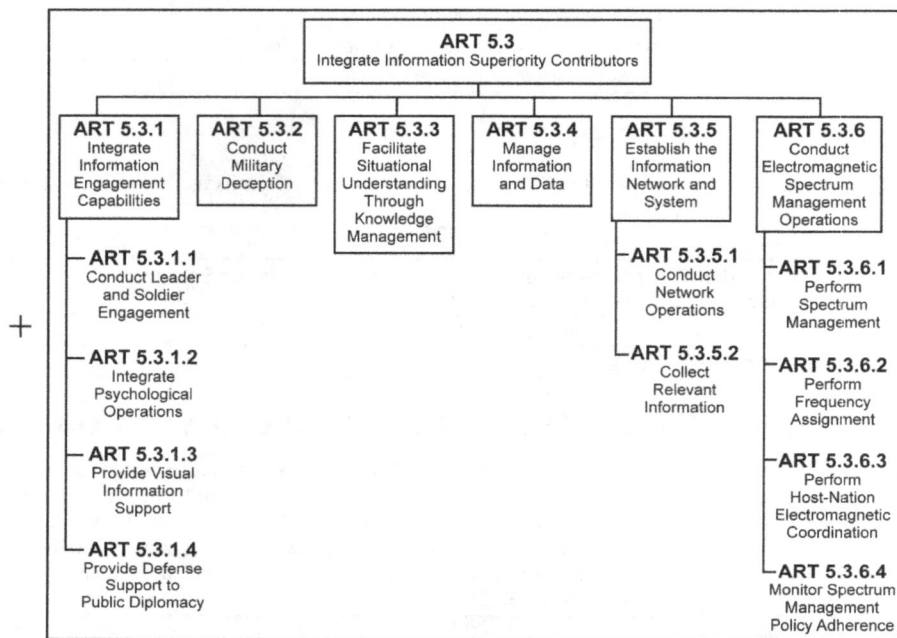

5-65. Commanders integrate four contributors to achieve information superiority during full spectrum operations. The four contributors are Army information tasks; intelligence, surveillance, and reconnaissance (ISR); knowledge management; and information management. ISR Army tasks are defined in ART 2.0 (The Intelligence Warfighting Function). (FM 3-0) (USACAC)

Note: The Army information tasks are defined within the appropriate warfighting functions.

Information engagement and military deception are included in this chapter.

Command and control warfare is in chapter 3, ART 3.0 (The Fires Warfighting Function).

Information protection and operations security are included in ART 6.0 (The Protection Warfighting Function).

ART 5.3.1 INTEGRATE INFORMATION ENGAGEMENT CAPABILITIES

5-66. *Information engagement* is the integrated employment of public affairs to inform U.S. and friendly audiences; psychological operations, combat camera, U.S. Government strategic communication and defense support to public diplomacy, and other means necessary to influence foreign audiences; and, leader and Soldier engagements to support both efforts. (FM 3-0) (USACAC)

No.	Scale	Measure
01	Yes/No	Unit identified information requirements in support of information engagement.
02	Yes/No	Unit developed information engagement objectives nested with commander's intent and strategic communications themes.
03	Yes/No	Unit developed information engagement measures of performance and measures of effectiveness.

No.	Scale	Measure
04	Yes/No	Unit determined information engagement role in achieving an operational advantage.
05	Yes/No	Unit participated in measures of performance and effects working groups.
06	Time	To integrate the analysis of socio-cultural, cognitive, physical, cyber, and electronic factors in the operational environment in support of information engagement operations.
07	Time	To provide input to base order for information engagement concept and prepare its annex with appendices.
08	Time	To integrate information engagement activities into the targeting process.
09	Time	To participate in the military decisionmaking process.
10	Time	To develop information engagement matrix and integrate with the target synchronization matrix.
11	Time	To maintain battle tracking and information engagement input to the common operational picture.
12	Time	To assess and analyze impact of information engagement activities in support of full spectrum operations.
13	Time	To assess measures of performance and measures of effectiveness.
14	Percent	Of information engagement activities not coordinated across staffs.
15	Percent	Of information engagement activities not coordinated with operations security, command and control warfare, information protection, and military deception.
16	Percent	Of information engagement working groups held to integrate information engagement in the operations process and synchronize information engagement activities.

ART 5.3.1.1 CONDUCT LEADER AND SOLDIER ENGAGEMENT

5-67. Face-to-face interaction by military leaders, Soldiers, Army civilians, and U.S. contractor personnel strongly influences the perceptions of the local populace. Carried out with discipline and professionalism, day-to-day interaction of Soldiers with the local populace among whom they operate has positive effects. Such interaction amplifies positive actions, counters enemy propaganda, and increases goodwill and support for the friendly mission. Likewise, meetings conducted by leaders with key communicators, civilian leaders, or others whose perceptions, decisions, and actions will affect mission accomplishment can be critical to mission success. These meetings provide the most convincing venue for conveying positive information, assuaging fears, and refuting rumors, lies, and misinformation. (FM 3-0) (USACAC)

No.	Scale	Measure
01	Yes/No	Unit commander prepared to conduct engagement and was familiar with the unit narrative, commander's scope of expertise, media personality profile and regulations concerning interviews.
02	Yes/No	Soldiers, Army civilians, and U.S. contractor personnel advised of the inevitability of media presence during military operations.
03	Yes/No	Soldiers, Army civilians, and U.S. contractor personnel briefed and acknowledged scope of information during interview.
04	Yes/No	Soldiers, Army civilians, and U.S. contractor personnel encouraged to represent themselves as soldiers and to speak about the jobs they perform for the Army.
05	Yes/No	Commander ensured media received maximum unrestricted disclosure of unclassified information in accordance with operational security guidance.
06	Yes/No	Units constructed and maintained media personality profiles to anticipate media objectives and were prepared to support those objectives.
07	Yes/No	Unit assessed outcomes of military leaders, Soldiers, Army civilians, and U.S. contractor personnel engagements and consolidated into some form of directive to conduct after action reviews.
08	Percent	Of local national actors and publics identified by unit.

No.	Scale	Measure
09	Percent	Of military leaders, Soldiers, Army civilians, and U.S. contractor personnel engagements coordinated across staffs.
10	Percent	Of local national actors and publics engaged by leaders and Soldiers.
11	Percent	Of military leaders, Soldiers, Army civilians, and U.S. contractor personnel engagements that successfully deliver unit narrative.
12	Time	Taken to issue friendly messages in response to crisis communications needs vs. adversary's propaganda timeline.

ART 5.3.1.2 INTEGRATE PSYCHOLOGICAL OPERATIONS

5-68. The Office of the Secretary of Defense reviews and approves all psychological operations (PSYOP) plans and programs. These include, at a minimum, PSYOP objectives, themes to stress and avoid, dissemination means and conduits, concept of operations, target audiences, and funding sources. Once PSYOP programs are approved, PSYOP planning occurs with the development of supporting PSYOP objectives, potential target audiences, and initial assessment criteria. PSYOP staff planners provide input to the military decisionmaking process, and incorporate PSYOP into the operation plan and order. (FM 3-05.30) (USAJFKSWCS)

No.	Scale	Measure
01	Yes/No	Unit reviewed U.S. laws, policies, regulations, and international treaties regarding the execution of PSYOP.
02	Yes/No	Unit reviewed approved PSYOP Office of the Secretary of Defense programs including PSYOP objectives, supporting PSYOP objectives, potential target audience lists, and initial assessment criteria.
03	Yes/No	Unit established approval authorities for PSYOP Series.
04	Yes/No	Unit obtained appropriate authorizations for the dissemination of PSYOP.
05	Yes/No	Unit incorporated programs approved by the Office of the Secretary of Defense into its operation plan and order.

ART 5.3.1.3 PROVIDE VISUAL INFORMATION SUPPORT

5-69. Create a record of unit activities. Still and motion video recordings include friendly forces, equipment, and positions before, during, and after engagements; terrain features in current or projected operational areas; battle damage to friendly, enemy, or host-nation property; any essential element of friendly information that assists a commander in conducting (planning, preparing, executing, and assessing) operations. (FM 6-02.40) (USASC&FG)

No.	Scale	Measure
01	Yes/No	Visual information support assisted the commander with the operations process.
02	Time	To provide combat camera detachment commander when determined necessary.
03	Time	To process media and produce imagery for short-notice tasking.
04	Time	To provide finished imagery products to customers.
05	Time	To provide finished imagery products to customers in the United States.
06	Time	To respond to and be on scene for short notice tasking.
07	Time	To review selected combat camera materials (for release, until release, including products to be released by public affairs representatives, before delivery to the releasing agency).
08	Percent	Of photographic images and electronic documentation available.
09	Percent	Of presented coverage deemed suitable by customer (such as audience attention and share).
10	Percent	Of subject coverage requests filled.

ART 5.3.1.4 PROVIDE DEFENSE SUPPORT TO PUBLIC DIPLOMACY

5-70. Defense support to public diplomacy is those activities and measures taken by the Department of Defense components to support and facilitate public diplomacy efforts of the U.S. Government. Defense support to public diplomacy is the military's role in supporting the U.S. Government's strategic communication program. It includes peacetime military engagement activities conducted as part of combatant commanders' theater security cooperation plans. (FM 3-0) (USACAC)

ART 5.3.2 CONDUCT MILITARY DECEPTION

5-71. Execute actions to mislead enemy or adversary military decisionmakers deliberately as to friendly military capabilities, intentions, and operations, thereby causing the enemy or adversary to take specific actions (or inactions) that contribute to accomplishing the friendly mission. Military deception includes simulative deception, simulative electronic deception, imitative electronic deception, and manipulative electronic deception. (FM 3-13) (USACAC)

No.	Scale	Measure
01	Yes/No	Unit achieved military deception objective.
02	Yes/No	Unit integrated military deception effort with other operations.
03	Yes/No	Military deception effort conformed to instructions from higher headquarters, statutory requirements, and international agreements.
04	Yes/No	Military deception effort supported the commander's intent.
05	Yes/No	Unit identified and then exploited target biases.
06	Yes/No	Unit maintained operations security throughout the process.
07	Yes/No	Unit used various sources—physical, electronic, imitative, simulative, and manipulative—to transmit the military deception story to the enemy.
08	Yes/No	Unit identified the deception target appropriate to the level of deception operation.
09	Time	To provide a military deception plan to support a branch or sequel.
10	Time	For enemy to identify military deception after commencing operations.
11	Time	To implement preplanned military deception plan.
12	Time	To identify key enemy decisionmakers.
13	Time	To identify enemy critical intelligence indicators.
14	Time	To assess operations security measures protecting the military deception operation from hostile detection and unwitting disclosure to friendly elements.
15	Percent	Of enemy force decoyed away from the area of the unit's decisive operation.
16	Percent	Of operation plans and orders that contain a military deception appendix to the information operations annex.
17	Percent	Of friendly military deception operations resulting in the enemy reallocating its resources.
18	Percent	Of enemy critical intelligence indicators successfully supplied with false or misleading information.
19	Percent	Of military deception events executed at the time and location specified.
20	Percent	Of physical environment and other conditions of the operational environment effectively used to support military deception effort.
21	Percent	Of false information contained in the military deception story.
22	Number	Of physical, technical, and administrative means available to convey information to the military deception target.

ART 5.3.3 FACILITATE SITUATIONAL UNDERSTANDING THROUGH KNOWLEDGE MANAGEMENT

5-72. Create, organize, apply, and transfer knowledge to facilitate situational understanding and decisionmaking. Knowledge management supports improving organizational learning, innovation, and performance. Knowledge management processes ensure that knowledge products and services are relevant, accurate, timely, and useable to commanders and decisionmakers. (FM 3-0) (USACAC)

No.	Scale	Measure
01	Yes/No	Knowledge management applied analysis and evaluation to information to create knowledge.
02	Yes/No	Commander's critical information requirements focused knowledge management product development.
03	Yes/No	Knowledge management narrowed the gap between relevant information required and information available.
04	Yes/No	Unit developed knowledge management plan.
05	Yes/No	Knowledge management plan addressed knowledge and information flow.
06	Yes/No	Knowledge management plan developed criteria for displaying the common operational picture.
07	Yes/No	Knowledge management supported developing situational awareness and situational understanding.
08	Yes/No	Knowledge management enabled rapid, accurate retrieval of previously developed knowledge to satisfy new requirements.
09	Yes/No	Knowledge management routed products to the appropriate individuals in a readily understood format.
10	Yes/No	Knowledge management kept commander and staff from being overwhelmed by information.

+ ART 5.3.4 MANAGE INFORMATION AND DATA

5-73. Direct relevant information to the right person at the right time in a usable format to facilitate situational understanding and decisionmaking. Use procedures and information systems to collect, process, display, store, and disseminate data and information. (FM 6-0) (USACAC)

No.	Scale	Measure
01	Yes/No	Information and data collected, processed, displayed, stored, and disseminated directed relevant information to the right persons at the right time in a usable format to facilitate situational understanding and decisionmaking.
02	Time	To direct, establish, and control the means by which the various staffs and forces send and receive operationally significant data or information, to minimize operations delayed or affected because of lack of information.
03	Time	For force common operating picture to reflect real-world situation.
04	Percent	Of critical information acquired and disseminated to subordinate commanders and appropriate members of subordinate staffs.
05	Percent	Of time that data was presented to the decisionmaker in the requested format.

+ ART 5.3.5 ESTABLISH THE INFORMATION NETWORK AND SYSTEM

5-74. Install, operate, and maintain networks that ensure information systems and personnel can perform information management activities. ART 5.3.5 includes network operations, functions (enterprise systems management, content staging and information dissemination management, and information assurance and computer network defense), and repositioning information systems (such as communications nodes) to maintain continuity of command and control (C2). (FM 6-0) (USACAC)

No.	Scale	Measure
01	Yes/No	Operations of the information network, systems, and personnel effectively performed information management activities in support of the commander's decisionmaking.
02	Time	To deliver high precedence messages.
03	Time	To integrate new headquarters into existing SECRET Internet Protocol Router Networks (SIPRNETs).
04	Time	To establish integrated communications system.
05	Time	To establish both data and voice communications with combatant command and its components.
06	Time	To submit frequency requests.
07	Percent	Of subordinate commanders and supporting staffs in communication with force headquarters.
08	Percent	Of C2, communications, and computer support systems that are operational.
09	Percent	Of C2, communications, and computer support systems that meet command reliability standards.
10	Percent	Of C2, communications, and computer support systems equipment down for unscheduled maintenance.
11	Percent	Of traffic sent on nondedicated or non-Department of Defense lines or circuits.
12	Percent	Of communications equipment, circuits, and connectivity with status accurately displayed.
13	Percent	Of communications outages with adequate back-up communications paths.
14	Percent	Of time essential C2, communications, and computer systems accessible from all locations.
15	Percent	Of force headquarters local area networks capable of interoperating.
16	Percent	Of time communications connectivity maintained with all units.
17	Percent	Of time voice and data communications (unsecure and secure) maintained.
18	Percent	Of equipment interoperability problems that result in late or lost data.
19	Percent	Of resources requested to fill all shortfalls
20	Percent	Of critical C2, communications, computer, and intelligence architecture nodes identified in the operation plan.
21	Number	Of instances of delay, disruption, or corruption of operational C2, communications, and computer systems due to failure to follow established policies and procedures.

ART 5.3.5.1 CONDUCT NETWORK OPERATIONS

5-75. Perform the essential tasks to support situational awareness and command and control that commanders will use to operate and defend the network. (FMI 6-02.60) (USASC&FG)

No.	Scale	Measure
01	Yes/No	The G-6 prepared and maintained information systems estimates, plans, and orders.
02	Yes/No	The G-6 recommended command post locations based on the information environment.
03	Yes/No	Unit established automation systems administration procedures for all information systems.
04	Yes/No	Unit established procedures for collecting, processing, displaying, storing, and disseminating data and information within the headquarters, staff sections, and major subordinate commands throughout the operations process.
05	Time	To manage and control information network capabilities and services.
06	Time	To coordinate, plan, and direct all information systems support interfaces with joint and multinational forces, including host-nation forces.
07	Time	To coordinate the availability of commercial information systems and information services.

No.	Scale	Measure
08	Time	To coordinate unit commercial and military satellite communications requirements.
09	Percent	Of time monitoring and making recommendations on all technical information systems activities.
10	Percent	Of time assessing information systems vulnerability and risk management (with the G-2 and G-7).
11	Percent	Of time recommending information systems network priorities for battle command.
12	Percent	Of time ensuring that redundant communications means are planned and available to pass time-sensitive information.

ART 5.3.5.2 COLLECT RELEVANT INFORMATION

5-76. Continually collect relevant information about the factors of mission, enemy, terrain and weather, troops and support available, time available, and civil considerations from the information environment by any means for processing, displaying, storing, and disseminating to support conducting current and future operations. (FM 6-0) (USACAC)

> *Note:* The distinction between ART 5.3.5.2 and ART 2.3 (Perform Intelligence, Surveillance, and Reconnaissance) is that ART 2.3 involves collecting information from the operational environment while ART 5.3.5.2 involves integrating intelligence information and products that exist within the command and control systems.

No.	Scale	Measure
01	Yes/No	Information collected supported military decisionmaking.
02	Time	Since latest relevant information collected.
03	Percent	Of decisions delayed awaiting collection of appropriate data.
04	Percent	Of requested information collected within time desired.
05	Percent	Of collected information that was high quality.

ART 5.3.5.2.1 Collect Friendly Force Information Requirements

5-77. Collect data about friendly forces from the information environment for processing, displaying, storing, and disseminating to support command and control functions. (FM 6-0) (USACAC)

No.	Scale	Measure
01	Yes/No	Information collected about friendly forces supported decisionmaking.
02	Yes/No	Information collected was current.
03	Percent	Of accuracy of information on essential logistics, maintenance, and personnel requirements and reports.
04	Percent	Of accuracy of information regarding subordinate unit deployments.
05	Percent	Of accuracy of data used by operations staff.
06	Percent	Of accuracy of subordinate unit status.
07	Percent	Of friendly units or organizations and personnel with current status known.

ART 5.3.5.2.2 Integrate Intelligence Products

5-78. Collect intelligence products from intelligence sources and combine them with friendly force information requirements for use in command and control functions. (FM 6-0) (USACAC)

Note: ART 2.3 (Perform Intelligence, Surveillance, and Reconnaissance) involves collecting information about the enemy or adversary and other conditions of the operational environment.

No.	Scale	Measure
01	Yes/No	Intelligence products integrated by information management supported decisionmaking.
02	Time	To establish connectivity with component intelligence agencies, combatant command, and national intelligence agencies (after arrival).
03	Time	To integrate intelligence products by users or common operational picture.
04	Percent	Of accuracy of intelligence situation displays integrated with other mission-essential information.
05	Percent	Of intelligence products properly integrated with other information.

ART 5.3.5.2.3 Assess Accuracy, Timeliness, Usability, Completeness, and Precision of Collected Information

5-79. Apply the criteria of accuracy, timeliness, usability, completeness, and precision to evaluate the quality of relevant information collected. (FM 6-0) (USACAC)

No.	Scale	Measure
01	Yes/No	Relevant information that meets the quality criteria served the commander's needs.
02	Time	To assess collected relevant data.
03	Percent	Of available information examined and considered in latest status reporting.
04	Percent	Of accuracy of data transmitted and disseminated.
05	Percent	Of time information passed by specified time.
06	Percent	Of time information on commander's critical information requirements passed by specified time.
07	Percent	Of time mission-essential information and threat assessments passed by specified time.
08	Percent	Of reports with no significant errors.

ART 5.3.5.2.4 Process Relevant Information to Create a Common Operational Picture

5-80. Add meaning to relevant information by formatting, plotting, translating, correlating, aggregating, organizing, categorizing, analyzing, and evaluating it to create the common operational picture (COP). (FM 6-0) (USACAC)

Note: The COP is an operational picture tailored to the user's requirements based on common data and information shared by more than one command. The COP facilitates collaborative planning and helps all echelons to achieve situational understanding.

No.	Scale	Measure
01	Yes/No	Meaning added to relevant information to create the COP supported decisionmaking.
02	Time	To compile appropriate reports.
03	Time	To provide analysis and evaluation to information.
04	Percent	Of time that unit maintained accurate and current COP in the area of operations.
05	Percent	Of reports processed within time limits.

ART 5.3.5.2.5 Display a Common Operational Picture Tailored to User Needs

5-81. Present relevant information in audio or visual formats that convey the common operational picture (COP) for decisionmaking and exercising other command and control functions. The COP format should be easily understandable to the user and tailored to the needs of the user and the situation. (FM 6-0) (USACAC)

No.	Scale	Measure
01	Yes/No	COP was tailored to the user's needs and the situation was easily understandable.
02	Time	Between the force COP and the real-world situation to maintain operational information, force status, and the capability to respond to an emerging situation and project branches or sequels.
03	Time	To display shared local databases.
04	Time	For decisionmaker to understand display in decisionmaking.
05	Percent	Of accurate mission-essential information maintained on situation displays.
06	Percent	Of current operational data displays.

ART 5.3.5.2.6 Store Relevant Information

5-82. Retain relevant information in any form that allows processing, displaying, or disseminating to authorized users when and as required to exercise command and control. ART 5.3.5.2.6 includes maintaining distributed or common databases. (FM 6-0) (USACAC)

No.	Scale	Measure
01	Yes/No	Stored relevant information was readily available for processing, displaying, or disseminating when and as required to support exercise of command and control.
02	Yes/No	Unit preserved and retained relevant information for historians to collect.
03	Time	To post unit reports to appropriate databases or pass to work centers (from receipt).
04	Time	To access and display shared remote databases.
05	Time	To enter most current information on force status.
06	Time	To access shared local databases.
07	Percent	Of relevant information required already in databases.

ART 5.3.5.2.7 Disseminate Common Operational Picture and Execution Information

5-83. Convey relevant information of any kind from one person or place to another by any means to improve understanding or to initiate or govern action, conduct, or procedure. (FM 6-0) (USACAC)

No.	Scale	Measure
01	Yes/No	Common operational picture and execution information disseminated between headquarters in time and with sufficient quality and quantity to allow those headquarters to initiate actions that met the commander's intent.
02	Yes/No	After approval, all orders and plans received by components and adjacent units.
03	Time	To process and disseminate status information (to subordinate units).
04	Time	To communicate all approved orders and plans to subordinate and adjacent units.
05	Time	To prepare and forward situation reports.
06	Percent	Of organizations or units receiving latest information.
07	Percent	Of command plans, reports, and other information passed error free.
08	Percent	Of addressees that received message.
09	Percent	Of addressees that received critical planning messages (such as warning orders).
10	Percent	Of messages sent outside normal communications channels.
11	Percent	Of accurate orders and requirements transmitted or disseminated within reporting criteria.
12	Percent	Of information that gets to appropriate people.
13	Percent	Of reports disseminated to all agencies within time limits.

No.	Scale	Measure
14	Percent	Of time mission-essential information passed within established criteria.
15	Number	Of instances where incoming information (which could affect outcome of operation) is not received by the person responsible for action.

ART 5.3.5.2.8 Communicate with Non-English Speaking Forces and Agencies

5-84. Communicate verbally, nonverbally, in writing, or electronically in the appropriate language of multinational, host-nation, and nongovernmental forces and agencies to meet all command and control (C2) requirements. This task requires U.S. personnel to establish and maintain effective rapport with the leaders and staff of multinational or host-nation forces. Information exchange with multinational forces is achieved through the establishment of a multinational information sharing enterprise called the Combined Enterprise Regional Information Exchange System. (FM 6-0) (USACAC)

Note: Classified and sensitive information is not passed to multinational partners in violation of policy guidance.

No.	Scale	Measure
01	Yes/No	Communications with non-English speaking personnel sufficiently met all C2 requirements to allow mission accomplishment effectively by force as a whole.
02	Yes/No	Unit protected classified and sensitive information when communicating with multinational partners.
03	Time	To communicate information or orders from one language to another orally or in writing.
04	Time	To develop interpersonal relationships.
05	Percent	Of needed information, not passed to or received by allies (due to lack of equipment interoperability).
06	Percent	Of needed information not passed to allies due to classification constraints.
07	Percent	Of accuracy of communication with non-English-speaking forces and agencies.
08	Percent	Of required linguist support provided.
09	Number	Of missions not accomplished and commitments not met due to faulty or lacking language support.
10	Number	Of cultural mistakes occurring while dealing with non-English-speaking forces or agencies.

ART 5.3.6 CONDUCT ELECTROMAGNETIC SPECTRUM MANAGEMENT OPERATIONS

5-85. Electromagnetic spectrum (EMS) operations is the overarching concept that incorporates spectrum management, frequency assignments, policy implementation, and host-nation coordination that enables the commander's effective use of the electromagnetic spectrum for full spectrum operations. Electromagnetic spectrum operations enable and support the six warfighting functions of command and control, intelligence, fires, movement and maneuver, protection, and sustainment. EMS operations consists of planning, operating, and coordinating the use of the electromagnetic spectrum through operational, engineering, administrative, and policy implementation procedures. The objective of EMS operations is to enable electronic systems that rely on wireless connectivity to perform their functions in the intended environment without causing or suffering unacceptable frequency interference. (JP 6-0) (USJFCOM JWFC)

ART 5.3.6.1 PERFORM SPECTRUM MANAGEMENT

5-86. Spectrum management consists of evaluating and mitigating electromagnetic environmental effects; managing frequency records and databases; de-conflicting frequencies, frequency interference resolution;

allotting frequencies; and coordinating electronic warfare to ensure electromagnetic dependent systems operate as intended. (JP 6-0) (USJFCOM JWFC)

No.	Scale	Measure
01	Yes/No	Spectrum was managed to satisfy mission requirements and met commander's intent.
02	Yes/No	All systems operated with no interference.
03	Time	To convert raw data to a useable format.
04	Time	To maintain databases.
05	Time	To deconflict spectrum assignments.
06	Time	To evaluate environmental effects.
07	Time	To resolve frequency interference.
08	Time	To coordinate electronic warfare issues.
09	Time	To process interference report.
10	Time	To coordinate, develop, and publish a joint restricted frequency list.
11	Percent	To coordinate with network managers.
12	Number	Of frequency assignments managed.
13	Number	Of systems requiring spectrum.

ART 5.3.6.2 PERFORM FREQUENCY ASSIGNMENT

5-87. Frequency assignment is the requesting and issuance of authorization to use frequencies for specific equipment such as combat net radio and frequencies for Army wireless networks. This also includes the planning necessary for these systems. (JP 6-0) (USJFCOM JWFC)

No.	Scale	Measure
01	Yes/No	Sufficient frequencies were requested to meet mission requirements.
02	Yes/No	Sufficient frequencies were available to meet mission requirements.
03	Yes/No	Frequency re-use plans were utilized to efficiently use spectrum.
04	Time	To generate radio loadsets and frequency plans.
05	Time	To build a communications-electronics operating instructions.
06	Time	Between request for frequencies.
07	Time	To obtain frequency approval.
08	Time	To design the frequency plan for area networks.

ART 5.3.6.3 PERFORM HOST-NATION ELECTROMAGNETIC COORDINATION

5-88. Each nation has sovereignty over its electromagnetic spectrum. The use of electromagnetic spectrum must be negotiated on a case-by-case basis. Approval to transmit within a country is based on the sovereignty of that country and their regulatory processes that evaluate the Department of Defense request for use of spectrum perceived potential for electromagnetic interference to local receivers. Use of military or commercial electromagnetic spectrum systems in host nations requires coordination and negotiation that result in formal approvals and certifications. (JP 6-0) (USJFCOM JWFC)

No.	Scale	Measure
01	Yes/No	Unit established relationship with host-nation agency responsible for radio frequency spectrum.
02	Yes/No	Command informed of host-nation restrictions on spectrum use.
03	Time	To process frequency request.
04	Time	To obtain frequency approval.
05	Percent	Of frequency requests filled.
06	Percent	Of mission degradation due to inadequate spectrum.

ART 5.3.6.4 MONITOR SPECTRUM MANAGEMENT POLICY ADHERENCE

5-89. International use of the electromagnetic spectrum is coordinated globally through the International Telecommunications Union). The United States Military Communications-Electronics Board (MCEB) is the main coordinating body for signal matters among Department of Defense (DOD) components. The main enforcement mechanism for DOD systems is the Spectrum Certification process (the frequency allocation to equipment process utilizing the DD Form 1494 [Application for Equipment Frequency Allocation]), which must be completed for all systems and equipment that emit or receive Hertzian waves. (JP 6-0) (USJFCOM JWFC)

No.	Scale	Measure
01	Yes/No	Unit observed all spectrum policies, regulations, and rules.
02	Yes/No	Commander was aware of policy, regulations, and rules affecting operations.
03	Time	To complete DD Form 1494.
04	Time	To get MCEB approval.
05	Time	To disseminate spectrum supportability guidance.
06	Number	Of systems not meeting Spectrum Certification compliance.

SECTION IV – ART 5.4: CONDUCT CIVIL-MILITARY OPERATIONS

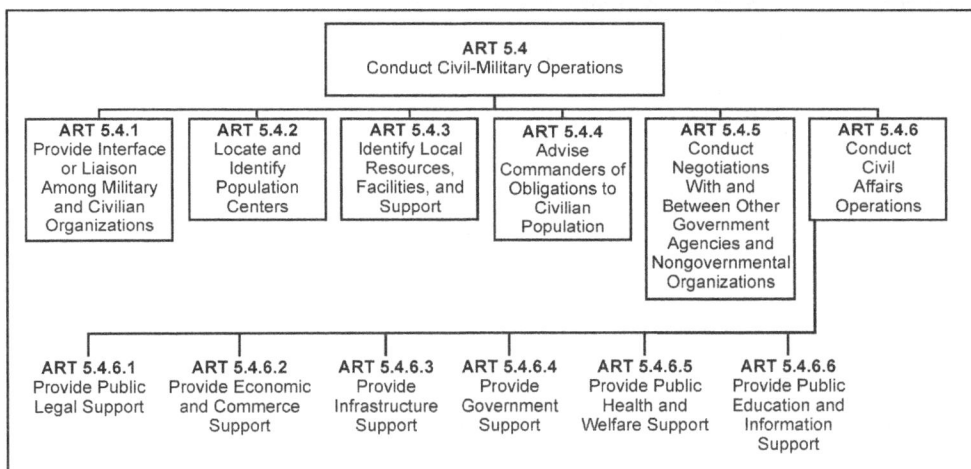

```
                                    ART 5.4
                        Conduct Civil-Military Operations

   ART 5.4.1      ART 5.4.2     ART 5.4.3      ART 5.4.4      ART 5.4.5      ART 5.4.6
   Provide        Locate and    Identify Local Advise         Conduct        Conduct
   Interface      Identify      Resources,     Commanders of  Negotiations   Civil
   or Liaison     Population     Facilities, and Obligations to With and      Affairs
   Among Military Centers       Support        Civilian       Between Other  Operations
   and Civilian                                Population      Government
   Organizations                                              Agencies and
                                                              Nongovernmental
                                                              Organizations

   ART 5.4.6.1    ART 5.4.6.2   ART 5.4.6.3    ART 5.4.6.4    ART 5.4.6.5    ART 5.4.6.6
   Provide Public Provide Economic Provide      Provide        Provide Public Provide Public
   Legal Support  and Commerce  Infrastructure Government      Health and     Education and
                  Support       Support        Support        Welfare Support Information
                                                                             Support
```

5-90. Civil-military operations (CMO) are the activities of a commander that establish, maintain, influence, or exploit relations between military forces, governmental and nongovernmental civilian organizations and authorities, and the civilian populace in a friendly, neutral, or hostile operational area in order to facilitate military operations to consolidate and achieve U.S. objectives. CMO may include performance by military forces of activities and functions normally the responsibility of local, regional, or national government. These activities may occur before, during, or after other military actions. They may also occur, if directed, in the absence of other military operations. CMO may be performed by designated civil affairs (CA), by other military forces, or by a combination of CA and other forces. (FM 3-05.40) (USAJFKSWCS)

No.	Scale	Measure
01	Yes/No	The conduct of CA operations in the area of operations (AO) supported mission accomplishment by Army, multinational, and special operations forces.
02	Yes/No	Army forces facilitated and mediated negotiations among opposing ethnic, cultural, religious, and social groups in the AO.

No.	Scale	Measure
03	Yes/No	Environmental regulations, laws, and considerations were taken into account during planning and present in procedures being followed.
04	Yes/No	Unit conducted civil reconnaissance focusing specifically on the civil component: area, structures, capabilities, organizations, people, and events.
05	Yes/No	Unit established measures of effectiveness with a mechanism to monitor and assess those measures.
06	Yes/No	Unit developed a concept to execute CA operations that were by, through, or with military and civilian organizations.
07	Yes/No	Unit ensured the CA plan and its execution addressed the management of civil information.
08	Time	To execute the CA operation plan or order.
09	Time	To refine CA program or produce CMO annex to operation order after receipt of warning order.
10	Time	To establish civil-military operations center.
11	Time	For legal authority to review CA plan.
12	Time	To deploy CA personnel in support of operations.
13	Time	To coordinate with local authorities on local crowd control measures.
14	Time	To assess situation and define requirements.
15	Time	To coordinate and liaise with host-nation populations and institutions, intergovernmental organizations, nongovernmental organizations, other government agencies, and military units represented in the AO.
16	Time	To identify and integrate civil affairs support and appropriate CA essential elements of information into command intelligence programs in the AO.
17	Time	To submit the CA plan to the supported element for review and approval.
18	Time	To identify cultural, ethnic, social, and religious characteristics of the local populace.
19	Time	To provide the supported element with necessary relevant cultural information to mitigate acts contrary to local culture and norms.
20	Percent	Of local population able to maintain normal day-to-day activities.
21	Percent	Of local population able to remain in their homes.
22	Percent	Of U.S. military supplies and resources in the AO used to perform CA operations.
23	Number	Of instances of supportive and hostile actions directed toward civilians in the AO.
24	Number	And types of restrictions imposed on the use of cultural property.
25	Number	Of instances of supportive and hostile actions directed toward the military in the AO.

ART 5.4.1 PROVIDE INTERFACE OR LIAISON AMONG MILITARY AND CIVILIAN ORGANIZATIONS

5-91. Provide interface among U.S. forces, host-nation population and institutions, nongovernmental organizations, intergovernmental organizations, and other government agencies. Minimize the likelihood and effects of civil interference with military operations. Prepare and transition liaison activities to follow-on military and civilian units and organizations as appropriate. (FM 3-05.40) (USAJFKSWCS)

No.	Scale	Measure
01	Yes/No	Liaison performed in the area of operations (AO) contributed to accomplishment of the mission.
02	Yes/No	Unit engaged host-nation forces capable of promoting stability.
03	Yes/No	Environmental considerations planning and procedures were present and being followed.
04	Time	To identify key civilian agencies and officials in the AO.

No.	Scale	Measure
05	Time	To establish liaison with military and civilian organizations.
06	Time	To establish information system interconnectivity between organizations in the AO.
07	Number	Of persons aided by interagency requests for assistance met in the AO.
08	Number	Of U.S. units in the AO assigned with host-nation liaison officers.
09	Number	Of different organizations collaborating on projects in the AO.
10	Number	Of instances of insufficient support provided to and from other nations, groups, and agencies.
11	Number	Of instances of liaison activities with military and civilian organizations initiated by the supported element.
12	Number	Of instances of liaison activities with the supported element initiated by military and civilian units and organizations.
13	Number	And types of information systems used to maintain information flow between liaison teams and supported or supporting organizations.
14	Cost	Of establishing and maintaining effective liaison activities in the AO.

ART 5.4.2 LOCATE AND IDENTIFY POPULATION CENTERS

5-92. Locate and identify population centers in the area of operations that may have impact on military operations or where military operations may significantly impact the civilian population. Anticipate civilian reaction to military operations in or near population centers. (FM 3-05.40) (USAJFKSWCS)

No.	Scale	Measure
01	Yes/No	The location and identification of population centers in the area of operations (AO) supported mission accomplishment by U.S. forces.
02	Yes/No	Civil affairs unit conducted area assessment in accordance with command guidance and mission requirements.
03	Yes/No	Advised the commander regarding the impact of operations on the civilian population.
04	Yes/No	Identified local areas likely to require protection from military activities.
05	Time	To identify the impact of military operations on the civilian population.
06	Time	To identify population centers that might interfere with combat operations in the AO.
07	Time	To plan, in conjunction with military police and local authorities, for the orderly movement of local dislocated civilians during the conduct of combat operations in AO.
08	Percent	Of operations anticipated to be degraded, delayed, or modified due to dislocated civilians in the AO.
09	Percent	Of resources in the AO required to control the movement of displaced civilians.
10	Percent	Of dislocated civilian control provided by the host nation.
11	Number	Of civilian casualties sustained during the conduct of military operations in the AO.
12	Number	Of dislocated civilian centers established in the AO.

ART 5.4.3 IDENTIFY LOCAL RESOURCES, FACILITIES, AND SUPPORT

5-93. Identify, locate, and help acquire local resources, civilian labor, facilities, and other support that tactical organizations require to accomplish their missions. Assist in the coordination and administration of host-nation agreements and other forms of support. (FM 3-05.40) (USAJFKSWCS)

No.	Scale	Measure
01	Yes/No	The identification of local resources, facilities, and other support assisted in accomplishing the mission of Army forces in the area of operations (AO).
02	Yes/No	Civil affairs unit completed area assessment per command guidance and mission requirements.
03	Yes/No	The diversion of local resources, facilities, and other support from civil to military units did not affect the local economy and civilian community in an unacceptable manner.
04	Time	To coordinate host-nation support agreements before or after unit arrival in an AO.
05	Time	To identify sources of dietary items necessary to support the host-nation population and other personnel operating in the AO.
06	Time	To identify sources of nondietary items necessary to support host-nation and U.S. personnel operating in the AO.
07	Time	To identify sources of transportation assets that can be used to support the mission.
08	Time	To identify housing facilities that can be used to support the mission of Army forces.
09	Time	To identify local people—interpreters, skilled craftsmen, professionals, and laborers—who can be hired to support the mission.
10	Percent	Of supplies and services used by U.S. units procured from host-nation sources.
11	Percent	Of sustainment and logistic efforts in the AO provided by host nation.
12	Percent	Of local economy that will be affected by the acquisition of local resources, facilities, and other support.
13	Number	And types of facilities used by U.S. units provided by host nation.
14	Number	And types of host-nation support agreements in effect in the AO.
15	Cost	Of identifying local resources, facilities, and support in the AO.

ART 5.4.4 ADVISE COMMANDERS OF OBLIGATIONS TO CIVILIAN POPULATION

5-94. Develop, with the staff judge advocate, requirements, and guidance for military personnel concerning the treatment of civilians. Provide authoritative guidance to military personnel on the customary and treaty law applicable to the conduct of land warfare as it applies to civilians. (FM 27-100) (TJAGLCS)

No.	Scale	Measure
01	Yes/No	Commander was advised on civil laws, common practices, and local customs that conflict with U.S. law, international agreements, or internationally recognized individual human rights.
02	Yes/No	Before mission start, the supporting G-9 or S-9 analyzed, formulated, and presented information about the civil aspects of the area of operations (AO). The format for this briefing package followed the supported command's standing operating procedures and was modified throughout the conduct of the operation as conditions changed.
03	Yes/No	Unit G-9 or S-9 advised the commander on essential goods and services—food, shelter, health support—necessary to sustain life for civilians.
04	Time	For the unit G-9 or S-9 to coordinate with appropriate agencies in the AO, such as the staff judge advocate, U.S. country team, and host-nation government legal officials.
05	Time	For the unit G-9 or S-9 to analyze the impact of the mission on civilians and to predict civilian casualties, property destruction, and infrastructure dislocation.
06	Time	For the unit G-9 or S-9 to consider religious aspects, customs, and cultures in the AO and their effect on operations.
07	Percent	Of difference between planned and actual requirements to support operations in AO.
08	Cost	Of meeting the commander's obligations for local populace.

ART 5.4.5 CONDUCT NEGOTIATIONS WITH AND BETWEEN OTHER GOVERNMENT AGENCIES AND NONGOVERNMENTAL ORGANIZATIONS

5-95. Conduct negotiations between the U.S. and local, military and civilian organizations. These negotiations can range widely in size and scope. (FM 3-07) (USACAC)

No.	Scale	Measure
01	Yes/No	The outcome of negotiations supported accomplishing the mission of U.S. forces in the area of operations (AO).
02	Yes/No	U.S. forces facilitated and mediated negotiations between opposing ethnic, cultural, environmental, commercial, religious, and social groups in the AO.
03	Yes/No	Negotiating team conditionally agreed on outcome of the negotiations before the start of negotiations.
04	Yes/No	Negotiating team avoided making the initial offer in the negotiations.
05	Yes/No	Negotiating team kept the objective of the negotiations in mind and focused on the big picture in lieu of single issues.
06	Yes/No	Negotiating team discovered additional variables, concessions, or bargaining chips during negotiations.
07	Yes/No	Negotiating team had a thorough understanding of U.S. interests and goals for negotiations.
08	Yes/No	Negotiating team kept accurate notes on progress of negotiations that may be referred to if the other party forgets, misunderstands, or attempts to distort interpretations of what was discussed and agreed.
09	Yes/No	Negotiating team summarized and confirmed the understandings of all parties continually.
10	Yes/No	Negotiating team had the skills to conduct negotiations. This included good communications skills; ability to use the dynamics of conflict; and knowledge of the area, culture, economies, political philosophy, language, customs, history, wants, needs, goals, probable assumptions, and communications or negotiation styles of the other parties to the negotiations.
11	Yes/No	Negotiating team's higher headquarters approved the negotiating team's concessions and positions.
12	Yes/No	Negotiation team negotiated terms for exchange of prisoners of war.
13	Yes/No	Negotiation team negotiated arrangements with belligerents.
14	Yes/No	Negotiation team negotiated or modified regional security arrangements with all interested parties.
15	Yes/No	Negotiation team negotiated the enhancement of cross border controls and security.
16	Yes/No	Negotiation team briefed the supported element on issues in negotiation.
17	Yes/No	Negotiating team gained preauthorization for expected negotiations outcome and concessions.
18	Yes/No	Negotiating team did not obligate the supported element on terms not previously approved.
19	Yes/No	Environmental regulations, laws, and considerations were taken into account during planning and present in procedures being followed.
20	Time	To complete background information on the parties involved in the negotiations to identify needs and interests to include personal and emotional aspects.
21	Time	To establish the conditions necessary for the conduct of successful negotiations. This includes building trust, rapport, and empathy with the other individuals involved in the negotiations.
22	Time	To determine negotiating concessions and bargaining chips that can be exploited during the conduct of negotiations.

No.	Scale	Measure
23	Time	To complete a legal review of agreements reached during negotiations.
24	Percent	Of desired objectives obtained during negotiations.
25	Percent	Of time U.S. force gets something in exchange for some type of concession in its negotiating position.
26	Number	And types of negotiations currently ongoing in the AO.
27	Cost	Of conducting negotiations with and between other government agencies and nongovernmental organizations.

ART 5.4.6 CONDUCT CIVIL AFFAIRS OPERATIONS

5-96. Conduct civil affairs operations by providing public legal support, economic and commerce support, infrastructure support, government support, public health and welfare support, and public education and information support. (FM 3-05.40) (USAJFKSWCS)

ART 5.4.6.1 PROVIDE PUBLIC LEGAL SUPPORT

5-97. Establish the fair, competent, and efficient application and fair and effective enforcement of the civil and criminal laws of a society through impartial legal institutions and competent police and corrections systems. This functional area includes judge advocates trained in international and comparative law as well as civil affairs specialists in related subjects. (FM 3-05.40) (USAJFKSWCS)

No.	Scale	Measure
01	Yes/No	The civilian public legal system operating in the area of operations (AO) provided for the rule of law and justice to the civilian population and supported the mission of the U.S. force.
02	Yes/No	The unit established liaison, evaluated the current judicial system, and developed plans that will improve the existing judicial system of the host nation.
03	Yes/No	The supporting civil affairs staff finalized the courses of action and briefed the supported commander and staff.
04	Time	To establish liaison with local judicial officials.
05	Time	To evaluate the judicial system in the AO. This includes determining effectiveness of the civilian court system, judiciary reliability, quantity and quality of legal resources available to civilians. It also includes finding out which areas need legal assistance; the range of criminal and civil legislation and statutes; and methods used to record, report, and publish laws and decisions, and deficiencies on government operations.
06	Time	To determine to reopen local civilian tribunals.
07	Time	To develop plans for providing legal support, such as establishing tribunals and other judicial and administrative agencies and determining the number, types, jurisdiction, procedures, and delegation of appointing authority of the tribunals and other judicial and administrative agencies.
08	Time	To establish a prison and institution administration review system.
09	Time	To develop an advisory program to upgrade the judiciary and other legal agencies.
10	Time	On average that individuals wait to settle civil and criminal charges.
11	Percent	Of available legal support in the AO devoted to supporting public legal support.
12	Percent	Of legal actions occurring in the AO conducted solely by local civilians.
13	Number	Of judge advocate general personnel supporting the conduct of public legal support.
14	Number	Of other Army personnel supporting the conduct of public legal support.
15	Number	Of local civilian lawyers, judges, and legal assistants cleared to conduct public legal support.
16	Number	Of criminal and administrative legal cases in the AO.
17	Cost	To provide public legal support in the AO.

ART 5.4.6.1.1 Provide Support to Indigenous Judicial Systems

5-98. Provide support to local judiciary system, establish civil administration courts, and help in preparing or enacting necessary laws for the enforcement of U.S. policy and international law. Prepare and conduct the transition of public legal support to follow-on military and civilian agencies as appropriate. (FM 3-05.40) (USAJFKSWCS)

No.	Scale	Measure
01	Yes/No	The civilian public legal system operating in the area of operations (AO) provided for the rule of law and justice to civilians and supported the mission of the U.S. forces.
02	Yes/No	The unit established liaison, evaluated the current judicial system, and developed plans to improve the existing judicial system of the host nation.
03	Yes/No	The supporting G-9 or S-9 finalized the courses of action and briefed the supported commander and staff.
04	Yes/No	Environmental considerations planning and procedures were present and being followed.
05	Time	To establish liaison with local judicial officials.
06	Time	To evaluate the judicial system in the AO. This includes determining effectiveness of the civilian court system, judiciary reliability, quantity and quality of legal resources available to civilians. It also determined which areas needed legal assistance; a range of criminal and civil legislation and statutes; and procedures to record, report, and publish laws, decisions, and deficiencies on government operations.
07	Time	To determine to reopen local civilian tribunals.
08	Time	To develop plans for providing legal support. Support includes establishing tribunals and other judicial and administrative agencies and determining the number, types, jurisdiction, procedures, and delegation of appointing authority of the tribunals and other judicial and administrative agencies.
09	Time	To establish a prison and institution administration review system.
10	Time	To develop an advisory program to upgrade the judiciary and other legal agencies.
11	Time	On average that individuals wait to settle civil and criminal charges.
12	Percent	Of available legal support in the AO devoted to supporting public legal support.
13	Percent	Of legal actions occurring in the AO conducted solely by local civilians.
14	Number	Of judge advocate general personnel supporting the conduct of public legal support.
15	Number	Of U.S. and multinational personnel supporting the conduct of public legal support.
16	Number	Of local civilian lawyers, judges, and legal assistants cleared to conduct public legal support.
17	Number	Of criminal and administrative legal cases in the AO.
18	Cost	To provide public legal support in the AO.

ART 5.4.6.1.2 Provide Property Control Support

5-99. Identify private and public property and facilities available for military use, and recommend policies and procedures to obtain them. Coordinate military acquisition of civilian property and facilities needed by military forces. Establish policies and procedures concerning custody and administration of private and public property. Control negotiable assets and resources of potential military use that are not otherwise under the supervision of other agencies. Prepare and conduct the transition of property controls to follow-on military and civilian agencies as appropriate. (FM 3-05.40) (USAJFKSWCS)

No.	Scale	Measure
01	Yes/No	The property control system in the area of operations (AO) supported the mission of U.S. forces.
02	Yes/No	The unit established liaison with civilian and military engineers or property control agencies, evaluated the current property control system, and developed plans to improve the existing property control system of the host nation.
03	Yes/No	The supporting G-9 or S-9 finalized courses of action and briefed the supported commander and staff.
04	Yes/No	Environmental considerations planning and procedures were present and being followed.
05	Time	To prepare property acquisition plans. Property acquisition plans classify the property to be acquired; identify and coordinate the acquisition of property for military use; prescribe a recording system for property acquired by the military from civilian sources; prescribe measures to protect and preserve civilian ownership records; prescribe measures to safeguard and properly manage the acquired property; plan for scheduling the acquisition of property; and plan for controlling negotiable assets and resources of potential military use that are not supervised by other agencies.
06	Time	To review property control support plans for compliance with international laws, treaties, and agreements.
07	Time	To establish a claims process for return and compensation for seized property.
08	Percent	Of difference between planned and actual requirements for property acquisition and control in the AO.
09	Percent	Of planned property acquisitions in the AO completed.
10	Percent	Of AO acquired for U.S. military use.
11	Number	Of U.S. property control teams operational in the AO.
12	Number	Size and types of property used to support military operations in the AO.
13	Number	Of instances when U.S. forces are not able to establish property control over civilian property acquired in the AO.
14	Cost	To establish and maintain property controls over civilian property acquired in the AO.

ART 5.4.6.2 PROVIDE ECONOMIC AND COMMERCE SUPPORT

5-100. Determine the availability of local resources for military and civilian use. Determine the economic controls and the governmental structure related to economics and commerce. Help in developing and executing price control and rationing programs. Direct support to keep key industries operating. Advise, assist, or direct governmental economic and commercial agencies. Develop and implement plans to prevent black-market activities. (FM 3-05.40) (USAJFKSWCS)

No.	Scale	Measure
01	Yes/No	The public economic and commercial system supported the mission of U.S. forces in the area of operations (AO).
02	Yes/No	The unit established liaison, evaluated the current economic and commerce system, and developed plans that will improve the existing economic and commerce system of the host nation.
03	Yes/No	The supporting civil affairs staff finalized the courses of action and briefed the supported commander and staff.
04	Time	To assess the operation of economic and commercial agencies in the AO. Includes developing a census of key economic and commercial industries, establishing communications links among other government agencies, university, and industry personnel. It includes preparing surveys to determine means of production, distribution channels, marketing methods, locations of raw materials, assistance programs available, food and fiber production and requirements, food and agricultural processing and storage facilities, and types and volumes of commodities entering trade.
05	Time	To determine the feasibility of establishing new industries.

No.	Scale	Measure
06	Time	To evaluate the effectiveness of monetary and fiscal systems and policies and make recommendations. This includes reviewing revenue producing systems, budgetary systems, treasury operations, central banking operations, and commercial banking operations.
07	Time	To prepare and coordinate plans to assist the country's economic and commercial agencies.
08	Time	To assess the availability of civilian resources for civilian and military use.
09	Time	To review economic and commerce support plans for compliance with international laws, treaties, and agreements.
10	Time	To develop plans to prevent or limit black market activities to include rationing and price control programs, controlling methods of distribution of critical commodities, and providing security measures to protect storage facilities and distribution points.
11	Percent	Of economy operating on a cash or credit versus barter.
12	Percent	Of stocks of critical supplies diverted to black market.
13	Percent	Of economy dependent on foreign investment and aid.
14	Percent	Of inflation rate in the AO.
15	Percent	Of individual civilian savings rate in the AO.
16	Percent	Of unemployment or underemployment in the AO.
17	Number	And types of commercial facilities restored to operating condition or improved in the AO.
18	Number	And types of new industries and services located in the AO.
19	Number	And types of critical commodities available primarily through the black market.
20	Number	Of metric tons per day of each type of critical commodity diverted to the black market.
21	Number	Of cases per day of individuals arrested for black market activities.
22	Cost	Of direct support or subsidies to keep key industries operating.

ART 5.4.6.2.1 Provide Food and Agriculture Support

5-101. Provide advice and assistance in establishing and managing crop improvement programs, agricultural training, use of fertilizers and irrigation, livestock improvement, and food processing, storage, and marketing. Direct the governmental food and agricultural agencies. Identify areas of staple crops and areas of surplus and deficit foodstuffs, and devise a means to distribute the surplus and eliminate the deficit. Identify locations and capacities of livestock, food processing, storage, and marketing areas. (FM 3-05.40) (USAJFKSWCS)

No.	Scale	Measure
01	Yes/No	The food and agricultural support system in the area of operations (AO) supported the mission of U.S. forces in the AO.
02	Yes/No	AO became self-sufficient in food and agricultural products or because of productivity of its economy and commerce could afford to import the shortfall in its required food and agricultural products.
03	Yes/No	The unit established liaison, evaluated the current food and agriculture system, and developed plans that will improve the existing food and agriculture system of the host nation.
04	Yes/No	The supporting civil affairs staff finalized the courses of action and briefed the supported commander and staff.
05	Time	To identify food processing and distribution requirements. This includes surveys to determine food production capability; limiting factors affecting the production, processing, and storage of food supplies; the transportation needs to complete the mission; and the development of regulatory measures or incentives that encourage the production, safeguard, and orderly distribution of agricultural products.

No.	Scale	Measure
06	Time	To develop plans for government food and agricultural programs. This requires determining the key personnel involved in agriculture and food production in the AO; providing interface with technical specialists working in food and agriculture; and conducting surveys of food and fiber requirements, processing and storage facilities and capabilities, marketing systems, credit availability, agricultural education, and land tenure. It also requires determining the location of food supplies, existing and projected shortages in food and agricultural products, and preparing studies on the population's dietary habits and nutritional requirements.
07	Time	To develop a plan to prevent black market activities. Includes establishing procedures to determine critical commodities in short supply or in danger of being in short supply; establishing a system to inventory available resources, requiring holders of excess resources to declare those stocks and surrender them against a receipt; determining if rationing or price control programs exist or need to exist; developing methods of distribution; issuing guidance on preparing and posting ordinances, laws, and proclamations for distribution to the local population; establishing a program to educate civilians on violations and resulting punishment; and establishing security measures to protect food storage facilities and distribution points.
08	Time	To assess the availability of food and agricultural resources from the AO for use by U.S. forces.
09	Time	To survey and supervise civilian farming methods in the AO. This includes establishing a farm credit system and technical and engineering advice programs. It also includes coordinating the procurement of farm labor during critical planting and harvesting seasons and estimating requirements for food, fertilizer, farm machinery, and other resources.
10	Time	To develop plans for the use and conservation of land, food, and marine resources to include setting procedures for seizing and maintaining all agricultural, fishing, and forestry records, and recommending plans for restoring, using, and conserving land, forests, and marine resources.
11	Time	To review food and agriculture support plans for compliance with international laws, treaties, and agreements.
12	Percent	Of the AO self-sufficient in food and agricultural products.
13	Percent	Of U.S. force requirements for food and agricultural products provided by the AO from its surplus stocks.
14	Percent	Of U.S. force committed to food and agricultural support.
15	Percent	Of population in the AO involved in food and agricultural production.
16	Percent	Of individuals in the AO involved in food production who own their farms.
17	Number	And types of agricultural exhibition projects conducted in the AO to introduce new farming equipment, techniques of farming, agricultural crops, and livestock.
18	Number	Of civilians in the AO involved in educational programs designed to improve their capability to farm and manage farm-related businesses.
19	Number	Of metric tons per day and types of food and agriculture products imported into the AO to meet the needs of its inhabitants.
20	Number	Of metric tons per day and types of food and agriculture products exported from the AO that are surplus to its requirements.
21	Cost	Of direct support or subsidies to provide necessary food and agriculture support to sustain the civilians in an AO.

ART 5.4.6.2.2 Provide Civilian Supply Support

5-102. Determine the availability of local supplies for civil and military use. Coordinate military needs for local resources and coordinate their acquisition. Determine the needs of the populace for emergency supplies and arrange for distribution in accordance with policy. Coordinate the movement of essential civilian supplies. Plan and supervise rationing programs. Arrange salvage of captured supplies that can be used by the civilians. Advise and assist host-nation governments in these activities, when appropriate. Direct governmental and commercial supply activities. (FM 3-05.40) (USAJFKSWCS)

No.	Scale	Measure
01	Yes/No	Public supply system in the area of operations (AO) supported the mission of U.S. forces.
02	Yes/No	The unit established liaison, evaluated the current civilian supply system, and developed plans that will improve the existing civilian supply system of the host nation.
03	Yes/No	The supporting civil affairs staff finalized the courses of action and briefed the supported commander and staff.
04	Yes/No	Authority reviewed the civilian supply plan for compliance with international laws, treaties, and agreements.
05	Time	To review civilian supply plan for compliance with international laws, treaties, and agreements.
06	Time	To plan, coordinate, and move supplies from current locations to distribution points.
07	Time	To identify resources to support civilian and military operations. Includes establishing procedures to survey civilian supply and associated accounting procedures. It also includes determining the adequacy of civilian supplies, agricultural and industrial patterns, and the effect redistributing resources would have on the populace.
08	Time	To develop plan to prevent black market activities. Includes establishing procedures to determine critical commodities in short supply or in danger of being in short supply; establishing a system to inventory available resources, requiring holders of excess resources to declare those stocks and surrender them against a receipt; determining if rationing or price control programs exist or need to exist; developing methods of distribution; issuing guidance on preparing and posting ordinances, laws, and proclamations for distribution to local civilians; establishing program to educate civilians on violations and resulting punishment; and establishing security measures to protect storage facilities and distribution points.
09	Percent	Of required civilian supplies provided internal to the AO.
10	Percent	Of required civilian supplies provided by salvaging captured enemy supplies.
11	Percent	Of required civilian supplies provided by U.S. forces or other agencies.
12	Percent	Of U.S. military supplies that can be provided from sources in the AO.
13	Percent	Of available U.S. military resources to conduct civilian supply support in the AO.
14	Percent	Of black market activity detected and suppressed by law enforcement operations.
15	Number	Of tons per day and types of emergency supplies provided to the populace.
16	Number	Of individuals per day provided emergency supplies.
17	Number	Of tons and types of enemy supplies salvaged for use by civilians of the AO within a given time.
18	Number	Of metric tons per day of each type of critical commodity diverted to the black market.
19	Number	Of cases per day of individuals arrested for black market activities.
20	Cost	Of civilian supply support.

ART 5.4.6.3 PROVIDE INFRASTRUCTURE SUPPORT

5-103. Provide infrastructure supports includes providing public communications support, public transportation support and public works and facilities support.

ART 5.4.6.3.1 Provide Public Communications Support

5-104. Manage communications resources to include postal services, telephone, telegraph, radio, television, computer systems, and public warning systems. Coordinate the use of communications resources for the military. Provide technical advice and assistance on communications systems. Recommend the allocation of civilian communications resources for civilian and military use. Direct civil communications agencies and provide advice, assistance, and supervision as required. Prepare and conduct the transition of public communications support to follow-on civilian and military units and organizations as appropriate. (FM 3-05.40) (USAJFKSWCS)

No.	Scale	Measure
01	Yes/No	Public communications systems in the area of operations (AO) supported the mission of U.S. forces.
02	Yes/No	The supporting G-9 or S-9 finalized courses of action and briefed the supported commander and staff.
03	Yes/No	Unit established liaison, evaluated the current public communications system, and developed plans to improve the existing public communications system of the host nation per international law, treaties, and agreements.
04	Time	To establish liaison with local public communications agencies.
05	Time	To identify available public mass communications resources and determine their conditions and capabilities.
06	Time	To plan the use and supervision of government and public communications systems in the AO.
07	Time	To conduct legal review of proposed changes to public communications system and regulations.
08	Time	From making changes to existing laws and ordinances to when the population is informed of the changes.
09	Time	To conduct civilian censorship program.
10	Time	To screen broadcast personalities and mass communications support personnel for political beliefs.
11	Percent	Of public mass communications resources in the AO under U.S. control.
12	Percent	Of civilians with access to electronic communications systems by type (radio, broadcast television, cable television, satellite broadcasts, short-wave, and Internet).
13	Percent	Of civilians whose only source of electronic communications information is under U.S. control.
14	Number	And types of mass communications systems operable in the AO.
15	Number	Of broadcast personalities and mass communications support personnel removed from their positions because of possible disagreements with U.S. policy in the AO.
16	Cost	To provide public communications support in the AO.

ART 5.4.6.3.2 Provide Public Transportation Support

5-105. Identify the modes and capabilities of transportation systems available in the civilian sector. Coordinate the use of locally available assets—to include, railroads, highways, ports, airfields, and motor vehicles—to support military operations. Prepare plans for the use of available civilian and military transportation assets for emergency civilian evacuation from combat areas or transportation of relief supplies. Provide advice and assistance in establishing and operating transportation facilities. Direct civilian transport agencies and functions. (FM 3-05.40) (USAJFKSWCS)

No.	Scale	Measure
01	Yes/No	Public transportation system in the area of operations (AO) supported the mission of U.S. forces.
02	Yes/No	Unit established liaison, evaluated the current public transportation system, and developed plans that will improve the existing public transportation system of the host nation.
03	Yes/No	The supporting civil affairs staff finalized the courses of action and briefed the supported commander and staff.
04	Time	To establish liaison with transportation agencies.
05	Time	To evaluate existing public transportation system capabilities in the AO. This includes the number and operating conditions of mass transportation assets, bulk carriers, and specialty transportation assets.

No.	Scale	Measure
06	Time	To identify civilian and military transportation needs and capabilities in the AO.
07	Time	To review public transportation plans for compliance with international laws, treaties, and agreements.
08	Percent	Of difference between planned and actual public transportation requirements in the AO.
09	Percent	Of civilians that has access to public transportation in the AO.
10	Percent	Of AO covered by public transportation systems.
11	Percent	Of U.S. forces in the AO involved in providing public transportation support.
12	Percent	Of U.S. military transportation requirements that can be met by public transportation assets in the AO.
13	Number	And types of U.S. forces needed to provide public transportation support.
14	Number	And types of mass transportation assets available in the AO.
15	Number	Of metric tons per day that can be moved by public transportation assets.
16	Number	Of passengers per day that can be moved by public transportation assets.
17	Cost	To provide public transportation support in the AO.

ART 5.4.6.3.3 Provide Public Works and Facilities Support

5-106. Coordinate public works and utilities support for military operations. Advise and assist in the construction, operation, and maintenance of public works and utilities. Direct public works and utilities operations. (FM 3-05.40) (USAJFKSWCS)

No.	Scale	Measure
01	Yes/No	Public works and facilities in the area of operations (AO) supported the mission of U.S. forces.
02	Yes/No	The unit established liaison, evaluated the current public works and facilities system, and developed plans that will improve the existing public works and facilities system of the host nation.
03	Yes/No	The supporting civil affairs staff finalized the courses of action and briefed the supported commander and staff.
04	Yes/No	Unit constructed the public works systems to plan specifications within allotted time.
05	Yes/No	Public works systems safeguarded the health of noncombatants in the AO.
06	Time	To establish liaison with public works and facilities agencies in the AO.
07	Time	To review public works and facilities in the AO. This includes determining the functions and authority of existing public works and facilities regulatory agencies.
08	Time	To develop public works program in the AO. This includes determining the type of support needed, the concerned civil agencies, and the dependability of local agencies and area residents, and developing security plans to protect public works facilities.
09	Time	To perform reconnaissance of selected sites for new public works facilities to evaluate the sites for suitability and conditions, identify construction problems and possible courses of action, and update or revise the engineer estimate.
10	Time	To prepare construction directives for public work facilities and issue it to the construction unit or company.
11	Time	To coordinate for and receive additional assets to perform public works functions.
12	Time	To monitor construction and conduct quality assurance inspections.
13	Time	To verify accuracy of public work construction plans and specifications to include ensuring that the bill of materials includes all required materials to complete construction.
14	Time	To design new construction requirements.
15	Time	To manage and administer public works program in the AO.

No.	Scale	Measure
16	Time	To conduct environmental baseline surveys on real estate being considered for acquisition or use by public works facilities.
17	Time	To inventory equipment located within and supporting public work facilities.
18	Time	To maintain public works records.
19	Percent	Difference between planned and actual requirements for public works in the AO.
20	Percent	Of planned public works facilities acquired or constructed in the AO.
21	Percent	Of existing public works facilities modernized in the AO.
22	Percent	Of existing public works facilities that can be used in their current condition.
23	Percent	Of existing public works facilities damaged by combat actions or natural disaster.
24	Percent	Of civilian nonbattle injuries and disease in the AO, attributable to inadequate public works support.
25	Percent	Of U.S. effort in the AO used to provide public works support.
26	Number	Of metric tons or liters and types of waste, refuse, and hazardous material produced per day in the AO.
27	Number	And types of public works facilities constructed or acquired.
28	Number	And types of U.S. forces used to provide public works support in the AO.
29	Cost	Of providing public works support in the AO.

ART 5.4.6.4 PROVIDE GOVERNMENT SUPPORT

5-107. Provide government support consists of providing public safety support and public administration support. (FM 3-05.40) (USAJFKSWCS)

ART 5.4.6.4.1 Provide Public Safety Support

5-108. Coordinate public safety activities for the military force. Provide liaison between the military forces and public safety agencies and coordinate the control of civilian movement. Advise, assist, or supervise local police, fire fighters, rescue agencies, and penal institutions. Supervise the enforcement of all laws and ordinances after the populace has been duly informed. Prepare and conduct the transition of public safety support to follow-on civil affairs units; other military units; host-nation assets; United Nations, intergovernmental, and nongovernmental organizations; and other government agencies as appropriate. (FM 3-05.40) (USAJFKSWCS)

No.	Scale	Measure
01	Yes/No	The supporting G-9 or S-9 finalized the courses of action and briefed the supported commander and staff.
02	Yes/No	Public safety system in the area of operations (AO) supported the mission of U.S. forces.
03	Yes/No	Unit mentored host-nation police forces.
04	Yes/No	Unit transferred public security responsibilities to host-nation police forces.
05	Yes/No	Unit promoted mine awareness.
05	Yes/No	Unit established liaison, evaluated the current public safety system, and developed plans that will improve the existing public safety system of the host nation.
06	Yes/No	Environmental considerations were planned and procedures were present and being followed.
07	Time	To establish liaison with public safety agencies operating in the AO.
08	Time	To evaluate public safety system in the AO. This includes determining availability, capabilities, effectiveness, and resources of existing public safety agencies, such as police, firefighting, prison, and emergency rescue agencies.

No.	Scale	Measure
09	Time	To develop plans to provide public safety assistance. This requires reviewing current civilian public safety plans, informing civilians of new and revised laws and ordinances, and integrating and supervising the enforcement of laws and ordinances to include civilian movement control restrictions.
10	Time	To review public safety plans for compliance with international laws, treaties, and agreements.
11	Percent	Of difference between planned public safety requirements and actual requirements in the AO.
12	Percent	Of planned public safety support gained in the AO.
13	Percent	Of civilians in the AO provided protection by public safety agencies and organizations.
14	Percent	Of public safety personnel who have passed screening tests for human rights abuses, criminal activities, and political views.
15	Percent	Of public safety personnel licensed or trained to perform their jobs.
16	Percent	Of public safety effort in the AO provided by U.S. forces.
17	Number	And types of public safety facilities located in the AO.
18	Number	And types of U.S. forces used to provide public safety in the AO.
19	Number	And types of criminal activities occurring in the AO within a given time.
20	Number	And types of fires and hazardous materials incidents in the AO within a given time.
21	Number	And types of civilian public safety personnel in the AO.
22	Number	Of civilian personnel detained in prisons and jails in the AO for civil offenses.
23	Cost	To support public safety in the AO.

ART 5.4.6.4.2 Provide Public Administration Support

5-109. Provide liaison to the military forces. Survey and analyze the operation of local government agencies: their structure, centers of influence, and effectiveness. Advise, assist, supervise, or direct government agencies. Recommend and, within the limits of authority, implement governmental functions, policies, and procedures for the conduct of government. Identify officials whose continued service would be adverse to U.S. interests and remove them from office. Recommend for appointment to key offices, individuals who are respected civilians of the area and who would best serve the interests of the United States. Individuals must be cleared by military intelligence prior to nomination. (FM 3-05.40) (USAJFKSWCS)

No.	Scale	Measure
01	Yes/No	Public administration in the area of operations (AO) supported the mission of U.S. forces.
02	Yes/No	The unit established liaison, evaluated the current public administration system, and developed plans that will improve the existing public administration system of the host nation in accordance with international law, treaties, and agreements.
03	Yes/No	The supporting civil affairs staff finalized the courses of action and briefed the supported commander and staff.
04	Time	To establish liaison with civilian public administration agency officials in the AO.
05	Time	To evaluate the operations of civilian public administration agency in the AO. This includes determining the availability and scope of public administration resources, reviewing public administration policies and regulations, and determining their effectiveness.
06	Time	To plan the use of public administration agencies resources to support U.S. operations.
07	Time	To develop plans for providing public administration assistance. This includes analyzing the areas and degree of need for public administration assistance and looking at existing public administration agencies to determine if they require restructuring.

No.	Scale	Measure
08	Time	To develop plans to restore civilian authority during and after military operations. This includes identifying institutions requiring restoration of their authority, evaluating local personnel to determine their loyalty and values, identifying individuals to be placed in authority positions, and removing individuals presumed subversive to U.S. objectives in the AO.
09	Time	To conduct legal review of planned changes to public administration system and regulations.
10	Percent	Of local control of public administration in the AO.
11	Percent	Of U.S. forces in the AO providing support to public administration.
12	Percent	Of local population satisfied with public administration services provided them in the AO.
13	Percent	Of local population complying with public administration registration, licensing, regulation, and fee or taxation requirements imposed on them.
14	Percent	Of public service administrators in the AO cleared by U.S. forces to perform their functions.
15	Number	Of public administration offices or facilities operating in the AO.
16	Number	Of civilians in the AO affected by public administration support.
17	Number	And types of units provided public administration support in the AO.
18	Number	Of individuals removed or prevented from assuming public office because their presumed beliefs are subversive to U.S. objectives in the AO.
19	Cost	To provide public administration support in the AO.

ART 5.4.6.5 PROVIDE PUBLIC HEALTH AND WELFARE SUPPORT

5-110. Determine the type and amount of welfare supplies needed for emergency relief. Plan and coordinate for the use of welfare supplies from all sources. Advise and assist the host-nation government. Establish and supervise emergency centers for distributing supplies and for housing and feeding civilians. (FM 3-05.40) (USAJFKSWCS)

No.	Scale	Measure
01	Yes/No	Public welfare system in the area of operations (AO) supported the mission of U.S. forces.
02	Yes/No	The unit established liaison, evaluated the current public welfare system, and developed plans that will improve the existing public welfare system of the host nation.
03	Yes/No	The supporting civil affairs staff finalized the courses of action and briefed the supported commander and staff.
04	Time	To establish liaison with public welfare agencies in the AO.
05	Time	To evaluate the public welfare system. This requires determining the extent of the welfare problem, number and location of civilian welfare organizations available, the resources that they have available, and their effectiveness in providing services to those who need them.
06	Time	To plan public welfare assistance. This includes the review of existing public welfare laws and programs and determining the numbers of needy civilians to be serviced, and the types of assistance they need (food and clothing), and their availability in the AO. It includes planning for the supervision of emergency shelters and feeding centers to include the recruitment and screening of public welfare personnel, the protection and evacuation of welfare storage and operating facilities, and the distribution of welfare supplies.
07	Time	To procure and transport public welfare supplies to storage or distribution centers.
08	Time	To conduct public welfare assistance.
09	Time	To educate civilians in the AO on public welfare support available to them.

No.	Scale	Measure
10	Percent	Of difference between planned public welfare requirements and actual requirements in the AO.
11	Percent	Of planned public welfare support achieved in the AO.
12	Percent	Of U.S. forces in the AO involved in the conduct of public welfare.
13	Percent	Of public welfare support in the AO provided by civilian organizations.
14	Percent	Of civilians in the AO with access to public welfare facilities.
15	Number	And types of U.S. forces providing public welfare support.
16	Number	Of civilian deaths resulting from an inability to access public welfare within a given time.
17	Number	Of civilians in the AO provided public welfare.
18	Number	And types of other government agencies and nongovernmental organizations providing public welfare in the AO.
19	Cost	To provide public welfare support in the AO.

ART 5.4.6.5.1 Provide Public Health Support

5-111. Estimate needs for additional medical support required by the civilian sector. Coordinate acquisition of medical support from voluntary agencies or U.S. military sources. Coordinate the use of civilian medical facilities and supplies by U.S. forces. Aid in the prevention, control, and treatment of endemic and epidemic diseases of the civilian populace. Survey and provide assistance with civilian health care (medical personnel, facilities, training programs), and provide guidance for provision of emergency services by U.S. personnel. Analyze, survey, supervise, or direct civilian public health and sanitation services, personnel, organizations, and facilities. (FM 3-05.40) (USAJFKSWCS)

No.	Scale	Measure
01	Yes/No	Public health system in the area of operations (AO) supported the mission of U.S. forces.
02	Yes/No	Civilian medical conditions posed a threat to U.S. forces in the AO.
03	Yes/No	The unit established liaison, evaluated the current public health system, and developed plans that will improve the existing public health system of the host nation.
04	Yes/No	The supporting civil affairs staff finalized the courses of action and briefed the supported commander and staff.
05	Time	To establish liaison with public health agencies in the AO.
06	Time	To evaluate the public health system. This includes determining public health resources available—personnel, facilities, and supplies—their condition and their capability to meet the medical requirements of military forces and civilians in the AO. It also includes evaluating the availability of these resources to civilians and the effectiveness of existing public health programs.
07	Time	To prepare plans to prevent and control communicable diseases. Plans require educating host-nation personnel on personal hygiene and sexually transmitted diseases, ensuring local ordinances informing civilians of medical and sanitary measures are made public and enforced, coordinating with public works to ensure that human and other hazardous wastes are safely disposed, protecting food consumed in their raw states, reducing breeding places for disease-carrying insects and animal vectors, preventing hazards to animal and marine life, and containing diseases endemic to the AO.
08	Time	To provide public health assistance. This requires analyzing the organization and functions of existing public health and sanitation agencies, controlling, treating, and preventing communicable diseases, protecting food and water supplies, and supervising the maintenance of public health facilities and records.
09	Time	To establish preventive medicine inoculation plan on activation of the AO.
10	Percent	Of difference between planned public health requirements and actual requirements in the AO.

No.	Scale	Measure
11	Percent	Of planned public health preventive medicine support achieved in the AO.
12	Percent	Of civilians inoculated for disease prevention in the AO.
13	Percent	Of civilians in the AO provided health education and training.
14	Percent	Of civilians with access to doctors, dentists, and ophthalmologists.
15	Percent	Of doctors, dentists, and ophthalmologists to civilians in the AO.
16	Percent	Of individuals providing public health services in the AO currently licensed or certified to perform those services.
17	Number	Of civilians living in the AO that must be provided public health support.
18	Number	And types of public health facilities available in the AO to include their bed capacity.
19	Number	Of currently licensed or certified doctors, physician assistants, nurses, midwives, paramedics, and emergency medical technicians providing public health services in the AO.
20	Number	And types of communicable and sexually transmitted disease cases in the AO.
21	Number	Of sanitary inspections conducted in the AO within a given time.
22	Number	Of instances of civilian restaurants and food service facilities shut down for sanitation violations.
23	Cost	To provide public health support in the AO.

ART 5.4.6.5.2 Provide Cultural Relations Support

5-112.   Provide information to military forces on the social, cultural, religious, and ethnic characteristics of the local populace. Develop codes of behavior to educate U.S. forces to reduce acts contrary to local customs and practice. Locate and identify religious buildings, shrines, and consecrated places, and recommend restrictions on their use. Act as a disinterested third party in negotiations between opposing ethnic, cultural, religious, and social groups in the area of operations (AO). Function in a liaison capacity between U.S. commanders and leaders of local social, cultural, religious, and ethnic groups. Consistent with mission requirements, recommend methods and techniques of operation that will be most acceptable to the local population. (FM 3-05.40) (USAJFKSWCS)

No.	Scale	Measure
01	Yes/No	The cultural affairs support system in the AO supported mission of U.S. forces.
02	Yes/No	The unit established liaison; assessed the host nation's social, cultural, religious, and ethnic characteristics; and developed plans that will improve the existing civil print information systems of the host nation.
03	Yes/No	The supporting civil affairs staff finalized the courses of action and briefs the supported commander and staff.
04	Yes/No	U.S. forces acted as the disinterested party in negotiations between opposing ethnic, cultural, religious, and social groups in the AO.
05	Time	To educate U.S. forces to reduce acts contrary to local customs and practice.
06	Time	To develop, coordinate, and enforce codes of behavior that complement local customs.
07	Time	To conduct liaison with local social, cultural, religious, and ethnic leaders.
08	Number	And types of restrictions placed on the use of religious buildings, shrines, consecrated places, and property by civilians and military forces.
09	Number	And types of restrictions placed on the use of historical buildings and property by the civilian population and military forces.
10	Number	And types of restrictions placed on the use of social gathering places by civilians and military forces.
11	Cost	To provide cultural affairs support in the AO.

ART 5.4.6.5.3 Resettle Dislocated Civilians

5-113. Estimate the number of dislocated civilians, their points of origin, and anticipated direction of movement. Plan movement control measures, emergency care, and evacuation of dislocated civilians so that they do not interfere with U.S. military operations. Coordinate with military forces for transportation, military police support, military intelligence screening or interrogation, and medical activities, as needed. Advise on or establish and supervise the operation of temporary or semi-permanent camps for dislocated civilians. Resettle or return dislocated civilians to their home in accordance with U.S. policy and objectives. Advise and assist host-nation and government agencies on camps and relief measures for dislocated civilians. Supervise the conduct of movement plans for dislocated civilians. (FM 3-05.40) (USAJFKSWCS)

No.	Scale	Measure
01	Yes/No	The dislocated civilian program in the area of operations (AO) supported accomplishment of the mission of U.S. forces.
02	Yes/No	The unit established liaison, coordinated activities, and developed plans in accordance with international law, treaties, and agreements.
03	Yes/No	The unit operations staff, with help from special functions teams, established care and control policies for dislocated civilians that minimized civilian interference with military operations and provided the assistance in accordance with international law.
04	Time	To review current situation and identify dislocated civilian requirements.
05	Time	To establish liaison with national authorities, relief agencies, and voluntary agencies involved with dislocated civilians.
06	Time	To determine impact of dislocated civilians on military operations.
07	Time	To develop, coordinate, and implement control measures for dislocated civilians to include movement control policy, collection points, and assembly areas for dislocated civilians.
08	Time	To develop policy for providing minimum, essential support requirements to dislocated civilians.
09	Time	To prepare plans for the establishment of dislocated civilian assembly points and camps. This includes determining transportation requirements, availability of local and military resources that can be used to support dislocated civilians, developing camp in-processing, administration, and security procedures.
10	Time	To establish policy for final disposition of dislocated civilians to include guidelines for their release to return to their homes, transition of U.S. control over their camps to host-nation, multinational, and nongovernmental organizations.
11	Time	To review dislocated civilian plans for compliance with international laws, treaties, and agreements.
12	Percent	Of difference between planned dislocated civilian support requirements and actual requirements.
13	Percent	Of planned dislocated civilian support capabilities currently available in the AO.
14	Percent	Of U.S. forces in the AO involved in the conduct of dislocated civilian operations.
15	Number	Of dislocated civilians in the AO.
16	Number	And types of forces involved in the conduct of dislocated civilian operations in the AO.
17	Number	And capacity of dislocated civilian camps and centers established in the AO.
18	Cost	Of dislocated civilian operations in the AO.

ART 5.4.6.5.4 Provide Arts, Monuments, and Archives Support

5-114. Prepare a list and map overlay showing the location of significant cultural properties requiring special protection. Include the name and significance of the persons or organizations having custody. Provide information for use in public affairs command information programs to inform all military personnel of directed actions concerning arts, monuments, and archives. Prepare plans and directives for the protection of arts, monuments, archives, and other cultural properties. Coordinate military support for the chemical, biological, radiological, and nuclear decontamination of cultural properties. Advise, assist, or direct the restoration of cultural properties that have been damaged. Help in locating, identifying, determining ownership, and safeguarding arts, monuments, and archives. (FM 3-05.40) (USAJFKSWCS)

No.	Scale	Measure
01	Yes/No	The arts, monuments, and archives program supported U.S. forces mission accomplishment.
02	Yes/No	The unit established liaison, coordinated activities, and developed plans for arts, monuments, and archives support per international law, treaties, and agreements.
03	Yes/No	The supporting civil affairs staff finalized courses of action and briefed the supported commander and staff.
04	Yes/No	Environmental considerations planning and procedures were present and being followed.
05	Time	To review or formulate arts, monuments, and archives policies and ensure these policies comply with U.S. goals and objectives.
06	Time	To conduct census of art, monument, and archive objects located in the area of operations (AO) and determine their condition. This includes marking these objects to ensure their identification in case of theft.
07	Time	To identify the facilities and security procedures available for protecting arts, monuments, and archives in the AO. This may involve designating collection points for art, monument, and archive objects.
08	Time	To establish liaison with local arts, monuments, and archive agencies.
09	Time	To advise local custodian in restoring, inventorying, and safeguarding arts.
10	Time	To review art, monument, and archive plans for compliance with international laws, treaties, and agreements.
11	Percent	Of art, monument, and archive objects in the AO that are lost to theft or destroyed or damaged by military operations or natural disaster.
12	Number	And types of U.S. forces involved in the protection of art, monument, and archive objects in the AO.
13	Number	And types of art, monument, and archive objects in the AO that are lost to theft or destroyed or damaged by military operations or natural disaster.
14	Cost	To provide arts, monuments, and archives support in the AO.

ART 5.4.6.6 PROVIDE PUBLIC EDUCATION AND INFORMATION SUPPORT

5-115. The conduct of civil affairs operations includes providing public education and information support. (FM 3-05.40) (USAJFKSWCS)

ART 5.4.6.6.1 Provide Public Education Support

5-116. Provide technical advice and help in planning and implementing needed education programs. Supervise schools and screen personnel and materials in the education system for compatibility with U.S. objectives and interests. Prepare and conduct the transition of public education support to follow-on civilian and military forces and agencies as appropriate. (FM 3-05.40) (USAJFKSWCS)

No.	Scale	Measure
01	Yes/No	Public education system in the area of operations (AO) supported the mission.
02	Yes/No	Unit established liaison, evaluated the current public education system, and developed plans to improve the existing public education system of the host nation.
03	Yes/No	The supporting civil affairs staff finalized courses of action and briefed the supported commander and staff.
04	Yes/No	Graduates of public education system were productive citizens.
05	Time	To establish liaison with local education officials.

No.	Scale	Measure
06	Time	To evaluate the local education system in the AO. This includes determining the availability, size, and quantity of education resources; the effectiveness of the education system; and extent of damage to educational facilities resulting from military operations or natural disaster.
07	Time	To develop plans to provide education assistance. This includes determining the need for public education assistance, reviewing school curriculums for resourcing require-ments and aligning them with desired outcomes, testing cycles, restoring or reopening existing schools, determining security requirements for educational facilities, and determining legal considerations regarding education and plans to administer schools.
08	Time	To conduct staff development of administrators and certified personnel.
09	Time	To acquire required educational supplies, such as textbooks, paper, pencils, laboratory equipment, and technology such as computers, educational programs, and test item databanks.
10	Time	To acquire vocational educational supplies, such as power tools, woodworking supplies, welding supplies, and automotive diagnostic equipment.
11	Percent	Alignment between desired or tested outcomes and curriculums.
12	Percent	Of population that can access the public education facilities on a regular basis.
13	Percent	Of student population achieving at or above established educational goals.
14	Percent	Of student population with special educational or occupational needs.
15	Percent	Of educational facilities in the AO that meet the educational needs of the students.
16	Percent	Of required educational supplies and technology in the hands of students.
17	Percent	Of students attending class on a given day.
18	Percent	Of students who graduate.
19	Percent	Of students obtaining postsecondary education.
20	Number	Of students, staff, faculty, and administrators in the AO.
21	Number	Of educational facilities in the AO.
22	Number	Of educational facilities closed because of military operations or natural disaster.
23	Number	Of administrative, certified, and classified personnel within the public education system removed because of presumed or proven subversive beliefs.
24	Number	Of administrative, certified, and classified personnel within the public education system removed because of incompetence.
25	Number	Of acts of violence against students and staff within a given time.
26	Cost	To provide public education support in the AO.
27	Ratio	Of instructors to students.

ART 5.4.6.6.2 Conduct Civil Information Support

5-117. Collect civil information. This information covers civil areas, structures, capabilities, organizations, people, and civil events within the commander's operational environment. Enter information into a central database and internally fuse it with the supported element, higher headquarters, and other agencies and organizations. This fusion ensures the timely availability of information for analysis and the widest possible dissemination of the raw and analyzed civil information to military and nonmilitary partners throughout the area of operations (AO). (FM 3-05.40) (USAJFKSWCS)

No.	Scale	Measure
01	Yes/No	The collection of civil information did not violate U.S. Code and applicable Department of Defense and Army regulations against collecting intelligence on U.S. citizens.
02	Yes/No	Civil information collected allowed the unit to determine civil centers of gravity.
03	Yes/No	Unit updated civil information products as additional information and data was collected.

No.	Scale	Measure
04	Yes/No	Civil information and data collected, processed, displayed, stored, and disseminated directed relevant information to the right persons at the right time in a usable format to facilitate situational understanding and decisionmaking.
05	Yes/No	Civil information collected supported decisionmaking.
06	Yes/No	Liaison maintained with host-nation authorities, military and civilian agencies, and other organizations as required by the factors of mission, enemy, terrain and weather, troops and support available, time available, civil considerations.
07	Yes/No	Unit disseminated civil information to appropriate headquarters and agencies.
08	Yes/No	Unit integrated information into force common operational picture.
09	Time	To coordinate with the host nation for civil information.
10	Time	To coordinate civil information collection by organic assets.
11	Time	To coordinate with staff intelligence officer for required support.
12	Time	To develop civil information requirements.
13	Time	To perform civil reconnaissance patrols designed to obtain civil information.
14	Percent	Of terrorist- or threat-related activities reported by civilians before they occur.
15	Percent	Of individuals engaged in terrorist- or threat-related activities reported by civilians after their occurrence.
16	Percent	Of destabilizing and other potentially disruptive elements in an AO identified before they negatively impact civil society.
17	Percent	Of civil information—such as population centers and the location of significant arts, monuments, and archives—included in the unit's database.

SECTION V – ART 5.5: EXECUTE COMMAND PROGRAMS

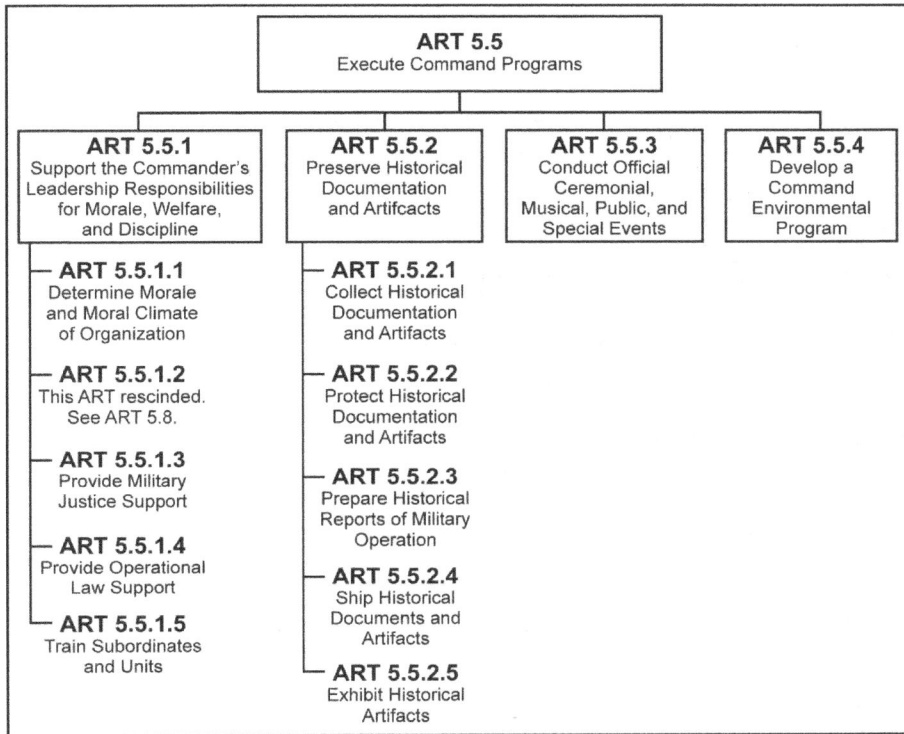

5-118.   Command programs are programs required by U.S. Code and Army regulations. In some cases doctrine also addresses aspects of these programs. Command programs include the following tasks: support commander's leadership responsibilities for morale, welfare, and discipline; preserve historical documentation and artifacts; train subordinates and units; and develop a command environmental program.

ART 5.5.1 SUPPORT THE COMMANDER'S LEADERSHIP RESPONSIBILITIES FOR MORALE, WELFARE, AND DISCIPLINE

5-119.   Support the commander's leadership influencing, operating, and improving activities that fulfill responsibilities for the morale, welfare, and discipline of Soldiers and Army civilians. (FM 6-22) (USACAC)

No.	Scale	Measure
01	Yes/No	Commander fulfilled responsibilities for the morale, welfare, and discipline of Soldiers and Army civilians.
02	Yes/No	Unit's mission was degraded, delayed, or disrupted due to stress-related illness or injury.
03	Percent	Of Soldiers in the command involved in disciplinary cases.
04	Percent	Of leaders trained or educated appropriate to rank.
05	Number	Of unit and personal commendations for achievement.
06	Number	Of unit and personal commendations for bravery.
07	Number	Of incidents of disrespect to authorities or unit.

ART 5.5.1.1 DETERMINE MORALE AND MORAL CLIMATE OF ORGANIZATION

5-120. Morale, the human dimension's most important intangible element, is an emotional bond that impacts the quality of organizational cohesion in the accomplishment of missions. Moral climate deals with the ability of a unit to do what is right even when there is pressure to do something else. (FM 6-22) (USACAC)

No.	Scale	Measure
01	Yes/No	The morale of Soldiers and Army civilians and moral climate of organization accurately and regularly determined.
02	Percent	Of crimes reported involving military behavior in violation of statutes of the Uniform Code of Military Justice.
03	Percent	Of reports from unit ministry team on level of morale that indicates the level is below average.
04	Number	Of morale incidents reported through medical channels.

+ ART 5.5.1.2 ESTABLISH AND MAINTAIN DISCIPLINE

5-121. ART 5.5.1.2 is rescinded. See ART 5.8 (Establish and Maintain Discipline).

ART 5.5.1.3 PROVIDE MILITARY JUSTICE SUPPORT

5-122. Advise and assist the commander in the administration of the uniformed code of military justice, to include the disposition of alleged offenses by courts-martial or nonjudicial punishment, appeals of nonjudicial punishment, and action on courts-martial findings and sentences. Supervise the administration and prosecution of courts-martial, preparation of records of trial, the victim-witness assistance program, and military justice training for all Army personnel. (FM 27-100) (TJAGLCS)

No.	Scale	Measure
01	Yes/No	Military justice administration supported the commander's leadership and discipline responsibilities.
02	Yes/No	Commanders briefed Soldiers on General Order No. 1 before deployment.
03	Time	To prepare a DA Form 2627 (Record of Proceedings under Article 15, UCMJ).
04	Time	To prepare a record of trial by courts-martial.
05	Time	To conduct Article 32 investigation.
06	Time	Between requests for military justice briefings and actual presentations.
07	Time	Between prereferral and referral of a case.
08	Time	Between referral and trial of a case.
09	Time	Between the occurrence of crime and its final disposition.
10	Percent	Of cases in which unlawful command influence occurs.
11	Percent	Of cases overturned on appeal.
12	Percent	Of cases requiring the detail of outside counsel.
13	Percent	Of victims and witnesses having an appointed victim-witness liaison.
14	Percent	Of victims and witnesses who receive a victim-witness information packet.
15	Percent	Of unit personnel who receive military justice training.
16	Percent	Of drafted specifications that properly state an offense.
17	Percent	Of drafted specifications that are supported by probable cause.
18	Percent	Of courts-martial cases co-chaired by a senior judge advocate.

ART 5.5.1.4 PROVIDE OPERATIONAL LAW SUPPORT

5-123. Support the command and control of military operations (the military decisionmaking process and conduct of operations) by performing mission analysis; preparing legal estimates; designing the operational legal support structure; writing legal annexes; assisting in the development and training of rules of engagement (ROE); reviewing all operation plans and orders; maintaining situational understanding; and advising on the legal aspects of targeting, internment and resettlement activities (including detainee operations), stability operations or civil support operations, and applying ROE, civil affairs operations, and information engagement. (FM 27-100) (TJAGLCS)

No.	Scale	Measure
01	Yes/No	Operational law support enhanced the chances of accomplishing the mission without violating the laws of war or rules of engagement.
02	Yes/No	Environmental considerations planning and procedures were present and being followed.
03	Yes/No	Unit determined ROE and clearly defined roles and responsibilities (including custody and transfer of detainees).
04	Time	To assist drafting and reviewing of the ROE for the operation.
05	Time	Between commander's requests for and receipt of legal advice or support.
06	Time	To prepare legal estimates.
07	Time	To draft legal annexes.
08	Time	Between requests for briefings on ROE or law of war and actual presentation of the briefings.
09	Time	To review existing interagency and multinational agreements.
10	Time	To review operation plans.
11	Time	To advise on legal aspects of internment and resettlement activities (including detainee operations).
12	Time	To advise on the legal aspects of stability operations.
13	Time	To advise on the legal aspects of support operations.
14	Percent	Of unit personnel who receive ROE briefings before deployment.
15	Percent	Of operational law judge advocates and support personnel with working knowledge of available Army information systems.
16	Percent	Of operational law judge advocates and support personnel with access to Army information systems.
17	Percent	Of operational law judge advocates and support personnel with access to the legal automation Army-wide system.
18	Percent	Of judge advocates who deploy with radio data links and radios.
19	Percent	Of judge advocates with access to a global positioning device.
20	Percent	Of judge advocates proficiently trained to use a global positioning device.
21	Percent	Of units or Soldiers that receive legal briefings on ROE or law of war, status-of-forces agreements, and host-nation law before deployment.
22	Percent	Of operational cells with a judge advocate detailed.
23	Percent	Of missions where a judge advocate participates in mission analysis.
24	Percent	Of targets reviewed by a judge advocate.
25	Percent	Of entities requiring legal liaison that have a designated judge advocate liaison.
26	Percent	Of crisis management team meetings attended by a judge advocate.
27	Percent	Of issues correctly identified, analyzed, and resolved to support the command and control mission.
28	Percent	Of legal opinions that reflect an accurate view of the law.
29	Percent	Of legal opinions that answer the client's questions clearly and concisely.
30	Percent	Of legal opinions in a form that is useful to the client.
31	Percent	Of opinions reviewed by a supervisor before release.

ART 5.5.1.5 TRAIN SUBORDINATES AND UNITS

5-124. Instruct military personnel to improve their capacity individually and collectively to perform specific military functions and tasks. Training prepares Soldiers, leaders, and units to conduct tactical operations and win. (FM 7-0) (USACAC)

No.	Scale	Measure
01	Yes/No	Training prepared Soldiers, leaders, and units to accomplish their actions and missions.
02	Time	For unit and personnel to perform training to standard.
03	Percent	Of units able to accomplish missions.
04	Percent	Of personnel able to perform assigned duties in missions.
05	Percent	Of mission-essential tasks trained to standard under prescribed conditions.
06	Percent	Of mission-essential tasks performed to standard in operations.

ART 5.5.1.5.1 Develop Mission-Essential Task List

5-125. Compile collective mission-essential tasks that must be performed successfully if an organization is to accomplish its operational missions. (FM 7-0) (USACAC)

No.	Scale	Measure
01	Yes/No	Unit developed a mission-essential task list (METL) that reflects the higher command METL and guidance.
02	Percent	Of mission-essential tasks selected corresponding to missions.
03	Percent	Of increase in effectiveness of unit and individuals on mission-essential tasks as a result of training relative to before training.
04	Number	Of missions that a unit receives that are not on its METL.

ART 5.5.1.5.2 Plan Training

5-126. Identify a desired outcome, develop effective ways of achieving it, recommend the most effective one, and produce a sequence of activities that achieve expected results. ART 5.5.1.3.2 includes assessing training proficiency, articulating a training vision, issuing training guidance, managing time, establishing training events, and allocating training resources to activities and events. (FM 7-0) (USACAC)

No.	Scale	Measure
01	Yes/No	Training plan reflected command and doctrinal guidance and the existing factors of mission, enemy, terrain and weather, troops and support available, time available, civil considerations that if executed will achieve the desired performance in operations.
02	Yes/No	Environmental regulations, laws, and considerations were taken into account during planning and present in procedures being followed.
03	Time	To establish required training program (from mission change).
04	Time	To prepare training plan.

ART 5.5.1.5.3 Prepare for Training

5-127. Initiate and perform activities by the unit before execution to improve its ability to train successfully. ART 5.5.1.3.3 includes rehearsals of trainers; preexecution checks of individuals and units to be trained, trainers to execute and evaluate training, and training support required; assembly of allocated training resources; and elimination of potential training distracters. (FM 7-0) (USACAC)

No.	Scale	Measure
01	Yes/No	Trainers and training resources were available and ready for the training audience when they arrived.
02	Yes/No	Environmental regulations, laws, and considerations were taken into account during planning and present in procedures being followed.
03	Time	For units and personnel to begin training.

No.	Scale	Measure
04	Time	For rehearsals of trainers.
05	Time	To assemble and position training resources.
06	Percent	Of required training resources provided and assembled.

ART 5.5.1.5.4 Execute Training

5-128. Put a plan into action by applying training resources to accomplish the training mission or objective. ART 5.5.1.5.4 includes effective presentation and practice during execution that is accurate, well structured, efficient, realistic, safe, and effective. (FM 7-0) (USACAC)

No.	Scale	Measure
01	Yes/No	Training conducted to established standards accomplished training mission.
02	Yes/No	Environmental regulations, laws, and considerations were taken into account during planning and present in procedures being followed.
03	Time	To complete required training.
04	Percent	Of units and assigned personnel attending and participating in training.
05	Number	Of accidents during training.
06	Number	Of times that planned training activities are modified for safety considerations.

ART 5.5.1.5.5 Assess Training

5-129. Evaluate the demonstrated ability of individuals, leaders, and units against specified training standards. Training may be evaluated against the training objectives or against the performance standards of the tasks for which being trained. (FM 7-0) (USACAC)

No.	Scale	Measure
01	Yes/No	Commander knew which tasks were performed at or above standard and which tasks did not meet standards.
02	Yes/No	Training evaluated against training objectives or performance standards.
03	Yes/No	The opposing force, training conditions, and observers and controllers were sufficient to trigger realistic training events.
04	Yes/No	Unit completed at least one after action review per major training event.
05	Yes/No	Environmental regulations, laws, and considerations were taken into account during planning and present in procedures being followed.
06	Time	To devise and complete training assessments.
07	Percent	Of personnel and units meeting standard.
08	Percent	Of mission-essential task list trained to standard.
09	Percent	Of mission-essential tasks meeting mission requirements.

ART 5.5.2 PRESERVE HISTORICAL DOCUMENTATION AND ARTIFACTS

5-130. To collect and safeguard, paper, photographic images, electronic documentation, and artifacts of key events, decisions, and observations of joint or combined operations or campaigns to support lessons learned analysis, public affairs efforts, doctrine development, and historical retention and writing. This task includes attending key briefings and meetings and interviewing key personnel to gather first-hand observations, facts, and impressions. This task further includes keeping a daily journal of personal observations and key events and packaging and forwarding collected information to appropriate agencies. Prepare and submit contingency historical reports that include required data. (FM 1-20) (Center of Military History)

No.	Scale	Measure
01	Yes/No	Artifacts were moved from the theater of operations and forwarded to an Army museum under the direction of the Center of Military History as soon as possible.
02	Yes/No	Historical team used appropriate technology to accomplish the mission.
03	Yes/No	Supported units received timely, accurate, and relevant historical products and Services.
04	Yes/No	Historical team anticipated and responded to the supported commander's needs.
05	Yes/No	Commander and staff of the supported unit were aware of and had access to historical products and services.
06	Yes/No	Soldiers within supported unit were aware of Army and unit heritage, customs, and traditions.
07	Yes/No	Supported unit exploited history to build cohesion and communicate Army values.
08	Yes/No	Military history was incorporated into supported unit training.
09	Yes/No	Proper policy exception authority was obtained for the consumptive use or alteration of artifacts by the supported unit.
10	Yes/No	All explosive material was inspected by explosive ordnance disposal and rendered inert, as necessary.
11	Yes/No	All retained explosive artifacts were fully documented and recorded as inert.
12	Yes/No	Items of historical significance (mission statement, after action reviews, or lesson learned summaries, and unit briefing slides) were maintained and submitted as part of the unit historical report per AR 25-400-2.
13	Yes/No	Historical team created and maintained organizational history files.
14	Yes/No	All artifacts sent from active operations were accompanied with complete documentation for provenance and historical significance.
15	Time	From termination of major event until all key personnel interviewed.
16	Time	Before documents are available for lessons learned analysis.
17	Time	Before initial status reports are submitted to higher echelons.
18	Time	Before contingency historical reports and supporting documents are sent to a central collection point.
19	Time	To respond to and be on scene for short notice tasking.
20	Time	Before after action reviews are submitted to higher echelons.
21	Percent	Of decision documents with predecisional material available.
22	Percent	Of key leaders interviewed after the event.
23	Percent	Of key staff members interviewed after the event.
24	Percent	Of properly captioned photographic images and electronic documentation available.
25	Percent	Of official documentation—such as maps, orders, photos, web pages—preserved in historical accounts.
26	Percent	Of operations that have enemy versions or accounts available for cross-referencing.
27	Percent	Of records retired or retained per G-1 records retirement system (rather than destroyed).
28	Percent	Of SFs 135 (Records Transmittal and Receipt) properly completed.
29	Percent	Of battles and engagements with photographic images and electronic documentation available.
30	Number	Of oral history interviews completed.
31	Number	Of end-of-tour interviews completed.

ART 5.5.2.1 COLLECT HISTORICAL DOCUMENTS AND ARTIFACTS

5-131. Collect documents, maps, photographs, video and audio recordings, artifacts, and other historical material that might not be preserved by retiring records. This task includes performing after-action interviews and special projects for subjects not treated in regular reports and documents. It also includes the collection of specific information for historical research projects being undertaken by higher historical offices when tasked to do so through command channels. (FM 1-20) (Center of Military History)

No.	Scale	Measure
01	Yes/No	Documents and artifacts relating to military operations were collected and preserved.
02	Yes/No	Historical team used appropriate technology to accomplish the mission.
03	Yes/No	All collected artifacts were reported to the chief of military history.
04	Yes/No	Historical documents and artifacts were properly classified and secured per their classification level.
05	Yes/No	Research collection was established to provide supported units with historical information relating to their current operations.
06	Yes/No	Historians and archivists coordinated with staff division chiefs, action or project officers, and other key personnel to ensure documents, oral interviews, visual images, and other source materials pertaining to historically significant developments and events that took place in the command were placed in the historical research collection.
07	Yes/No	Historical research collections were established and maintained per AR 870-5.
08	Yes/No	Support history detachment advised and assisted the commander and the command's record managers to ensure proper records management regarding documents designated as permanent per AR 25-400-2 (daily journals, plans, files, and so forth).
09	Yes/No	Unit and command histories were regularly prepared and transmitted per regulatory guidance.
10	Yes/No	Supporting military history detachment established working relationships with organization or installation records managers, librarians, and museum curators during the selection of documents for the collection to ensure that all source materials were available to the command.
11	Yes/No	Historians and archivists coordinated with museum personnel to differentiate between historical documents held by museums in support of their collections (as defined in AR 870-20).
12	Time	Before documents are available for lessons learned analysis.
13	Time	Before contingency historical reports and supporting documents are sent to a central collection point.
14	Percent	Of decision documents with predecisional material available.
15	Percent	Of photographic images and electronic documentation available.
16	Percent	Of official documentation—such as maps, orders, photos, web pages—preserved in historical accounts.
17	Percent	Of records retired or retained (rather than destroyed).
18	Percent	Of SFs 135 (Records Transmittal and Receipt) properly completed.
19	Percent	Of battles and engagements with photographic images and electronic documentation available.
20	Percent	Of artifacts unaccounted for after 100-percent inventory.
21	Percent	Of artifact records with an incorrect entry in the "location" block.
22	Number	Of end-of-tour interviews completed.

ART 5.5.2.2 PROTECT HISTORICAL DOCUMENTS AND ARTIFACTS

5-132. Preventing the damage of historical documents and artifacts by either external forces such as mishandling, an unstable environment, or by the intrinsic nature of the materials used to make them. Some historical records and artifacts materials are much more fragile than others are and may have special requirements for care. (FM 1-20) (Center of Military History)

No.	Scale	Measure
01	Yes/No	Historical documents and artifacts were maintained as much as possible in a manner that prevents further deterioration while still allowing their use in historical and operational studies and education.
02	Yes/No	Paper documents were not exposed to excessive amounts of light. Lights were turned off in rooms that are not in use. Daylight was blocked by the use of curtains, shades, or plastic filtering films.
03	Yes/No	Paper documents were not exposed to rapid changes or extremes in temperature and humidity. Temperature was maintained at 68-degrees (within a range of plus or minus 5 degrees) Fahrenheit and 50-percent (within a range of plus or minus 5 percent) relative humidity in book rooms that were in regular use.
04	Yes/No	Paper documents were shelved correctly.
05	Yes/No	Paper documents were regularly checked for pest infestations.
06	Yes/No	Electronic documents were stored in formats that can be read by successive software programs per Department of Defense military standards.
07	Yes/No	Photographs were properly labeled to identify who, what, how, when, and where and who took the photograph on the back of the photo or on the sheet of paper containing the photo.
08	Yes/No	Artifacts containing organic materials—such as leather, fur, horn, feathers, ivory, wool, paper, and cotton—were protected as much as possible from environmental damage such as light, humidity, temperature variations, air pollution, pests, and destructive handling.
09	Yes/No	Artifacts composed of inorganic materials such as metal, stone, glass, and ceramics were protected as much as possible from environmental damage and destructive handling.
10	Percent	Of historical documents and artifact collection that have been properly treated to ensure preservation.

ART 5.5.2.3 PREPARE HISTORICAL REPORTS OF MILITARY OPERATIONS

5-133. Provide well-researched studies and analyses, accurate historical information, institutional memory, historical perspective, and input to lessons learned to support commanders and staffs in problem solving and decisionmaking. (FM 1-20) (Center of Military History)

No.	Scale	Measure
01	Yes/No	Echelon headquarters had a history program that provided studies and analyses, accurate historical information, institutional memory, and historical perspectives to its constituent commanders and staffs to support problem solving and decisionmaking.
02	Yes/No	Echelon headquarters exercised staff supervision over subordinate unit and organizational history programs and activities, including biennial staff assistance visits to subordinate command history offices and review and evaluation of the professional historical credentials and qualifications of all candidates for command historian positions in subordinate commands.
03	Yes/No	Historian prepared historical reports supported leader development.
04	Yes/No	Command historian assisted in the planning and preparation of historical reports, including the command report.
05	Yes/No	Short historical studies of immediate use to the command were prepared to provide insights concerning special tactics, techniques, and battlefield improvisations.
06	Time	Required to establish command historical programs to include monographs, doctrinal and special studies, histories, documentary collections, oral history interviews, and studies on topics and events of historical significance to the command and the Army.
07	Time	To prepare historical manuscripts for publication.
08	Time	To respond to historical inquiries from within or external to the supported unit.
09	Time	Required to perform instruction or teaching duties in military history when assigned to a service school.

No.	Scale	Measure
10	Percent	Of available historian resources devoted toward performing instruction or teaching in military history.
11	Number	Of doctrinal and special studies prepared.
12	Number	Of biennial staff assistance visits to subordinate command history offices.
13	Number	Of oral history interviews.
14	Number	Of historical inquiries for which a response was prepared.
15	Number	Of monographs prepared on selected operations, battles, activities, or problems.

ART 5.5.2.4 SHIP HISTORICAL DOCUMENTS AND ARTIFACTS

5-134. Preventing damage to historical records and artifacts during shipment to and from historical collections by external forces such as mishandling or an unstable environment. (FM 1-20) (Center of Military History)

No.	Scale	Measure
01	Yes/No	When they were no longer needed locally for research or references, forward materials were produced or collected through the theater historian to the Army Center of Military History.
02	Yes/No	Unit commanders were required to maintain organizational history files by shipping those files to storage when the unit was inactivated, disbanded, or reduced to zero strength, while in a combat zone, or when otherwise unable to care for them.
03	Yes/No	Commanders of reactivated units or active units with stored organizational history files addressed their requests for the return of those files to the Center of Military History.
04	Yes/No	Questions concerning the proper packing of historical artifacts or art were referred directly to the Center of Military History or to a certified Army museum.
05	Yes/No	Historical artifacts and works of art were shipped commercially or by mail and packed to preclude any damage as described in FM 38-700.
06	Yes/No	The same level of care was taken to protect artifacts and art in transit as was used in their storage, including the use of museum-safe materials.
07	Yes/No	Historical artifacts and works of art were sent by registered mail (return receipt requested) when size and weight met commercial carrier requirements and when economically advantageous.
08	Yes/No	The artifact responsible officer obtained the chief curator's prior approval in writing (to include a document number assigned by the artifact accountable officer) before shipment of any item.
09	Yes/No	The artifact responsible officer provided a copy of the written approval, as part of the shipment, to the receiving museum, organization, or activity.
10	Yes/No	Shipments of artifacts containing hazardous material or restricted material were shipped per current regulations and directives.
11	Number	And types of historical records and artifacts shipped to and from the headquarters.

ART 5.5.2.5 EXHIBIT HISTORICAL ARTIFACTS

5-135. This task encompasses the visual means by which the interpretation of Army history is met and involves the exhibition of a portion of the collection of a museum or historical collection. It is the technique by which the majority of the institution's audience is reached. Museum exhibits encompass permanent, temporary, traveling, and remote displays. (FM 1-20) (Center of Military History)

No.	Scale	Measure
01	Yes/No	Historical exhibits conformed to professional standards.
02	Yes/No	Army museums and museum activities were organizationally aligned where they will be most effective as training, educational, and research institutions.
03	Yes/No	Exhibits and historical documents and artifact collections supported military training,

No.	Scale	Measure
		education, research, and development.
04	Yes/No	Historical exhibit facilities and collections were maintained in a professional manner as directed in public law and Army regulations.
05	Yes/No	Historical artifacts and art held in custody by any Army agency or organization were accounted for, cataloged, preserved, transferred, and disposed of per AR 870-20.
06	Yes/No	Extreme care was taken to prevent the loss, damage, or destruction of historical artifacts or artwork.

ART 5.5.3 CONDUCT OFFICIAL CEREMONIAL, MUSICAL, PUBLIC, AND SPECIAL EVENTS

5-136. Conduct world-class Army and Department of Defense ceremonial, musical, memorial affairs, and public events, locally and world-wide on behalf of the Nation's civilian and military leaders. Provide Army military honors for private memorial services, religious and worship services, and Army ceremonial support. (FM 3-21.5) (USAIS)

Note: This task includes public wreath laying ceremonies, state funerals, and the honor guard at the Tomb of the Unknowns.

No.	Scale	Measure
01	Yes/No	Event support contributed to mission accomplishment either by enhancing unit cohesion and morale or by entertaining the civilian population.
02	Time	To rehearse the event required for the mission.
03	Time	To rehearse drill and ceremony required for the mission.
04	Time	To coordinate the performance of an assigned mission.
05	Time	To arrange logistic and administrative support for the band and other units participating in the event.
06	Time	To obtain recommendations and legal advice from the staff judge advocate.
07	Percent	Of authorized personnel required to perform the specific mission.
08	Percent	Of authorized musical equipment on hand and serviceable.

ART 5.5.4 DEVELOP A COMMAND ENVIRONMENTAL PROGRAM

5-137. Identify areas affected by environmental considerations. Determine specifics of a command program for a unit or organization that supports the Army program. Develop a command environmental program. Successfully incorporate environmental considerations into all operations by implementing the commander's environmental program. The commander may use a designated unit environmental compliance officer to assist in implementing this program. (FM 3-100.4) (USAES)

No.	Scale	Measure
01	Yes/No	Command environmental program prevented or mitigated the frequency of environmental incidents by Soldiers.
02	Yes/No	Environmental officer was trained and appointed on orders for all subordinate units per AR 200-1.
03	Yes/No	Unit standing operating procedures covered spill prevention and response, pollution prevention, and the use of the material safety data sheets.
04	Yes/No	All Soldiers received and are current in their required environmental training.
05	Yes/No	Unit environmental compliance officer conducted preoccupation environmental survey of all sites to be occupied by the unit.
06	Time	Of measures introduced to mitigate or eliminate the risk of hazardous material, petroleum, oils, and lubricants spills (or other types of releases).

No.	Scale	Measure
07	Time	To modify the command environmental program to include new environmental considerations, rules, or specific guidance.
08	Number	Of violations of the command environmental program occurring within a given time.
09	Number	Of environmental hazards not covered initially by the command environmental program.

SECTION VI – ART 5.6: INTEGRATE SPACE OPERATIONS

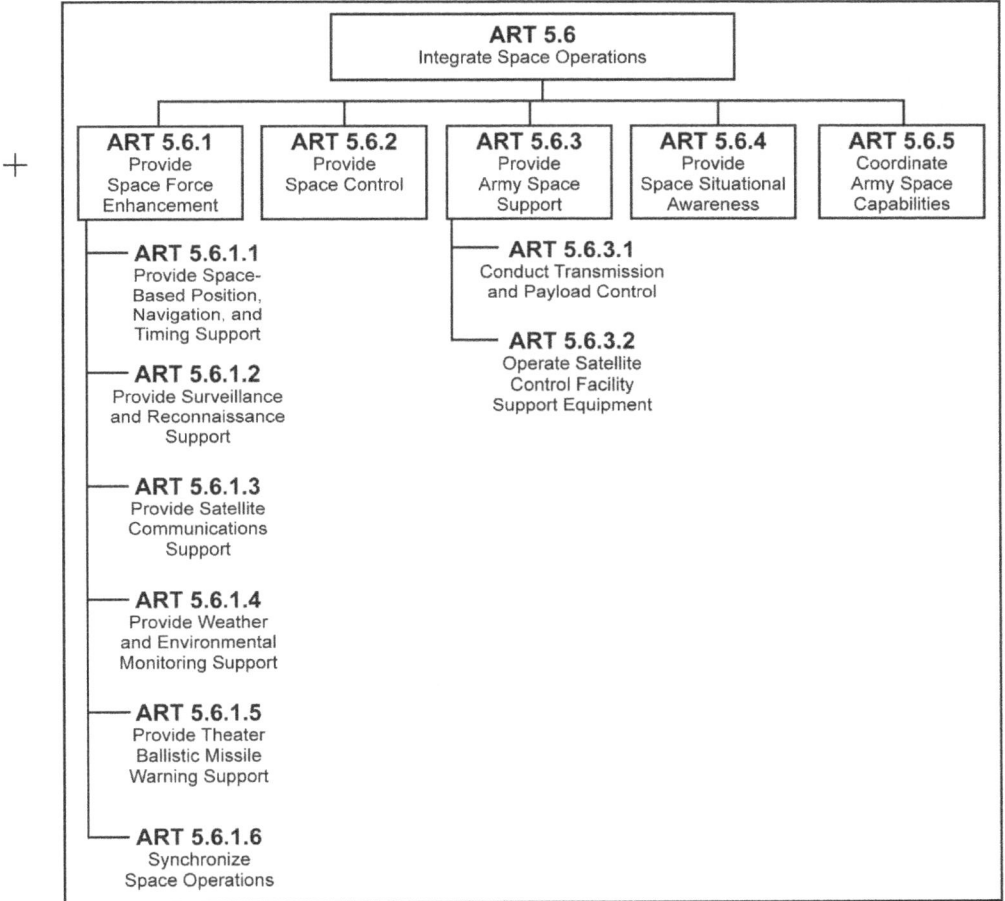

ART 5.6
Integrate Space Operations

ART 5.6.1
Provide
Space Force
Enhancement

ART 5.6.2
Provide
Space Control

ART 5.6.3
Provide
Army Space
Support

ART 5.6.4
Provide
Space Situational
Awareness

ART 5.6.5
Coordinate
Army Space
Capabilities

ART 5.6.1.1
Provide Space-
Based Position,
Navigation, and
Timing Support

ART 5.6.1.2
Provide Surveillance
and Reconnaissance
Support

ART 5.6.1.3
Provide Satellite
Communications
Support

ART 5.6.1.4
Provide Weather
and Environmental
Monitoring Support

ART 5.6.1.5
Provide Theater
Ballistic Missile
Warning Support

ART 5.6.1.6
Synchronize
Space Operations

ART 5.6.3.1
Conduct Transmission
and Payload Control

ART 5.6.3.2
Operate Satellite
Control Facility
Support Equipment

5-144. Space operations are integrated thoroughly into the force structure to enable Army operations, and are essential for mission accomplishment. Staffs down to brigade level integrate space capabilities and vulnerabilities into their mission analysis process. To ensure the maximum use of space, the Army integrates space capabilities into routine operations. (FM 3-14) (USASMDC)

ART 5.6.1 PROVIDE SPACE FORCE ENHANCEMENT

5-145. Provide space force enhancement to the commander, staff, and subordinate units support using space-based sensors and payloads. Space force enhancement support to the Soldier includes position navigation and timing, surveillance and reconnaissance, communication, weather and environmental monitoring, and integrated missile warning. (FM 3-14) (USASMDC)

No.	Scale	Measure
01	Yes/No	Unit identified space force enhancement areas that can affect mission.
02	Yes/No	Unit provided space force enhancement products and information to subordinate units.
03	Yes/No	Unit provided space force enhancement products and information to support current and future operations.
04	Yes/No	Unit received space planning products in sufficient time to incorporate into planning.

ART 5.6.1.1 PROVIDE SPACE-BASED POSITION, NAVIGATION, AND TIMING SUPPORT

5-146. Provide position, navigation, and timing support to assist the integration of the global positioning system (GPS) satellite constellation with user-level equipment. This task includes assessing the ability of both friendly and threat asset use, assessing and countering threats to friendly use, countering threat asset use and identifying requirements and coordinating for theater level enhanced coverage. (FM 3-14) (USASMDC)

No.	Scale	Measure
01	Yes/No	Unit identified friendly position navigation and timing support required, to include enhanced coverage requirements for systems and precision-guided weapons, and it integrated assessment into operational planning and execution.
02	Yes/No	Unit assessed threats to friendly position, navigation and timing accuracies; disseminated tactics, techniques, and procedures to users.
03	Yes/No	Unit assessed threat access and use of position, navigation, and timing assets and friendly abilities to counter or degrade use.
04	Yes/No	Unit identified constellation coverage and satellite status and disseminated times of decreased accuracy with sufficient planning time available.
05	Yes/No	Unit coordinated for additional operational capabilities to enhance coverage or counter known or suspected threat jammers.
06	Time	To process request for GPS-enhanced theater support.
07	Time	To assess and respond to reports of local GPS degradation.

ART 5.6.1.2 PROVIDE SURVEILLANCE AND RECONNAISSANCE SUPPORT

5-147. Provide intelligence, surveillance, and reconnaissance support to the Soldier by coordinating and using Department of Defense, national, and commercial space-based sensors and payloads and by coordinating with intelligence collection management personnel to enhance the G-2 collection capabilities. (FM 3-14) (USASMDC)

No.	Scale	Measure
01	Yes/No	Unit maintained situational awareness of satellites and space-based sensors and processors.
02	Yes/No	Unit integrated the capabilities of national technical means surveillance and reconnaissance assets into the surveillance and reconnaissance plan.
03	Yes/No	Unit integrated the capabilities of commercial surveillance and reconnaissance assets into the surveillance and reconnaissance plan.
04	Yes/No	Unit prepared and submitted requests for information through appropriate collection management process to obtain required surveillance and reconnaissance data.

ART 5.6.1.3 PROVIDE SATELLITE COMMUNICATIONS SUPPORT

5-148. Provide satellite communications support through coordination with regional satellite communications support centers, assessing satellite communications requirements and processing necessary requests for additional support as required. (FM 3-14) (USASMDC)

No.	Scale	Measure
01	Yes/No	Unit maintained situational awareness of satellites and satellite communications payloads.
02	Yes/No	Unit integrated available satellite communications resources in support of unit operations.
03	Yes/No	Unit monitored the operational status of available satellite communications resources and reported outages.
04	Yes/No	Tactical satellite communications network and systems supported allocation and apportionment decisions and instructions.
05	Percent	Of satellite communications nodes that possess required communications capabilities.

ART 5.6.1.4 PROVIDE WEATHER AND ENVIRONMENTAL MONITORING SUPPORT

5-149. Provide weather and environmental monitoring support to maintain situational awareness of space and terrestrial weather, solar events and other atmospheric events, assess their impacts on space-based sensors and payloads, and inform commanders and staff on mission impacts. (FM 3-14) (USASMDC)

No.	Scale	Measure
01	Yes/No	Unit maintained situational awareness of national, civil, and commercial weather and environmental monitoring satellites and satellite payloads.
02	Yes/No	Unit maintained situational awareness of predicted and assessed impact on supported unit operations.
03	Yes/No	Unit assessed sun conjunction events and their impact on unit operations.
04	Yes/No	Unit assessed terrestrial weather and its effects on space-based sensors and payloads.
05	Yes/No	Unit assessed space weather impacts on position, navigation, timing, and missile warning.
06	Yes/No	Unit assessed space weather impacts on communications and intelligence, surveillance, and reconnaissance space-based sensors and payloads.
07	Yes/No	Unit monitored national, civil, and commercial Web sites to ensure the most current terrestrial and space weather information was available.
08	Yes/No	Unit monitored the operational status of available environmental monitoring satellite resources and reported outages.
09	Yes/No	Unit prepared requests for information to obtain specialized space-based environmental monitoring products such as multi- and hyper-spectral imagery or changed detection products to support planning and mission execution.
10	Yes/No	Unit integrated available national, civil, and commercial environmental monitoring satellite resources in support of unit operations.

ART 5.6.1.5 PROVIDE THEATER BALLISTIC MISSILE WARNING SUPPORT

5-150. Provide theater ballistic missile warning support to the in-theater commander with fast, accurate theater ballistic missile launch, trajectories and impact location information. It provides advanced warning and targeting data to missile defense batteries for more accurate firing solutions. (FM 3-14) (USASMDC)

No.	Scale	Measure
01	Yes/No	Unit advised and updated commander and staff on theater ballistic missile warning architecture, processes, and dissemination methods.
02	Yes/No	Unit advised and updated commander and staff on theater event system processes and dissemination methods.
03	Yes/No	Unit assessed the impact of theater event system outages and advised the commander and staff of effects.
04	Percent	Of valid space-enabled theater ballistic missile alerts.

+ART 5.6.1.6 SYNCHRONIZE SPACE OPERATIONS

5-151. Provide Army space representation and support to the space coordinating authority. Assist the space support elements in ensuring Army space equities are recognized and incorporated into joint space operations. Assist in the joint space planning process and development of the space priorities. Coordinate space operations through the Army battlefield coordination detachment. (FM 3-14) (USASMDC)

No.	Scale	Measure
01	Yes/No	Unit conducted joint space planning.
02	Yes/No	Unit integrated space control.
03	Yes/No	Unit coordinated space requirements with director of space forces.
04	Yes/No	Unit synchronized Joint Functional Component Command (JFCC)-space and Combined Joint Task Force (CJTF) space control effects.
05	Yes/No	Unit coordinated CJTF space effects requirements.

ART 5.6.2 PROVIDE SPACE CONTROL

5-152. To ensure friendly unit access to space to enable maneuver forces to benefit from space force enhancement and to deny the enemy use of space to contribute to gaining and maintaining information superiority as an advantage to friendly maneuver operations. (FM 3-14) (USASMDC)

No.	Scale	Measure
01	Yes/No	Unit advised commander and staff on the capabilities and limitations of space control operations.
02	Yes/No	Unit integrated space control planning into its decisionmaking and effects planning.
03	Yes/No	Unit nominated targets and described desired effects of space control operations.
04	Percent	Of friendly systems impacted by threat application of space control operations.

ART 5.6.3 PROVIDE ARMY SPACE SUPPORT

5-153. Provide payload and network control of satellite communications system constellations for Department of Defense and maintain a backup contingency control capability through Army-maintained satellite operations centers. This task includes conducting the day-to-day telemetry, tracking, and commanding needed for optimal performance and health of assets as well as planning and coordinating the resolution of satellite anomalies. (FM 3-14) (USASMDC)

No.	Scale	Measure
01	Yes/No	Unit maintained situational awareness of satellites and satellite communications payloads.
02	Yes/No	Unit integrated available satellite communications resources in support of unit operations.
03	Yes/No	Unit monitored the operational status of available satellite communications resources and reported outages.
04	Yes/No	Unit prepared and submitted requests for additional satellite communications as required.
05	Yes/No	Operations of the tactical satellite communications network and systems supported allocation and apportionment decisions and instructions.
06	Percent	Of satellite communications nodes that possess required communications capabilities.
07	Percent	Of successful, uninterrupted satellite communications.

+ ART 5.6.3.1 CONDUCT TRANSMISSION AND PAYLOAD CONTROL

5-154. Provide transmission and payload control of satellite communications system constellations for Department of Defense and maintain a backup contingency control capability through Army-maintained satellite operations centers. Payload control is responsible for configuring and maintaining the satellite transponders at assigned level. (FM 3-14) (USASMDC)

No.	Scale	Measure
01	Yes/No	Unit performed Common Network Planning Software/ Defense Satellite Communication System Network Planning Software database procedures.
02	Yes/No	Unit performed trend analysis.
03	Yes/No	Unit performed Electromagnetic Interference resolution.
04	Yes/No	Unit reacted to an anomalous condition.
05	Yes/No	Unit accessed carriers.
06	Yes/No	Unit performed payload reconfiguration.
07	Yes/No	Unit performed satellite power management.
08	Yes/No	Unit reacted to emergency condition.
09	Yes/No	Unit managed ship movements across satellites.

+ART 5.6.3.2 OPERATE SATELLITE CONTROL FACILITY SUPPORT EQUIPMENT

5-155. Provide satellite the telemetry, tracking, and commanding needed for optimal performance and health of assets as well as planning and coordinating the resolution of satellite anomalies. (FM 3-14) (USASMDC)

No.	Scale	Measure
01	Yes/No	Unit performed transmit and receive subsystem operations.
02	Yes/No	Unit performed antenna subsystem operation.
03	Yes/No	Unit performed ancillary equipment operations.
04	Yes/No	Unit performed control subsystem maintenance.
05	Yes/No	Unit performed ancillary equipment maintenance.
06	Yes/No	Unit restored site power.
07	Yes/No	Unit restored host earth terminal communications.

ART 5.6.4 PROVIDE SPACE SITUATIONAL AWARENESS

5-156. Space situational awareness includes space intelligence, space surveillance, space reconnaissance, space and terrestrial weather monitoring, and space common operational picture (COP). In support of the COP, monitor, detect, and characterize authorized and unauthorized satellite access on key communications platforms maintaining information superiority as an advantage to friendly maneuver operations. In space surveillance, executed space tracking and space object identification in support of the space surveillance network. Provide space situational awareness in the commander's COP. (FM 3-14) (USASMDC)

No.	Scale	Measure
01	Yes/No	Unit supported the generation of the space order of battle.
02	Yes/No	Unit provided updates to the space catalog.
03	Yes/No	Unit provided observations on space systems.
04	Yes/No	Unit found, identified, tracked, and monitored space systems.
05	Yes/No	Unit monitored and assessed space systems for events and status changes.
06	Yes/No	Unit supported the characterization of space systems.
07	Yes/No	Unit advised commander and staff on the capabilities and limitations of space situational awareness mission.

No.	Scale	Measure
08	Yes/No	Unit integrated space situational awareness planning into its decisionmaking and effects planning.
09	Yes/No	Unit nominated targets and described desired effects of space situational awareness.
10	Yes/No	Mission was received via proper tasking authority and channel.
11	Time	To report intelligence to commander.

ART 5.6.5 COORDINATE ARMY SPACE CAPABILITIES

5-157.   Plan, coordinated, integrate, and control Army space capabilities and force structure to ensure the responsive application of space assets in support of the Soldier. Space capabilities include the mission areas of space force enhancement, space control, and space support and space situational awareness. (FM 3-14) (USASMDC)

No.	Scale	Measure
01	Yes/No	Unit compared organic space support assets in the mission to the need for augmentation.
02	Yes/No	Unit identified additional space forces required to execute mission.
03	Yes/No	Unit requested space augmentation support from the unit operations officer and commander.
04	Yes/No	Unit prepared and coordinated the request for space forces.
05	Time	To deploy space forces.

SECTION VII – ART 5.7: CONDUCT PUBLIC AFFAIRS OPERATIONS

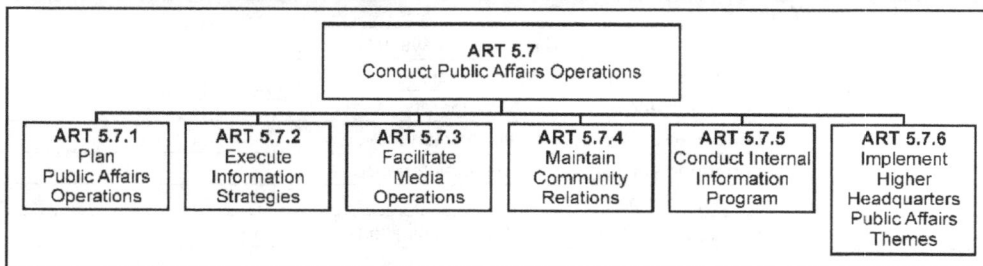

5-158.   Public affairs operations proactively informs and educates internal and external publics through public information, command information, and direct community engagement. Public affairs is a commander's responsibility to execute public information, command information, and community engagement directed toward both the external and internal publics with interest in the Department of Defense. (FM 3-0) (USACAC)

ART 5.7.1 PLAN PUBLIC AFFAIRS OPERATIONS

5-159.   Advise and assist the commander and command (or host nation) in public affairs planning. This includes developing information strategies and program, facilitating media operations, and performing community relations programs. (FM 46-1) (OCPA)

No.	Scale	Measure
01	Yes/No	Public affairs program in the area of operations (AO) supported mission accomplishment.
02	Yes/No	Public affairs officer developed contingency plans for conducting public affairs operations in the AO.
03	Yes/No	Public affairs officer provided contingency plans to the commander for conducting

No.	Scale	Measure
		public affairs operations in the AO.
04	Time	In advance to provide public affairs group for scheduled events.
05	Time	After event to release news.
06	Time	To provide an initial position on breaking news story.
07	Percent	Of plan phases have incorporated public affairs strategy.
08	Percent	Of public affairs guidance coordinated with operations, plans, and policy as needed.

ART 5.7.2 EXECUTE INFORMATION STRATEGIES

5-160. Identify affected internal and external audiences and their information requirements. Determine available communications channels to all audiences. Develop key command messages derived from the overall public affairs strategy. Acquire information to support messages. Process information for delivery through appropriate means. Protect information to meet operations security requirements and distribute information to audiences. (FM 46-1) (OCPA)

No.	Scale	Measure
01	Yes/No	Information strategies supported accomplishment of the unit mission in the area of operations (AO).
02	Yes/No	Public affairs officer developed contingency plans for conducting information strategies and programs, facilitating media operations, and leading community relations program in the AO.
03	Yes/No	Public affairs officer provided contingency plans to the commander for conducting information strategies and programs, facilitating media operations, and leading community relations program in the AO.
04	Time	To prepare for and hold first news conference on crisis or major event.
05	Time	Since last national media pool classified briefing.
06	Time	Before deployment to establish and disseminate media policy.
07	Time	To develop public affairs mission analysis.
08	Percent	Of press operational access rules and security procedures incorporated in operation plan.
09	Percent	Of releases error free.
10	Number	Of press releases per week.
11	Number	And types of information that is classified or withheld from press to avoid embarrassment.

ART 5.7.3 FACILITATE MEDIA OPERATIONS

5-161. Provide assistance to media that is covering operations. This includes assisting media on coverage ground rules; arranging interviews and briefings; coordinating unit visits and escorts; and providing assistance in arranging transportation, messing, billeting, communications support, protective equipment, and medical care. (FM 46-1) (OCPA)

No.	Scale	Measure
01	Yes/No	Unit conduct of media operations supported public affairs operations.
02	Time	To establish information bureau (following execute order).
03	Time	To provide public affairs guidance (after crisis event).
04	Time	To transmit print journalist stories during crisis or combat (from receipt).
05	Time	To close required media equipment (such as commercial television ground stations) to the area of operations.
06	Time	To get Department of Defense media pool into the area of operations.
07	Percent	Of media requests for access to key senior officials accepted.
08	Percent	Of media support requests answered.

ART 5.7.4 MAINTAIN COMMUNITY RELATIONS

5-162. Assist in conducting (planning, preparing, executing, and assessing) community relations programs as resources permit. It includes the conduct of programs to establish and sustain mutually beneficial relationships with the public, focusing on communities neighboring, or directly affected by Army activities. (FM 46-1) (OCPA)

No.	Scale	Measure
01	Yes/No	Community relations program supported unit mission accomplishment in the area of operations and at home base.
02	Time	To set up hometown news release program to publicize troops' successes.
03	Time	To develop public opinion baseline.
04	Percent	Of local customs, laws, and policies concerning presence of media researched and included in planning.
05	Percent	Of requests for information from organizations and private citizens answered.

ART 5.7.5 CONDUCT INTERNAL INFORMATION PROGRAM

5-160. Provide information of interest to Army forces to include data that commanders want their Soldiers to know. (FM 46-1) (OCPA)

No.	Scale	Measure
01	Yes/No	Soldiers knew the information that their commander wanted them to know.
02	Time	To establish newsletter or newspaper for deployed troops.
03	Percent	Of forces consuming internally prepared information.
04	Number	Of internal news releases per week.

ART 5.7.6 IMPLEMENT HIGHER HEADQUARTERS PUBLIC AFFAIRS THEMES

5-161. Transmit themes and information from higher headquarters to the U.S. military audience as part of an echelon's internal information program. (FM 46-1) (OCPA)

No.	Scale	Measure
01	Yes/No	Unit implemented higher headquarters public affairs themes to support mission accomplishment in the area of operations.
02	Time	To prepare higher headquarters public affairs themes for dissemination to forces.
03	Percent	Of higher headquarters public affairs themes disseminated to forces.
04	Percent	Of favorable response in forces to higher headquarters public affairs themes.

+ SECTION VIII – ART 5.8 ESTABLISH AND MAINTAIN DISCIPLINE

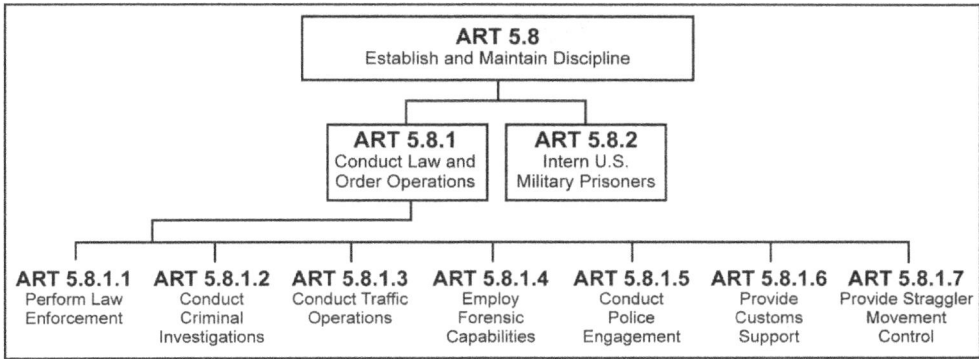

5-162. Establish and maintain discipline through military law enforcement, office of personnel management, regulations, justice, and confinement activities that regulate the force to comply with command policies and directives, ensure a lawful and orderly environment, and suppress criminal behavior. Lead the programs in such manner as to promote self-discipline and respect for authority. Internalize and practice Army values with minimum coercion. (FM 6-22) (USACAC)

No.	Scale	Measure
01	Yes/No	Soldiers and Army civilians exhibited respect for authority and internalized and practiced Army values. Organization had lawful and orderly environment.
02	Number	Of reported disciplinary events.
03	Number	Of incidents involving off-duty Soldiers that involve alcohol or drugs within a given time.
04	Number	Of accidents involving military vehicles within a given time.
05	Number	Of crimes reported within a given time.
06	Number	Of law and order incidents not covered by existing policy or standing operating procedures.

ART 5.8.1 CONDUCT LAW AND ORDER OPERATIONS

5-163. Law and order operations encompass policing and the associated law enforcement activities to control and protect populations and resources to facilitate the existence of a lawful and orderly environment. L&O operations and the associated skills and capabilities inherent in that function provides the fundamental base on which all other military police functions are frame and conducted. (FM 3-39) (USAMPS).

No.	Scale	Measure
01	Yes/No	Unit suppressed criminal behavior in the unit area of operations (AO).
02	Yes/No	Unit coordinated with the appropriate U.S. and host-nation authorities for law and order operational support.
03	Yes/No	Unit coordinated actions to remove conditions promoting crime.
04	Yes/No	Unit performed civilian police functions including investigating crimes and making arrests.
05	Yes/No	Unit conducted special police operations requiring formed units, including investigations and arrests.
06	Yes/No	Unit provided guidance on rules for use of force, rules of engagement and rules of interaction.
07	Yes/No	Unit developed plans and standing operating procedures concerning law enforcement operations.

No.	Scale	Measure
08	Percent	Of time military police subordinate elements conduct law and order missions.
09	Percent	Of criminal behavior suppressed in the AO.
10	Percent	Of military police assets distributed to conduct law and order operations per the plan.
11	Number	Of serious crimes—such as crimes against the United States, political crimes, and war crimes—referred to the Army Criminal Investigation Command for investigation within a given time.
12	Number	Of crimes reported within a given time.
13	Number	Of law and order incidents not covered by existing or established policy or standing operating procedures.
14	Number	Of military working dogs needed for patrol or the detection of narcotics and explosives

ART 5.8.1.1 Perform Law Enforcement

5-164.   Law enforcement (LE) are those activities performed by personnel authorized by legal authority to compel compliance with, and investigate violations of, laws, directives, and punitive regulations. LE occurs in support of governance and the rule of law; for LE to occur, a legal system must exit. Typically, LE is performed by personnel trained as police officers who are held directly accountable to the governmental source of their authority. (FM 3-39) (USAMPS)

No.	Scale	Measure
01	Yes/No	The conduct of law enforcement operations helped the commander maintain the efficiency of command.
02	Yes/No	Staff briefed and monitored military police sections.
03	Yes/No	Environmental regulations, laws, and considerations were taken into account during planning and present in procedures being followed.
04	Percent	Of crimes or incidents resolved within 30 days.
05	Percent	Of requests for rail and road movement escorts met.
06	Percent	Of security force requirement available to meet operational needs.
07	Number	Of serious crimes—such as crimes against the United States, political crimes, and war crimes—referred to the U.S. Army Criminal Investigation Command for investigation.
08	Number	Of military working dogs needed for patrol or the detection of narcotics and explosives.

ART 5.8.1.2 Conduct Criminal Investigations

5-165.   Investigate offenses against Army forces or property committed by persons subject to military law. This includes minor crimes and major incidents involving death, serious bodily injury, and war crimes. Conduct host-nation training and support pertaining to criminal investigations. (FM 3-19.13) (USAMPS)

Note: This task links to ART 7.3.2.3 (Perform Host-Nation Police Training and Support) and ART 7.4.2 (Provide Support to Civil Law Enforcement).

No.	Scale	Measure
01	Yes/No	Law enforcement agency personnel investigated and referred offenses against Army forces or property committed by persons subject to military law.
02	Yes/No	Law enforcement agency personnel monitored all ongoing investigations in the theater of operation and received final reports from subordinate elements.
03	Yes/No	Law enforcement agency personnel coordinated with U.S. Army Criminal Investigation Command for investigation of all major incidents (deaths, serious bodily injury, and war crimes).
04	Yes/No	Law enforcement agency personnel took control of crime scene.
05	Yes/No	Law enforcement agency personnel identified personnel involved in the crime.

No.	Scale	Measure
06	Yes/No	Senior law enforcement agency individual formulated investigative plan.
07	Yes/No	Law enforcement agency personnel processed crime scene.
08	Yes/No	Law enforcement agency personnel processed maintain chain of custody for all collected evidence.
09	Yes/No	Law enforcement agency personnel released crime scene to appropriate individuals.
10	Yes/No	Law enforcement agency personnel pursued immediate investigative leads.
11	Yes/No	Law enforcement agency personnel document all interviews, law enforcement interrogations, and observations conducted the investigation.
12	Yes/No	Law enforcement agency personnel modified standard interview and law enforcement interrogation techniques to overcome any language barriers and cultural differences.
13	Yes/No	Law enforcement agency personnel Investigators maintained a disciplined and systematic approach in their questioning during interviews and law enforcement interrogation.
14	Yes/No	Law enforcement agency personnel conduct follow-up investigations as necessary.
15	Yes/No	Law enforcement agency personnel coordinate investigative efforts with SJA.
16	Yes/No	Law enforcement agency personnel closed the case by preparing final report.
17	Time	To complete crime analysis.
18	Percent	Of investigations of minor reported incidents.
19	Percent	Of investigations conducted and reported per AR 195-2.
20	Percent	Of returns on deficient reports of investigations for corrective action or for further investigative activity.
21	Percent	Of crime analysis performed correctly.
22	Percent	Of case documents and required reports prepared per legal, regulatory, and standing operating procedure guidance.
23	Number	Of hotline complaints referred to criminal investigation division or military police investigation.

ART 5.8.1.3 Conduct Traffic Operations

5-166.   Develop and implement plans and policies concerning traffic flow, traffic safety, and enforcement of traffic laws. Conduct traffic accident investigation and prevention as well as implement traffic control studies, surveys, and necessary traffic enforcement programs. Assist in implementing traffic education and safety programs. In support of the deployed operational commander, maintain the security and viability of the strategic and tactical lines of communication so commanders can deploy and employ their forces. Support the commander's freedom of movement by operating traffic control posts, defiles, or mobile patrols; erecting route signs on main or alternate supply routes; or conducting a reconnaissance for bypassed or additional routes. Conduct traffic studies for the ground commander and provide recommendations to ensure traffic ability on critical routes including identification of reserved or restricted routes, traffic control measures, and other protection measures. Provide a permanent representative to the highway traffic headquarters in the theater of operations. Conduct host-nation training and support pertaining to traffic operations. (FM 3-39) (USAMPS)

Note: This task links to ART 7.3.2.3 (Perform Host-Nation Police Training and Support), ART 6.5 (Conduct Operational Area Security), ART 7.3.3.2 (Control Movement of Dislocated Civilians), ART 7.3.3.4 (Conduct Populace Resource Control), and ART 7.4.2 (Provide Support to Civil Law Enforcement).

No.	Scale	Measure
01	Yes/No	Unit had appropriate plans, policies and procedures for traffic control.
02	Yes/No	Law enforcement agency personnel prepared the traffic control plan.

No.	Scale	Measure
03	Yes/No	Law enforcement personnel coordinated traffic control activities with other headquarters, staff offices, and civil authorities.
04	Yes/No	Law enforcement agency personnel enforce traffic laws, regulations and orders.
05	Yes/No	Traffic control devices were recommended.
06	Yes/No	Unit assisted in accident prevention and vehicle safety programs.
07	Yes/No	Law enforcement personnel investigated and reported traffic accidents.
08	Yes/No	Law enforcement personnel operated traffic control points (on the installation) and main supply route control points (in the theater of operations).
09	Yes/No	Unit reported information concerning traffic, road usage, and progress of movements.
10	Yes/No	Unit performed route reconnaissance.
11	Yes/No	Unit reported any information of intelligence or law enforcement value.
12	Yes/No	Unit implemented alternate or temporary routes for traffic.
13	Percent	Of traffic escort and convoy escort duties.
14	Percent	Of other duties involving security, law enforcement, and crime prevention.
15	Number	Of traffic ability studies and assessments that were conducted.

ART 5.8.1.4 Employ Forensic Capabilities

5-167. Employ forensic capabilities, including the use of forensic evidence and biometric identification, through numerous modes such as fingerprints, Deoxyribonucleic acid (DNA), iris scans, firearm and tool mark analysis, and forensic examination of crime scenes and incident sites. Forensic and biometric evidence may be collected in support of criminal investigations involving crimes committed by persons against Army forces or property under the jurisdiction of military law enforcement personnel. This includes minor crimes and major incidents involving death, serious bodily injury, and war crimes. Forensic and biometric capabilities also can support tactical operations in an effort to provide forensic analysis of collected evidence, identification of personnel or other investigatory requirements in support of operational commanders in an effort identify threat personnel, organizations, and processes to enable appropriate engagement by military forces, Conduct host-nation training and support pertaining to forensic and biometric capabilities. (FM 3-39) (USAMPS).

> *Note:* This task links to ART 7.3.2.3 (Perform Host-Nation Police Training and Support) and ART 7.4.2 (Provide Support to Civil Law Enforcement).

No.	Scale	Measure
01	Yes/No	Unit conducted appropriate collection and handling procedures.
02	Yes/No	Force included personnel trained in collection and handling procedures.
03	Yes/No	Protective packages and containers were available to safely package and transport materials.
04	Yes/No	Unit located and safeguarded key witnesses, documents, and other evidence related to key ongoing or potential investigations and prosecutions.
05	Yes/No	Unit maintained chain of custody.
06	Number	Of items of evidence processed.
07	Number	Of processed items used to support criminal prosecution.
08	Time	To process collected evidence.

ART 5.8.1.5 Conduct Police Engagement

5-168. Police engagement is a type of information engagement that occurs between police personnel, organizations, or populations for the purpose of maintaining social order. Military police and USACIDC personnel engage local, host nation, and coalition police partners; police agencies; civil leaders; and local populations for critical information that can influence military operations or destabilize an area of operations. (FM 3-39) (USAMPS).

Note: This task links to ART 7.3.2.3 (Perform Host-Nation Police Training and Support) and ART 7.4.2 (Provide Support to Civil Law Enforcement).

No.	Scale	Measure
01	Yes/No	Police engagement did not violate U.S. Code and applicable Department of Defense and Army regulations against collecting intelligence on U.S. persons.
02	Yes/No	Public information venues were used to conduct police engagement with the local populace and community leaders.
03	Yes/No	Unit coordinated funds to establish and maintain a police informant operation.
04	Yes/No	Unit identified patterns or trends relevant to proactive law and order operations.
05	Yes/No	Unit analyzed police information and produced actionable police and criminal intelligence.
06	Yes/No	Unit recruited and developed law enforcement informants in the area of operations (AO).
07	Yes/No	Police engagement concept of operations and coordinating instructions developed for the operation plan and order as required.
08	Yes/No	Police engagement was actively employed with other government and nongovernment agencies.
09	Yes/No	All personnel were informed of specific informational themes to use when conducting informal police engagement during the course of normal interaction with the population.
10	Yes/No	All personnel were informed of specific police intelligence requirements.
11	Time	To identify and assess latest criminal information collected.
12	Time	To identify criminal information resources in the AO.
13	Time	To identify criminal trends and patterns developed in the AO.
14	Time	To analyze external police and criminal intelligence reports.
15	Time	To assess internally created police information.
16	Time	To produce criminal information bulletins and alert notices.
17	Percent	Of priority intelligence requirement collection efforts directed toward subordinate units.
18	Percent	Of available police intelligence resources in the AO.
19	Percent	Of known terrorist- and criminal-related activities reported by informants before their occurrence.
20	Percent	Of identified individuals engaged in terrorist- and criminal-related activities reported by informants after their occurrence.
21	Percent	Of accurately reported terrorist- and criminal-related activities.

ART 5.8.1.6 Provide Customs Support

5-169.   Perform tactical actions that enforce restrictions on controlled substances and other contraband violations that enter and exit an area of operations (AO). Customs support can be conducted in support of U.S. customs laws to ensure Department of Defense organizations comply with or support host-nation (HN) customs laws at HN points of entry or exit. Conduct HN training and support pertaining to customs support. (FM 3-39) (USAMPS)

Note: This task links to ART 7.3.1.3 (Conduct Border Control, Boundary Security, and Freedom of Movement).

No.	Scale	Measure
01	Yes/No	Unit established border security/customs enforcement.
02	Yes/No	Unit prevented controlled substances and other contraband items prohibited by law, regulation, or command policy from entering or exiting an AO subject to customs restrictions, policies, and laws.
03	Yes/No	Unit given military customs preclearance.
04	Time	To report custom inspection results to the U.S. border entry point and to the military customs staff advisor.
05	Time	To complete desired level of customs inspections.
06	Time	To consult with neighboring countries on border security plans.
07	Percent	Of redeploying unit personnel and equipment examined or inspected.
08	Percent	Of restricted items identified for U.S. Customs Service or Department of Agriculture review.
09	Percent	Of key customs concerns for the AO identified.
10	Percent	Of violations of border crossing laws and regulations reported to supported commander and appropriate law enforcement agencies.
11	Percent	Of shipments identified for inspection for the presence of U.S. Customs Service and Department of Agriculture restricted items.
12	Number	Of prohibited items barred from shipment within a given time.
13	Number	Of military working dogs needed for patrol or the detection of narcotics and explosives.

ART 5.8.1.7 Provide Straggler Movement Control

5-170. Straggler control involves the direction of uninjured stragglers to their parent unit or to a replacement unit as command policies dictate. Stragglers are military personnel who have become separated from their command by events on the battlefield. If stragglers are ill, wounded, or in shock, they are moved to the nearest medical facility. (FM 3-39) (USAMPS)

No.	Scale	Measure
01	Yes/No	Unit returned stragglers to military or unit control or provided medical assistance as soon as possible.
02	Yes/No	Unit located straggler control posts and collecting points on likely routes of straggler flow.
03	Yes/No	Unit confiscated and disposed of equipment, property, and documents on stragglers per the straggler control plan.
04	Yes/No	Unit diverted stragglers from main supply routes onto alternate routes or collecting points to alleviate congestion of the main supply routes.
05	Percent	Of unit that became stragglers.
06	Number	Of deliberate stragglers escorted back to their unit.
07	Number	Of instances in which deliberate stragglers are detained until they can be transported to a set location as designated by the straggler control plan.
08	Number	Of stragglers assisted, detained, or apprehended when they became separated from their units without authority.
09	Number	Of stragglers returned to military or unit control or provided medical assistance as soon as possible.
10	Number	Of stragglers diverted from main supply routes onto alternate routes or collecting points to alleviate congestion of the main supply routes.

ART 5.8.2 INTERN U.S. MILITARY PRISONERS

5-171. Detain, sustain, protect, and evacuate U.S. military prisoners. ART 5.8.3 includes the establishment of temporary detention facilities. It also includes the operation of long-term confinement facilities. (FM 3-19.40) (USAMPS)

Note: This task links to ART 4.4 (Conduct Internment and Resettlement Operations).

No.	Scale	Measure
01	Yes/No	Unit detained, sustained, and protected U.S. military prisoners until their trials were completed and their sentences were served.
02	Yes/No	Unit retained U.S. military prisoners in custody until trial or until transferred to a field confinement facility.
03	Yes/No	Unit evacuated convicted prisoners from the area of operations (AO) per policy or law.
04	Yes/No	Unit evaluated the operation to include identifying the location of temporary detention facilities for U.S. military prisoners.
05	Yes/No	Unit evacuated U.S. military prisoners from the forward battle area confinement facility to the support area confinement facility as conditions warranted.
06	Yes/No	Unit established a field detention facility in the AO.
07	Yes/No	The confinement facility staff protected and sustained U.S. military prisoners.
08	Yes/No	The confinement facility staff prevented the disruption of facility operations by enemy attack.
09	Number	Of U.S. military prisoners detained or evacuated from AO.
10	Number	Of cases where confinement facility medical section provided immediate problem solving and crisis intervention to U.S. military prisoners interned at the facility.
11	Number	Of prisoner movements or transfers following specified routes.
12	Number	Of U.S. military prisoners picked up and transported with proper written authorization.
13	Number	Of U.S. military prisoners sent to higher headquarter confinement facilities within 72 hours of arrival.
14	Number	Of those U.S. military prisoners requiring special billeting and accountability provided health and welfare items while still retaining proper custody and control.
15	Number	Of U.S. military prisoners escaping from confinement facilities in the AO.
16	Number	Of military working dogs needed for patrol or the detection of narcotics and explosives.

This page intentionally left blank.

Chapter 6

ART 6.0: The Protection Warfighting Function

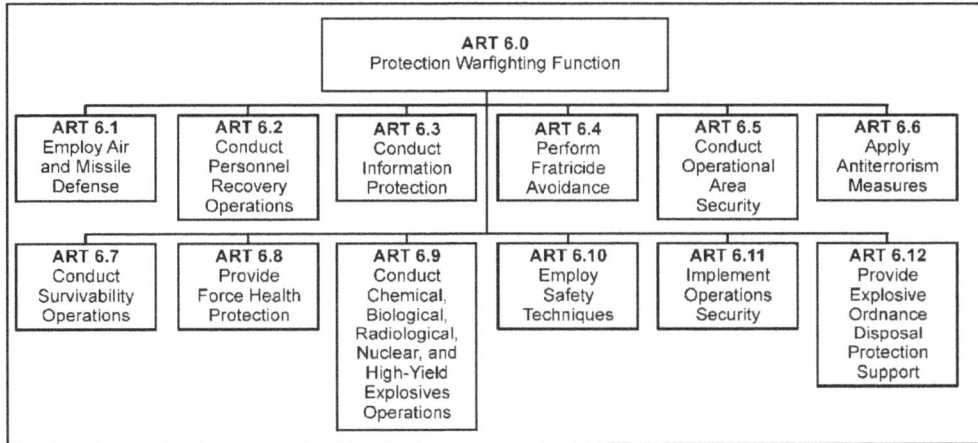

```
┌─────────────────────────────────────────────────────────────────────────┐
│                          ART 6.0                                          │
│                Protection Warfighting Function                            │
│                                                                           │
│  ART 6.1    ART 6.2    ART 6.3    ART 6.4    ART 6.5    ART 6.6          │
│  Employ Air Conduct    Conduct    Perform    Conduct    Apply            │
│  and Missile Personnel Information Fratricide Operational Antiterrorism  │
│  Defense    Recovery   Protection Avoidance  Area       Measures         │
│             Operations                       Security                     │
│                                                                           │
│  ART 6.7    ART 6.8    ART 6.9    ART 6.10   ART 6.11   ART 6.12         │
│  Conduct    Provide    Conduct    Employ     Implement  Provide          │
│  Survivability Force   Chemical,  Safety     Operations Explosive        │
│  Operations Health     Biological, Techniques Security  Ordnance         │
│             Protection Radiological,                     Disposal         │
│                        Nuclear, and                      Protection       │
│                        High-Yield                        Support          │
│                        Explosives                                         │
│                        Operations                                         │
└─────────────────────────────────────────────────────────────────────────┘
```

The *protection warfighting function* is the related tasks and systems that preserve the force so the commander can apply maximum combat power. Preserving the force includes protecting personnel (combatants and noncombatants), physical assets, and information of the United States and multinational military and civilian partners. The protection warfighting function facilitates the commander's ability to maintain the force's integrity and combat power. Protection determines the degree to which potential threats can disrupt operations and counters or mitigates those threats. Emphasis on protection increases during preparation and continues throughout execution. Protection is a continuing activity; it integrates all protection capabilities to safeguard bases, secure routes, and protect forces. (FM 3-0) (USACAC)

SECTION I – ART 6.1: EMPLOY AIR AND MISSILE DEFENSE

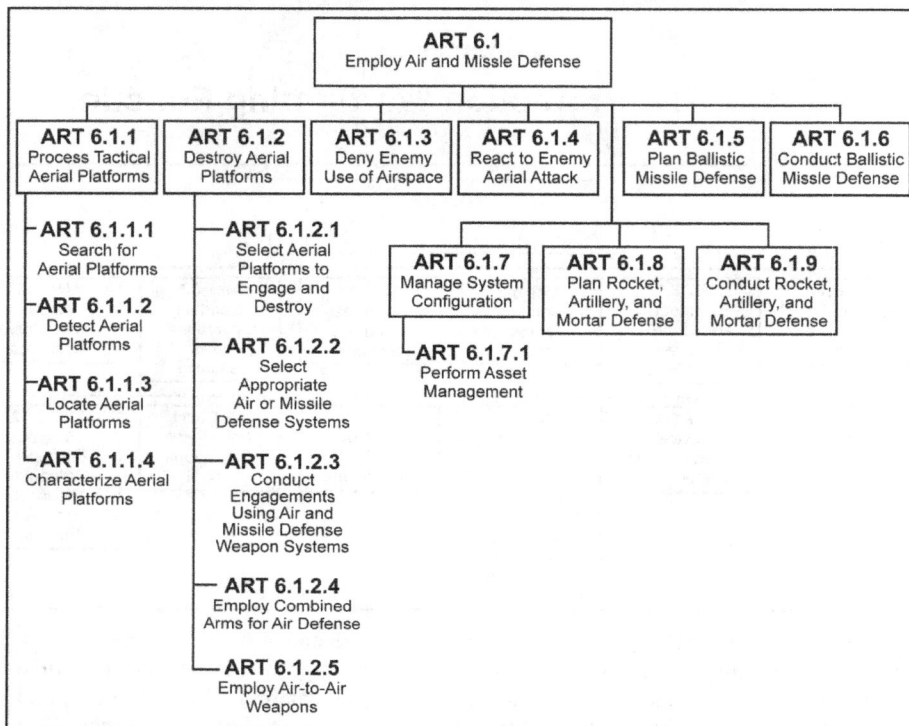

6-1. The air defense system protects the force from missile attack, air attack, and aerial surveillance by any of the following: interceptor missiles, ballistic missiles, cruise missiles, conventional fixed- and rotary-wing aircraft, and unmanned aircraft systems. It prevents enemies from interdicting friendly forces, while freeing commanders to synchronize movement and firepower. All members of the combined arms team perform air defense tasks; however, ground-based air defense artillery units execute most Army air defense operations. ART 6.1 includes fires at aerial platforms by both dedicated air defense systems and nondedicated weapon systems. (FM 3-27.10) (USASMDC)

ART 6.1.1 PROCESS TACTICAL AERIAL PLATFORMS

6-2. Provide advanced warning for all aerial platforms, select targets, and match the appropriate response to tactical aerial platforms including tactical ballistic missile, taking into account operational requirements and capabilities of systems and units. (FM 44-100) (USAADASCH)

No.	Scale	Measure
01	Yes/No	Unit detected, located, and identified all tactical aerial platforms in area of operations (AO).
02	Yes/No	Unit developed and refined early warning plan.
03	Yes/No	Unit employed all available means to detect tactical aerial platforms.
04	Yes/No	Unit received advanced warning of aerial platforms.
05	Yes/No	Unit processed advanced warning of aerial platforms.
06	Yes/No	Unit transmitted advanced warning of aerial platforms.
07	Time	In advance of air attack that advanced warning was provided in AO.
08	Percent	Of enemy offensive air sorties against which friendly air defense assets are assigned.

No.	Scale	Measure
09	Percent	Of enemy air attacks in AO for which early warning provided.
10	Percent	Of force in AO provided early warning of incoming air threat in time to allow them to initiate passive air defense.
11	Percent	Of time that early warning system is operational.
12	Percent	Of air threat warning estimates concerning attack timing and numbers considered accurate by maneuver units.
13	Percent	Of losses caused by hostile air attack and surveillance.
14	Percent	Of AO covered by early warning system.

ART 6.1.1.1 SEARCH FOR AERIAL PLATFORMS

6-3.   Systematically conduct surveillance of a defined area so that all parts of a designated airspace are visually inspected or searched by sensors. (FM 44-100) (USAADASCH)

No.	Scale	Measure
01	Yes/No	Sensor system used to conduct surveillance detected aerial platforms in its current mode of operations from its current location.
02	Time	To refine air defense plan.
03	Time	To revisit each part of the airspace over the area of operations (AO)—how long does it take the sensor to conduct a 6400-mil sweep.
04	Percent	Of AO covered by air defense sensors that can detect projected enemy aerial platforms.
05	Percent	Of air defense sensors functioning in AO.
06	Percent	Of air threat warning estimates concerning attack timing and numbers considered accurate by maneuver units.
07	Percent	Of effectiveness of the system conducting the aerial surveillance given current environmental conditions, such as weather and characteristics of the surrounding terrain to include interference or restrictions placed on its operations resulting from its proximity to other military or civilian equipment.
08	Number	Of air defense sensors functioning in AO.

ART 6.1.1.2 DETECT AERIAL PLATFORMS

6-4.   Determine or detect an aerial platform of possible military significance but cannot confirm it by recognition. (FM 44-100) (USAADASCH)

No.	Scale	Measure
01	Yes/No	Unit detected all aerial platforms in the area of operations (AO).
02	Time	To refine air defense plan.
03	Time	To report the direction of movement, altitude, and rate of movement, and to estimate if platform is a possible enemy aerial platform and target.
04	Percent	Of potential aerial platforms determined to be friendly aircraft by an identification, friend or foe (IFF) system.
05	Percent	Of potential aerial platforms determined to be friendly aircraft by other than an IFF system.
06	Percent	Of potential aerial platforms in AO detected by air defense sensors.
07	Percent	Of enemy aircraft in AO that are not detected by friendly air defense sensors.
08	Percent	Of air threat warning estimates concerning attack timing and numbers considered accurate by maneuver units.
09	Number	Of potential aerial platforms in AO detected by air defense sensors.
10	Number	Of potential aerial platforms in AO detected by ground observers.
11	Number	Of enemy aircraft in AO that are not detected by air defense sensors.

ART 6.1.1.3 LOCATE AERIAL PLATFORMS

6-5. Confirm the track of an aircraft or missile platform in flight. (FM 44-100) (USAADASCH)

No.	Scale	Measure
01	Yes/No	Unit located all detected aerial platforms in the area of operations (AO).
02	Time	To refine air defense plan.
03	Percent	Of aerial platforms in AO correctly located with targetable accuracy by air defense sensors.
04	Percent	Of aerial platforms in AO not located with targetable accuracy by air defense sensors.
05	Percent	Of air defense sensors mission capable in the AO.
06	Percent	Of AO covered by air defense sensors.
07	Percent	Of air threat warning estimates concerning attack timing and numbers considered accurate by maneuver units.
08	Number	Of aerial platforms in AO correctly located with targetable accuracy by air defense sensors.
09	Number	Of aerial platforms in AO not located with targetable accuracy by air defense sensors.
10	Number	Of air defense sensors available in the AO.

ART 6.1.1.4 CHARACTERIZE AERIAL PLATFORMS

6-6. Classifies, types, and identifies tracks and evaluates engageable tracks to determine the threat they pose to friendly assets and areas. (FM 44-100) (USAADASCH)

No.	Scale	Measure
01	Yes/No	Unit identified all detected and located tactical aerial platforms in the area of operations (AO).
02	Time	To refine air defense plan.
03	Time	To identify aerial platform as friendly, hostile, or unknown once it has been detected and located.
04	Percent	Of aerial platforms in AO correctly identified.
05	Percent	Of enemy aircraft in AO that penetrate the air defense sensor network undetected.
06	Percent	Of air threat warning estimates concerning attack timing and numbers considered accurate by maneuver units.
07	Number	Of aerial platforms in AO correctly identified.
08	Number	Of enemy aerial platforms in AO that penetrate the air defense sensor network undetected.

ART 6.1.2 DESTROY AERIAL PLATFORMS

6-7. Destroy all types of enemy aerial platforms in flight in the area of operations. (FM 44-100) (USAADASCH)

No.	Scale	Measure
01	Yes/No	Unit changed its objectives, plan, or operational timetable due to enemy air attack.
02	Yes/No	Destruction of an aerial platform was done per established rules of engagement.
03	Time	To assign a specific air defense weapon system to a specific target.
04	Time	For air defense weapon system to acquire, track, and engage as necessary a specific aerial target once assigned responsibility for engaging that specific aerial platform.
05	Time	For air defense weapon system to recycle or reload so that it is capable of engaging another aerial platform.
06	Time	To determine weapons control status.
07	Percent	Of losses caused by hostile air attack and surveillance.

No.	Scale	Measure
08	Percent	Of available systems directed against declared hostile aerial platforms.
09	Percent	Of enemy aerial platforms destroyed.
10	Percent	Of detected enemy aerial platforms against which air defense weapons are assigned.
11	Percent	Of target cueing information received by air defense weapon systems.
12	Number	Of enemy aerial platforms destroyed.
13	Number	Of different enemy aerial platforms that can be engaged simultaneously.

ART 6.1.2.1 SELECT AERIAL PLATFORMS TO ENGAGE AND DESTROY

6-8.   Analyze each aerial platform to determine if and when it should be destroyed or engaged according to the threat posed, the tactical benefit, and the commander's guidance. ART 6.1.2.1 requires building and maintaining a complete, accurate, and relevant integrated air picture and having current control information. (FM 44-100) (USAADASCH)

No.	Scale	Measure
01	Yes/No	Unit selected aerial platforms that met the commander's guidance for engagement and destruction.
02	Time	To select aerial platforms to engage and destroy.
03	Time	To determine targeting solution after making decision to engage selected aerial platform.
04	Percent	Of enemy aerial platforms in the area of operations (AO) correctly identified and attacked by air defense systems.
05	Percent	Of mission capable air defense systems in AO.
06	Percent	Of available systems directed against declared hostile aerial platforms.
07	Percent	Of aerial targets in the AO that meet the commander's guidance for engagement and destruction.
08	Number	Of enemy aircraft in AO correctly identified and attacked by air defense systems.
09	Number	Of air defense systems in AO.
10	Number	Of fratricide incidents.

ART 6.1.2.2 SELECT APPROPRIATE AIR OR MISSILE DEFENSE SYSTEMS

6-9.   Determine the appropriate air or missile defense systems for engaging a particular aerial platform. (FM 44-100) (USAADASCH)

No.	Scale	Measure
01	Yes/No	Unit selected system capable of engaging and destroying targeted aerial platform.
02	Time	To select and assign attack system once an enemy aerial platform is identified and located with targeting accuracy.
03	Percent	Of probability of selected air defense system hitting targeted aerial platform.
04	Percent	Of enemy aerial platforms selected for attack by dedicated air defense systems.
05	Percent	Of enemy aerial platforms engaged by unit small-arms air defense.
06	Percent	Of enemy aerial platforms not selected for attack by dedicated air defense systems.
07	Percent	Of enemy aerial platforms engaged by more than one air defense system.
08	Percent	Of available dedicated air defense systems directed against declared hostile aerial platforms.
09	Number	Of available dedicated air defense systems directed against declared hostile aerial platforms.

ART 6.1.2.3 CONDUCT ENGAGEMENTS USING AIR AND MISSILE DEFENSE WEAPON SYSTEMS

6-10. Use air and missile defense weapon systems to destroy aerial platforms and protect the force. (FM 44-100) (USAADASCH)

No.	Scale	Measure
01	Yes/No	Friendly course of action changed as a result of enemy air attack.
02	Yes/No	Air and missile defense weapons were used per established rules of engagement.
03	Time	To warn dedicated air defense units after identifying inbound enemy aerial platforms.
04	Time	For air and missile defense weapon system to acquire, track, and engage as necessary a specific aerial platform once assigned responsibility for the engagement of that specific aerial platform.
05	Time	To determine weapons control status.
06	Time	For air defense weapon system to recycle or reload so that it is capable of engaging another aerial platform.
07	Percent	Of enemy aerial platforms able to penetrate air defense network to deliver ordnance or accomplish mission.
08	Percent	Of all air defense systems positioned to engage the enemy aerial platforms.
09	Percent	Of friendly courses of action that must be changed because of enemy air attack.
10	Percent	Of enemy aerial platforms engaged that are destroyed by each air defense weapon system.
11	Percent	Of friendly casualties attributed to enemy aerial platforms.
12	Percent	Of enemy engaged aerial platforms deterred from delivering their ordnance on target.
13	Percent	Of available air defense systems directed against declared hostile aerial platforms.
14	Number	Of destroyed enemy aerial platforms by air defense weapon system.

ART 6.1.2.3.1 Determine Air and Missile Defense Weapon System Capability for Engagement of Aerial Platforms

6-11. Determine the air and missile defense weapon system that can provide the required results. Includes consideration of weapons engagement zones and system characteristics and capability to defeat target. (FM 44-100) (USAADASCH)

No.	Scale	Measure
01	Yes/No	Available air and missile defense weapons engaged their intended targets.
02	Time	To determine system capability for engaging an identified enemy aerial platform.
03	Percent	Of enemy aerial platforms allocated to each air defense weapon system.
04	Percent	Of aerial platforms engaged within the effective range of the selected weapon system.
05	Number	Of enemy aerial platforms allocated to each weapon system.

ART 6.1.2.3.2 Determine Air and Missile Defense Weapon System Availability for Aerial Engagement

6-12. Determine the air and missile defense weapon system available for executing operation. ART 6.1.2.3.2 includes consideration of weapons control status and determination of physical environment restrictions and engagement criticality. (FM 44-100) (USAADASCH)

No.	Scale	Measure
01	Yes/No	Selected air and missile defense weapon system was mission capable. Rules of engagement allowed it to be used, and it could be brought to a position where it could engage the intended target.
02	Yes/No	Enemy aerial platform was engaged to accomplish mission.

No.	Scale	Measure
03	Time	To determine weapons control status.
04	Time	To analyze any restriction caused by physical environment.
05	Time	To determine engagement criticality—the enemy aerial platform must be engaged to accomplish mission.
06	Percent	Of available air defense systems in the area of operations that have the capability to engage a specific hostile aerial platform.
07	Percent	Of targets not attacked in a timely manner due to nonavailability of appropriate air defense weapon system.

ART 6.1.2.3.3 Designate Air and Missile Defense Weapon System for Aerial Engagement

6-13. Designate air and missile defense weapon system to perform an engagement. (FM 44-100) (USAADASCH)

No.	Scale	Measure
01	Yes/No	Selected system accomplished the desired task.
02	Time	To determine available systems for engaging target on identification of enemy aerial platform.
03	Time	To select appropriate weapon system.
04	Time	For air defense weapon system to acquire, track, and engage as necessary a specific aerial target once assigned responsibility for the engagement of that specific aerial target.
05	Time	For air defense weapon system to recycle or reload so that it is capable of engaging another aerial platform.
06	Percent	Of available air defense weapon systems.
07	Percent	Of aerial targets not attacked in a timely manner due to nonavailability of appropriate air defense weapon systems.
08	Percent	Of reliability of air defense weapon system.
09	Percent	Of enemy aerial platforms able to penetrate air defense network to deliver ordnance or accomplish mission.
10	Percent	Of all air defense systems positioned to engage the enemy aerial platforms that engage the enemy aerial platforms.
11	Percent	Of friendly courses of action that must be changed because of enemy air attack.
12	Number	Of air defense weapon systems used in engagement of aerial platform.

ART 6.1.2.4 EMPLOY COMBINED ARMS FOR AIR DEFENSE

6-14. Use weapon systems other than dedicated ground based air defense systems—such as small arms, tank cannons, and antiarmor missiles—to destroy aerial targets. (FM 44-8) (USAADASCH)

No.	Scale	Measure
01	Yes/No	Commander modified course of action because of enemy air attack.
02	Yes/No	Employment of combined arms for air defense was done per established rules of engagement.
03	Time	To warn nonair defense units after identifying inbound enemy aerial platforms.
04	Time	To determine weapons control status.
05	Time	For unit weapon systems to acquire, track, and engage as necessary a specific aerial target once assigned responsibility for engaging that specific aerial target.
06	Time	For unit weapon systems conducting self-defense against air attack to recycle or reload so that they are capable of engaging another aerial target.
07	Percent	Of enemy aerial platforms able to penetrate air defense network to deliver ordnance or accomplish mission.

No.	Scale	Measure
08	Percent	Of all unit weapon systems positioned to engage enemy aerial platforms that engage the enemy aerial platforms.
09	Percent	Of friendly courses of action that must be changed because of enemy air attack.
10	Percent	Of enemy aerial platforms engaged destroyed by each weapon system.
11	Percent	Of friendly casualties attributed to enemy aerial platforms.
12	Percent	Of enemy engaged aerial platforms deterred from delivering their ordnance on target.
13	Percent	Of weapon system operators using correct aerial engagement aiming techniques.
14	Percent	Of reduced the effectiveness of the enemy's attack as a result of the unit conducting self defense against air attack.
15	Number	Of destroyed enemy aerial platforms by unit weapon system.
16	Number	Of friendly casualties attributed to enemy aerial attack.

ART 6.1.2.5 EMPLOY AIR-TO-AIR WEAPONS

6-15. Use weapon systems carried on aircraft to destroy aerial targets. Army aircraft normally do this in self-defense. (FM 3-04.126) (USAAWC)

No.	Scale	Measure
01	Yes/No	Air-to-air weapons destroyed intended targets.
02	Yes/No	Friendly aircraft maintained mutual support throughout engagement.
03	Yes/No	Use of air-to-air weapons was accomplished per established rules of engagement.
04	Time	To warn friendly aerial platforms of inbound enemy aircraft after identification as hostile.
05	Time	Available to friendly aerial platforms to prepare to engage inbound enemy aircraft.
06	Time	To designate and initiate selected air-to-air combat drill.
07	Time	To get ordnance on target after initiation of task.
08	Time	To complete air-to-air combat after target identification.
09	Percent	Of total number of air-to-air combat engagements within a given time where friendly system remains undetected while maneuvering into position where it can effectively engage the enemy aerial platform.
10	Percent	Probability of a hit.
11	Percent	Probability of a kill given a hit.
12	Percent	Of enemy air attacks detected early enough to allow engagement.
13	Percent	Of casualties of friendly aerial platforms conducting air-to-air combat.
14	Percent	Of engaged enemy aerial platforms destroyed by friendly aerial platforms.
15	Percent	Of enemy aerial platforms not engaged by available friendly aerial platforms.
16	Percent	Of available systems directed against declared hostile aerial platforms.
17	Number	Of enemy aerial platforms destroyed by friendly aerial platforms.

ART 6.1.3 DENY ENEMY USE OF AIRSPACE

6-16. Take passive air defense measures to prevent aircraft from effectively engaging the unit. Passive air defense measures, when the unit is not in the path or target of the enemy aircraft, include moving to cover and concealment and preparing to engage the attacking or any follow-on aircraft. (FM 44-100) (USAADASCH)

Note: ART 6.1.2.4 (Employ Combined Arms for Air Defense) addresses active self-defense measures taken against aerial attack by combined arms units.

No.	Scale	Measure
01	Yes/No	Commander modified course of action because of enemy air attack.
02	Time	To warn combined arms units after identifying inbound enemy aerial platforms.
03	Time	To move to covered and concealed positions.
04	Percent	Of enemy aerial platforms able to penetrate the air defense network to deliver ordnance or accomplish mission.
05	Percent	Of friendly courses of action that must be changed due to enemy air attack.
06	Percent	Of enemy aerial platforms unable to acquire friendly personnel and equipment to attack.
07	Percent	Of friendly casualties attributed to enemy aerial platforms.
08	Number	Of friendly casualties attributed to enemy aerial attack.

ART 6.1.4 REACT TO ENEMY AERIAL ATTACK

6-17. Prevent or degrade enemy use of airspace through fire potential or other means, such as smoke, not involving directly attacking aerial platforms. (FM 44-100) (USAADASCH)

No.	Scale	Measure
01	Yes/No	The enemy did not attempt to use designated portions of the airspace.
02	Yes/No	The denial of enemy use of airspace was done per established rules of engagement.
03	Time	To refine air defense plan.
04	Time	Since last enemy aerial attack.
05	Time	To warn all units in the area of operations after identification of inbound enemy aerial platforms.
06	Time	To report the direction of movement, altitude, rate of movement, and estimated target of enemy aerial platforms.
07	Percent	Of enemy aerial platforms attempting to penetrate into friendly airspace deterred from doing so by friendly fire potential.
08	Percent	Of potential target areas obscured by smoke.
09	Percent	Of losses caused by hostile air attack and surveillance.
10	Percent	Of low level flight corridors rendered unusable by the friendly fire potential.
11	Number	Of errors in the performance of the air defense sensor network in a given time.
12	Number	Of enemy aerial platforms attempting to penetrate into the airspace over friendly ground forces.

ART 6.1.5 PLAN BALLISTIC MISSILE DEFENSE

6-18. Plan and coordinate protection of the homeland and operational forces from ballistic missile attack by direct defense and by destroying the enemy's missile capacity. Determine essential requirements to achieve mission accomplishment using ballistic missile defense weapons in a defensive role to defend, detect, defeat, deter, and protect. Centralized planning for missile defense includes the protection of operational forces in the combatant commander's area of responsibility and destroying ballistic missile platforms in flight. Missile defense operations include all forces and activities that support active defense, passive defense, and attack operations. (FM 3-27.10) (USASMDC)

No.	Scale	Measure
01	Yes/No	Unit planned ballistic missile defense.
02	Yes/No	Unit determined command and control for ground based missile defense.
03	Yes/No	Unit defined rules of engagement according to threat capability parameters and national security objectives.
04	Yes/No	Missile defense support was available to assist the staff elements.
05	Yes/No	Unit conducted detailed planning that maximizes Service capabilities.

No.	Scale	Measure
06	Yes/No	Unit linked with the planning of other air and missile defense capabilities.
07	Yes/No	Defended assets list affected configuration of the ground missile defense system and its defensive task plans.
08	Time	To develop and coordinate combatant command's defended assets list.
09	Time	To allocate interceptors in a timely manner to defend the homeland and operational forces.
10	Time	To respond to reports of outages and degradation.
11	Percent	In degrees in which ongoing or planned operations adversely affected by threat defenses.
12	Number	Of available systems directed against declared threatening missile platforms.

ART 6.1.6 CONDUCT BALLISTIC MISSILE DEFENSE

6-19. Defend an assigned area of interest, friendly forces, and infrastructure from ballistic missile attack. Conduct active ballistic missile defense operations to negate significant enemy missile attack. Missile defense activities are designed to destroy attacking enemy missiles exo- or endo-atmosphere, or to nullify or reduce the effectiveness of such attack. Conduct destruction of hostile missiles per rules of engagement. Provide ballistic missile warning support to the Soldier using contributing sensors. (FM 3-27.10) (USASMDC)

No.	Scale	Measure
01	Yes/No	Unit employed combined arms for ballistic missile defense.
02	Yes/No	Unit conducted ballistic missile warning.
03	Yes/No	Space-based sensors maintained constant surveillance of potential adversary launch activities.
04	Yes/No	Unit used space-based sensors to continue tracking missile until booster burnout.
05	Yes/No	Unit evaluated a threat launch of a ballistic missile capable of striking the defenses assets.
06	Yes/No	Unit selected battle plan to match the decisions for missile's trajectory.
07	Yes/No	Unit determined the state vectors, predicted an impact point, assessed whether the defended area was threatened, and alerted missile defense systems.
08	Yes/No	Unit verified rules of engagement were met.
09	Yes/No	Unit processed ballistic missile targets.
10	Yes/No	Unit conducted effective decentralized engagements.
11	Yes/No	Unit launched interceptors singly or in multiples (ripple fire) according to established firing doctrine.
12	Yes/No	Unit intercepted, engaged, and neutralized threat missiles.
13	Yes/No	Unit conducted kill assessment for engagements and made recommendations for additional firings.
14	Time	To assign a specific system to a specific target.
15	Percent	Of threat warning estimates concerning attack timing and numbers considered accurate.
16	Percent	Of hostile missiles engaged and destroyed.
17	Percent	Of errors in performance of surveillance, identification, and track monitor procedures.

ART 6.1.7 MANAGE SYSTEM CONFIGURATION

6-20. Maintain optimal system configuration to support ballistic missile defense operations necessary to respond operationally and tactically. Identify and minimize degrading effects on readiness and maintain directed readiness condition. Be familiar with system element's (to include supporting sensor's) capabilities and limitations so the system configuration can be adjusted to ensure the optimal defense. Assess preplanned maintenance, test, and exercise requests to modify system configuration. Determine impact on the ballistic missile defense mission. Approve or disapprove requested modification. Ensure approved preplanned configuration changes are executed on time. Assess real-time (unplanned) system element outage impact on the ballistic missile defense mission. Decide or direct relevant information to the right person, at the right time, in a usable format, to facilitate situational understanding and decisionmaking. (FM 3-27.10) (USASMDC)

No.	Scale	Measure
01	Yes/No	Unit maintained optimal system configuration necessary to respond operationally and tactically without negative effects on the defense.
02	Yes/No	Unit maintained optimal system configuration necessary to maintain readiness and readiness condition.
03	Yes/No	Unit was familiar with system element's capabilities and limitations.
04	Yes/No	Unit assessed preplanned maintenance, test, and exercise requests to modify system configuration, determined impact on the mission, and made decision.
05	Yes/No	Unit ensured approved preplanned configuration changes were executed on time.
06	Yes/No	Unit assessed real-time (unplanned) system element outages impact on the mission, determined impact on the mission, and made or recommended decision.
07	Yes/No	For recommendations, relevant readiness information was disseminated to the right person, at the right time, in a usable format, to facilitate situational understanding and decisionmaking.
08	Time	Of lag between approved real-time system reconfiguration and commencement of system reconfiguration.
09	Number	Of maintenance, test, and exercise operations adversely affected by disapprovals.
10	Number	Of incidents of critical information not reaching person responsible for decision in a timely manner.

+ ART 6.1.7.1 Perform Asset Management

6-20.1. Manage changes to assets supporting the Ballistic Missile Defense System to support ballistic missile defense operations necessary to respond operationally and tactically. Identify and minimize degrading effects of scheduled and unscheduled changes on readiness and maintain directed readiness condition. Be familiar with the system element's (to include supporting sensors) capabilities and limitations so the system configuration can be adjusted to ensure the optimal defense. Assess preplanned maintenance, test, and exercise requests to modify system configuration. Recommend approval or disapproval of requested modification through proper channels for combatant command. Assess real-time (unplanned) system element outage impact on the ballistic missile defense mission. Make appropriate changes to the Ballistic Missile Defense System health and status or operational capability as required. Advise combatant command in determining the Ballistic Missile Defense System capability. Decide or direct relevant information to the right person, at the right time, in a usable format, to facilitate situational understanding and decisionmaking. (FM 3-27.10) (USASMDC)

No.	Scale	Measure
01	Yes/No	Unit assessed preplanned maintenance, test, and exercise requests to modify system configuration (assets), determined impact on the mission, and made decision.
02	Yes/No	Unit assessed real-time (unplanned) system asset outages impact on the mission, determined impact on the mission, and made or recommended decision.
03	Yes/No	Unit was familiar with system asset capabilities and limitations.
04	Yes/No	For recommendations, relevant readiness information was disseminated to the right person, at the right time, in a usable format, to facilitate situational understanding and decisionmaking.
05	Time	To respond to reports of outages and degradation.
06	Time	To identify asset status change and take appropriate action to minimize degrading effects or maximize enhancement of that change on the overall system capabilities.
07	Time	Change in asset status for which it was not accounted.

No.	Scale	Measure
08	Number	Of maintenance, test, and exercise operations adversely affected by disapprovals.
09	Number	Of incidents of critical asset information not reaching the person responsible for a decision in a timely manner.

ART 6.1.8 PLAN ROCKET, ARTILLERY, AND MORTAR DEFENSE

6-21. Plan and coordinate to protect operational forces, forward operating bases, and aerial ports and seaports of debarkation from rocket, artillery, and mortar (RAM) attack by direct defense and by destroying the enemy's RAM capability. Determine essential requirements to achieve mission accomplishment using RAM defense weapons in a defensive role to deny, sense, warn, intercept, shape & respond and protect. RAM defense operations include all forces and activities that support active defense, passive defense, and counterfire operations. (FMI 3-01.60) (USAADASCH)

No.	Scale	Measure
01	Yes/No	Unit planned RAM defense.
02	Yes/No	Unit conducted RAM attack warning.
03	Yes/No	Unit defined rules of engagement according to threat capability parameters.
04	Yes/No	RAM attack support was available.
05	Yes/No	Unit conducted detailed planning that maximizes Service capabilities.
06	Yes/No	Defended asset list affected defense configuration.
07	Time	To develop and coordinate the commander's defended asset list.
08	Time	To allocate assets in support of the defended asset list.
10	Percent	In degrees in which ongoing or planned operations adversely affected by threat defenses.
11	Number	Of available systems directed against threatening RAM targets.

ART 6.1.9 CONDUCT ROCKET, ARTILLERY, AND MORTAR DEFENSE

6-22. Defend an assigned asset, friendly forces, and infrastructure against rocket, artillery, and mortar (RAM) attack. Conduct RAM defense to negate RAM attacks. RAM defenses are designed to destroy attacking enemy RAM or to nullify or reduce the effectiveness of such attack. Conduct destruction of hostile RAM targets per rules of engagement. Minimize collateral damage and reduce probability of casualties within adjacent populated areas. This task includes detection, discrimination, and tracking of in-flight RAM to support cueing and handoff of targets for engagements. Provide RAM attack warning support to the Soldier using available sensors. (FMI 3-01.60) (USAADASCH)

No	Scale	Measure
01	Yes/No	Unit conducted timely RAM attack warning, providing time for protected forces and assets to prepare for impending attack.
02	Yes/No	Unit evaluated threat launch of RAM capable of striking the defended assets.
03	Yes/No	Unit predicted an impact point and assessed whether defended assets were threatened.
04	Yes/No	Unit verified rules of engagement were met.
05	Yes/No	Unit launched interceptors according to established firing doctrine.
06	Yes/No	Unit engaged, intercepted, and neutralized RAM targets beyond ranges or altitudes at which damage or destruction of forces, materiel, or infrastructure could occur.
07	Time	To assign a specific system to a specific target.
08	Percent	Of hostile targets engaged and destroyed.
09	Percent	Of errors in performance of surveillance and tracking procedures.

SECTION II – ART 6.2: CONDUCT PERSONNEL RECOVERY OPERATIONS

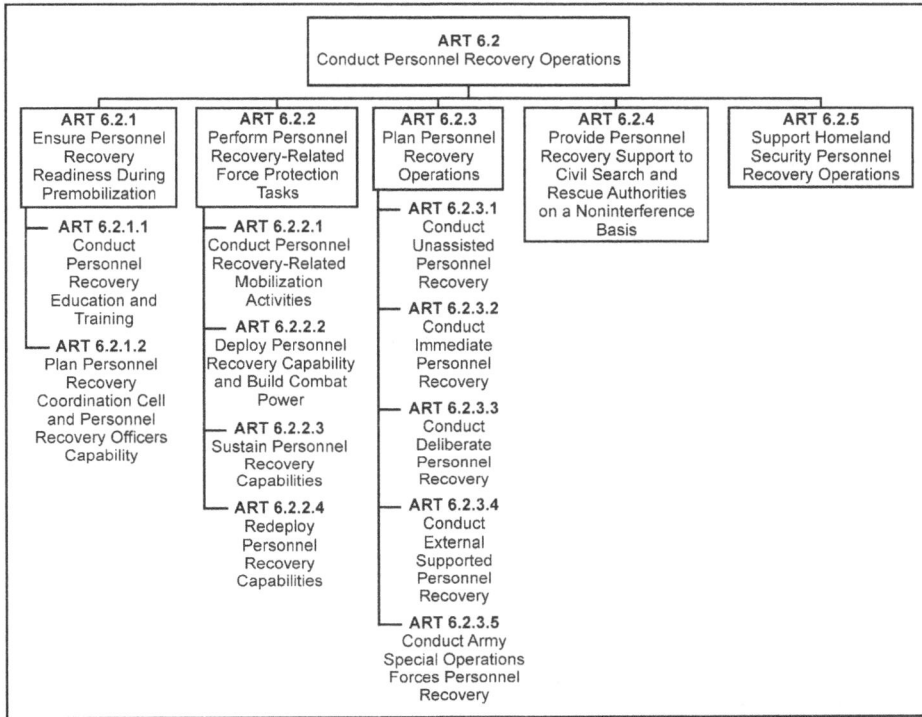

ART 6.2
Conduct Personnel Recovery Operations

ART 6.2.1
Ensure Personnel Recovery Readiness During Premobilization

- **ART 6.2.1.1** Conduct Personnel Recovery Education and Training
- **ART 6.2.1.2** Plan Personnel Recovery Coordination Cell and Personnel Recovery Officers Capability

ART 6.2.2
Perform Personnel Recovery-Related Force Protection Tasks

- **ART 6.2.2.1** Conduct Personnel Recovery-Related Mobilization Activities
- **ART 6.2.2.2** Deploy Personnel Recovery Capability and Build Combat Power
- **ART 6.2.2.3** Sustain Personnel Recovery Capabilities
- **ART 6.2.2.4** Redeploy Personnel Recovery Capabilities

ART 6.2.3
Plan Personnel Recovery Operations

- **ART 6.2.3.1** Conduct Unassisted Personnel Recovery
- **ART 6.2.3.2** Conduct Immediate Personnel Recovery
- **ART 6.2.3.3** Conduct Deliberate Personnel Recovery
- **ART 6.2.3.4** Conduct External Supported Personnel Recovery
- **ART 6.2.3.5** Conduct Army Special Operations Forces Personnel Recovery

ART 6.2.4
Provide Personnel Recovery Support to Civil Search and Rescue Authorities on a Noninterference Basis

ART 6.2.5
Support Homeland Security Personnel Recovery Operations

6-23. Conduct personnel recovery (PR) operations to recover and return own personnel, whether Soldier, Army civilian, selected Department of Defense (DOD) contractors, or other personnel as determined by the Secretary of Defense who are isolated, missing, detained, or captured (IMDC) in an operational environment. IMDC personnel consist of U.S. forces, Army civilians, and DOD contractors who deploy with the force, or other personnel as designated by the President who are beyond the Army positive or procedural control of their unit, in an operational environment requiring them to survive, evade, resist, or escape. It is every unit's task to have procedures in place to be ready to recover personnel, whether Soldier, civilian, or contractor. (FM 3-50.1) (USACAC)

No.	Scale	Measure
01	Yes/No	The commander, staff, units, and individuals reviewed and developed PR guidance.
02	Yes/No	The commander, staff, and units acquired PR equipment.
03	Yes/No	The commander, staff, units, and individuals conducted PR-focused education and training.
04	Yes/No	Subordinate commands and units integrated PR into internal standing operating procedures consistent with guidance from higher headquarters.

ART 6.2.1 ENSURE PERSONNEL RECOVERY READINESS DURING PREMOBILIZATION

6-24. Ensure personnel recovery (PR) readiness during premobilization by including PR in all efforts during peacetime to organize, train, and equip to accomplish missions. Premobilization preparation actions are necessary for commanders and their staffs, units, and individuals and consist of reviewing and developing PR guidance, acquiring PR equipment, ensuring PR administrative requirements are met, and conducting PR-focused education and training. (FM 3-50.1) (USACAC)

No.	Scale	Measure
01	Yes/No	The commander, staff, units, and individuals reviewed and developed PR guidance.
02	Yes/No	The commander, staff, and units acquired PR equipment.
03	Yes/No	The commander, staff, units, and individuals conducted PR-focused education and training.
04	Yes/No	Subordinate commands and units integrated PR into internal standing operating procedures consistent with guidance from higher headquarters.

ART 6.2.1.1 CONDUCT PERSONNEL RECOVERY EDUCATION AND TRAINING

6-25. Ensure all required personnel recovery (PR) individual education and training requirements are conducted per established policies and directives. Coordinate code of conduct; survival, evasion, resistance, and escape; and PR education and training related to the country (theater). Department of Defense; Headquarters, Department of the Army; and combatant commanders establish education and training requirements. (FM 3-50.1) (USACAC)

No.	Scale	Measure
01	Yes/No	Unit reviewed policy, doctrine, regulations, and operation plans to develop guidance and intent for training.
02	Yes/No	Individual PR education and training requirements were included in individual training plans.
03	Yes/No	Units established training management systems to ensure all required training was accomplished.
04	Yes/No	Unit developed processes and procedures to ensure all incoming Soldiers, civilians, and contractors received required PR training.
05	Yes/No	All leaders and Soldiers received required PR training.
06	Yes/No	All Army civilians and Department of Defense contractors receive required training.

ART 6.2.1.2 PLAN PERSONNEL RECOVERY COORDINATION CELL AND PERSONNEL RECOVERY OFFICERS CAPABILITY

6-26. While the staff members collect personnel recovery (PR)-related information in their specialty areas, a central point for gathering the information from all the staff members is required to establish a usable common operational picture. The PR coordination cells at the division level and personnel recovery officers at brigade and below are the fusion points for the staffs' collaborative efforts to gather PR-related information for their respective units. PR coordination cells and officers coordinate with the individual staff members to collect, process, store, display, and disseminate this information. Personnel recovery coordination cells and officers are collectively referred to as PR cells. (FM 3-50.1) (USACAC)

No.	Scale	Measure
01	Yes/No	Unit ensured reliable communications with subordinate unit PR officers, other personnel recovery coordination cells, and the joint personnel recovery center.
02	Yes/No	Unit coordinated deliberate recoveries for the component.
03	Yes/No	Unit reviewed accountability and movement reporting procedures of subordinate units.
04	Yes/No	Unit assisted in immediate recoveries when requested by subordinate units.
05	Yes/No	Unit coordinated for component fire support to the operation.
06	Yes/No	Unit ensured subordinate units have access to standing operating procedures developed by the joint personnel recovery center.
07	Yes/No	Unit ensured subordinate units have sufficient evasion aids.
08	Yes/No	Unit coordinated for air and ground transportation and medical support.
09	Yes/No	Unit gathered PR-specific information developed by joint personnel recovery center and personnel recovery coordination cells and disseminated to subordinate units.
10	Yes/No	Unit identified subordinate unit PR equipment shortfalls to the personnel recovery coordination cell.

*ART 6.2.2 PERFORM PERSONNEL RECOVERY-RELATED FORCE PROTECTION TASKS

6-27. Perform personnel recovery (PR)-related tasks in planning and preparing for deployment. Unit PR personnel update and obtain equipment and materials necessary to perform assigned mission. Commanders and staffs ensure adequate PR capability is programmed into personnel and equipment flow to assure ability to support the force. PR personnel must be included in the time-phased force and deployment list and equipment included in the updated deployment equipment list. PR organizations should ensure that all unit personnel complete all necessary PR deployment requirements. (FM 3-50.1) (USACAC)

No.	Scale	Measure
01	Yes/No	Guidance and intent synchronized with the theater plan.
02	Yes/No	Processes, plans, and procedures integrated into the theater plan.
03	Yes/No	Maps and charts were standardized.
04	Yes/No	Special and personal staff integrated into family support plan.
05	Yes/No	Interoperability of location methods ensured.
06	Yes/No	Personnel and equipment were programmed into deployment plans per commander's intent.
07	Yes/No	Location techniques and systems synchronized.
08	Yes/No	All unit personnel completed PR-related training prior to deployment.
09	Yes/No	Personnel recovery coordination cells and personnel recovery officers received all necessary training.

ART 6.2.2.1 CONDUCT PERSONNEL RECOVERY-RELATED MOBILIZATION ACTIVITIES

6-28. Conduct personnel recovery (PR)-related mobilization activities by obtaining specific PR guidance from the supported combatant commander. Focus previous training and tailor existing guidance to the specific requirements of the supported combatant command. Examples of combatant command guidance include theater PR regulations, appendix 5 to annex C of joint operation plans and orders, theater PR standing operating procedures (SOPs), PR special instructions, and isolated personnel guidance. (FM 3-50.1) (USACAC)

No.	Scale	Measure
01	Yes/No	PR coordination cells and personnel recovery officers reviewed operation specific command and control procedures provided by theater specific guidance.
02	Yes/No	Training and rehearsals were scheduled against identified shortfalls in organization, equipment, and procedures.
03	Yes/No	Unit conducted integrated rehearsals for PR operations such as joint, combined arms, and warfighting function.
04	Yes/No	Unit operation plans and orders included PR guidance as PR appendix to annex C as required.
05	Yes/No	Unit updated all PR SOPs and other guidance to synchronize with theater PR guidance as necessary.

ART 6.2.2.2 DEPLOY PERSONNEL RECOVERY CAPABILITY AND BUILD COMBAT POWER

6-29. During deployment, commanders must determine who will be providing personnel recovery (PR) coverage for their arriving forces. Until commanders can build sufficient combat power to provide PR for their forces, other forces in theater must provide PR support. Build combat power through reception, staging, onward movement, and integration (RSOI). Early and continuous connectivity with the theater PR architecture enables commanders to report isolated, missing, detained, or captured (IMDC) incidents during deployment. This also allows commanders to report unit status when ready to assume responsibility for assigned PR missions. (FM 3-50.1) (USACAC)

No.	Scale	Measure
01	Yes/No	PR capability transitioned through the RSOI process.
02	Yes/No	Early deployment of PR capability was planned for in the time-phased force and deployment data.
03	Yes/No	Procedures for relief in place of the PR capability were developed.
04	Yes/No	PR capabilities were available until all forces, including contractors deploying with the force and Army civilians, redeployed.
05	Yes/No	Integrated rehearsals were planned and conducted.
06	Yes/No	PR cells capabilities to monitor PR asset status were established and maintained.
07	Yes/No	Unit implemented accountability procedures to identify actual IMDC events and preclude false reports.
08	Yes/No	Unit developed plans to identify personnel and equipment to maintain accountability and communications with movement serials.
09	Yes/No	Requirements to support command were implemented as required.
10	Yes/No	Unit built combat power and PR capability as planned.
11	Yes/No	Unit established training and rehearsal areas and ranges as planned.
12	Yes/No	Unit identified capabilities and shortfalls to provide recovery en route and during RSOI to supported command.

ART 6.2.2.3 SUSTAIN PERSONNEL RECOVERY CAPABILITIES

6-30. Sustain personnel recovery (PR) capabilities during the conduct of operations by having commander, staffs, units, and individuals refine their skills. Conduct rehearsals to exercise battle drills to ensure proficiency. New and replacement personnel will require training and equipment. Personnel recovery coordinating messages are prepared and disseminated per unit standing operating procedures. Commanders establish and maintain personnel accountability procedures. Communications architectures are established and maintained to ensure operational capabilities. (FM 3-50.1) (USACAC)

No.	Scale	Measure
01	Yes/No	Unit conducted rehearsals to exercise battle drills.
02	Yes/No	Battle drills were refined as changing battlefield conditions changed.
03	Yes/No	Unit established and maintained PR cell capability to monitor PR asset status.
04	Yes/No	Unit enforced accountability procedures.
05	Yes/No	Unit exercised reporting procedures within theater communications architecture.
06	Yes/No	Unit PR cells prepared and disseminated PR coordinating messages as required.
07	Yes/No	New personnel were trained in the PR procedures and provided equipment as necessary.

ART 6.2.2.4 REDEPLOY PERSONNEL RECOVERY CAPABILITIES

6-31. As units redeploy, either back to continental United States (CONUS) or to another area of operations (AO), actions accomplished for personnel recovery (PR) are similar to those undertaken during deployment. An important task is the transfer of PR responsibility, including the key task of transferring lessons learned gathered during after action reviews. When units redeploy to CONUS, PR responsibility for the AO transfers to incoming forces. If redeploying to another AO, units integrate into that AO's established PR architecture or build one if it does not exist. (FM 3-50.1) (USACAC)

No.	Scale	Measure
01	Yes/No	Unit ensured PR capabilities were available until all forces, including contractors and Army civilians, have redeployed.
02	Yes/No	Unit ensured counterintelligence debriefing of recovered personnel.

ART 6.2.3 PLAN PERSONNEL RECOVERY OPERATIONS

6-32. Preparation does not stop when employment begins. Rehearsals and battle drills continue and should become more demanding as skills increase. Unit plans are refined as after action reviews from personnel recovery (PR) operations identify changes required in task organization, command relationships, and recovery doctrine. New and replacement personnel will require training and equipment. Commanders and staffs, units, and individuals continue to refine their skills throughout the employment phase of operations. (FM 3-50.1) (USACAC)

No.	Scale	Measure
01	Yes/No	Unit enforced planning system to provide timely reporting; accurately reported validation and determined location; and rapidly disseminated the information to the entire PR architecture for coordinated response.
02	Yes/No	The system provided for accurate record keeping without degrading the PR effort.
03	Yes/No	The primary mission continued parallel to the recovery effort.
04	Yes/No	The goal was recovery of the isolated, missing, detained, or captured person.
05	Yes/No	Unit planned counterintelligence support to identify intelligence and security threats to PR in the area of operations.

ART 6.2.3.1 CONDUCT UNASSISTED PERSONNEL RECOVERY

6-33. Conduct unassisted personnel recovery (PR) to achieve own recovery without outside assistance. An unassisted recovery typically involves an evasion effort by isolated, missing, detained, or captured (IMDC) personnel to get back to friendly forces, or to a point where they can be recovered via another method. While the code of conduct requires IMDC personnel to make every effort to evade or escape, commanders must strive to recover these personnel via one or a combination of the other methods. (FM 3-50.1) (USACAC)

No.	Scale	Measure
01	Yes/No	Unit enforced planning system to provide timely reporting; accurately reported validation and determined location; and rapidly disseminated the information to the entire PR architecture for coordinated response.
02	Yes/No	The system provided for accurate record keeping without degrading the PR effort.
03	Yes/No	The primary mission continued parallel to the recovery effort.

ART 6.2.3.2 CONDUCT IMMEDIATE PERSONNEL RECOVERY

6-34. Conduct immediate personnel recovery (PR) operations to locate and recover isolated, missing, detained, or captured (IMDC) personnel by forces directly observing the isolating event or through the reporting process it is determined that IMDC personnel are close enough for forces to conduct a rapid recovery. Immediate recovery assumes that the tactical situation permits a recovery with the forces at hand without detailed planning or coordination. (FM 3-50.1) (USACAC)

No.	Scale	Measure
01	Yes/No	Unit enforced planning system to provide timely reporting; accurately reported validation and determined location; and rapidly disseminated the information to the entire PR architecture for coordinated response.
02	Yes/No	The system provided for accurate record keeping without degrading the PR effort.
03	Yes/No	The primary mission continued parallel to the recovery effort.

ART 6.2.3.3 CONDUCT DELIBERATE PERSONNEL RECOVERY

6-35. Conduct deliberate personnel recovery (PR) when an incident is reported and an immediate recovery is not feasible or was not successful. Weather, enemy actions, location of isolated, missing, detained, or captured personnel, and recovery force capabilities are examples of factors that may require the detailed planning and coordination of a deliberate recovery. (FM 3-50.1) (USACAC)

No.	Scale	Measure
01	Yes/No	Unit enforced planning system to provide timely reporting; accurately reported validation and determined location; and rapidly disseminated the information to the entire PR architecture for coordinated response.
02	Yes/No	The system provided for accurate record keeping without degrading the PR effort.
03	Yes/No	The primary mission continued parallel to the recovery effort.

ART 6.2.3.4 CONDUCT EXTERNAL SUPPORTED PERSONNEL RECOVERY

6-36. Conduct external supported personnel recovery (PR) when immediate or deliberate recovery is not feasible or was not successful. External supported personnel recovery is either the support provided by the Army to other joint task force components, interagency organizations, or multinational forces or the support provided by these entities to the Army. Close air support; intelligence, surveillance, and reconnaissance; and airborne command and control are examples of capabilities that may be required from different components to execute an external supported recovery. (FM 3-50.1) (USACAC)

No.	Scale	Measure
01	Yes/No	Unit enforced planning system to provide timely reporting; accurately reported validation and determined location; and rapidly disseminated the information to the entire PR architecture for coordinated response.
02	Yes/No	The system provided for accurate record keeping without degrading the PR effort.
03	Yes/No	The primary mission continued parallel to the recovery effort.
04	Yes/No	Unit provided the support required to support an external support requirement.
05	Yes/No	Unit transmitted a request for external support as necessary.

ART 6.2.3.5 CONDUCT ARMY SPECIAL OPERATIONS FORCES PERSONNEL RECOVERY

6-37. Conduct Army special operations force (ARSOF) personnel recovery (PR) missions to achieve specific, well-defined, and often sensitive results of strategic or operational significance. ARSOF PR missions are conducted in support of their own operations, when directed by the joint task force commander to support a PR operation, when the threat to the recovery force is high enough to warrant the conduct of a special operation, and when ARSOF is the only force available. Detailed planning, rehearsals, and in-depth intelligence analysis characterize ARSOF PR missions. This ART includes unassisted evasion and nonconventional assisted recovery. (FM 3-05.231) (USAJFKSWCS)

No.	Scale	Measure
01	Yes/No	Unit enforced planning system to provide timely reporting; accurately reported validation and determined location; and rapidly disseminated the information to the entire PR architecture for coordinated response.
02	Yes/No	The primary mission continued parallel to the recovery effort.
03	Yes/No	The isolated, missing, detained, or captured person was recovered.

ART 6.2.4 PROVIDE PERSONNEL RECOVERY SUPPORT TO CIVIL SEARCH AND RESCUE AUTHORITIES ON A NONINTERFERENCE BASIS

6-38. Department of the Army (DA) provides support to civil search and rescue (SAR), only when directed. The National Search and Rescue Plan, International Aeronautical and Maritime Search and Rescue manual, various international agreements, and Department of Defense and DA policies all provide the authoritative basis for military participation in civil SAR efforts. Military commanders, regardless of Service, may be requested to support civil SAR operations when they have the capability to do so. (FM 3-50.1) (USACAC)

No.	Scale	Measure
01	Yes/No	Unit identified critical command, control, and communications requirements with appropriate civil authorities and agencies.
02	Yes/No	Unit coordinated procedures to report, locate, support, and recover with appropriate civil authorities and agencies.
03	Yes/No	Unit coordinated equipment requirements with appropriate civil authorities and agencies.

ART 6.2.5 SUPPORT HOMELAND SECURITY PERSONNEL RECOVERY OPERATIONS

6-39. In cases where the President declares a "major disaster or emergency," a number of possible mechanisms are activated to support homeland security personnel recovery operations resulting from a major disaster or emergency. The Federal Emergency Management Agency of the Department of Homeland Security becomes the lead response agency in such cases as provided in the National Response Plan. The Department of Defense is signatory to the National Response Framework. The circumstances that exist before and after a disaster declaration may involve civil search and rescue operations carried out under the National Search and Rescue Plan, and may even involve mass rescue operations as discussed below. (FM 3-50.1) (USACAC)

No.	Scale	Measure
01	Yes/No	Unit identified critical command, control, and communications support requirements with appropriate civil authorities and agencies.
02	Yes/No	Unit coordinated procedures to report, locate, support, and recover with appropriate civil authorities and agencies.
03	Yes/No	Unit coordinated equipment requirements with appropriate civil authorities and agencies.

SECTION III – ART 6.3: CONDUCT INFORMATION PROTECTION

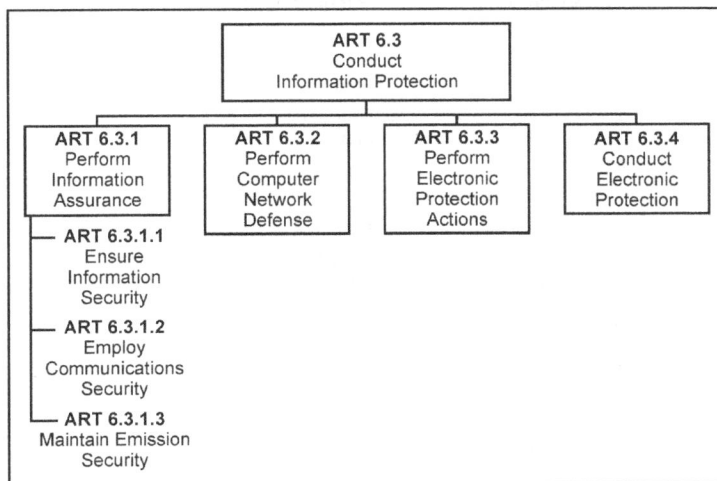

6-40. *Information protection* is active or passive measures that protect and defend friendly information and information systems to ensure timely, accurate, and relevant friendly information. It denies enemies, adversaries, and others the opportunity to exploit friendly information and information systems for their own purposes. Information protection comprises information assurance, computer network defense, and electronic protection. (FM 3-0) (USACAC)

No.	Scale	Measure
01	Yes/No	Unit course of action was not compromised by enemy information operations (IO).
02	Time	To develop and refine IO annex to operation order.
03	Time	To identify, determine appropriate response, and implement changes in response to a possible threat to friendly information systems.
04	Time	For friendly information and intelligence collection sensor system managers, operators, and emergency response teams and contact teams to respond, identify, and correct system failures attributed to enemy IO.
05	Percent	Of time units in area of operations (AO) are in restrictive information operations condition.
06	Percent	Of friendly emitters in AO known to have been exploited by an enemy.
07	Percent	Of information systems hardware, software components, and databases backed up by replacement components or backup files in case of failure or compromise.
08	Percent	Of information system software components and databases protected by firewalls and virus detection software.
09	Number	Of times to reprogram information system software in response to identified threats.
10	Number	Of instances of enemy IO disabling, corrupting, or compromising friendly information systems and intelligence collection sensors.
11	Number	Of instances of electronic fratricide in the AO.

ART 6.3.1 PROVIDE INFORMATION ASSURANCE

6-41. Plan, establish, and conduct programs and procedures to protect information and information systems. Implement safeguards and controls on data networks and computer systems. Ensure availability, integrity, authenticity, and security of information networks, systems, and data. Detect and react to compromises. Restore networks, systems, and data. (FM 3-13) (USACAC)

No.	Scale	Measure
01	Yes/No	Systems administrators and operators performed risk assessment of potential threats to friendly information systems and took appropriate action to respond to those risks.
02	Time	For information system emergency response teams to identify, respond, and correct information system failures attributed to adversary information operations or criminal mischief.
03	Percent	Of information systems not protected by firewalls, virus detection software, and other appropriate information protection measures.
04	Percent	Of information systems hardware components, software programs, and databases that have backups to replace or duplicate them in case of failure or corruption.
05	Percent	Of enemy or criminal attempts to disable, corrupt, or compromise friendly information system components, software, and databases that are successful.
06	Percent	Of enemy or criminal attempts to disable, corrupt, or compromise friendly information system components, software, and databases detected by system administrators and operators and automated protective systems, such as firewalls.
07	Percent	Of friendly information systems linked to the Internet.
08	Number	Of redundant communications paths available to connect information systems.
09	Number	Of attempts to disable, corrupt, or compromise friendly information system components, software, and databases.
10	Number	And types of friendly information systems linked to unsecured and secured Internet.

ART 6.3.1.1 ENSURE INFORMATION SECURITY

6-42. Deny the enemy access to electronic information (both communications and noncommunications) that could be used to identify friendly capabilities and intentions. (FM 3-13) (USACAC)

No.	Scale	Measure
01	Yes/No	Signal security compromises degraded, delayed, or modified unit operations.
02	Yes/No	Firewalls, virus protection software, or other information protection measures protected unit information systems.
03	Time	To refine and synchronize signal and information operations (IO) annexes to operation order.
04	Time	To complete operations security (OPSEC) assessment in the area of operations (AO).
05	Time	To identify improper occurrence of signal security.
06	Time	For appropriate information response teams to respond, identify, and correct information system failures attributed to enemy offensive IO or criminal activity.
07	Percent	Of increased or decreased number of security violations on combat net radios in the AO within a given time.
08	Percent	Of successful enemy attempted penetration of friendly information systems.
09	Percent	Of emitter system administrators and operators who have current OPSEC training.
10	Percent	Of enemy sensor coverage in AO known to friendly force.
11	Percent	Of identified friendly vulnerabilities in AO exploited by enemy actions.
12	Percent	Of electronic communications in AO encrypted or secured.
13	Percent	Of message traffic in AO exploited by enemy.
14	Percent	Of friendly emitters in AO exploited by enemy.
15	Percent	Of signal security measures previously assessed unsatisfactory that have improved based on assessment.
16	Percent	Of friendly operations conducted in a restrictive emission control environment.
17	Percent	Of units, installations, and agencies in AO operating from a common signal operation instruction.
18	Percent	Of unit communications systems required to maintain more than one encryption system.
19	Number	Of security violations on combat net radios in the AO.
20	Number	Of teams fielded to monitor friendly emitters.
21	Number	Of interceptions of friendly communications during planning and execution.
22	Number	Of instances when frequency allocation or frequency management fails to prevent signal fratricide.

ART 6.3.1.2 EMPLOY COMMUNICATIONS SECURITY

6-43. Deny the enemy information of value that might be derived from the possession and study of telecommunications. (FM 6-02.72) (USASC&FG)

No.	Scale	Measure
01	Yes/No	Communications security compromises degraded, delayed, or modified unit operations.
02	Yes/No	Unit executed controlling authority functions.
03	Time	To refine and synchronize signal annex to operation order.
04	Time	To complete communications security assessment in the area of operations (AO).
05	Time	To identify improper occurrences of communications security.
06	Percent	Of increased or decreased number of security violations on combat net radios in the AO within a given time.
07	Percent	Of enemy sensor coverage in AO known to friendly force.
08	Percent	Of successful enemy attempted penetration of friendly information systems.
09	Percent	Of information system administrators and operators who have current operations security training.
10	Percent	Of identified friendly communications vulnerabilities in AO exploited by enemy actions.
11	Percent	Of electronic communications in AO encrypted or secured.
12	Percent	Of message traffic in AO exploited by enemy.

No.	Scale	Measure
13	Percent	Of friendly information systems in AO exploited by enemy.
14	Percent	Of communications security measures previously assessed unsatisfactory that have improved based on assessment.
15	Percent	Of friendly operations conducted in a restrictive emission control environment.
16	Percent	Of units, installations, and agencies in AO operating from a common signal operation instructions.
17	Percent	Of unit communications systems requiring more than one encryption system.
18	Percent	Of communications systems using encryption.
19	Percent	Of systems that include communications security in communications network planning.
20	Number	Of communications security incidents reported.
21	Number	Of security violations on combat net radios in the AO.
22	Number	Of teams fielded to monitor friendly communications systems.
23	Number	Of interceptions of friendly communications during planning and execution.
24	Number	Of redundant communications paths available to connect operational information systems.

ART 6.3.1.3 MAINTAIN EMISSION SECURITY

6-44. Plan and implement measures to deny unauthorized persons information of value that might be derived from their interception and study of electromagnetic radiation. Select and control the use of electromagnetic, acoustic, or other emitters to optimize friendly operations and capabilities while minimizing detection by enemy sensors and mutual interference among friendly systems. (FM 6-02.72) (USASC&FG)

No.	Scale	Measure
01	Yes/No	Signal emission security compromises degraded, delayed, or modified unit operations.
02	Time	To refine and synchronize signal and IO annexes to operation order.
03	Time	To complete operations security (OPSEC) assessment in the area of operations (AO).
04	Time	To identify improper maintenance of emission security.
05	Percent	Of emitter system administrators and operators who have current OPSEC training.
06	Percent	Of enemy sensor coverage in AO known to friendly force.
07	Percent	Of identified friendly vulnerabilities in AO exploited by enemy actions.
08	Percent	Of electronic communications in AO encrypted or secured.
09	Percent	Of friendly emitters in AO exploited by enemy.
10	Percent	Of emission security measures previously assessed unsatisfactory that have improved based on new assessment.
11	Percent	Of friendly operations conducted in a restrictive emission control environment.
12	Percent	Of friendly courses of action that the enemy can determine by observing friendly emitters.
13	Number	Of emission security violations in the AO in a given time.
14	Number	Of teams fielded to monitor friendly emitters.
15	Number	Of interceptions of friendly emitters during planning and execution.
16	Number	Of instances when frequency allocation or frequency management fails to prevent signal fratricide.

ART 6.3.2 PERFORM COMPUTER NETWORK DEFENSE

6-45. Computer network defense is actions to defend against unauthorized activity within computer networks. Computer network defense includes monitoring, detection, analysis, response, and restoration activities. These activities are performed by multiple disciplines, such as operations, network administrators, intelligence, counterintelligence, and law enforcement. (FM 3-13) (USACAC)

No.	Scale	Measure
01	Yes/No	Enemy offensive information operations (IO) compromised unit course of action.
02	Time	To develop and refine IO annex to operation order.
03	Time	To identify, determine appropriate response, and implement changes in response to a possible threat to information systems.
04	Time	For friendly information and intelligence collection sensor system managers, operators, and emergency response teams or contact teams to respond, identify, and correct system failures attributed to enemy offensive IO.
05	Percent	Of time units in the area of operations (AO) are in restrictive emission control condition.
06	Percent	Of friendly emitters in the AO known to have been exploited by an enemy.
07	Percent	Of information systems hardware, software components, and databases backed up by replacement components or backup files in case of failure or compromise.
08	Number	Of times to reprogram information system software in response to identified threats.
09	Number	Of instances of enemy offensive IO disabling, corrupting, or compromising friendly information systems and intelligence collection sensors.
10	Number	Of instances of electronic fratricide in the AO.

ART 6.3.3 PERFORM ELECTRONIC PROTECTION ACTIONS

6-46. Plan and implement active and passive means to protect personnel, facilities, and equipment from any effects of friendly or enemy employment of electronic warfare (EW) that may degrade, neutralize, or destroy friendly combat capability. Electronic protection includes the hardening of equipment and facilities from the effects of EW; emission control procedures and measures as outlined in ART 6.3.1.3 (Maintain Emission Security); and the efficient management of the electromagnetic spectrum as outlined in ART 5.3.6 (Conduct Electromagnetic Spectrum Management Operations). (JP 6-0) (USJFCOM JWFC)

No.	Scale	Measure
01	Yes/No	EW mission spectrum requirements were de-conflicted with the unit spectrum manager.
02	Yes/No	All EW systems operated without interference.
03	Yes/No	EW system emission security compromises degraded, delayed, or modified unit operations.
04	Time	To evaluate EW-related frequency interference issues.
05	Time	To coordinate EW-related frequency interference issues.
06	Time	To resolve EW-related frequency interference issues.
07	Time	To identify improper maintenance of emission security.
08	Time	To respond to new threats through reprogramming of systems.
09	Time	To take appropriate measure against friendly or enemy EW system interference.
10	Percent	Of identified friendly vulnerabilities in the area of operations (AO) exploited by enemy actions.
11	Percent	Of friendly emitters in the AO exploited by enemy.
12	Percent	Of friendly operations conducted in a restrictive emission control environment.
13	Percent	Of emission control procedures that have improved from previous assessments
14	Percent	Of successful EW system reprogramming events.
15	Percent	Of friendly systems affected by friendly EW systems.
16	Percent	Of friendly systems affected by enemy EW systems.
17	Number	Of frequency interference issues.
18	Number	Of EW systems operating on assigned frequencies.
19	Number	Of EW systems detected by enemy sensors.
20	Number	Of emission security violations in the AO in a given time.
21	Number	Of instances when frequency allocation or frequency management fails to prevent signal fratricide.
22	Number	Of EW system reprogramming events.

No.	Scale	Measure
23	Number	Of instances when EW system reprogramming is unsuccessful.
24	Number	Of friendly systems affected by friendly or enemy EW systems.

ART 6.3.4 CONDUCT ELECTRONIC PROTECTION

6-47. Plan and implement actions such as communications avoidance or communications antijamming measures to protect personnel, facilities, and equipment from friendly and enemy employment of electronic warfare that degrade, neutralize, or destroy friendly combat capability. (FM 3-13) (USACAC)

No.	Scale	Measure
01	Yes/No	Unit course of action was not compromised by enemy offensive information operations (IO).
02	Time	To develop and refine IO annex to operation order.
03	Time	For friendly information and intelligence collection sensor system managers, operators, and emergency response teams or contact teams to respond, identify, and correct system failures attributed to enemy offensive IO.
04	Time	To identify, determine appropriate response, and implement changes in response to a possible threat to information systems.
05	Percent	Of time units in the area of operations (AO) are in restrictive information operations condition.
06	Percent	Of friendly emitters in the AO known to have been exploited by an enemy.
07	Percent	Of information systems hardware, software components, and databases backed up by replacement components or backup files in case of failure or compromise.
08	Number	Of times to reprogram information system software in response to identified threats.
09	Number	Of instances of enemy offensive IO disabling, corrupting, or compromising friendly information systems and intelligence collection sensors.
10	Number	Of instances of electronic fratricide in the AO.

SECTION IV – ART 6.4: PERFORM FRATRICIDE AVOIDANCE

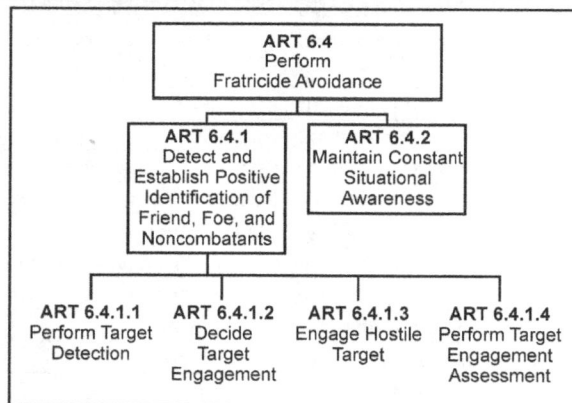

6-48. Fratricide is the unintentional killing of friendly personnel by friendly firepower. Responsibility for preventing fratricide is the responsibility of the commander, yet all Soldiers must avoid the reluctance to employ, integrate, and synchronize all the combat power at the critical time and place. (FM 3-20.15) (USAARMC)

ART 6.4.1 DETECT AND ESTABLISH POSITIVE IDENTIFICATION OF FRIEND, FOE, AND NONCOMBATANTS

6-49. Discretely and positively determine, by any means, the identity of tactical units, their equipment, and personnel, or of phenomena, such as communications-electronic patterns. Distinguish these forces from hostile or unknown forces and means, one from the other. (FM 5-19) (CRC)

No.	Scale	Measure
01	Yes/No	Unit was able to correctly identify other forces, equipment, and personnel in the area of operations (AO).
02	Time	To refine the force protection plan.
03	Time	Elapsed before enemy begins to mimic identification or recognition procedures.
04	Time	To confirm the identified or unidentified friendly unit or system.
05	Time	To confirm the identity of an unidentified target.
06	Time	To pass a target identity to the decisionmaker.
07	Time	To change codes in identification, friend or foe (IFF) systems.
08	Percent	Of force in the AO that has passive identification interrogation capability.
09	Percent	Of IFF systems operating correctly.
10	Percent	Of force in AO using their IFF systems.
11	Percent	Of friendly systems in the AO destroyed by friendly fire.
12	Percent	Of casualties in the AO from friendly fire.
13	Percent	Of positive identification false negatives (friendly identified as enemy) in the AO.
14	Percent	Of positive identification false positives (enemy identified as friendly) in the AO.
15	Percent	Of units accurately reporting their locations.
16	Number	Of penetrations in the AO by unknown targets.
17	Number	Of IFF systems operating correctly in the AO.

ART 6.4.1.1 PERFORM TARGET DETECTION

6-50. Characterize detected objects as friend, enemy, or neutral. In combat operations, discriminate between recognizable objects as being friendly, neutral, or enemy, or the name that belongs to the object as a member of a class. Identify a recognized object and the specific designation of that object to determine to the extent that firing or other tactical decisions concerning it could be made. This aspect of combat identification is time sensitive and directly supports a combatant's shoot or don't-shoot decision for detected objects on the battlefield. (FM 3-20.15) (USAARMC)

No.	Scale	Measure
01	Yes/No	Unit classified detected object as friend, foe, or neutral.
02	Yes/No	Unit determined threat level of object.
03	Percent	Of objects detected.
04	Percent	Of objects not detected.

ART 6.4.1.2 DECIDE TARGET ENGAGEMENT

6-51. Determine the appropriate application of military options and weapons resources on identified objects. Identify the process used to execute a course of action developed to engage a target determined by situational awareness and available weapons or resources. Select a course of action as the one most favorable to accomplish the mission. In an estimate of the situation, a clear and concise statement of the line of action intended to be followed by the commander as the one most favorable to successfully accomplishing the mission. (FM 3-20.15) (USAARMC)

No.	Scale	Measure
01	Yes/No	Unit determined target displays hostile intent.
02	Yes/No	Unit determined if target can be engaged within rules of engagement.
03	Yes/No	Unit determined if available resources can destroy the target.
04	Yes/No	Unit determined the most favorable course of action.

ART 6.4.1.3 ENGAGE A HOSTILE TARGET

6-52. Use military options or resources to engage a target with appropriate lethal or nonlethal weapons. Dominate targets and protect friendly and neutral personnel. (FM 3-20.15) (USAARMC)

No.	Scale	Measure
01	Yes/No	Unit used appropriate weapon to engage target.
02	Yes/No	Unit protected friendly and neutral personnel.
03	Time	To engage targets.
04	Percent	Of targets suppressed.
05	Percent	Of targets destroyed.
06	Percent	Of targets not engaged.
07	Percent	Of risk to friendly forces to engage targets.
08	Percent	Of risk to neutral personnel to engage targets.

ART 6.4.1.4 PERFORM TARGET ENGAGEMENT ASSESSMENT

6-53. Assess the applied resources applied and whether the action generated the desired effects in support of the commander's fires objectives. (FM 3-20.15) (USAARMC)

No.	Scale	Measure
01	Yes/No	Unit used effective process to assess existing or new threats.
02	Yes/No	Unit achieved desired engagement results.
03	Yes/No	Unit used information to detect additional threats.
04	Yes/No	Unit used appropriate level of force to engage target.
05	Time	To complete engagement assessment.
06	Time	To send assessment to higher headquarters.

ART 6.4.2 MAINTAIN CONSTANT SITUATIONAL AWARENESS

6-54. *Situational awareness* is immediate knowledge of the conditions of the operation, constrained geographically and in time. It is the ability to maintain a constant, clear mental picture of the tactical situation. This picture includes an understanding of both the friendly and enemy situations and of relevant terrain. It also includes relating events in time to form logical conclusions and make decisions that anticipate events. (FM 3-0) (USACAC)

No.	Scale	Measure
01	Percent	Of friendly forces following established procedures to identify themselves.
02	Percent	Of friendly forces detecting friendly objects or entities.

SECTION V – ART 6.5: CONDUCT OPERATIONAL AREA SECURITY

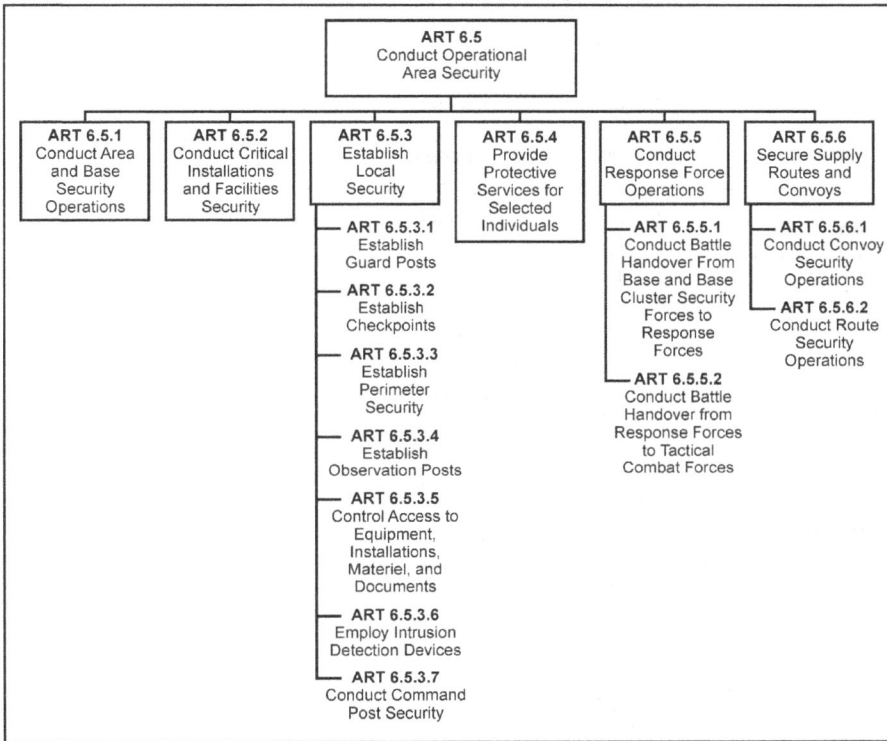

```
                          ART 6.5
                    Conduct Operational
                      Area Security
```

ART 6.5.1	ART 6.5.2	ART 6.5.3	ART 6.5.4	ART 6.5.5	ART 6.5.6
Conduct Area and Base Security Operations	Conduct Critical Installations and Facilities Security	Establish Local Security	Provide Protective Services for Selected Individuals	Conduct Response Force Operations	Secure Supply Routes and Convoys

ART 6.5.3.1 Establish Guard Posts

ART 6.5.3.2 Establish Checkpoints

ART 6.5.3.3 Establish Perimeter Security

ART 6.5.3.4 Establish Observation Posts

ART 6.5.3.5 Control Access to Equipment, Installations, Materiel, and Documents

ART 6.5.3.6 Employ Intrusion Detection Devices

ART 6.5.3.7 Conduct Command Post Security

ART 6.5.5.1 Conduct Battle Handover From Base and Base Cluster Security Forces to Response Forces

ART 6.5.5.2 Conduct Battle Handover from Response Forces to Tactical Combat Forces

ART 6.5.6.1 Conduct Convoy Security Operations

ART 6.5.6.2 Conduct Route Security Operations

6-55. Operational area security is a form of security operations conducted to protect friendly forces, installations, routes, and actions within an area of operations. Although vital to the success of military operations, operational area security is an economy of force mission designed to ensure the continued conduct of sustainment operations and to support decisive and shaping operations. (FM 3-90) (USACAC)

No.	Scale	Measure
01	Yes/No	The operations of the area security force provided the protected force or installation with sufficient reaction time and maneuver space.
02	Yes/No	Area security forces were in place not later than time specified in operation order.
03	Yes/No	Area security force prevented enemy ground observation of protected force or installation.
04	Yes/No	Collateral damage due to the conduct of area security operations was within acceptable limits.
05	Yes/No	Area security force provided early and accurate warning of enemy approach.
06	Yes/No	Area security force oriented its operations on the protected forces and facilities.
07	Yes/No	Area security force performed continuous reconnaissance.
08	Yes/No	Area security force maintained contact with enemy forces.
09	Yes/No	Area security force protected government-sponsored civilian stabilization and reconstruction personnel.
10	Yes/No	Area security force protected contractor and nongovernmental organization stabilization personnel and resources.

No.	Scale	Measure
11	Yes/No	Commander developed criteria for ending the area security operation.
12	Yes/No	Area commander established useful intelligence links with local authorities.
13	Time	To conduct reconnaissance of the area.
14	Time	To plan area security operation.
15	Time	To prepare for the area security operation including the conduct of troop movement.
16	Time	To execute the area security operation.
17	Time	To report enemy activities to appropriate headquarters.
18	Time	Before the secured force, installation, or route encounters enemy forces.
19	Time	To integrate host-nation or third-nation security forces and means into friendly area security operations.
20	Time	Between observation and surveillance of named areas of interest in the secured area.
21	Time	For a reaction force or tactical combat force to respond and reach an installation or facility under attack.
22	Time	Of increased required to transit an area due to enemy attacks on transportation facilities and road networks.
23	Percent	Of security force casualties during the area security operation.
24	Percent	Of casualties (secured force or installation and people using secured routes) during the area security operation.
25	Percent	Of unit combat power needed to provide desired degree of security.
26	Percent	Of decreased support capability of sustainment units due to enemy attacks.
27	Percent	Of decreased support capability of sustainment units due to the requirement to provide security forces from internal assets.
28	Percent	Of decreased transport capability of a line of communications or main supply route due to enemy attacks.
29	Percent	Of increased availability of area security forces through use of host-nation or third-nation security forces.
30	Percent	Of enemy reconnaissance and other forces destroyed or repelled by the area security force.
31	Percent	Of friendly operations judged as not compromised prior to or during execution.
32	Percent	Of operations not compromised (based on enemy prisoner of war interrogations or captured documents).
31	Percent	Of critical facilities in the area of operations hardened or protected by area security forces.
32	Percent	Of security measures completed for a given facility in the secured area.
33	Percent	Of the secured area that can be observed by visual observation or covered by sensors at any given time.
34	Percent	Of lines of communications and main supply routes in the area secured.
35	Percent	Of available military police effort in area used to provide area security, such as command post guards and reaction forces.
36	Percent	Of information system networks that have multiple paths over which to transmit data.
37	Percent	Of attempted enemy attacks—including terrorist attacks—that penetrate area security.
38	Number	And types of maneuver forces used to provide area security.
39	Number	And types of enemy forces operating in the area being secured.
40	Number	Of incidents where enemy forces affect the security of friendly units and facilities, such as terrorist attacks, snipping, and isolated mortar or rocket attacks.
41	Number	Of incidents where enemy forces compromise friendly courses of actions, level II and level III attacks, or terrorist attacks that penetrate into target area.

No.	Scale	Measure
42	Number	Of casualties incurred by the security force during the conduct of the area security operation.
43	Number	Of casualties incurred by the secured force or installation during the conduct of the area security operation.
44	Number	Of mobility corridors or avenues of approach that the area security force can observe.
45	Number	Of observation or guard posts that the area security force can establish.
46	Number	Of enemy reconnaissance and other forces destroyed during security operation.
47	Square Kilometers	Size of area being secured.

ART 6.5.1 CONDUCT AREA AND BASE SECURITY OPERATIONS

6-56. Area and base security operations are a specialized area security operation. It protects friendly forces, installations, and actions in the support area. It includes measures taken by military units, activities, and installations to protect themselves from acts designed to impair their effectiveness. (FM 3-90) (USACAC)

No.	Scale	Measure
01	Yes/No	Operations of the security forces provided the protected force or installation with sufficient reaction time and maneuver space.
02	Yes/No	Sustainment area and base security forces were in place not later than time specified in operation order.
03	Yes/No	Sustainment area and base security forces prevented enemy ground observation of protected force or installation.
04	Yes/No	Collateral damage was due to the conduct of sustainment area; base security operations were within acceptable limits.
05	Yes/No	Security force provided early and accurate warning of enemy approach toward base perimeter.
06	Yes/No	The security force oriented its operations on the protected facilities.
07	Yes/No	The security force performed continuous reconnaissance.
08	Yes/No	The security force maintained contact with enemy forces.
09	Yes/No	The area commander established useful intelligence links with local authorities.
10	Yes/No	Commander prioritized sustainment area and base security efforts to protect most critical resources.
11	Time	To conduct reconnaissance of the sustainment area.
12	Time	To plan sustainment area and base security operations.
13	Time	To prepare for the sustainment area and base security operations including the conduct of troop movement.
14	Time	To execute the sustainment area and base security operations.
15	Time	To report enemy activities to appropriate headquarters.
16	Time	Of warning before the secured force, installation, or route encounters enemy forces.
17	Time	To integrate host-nation or third-nation security forces and means into friendly sustainment area and base security operations.
18	Time	Between observation and surveillance of named areas of interest in secured area.
19	Time	For a reaction force or tactical combat force to respond and reach an installation or facility under attack.
20	Percent	Of security forces casualties during the sustainment area and base security operations.
21	Percent	Of casualties (secured force or installation and people using secured routes) during the sustainment area and base security operations.
22	Percent	Of unit combat power to provide desired degree of sustainment area and base security.

No.	Scale	Measure
23	Percent	Of decreased support capability of sustainment units due to the requirement to provide security forces from internal assets.
24	Percent	Of decreased support capability of sustainment units due to enemy attacks.
25	Percent	Of decreased transport capability of a line of communications or main supply route due to enemy attacks.
26	Percent	Of increased availability of sustainment area and base security forces through use of host-nation or third-nation security forces.
27	Percent	Of enemy reconnaissance and other forces destroyed or repelled by the base security forces.
28	Percent	Of friendly sustaining operations judged as not compromised before or during execution.
29	Percent	Of critical facilities in the area of operations hardened and protected by security forces.
30	Percent	Of security measures—such as perimeter fences, cleared fields of fire, and anti-intrusion detection devices—completed for a given facility in the sustainment area.
31	Percent	Of the sustainment area that can be observed by visual observation or covered by sensors at any given time.
32	Percent	Of lines of communications and main supply routes secured in the sustainment area.
33	Percent	Of available military police effort in area used to provide sustainment area security, such as reaction forces.
34	Percent	Of attempted enemy attacks—including terrorist attacks—that penetrate a base's perimeter security.
35	Number	And types of maneuver forces used to provide sustainment area and base security.
36	Number	And types of enemy forces operating in the echelon sustainment area.
37	Number	Of incidents where enemy forces affect the security of friendly bases, such as terrorist attacks, snipping, and isolated mortar or rocket attacks.
38	Number	Of incidents where enemy forces compromise friendly courses of action, level II and level III attacks, or terrorist attacks that penetrate into their target area.
39	Number	Of security force casualties during the sustainment area and base security operations.
40	Number	Of secured force or installation casualties during the sustainment area and base security operations.
41	Number	Of mobility corridors or avenues of approach that can be observed by the area security force.
42	Number	Of observation posts, guard posts, or checkpoints that can be established by the sustainment area security force.
43	Number	Of enemy reconnaissance and other forces destroyed during the conduct of sustainment area security operations.
44	Square Kilometers	Size of the echelon sustainment area.

ART 6.5.2 CONDUCT CRITICAL INSTALLATIONS AND FACILITIES SECURITY

6-57. Use protective measures to prevent or reduce the effects of enemy hostile acts (such as sabotage, insurgent actions, and terrorist attack) against unit critical facilities and systems designated as Site Security Level-A or Protection Level I. Protective measures include conducting local security operations, protecting individuals and systems, preparing fighting positions, preparing protective positions, employing protective equipment, reacting to enemy direct fire, reacting to enemy indirect fire, reacting to enemy aerial attack, reacting to a terrorist incident. (FM 3-27.10) (USASMDC)

No.	Scale	Measure
01	Yes/No	Security force protected installation or facility from damage.
02	Yes/No	Unit established executable antiterrorism program.
03	Yes/No	Unit established procedures to change force protection conditions.
04	Yes/No	Unit had procedures to respond to enemy use of chemical, biological, radiological, nuclear, and high-yield explosives weapons.
05	Yes/No	Antiterrorism and physical security plan was coordinated, approved, and executable.
06	Time	To refine base and base cluster defense plan.
07	Time	For a higher headquarters to assess base and base cluster defense plans.
08	Time	To coordinate additional assets for unit lines of communications.
09	Time	For reaction forces and response forces to respond to enemy threats to critical installations or facilities.
10	Time	To review counterintelligence plans for major tactical units in the area of operations.
11	Percent	Of unit to secure critical installations and provide facility security.
12	Percent	Of successful level I, level II, and terrorist attacks in echelon sustainment area.
13	Percent	Of decreased friendly installations and facilities capabilities due to successful attacks.
14	Percent	Of critical installations, facilities, and communications hardened against attack.
15	Percent	Of alternate path communications supporting operations in the area of operations.
16	Percent	Of friendly installations and unit having current counterterrorism or antiterrorism training programs in effect.
17	Percent	Of threat assessments passed within established criteria.
18	Percent	Of tactical units in the area of operations that have counterintelligence plans.
19	Number	Of level I, level II, and terrorist attacks attempted against critical installations and facilities in the unit sustainment area.
20	Number	Of friendly force actions that disrupt enemy intelligence collection efforts.
21	Number	Of enemy acts against friendly forces near the unit.
22	Number	Of instances of operations degraded, disrupted, delayed, or modified due to successful enemy penetration of critical installations and facilities.

ART 6.5.3 ESTABLISH LOCAL SECURITY

6-58. Take measures to protect friendly forces from attack, surprise, observation, detection, interference, espionage, terrorism, and sabotage. ART 6.5.3 enhances the freedom of action of tactical units in an area of operations by identifying and reducing friendly vulnerability to hostile acts, influence, or surprise. (FM 3-90) (USACAC)

No.	Scale	Measure
01	Yes/No	Effective local security existed in a 360-degree arc around the unit.
02	Yes/No	Commander adjusted unit levels of alert based on the factors of mission, enemy, terrain and weather, troops and support available, time available and civil considerations.
03	Time	To plan local security operations.
04	Time	To prepare for the conduct of local security operations.
05	Time	That local security will be maintained.
06	Time	To establish observation and guard posts.
07	Time	To conduct patrols of the local area.
08	Time	To emplace camouflage.
09	Time	Between observation and surveillance of dead space within direct fire range of the unit's perimeter.

No.	Scale	Measure
10	Time	For all unit personnel to occupy fighting and survivability positions on receipt of warning of enemy attack or operation order.
11	Time	To site and emplace protective obstacles, such as concertina wire and command detonated antipersonnel mines.
12	Time	To adjust local security measures in reaction to changes in environmental conditions, such as fog, rain, and nightfall.
13	Time	For unit reaction force to respond to enemy penetration of unit perimeter.
14	Time	To establish ambushes to provide local protection under limited visibility conditions.
15	Percent	Of unit observing stand-to time and procedures as outlined in unit standing operating procedures.
16	Percent	Of unit observing movement control restrictions.
17	Percent	Of unit observing unit noise and light discipline protocols.
18	Percent	Of available ground sensors, night vision devices, and daylight sights in operating condition.
19	Percent	Of local area around the unit under continuous observation or surveillance.
20	Percent	Of unit to provide local security.
21	Percent	Of decreased sustainment unit functional capabilities due to the requirement for those units to provide their own local security.
22	Number	And types of ground sensors, night vision devices, and daylight sights in operating condition.
23	Number	Of observation and guard posts established.
24	Number	Of patrols operating at any given time.
25	Number	Of ambushes operating at any given time.
26	Number	Of instances of enemy surveillance and reconnaissance attempts disrupted by friendly local security activities.
27	Number	Of level I and terrorist attacks directed against the unit.

ART 6.5.3.1 ESTABLISH GUARD POSTS

6-59. Delineate the organization and functions of interior and exterior guards to include orders, countersigns, parole words, and responsibility of the main guard; the duties of personnel; and methods of mounting the guard. (FM 3-90) (USACAC)

No.	Scale	Measure
01	Yes/No	Personnel manning guard posts took appropriate action per rules of engagement and special orders to prevent unauthorized entry or exit from protected facility.
02	Yes/No	Guard posts hardened against terrorist or level I attack.
03	Yes/No	Personnel manning guard posts allowed only authorized persons and vehicles access to the protected site.
04	Yes/No	Guard posts allowed adequate observation of mobility corridors and access routes leading into and out of the protected site.
05	Yes/No	Guard posts communicated with guardhouse, base defense operations center, and unit command post.
06	Yes/No	Method of mounting guard was per doctrine, regulations, and unit standing operating procedures.
07	Time	To assess the site—identify threat and vulnerabilities, review existing security arrangements, coordinate with facility commander, and conduct reconnaissance of the area.
08	Time	To develop guard post orders.
09	Time	To establish communication with guard house, base defense operations center, and unit command post.

No.	Scale	Measure
10	Time	To establish barrier control measure using available materials.
11	Time	To implement access controls, such as access rosters, badge systems, and duress codes.
12	Time	To establish challenge and password system.
13	Time	To emplace perimeter control measures to include concertina wire, mines, trenches, barricades, fences, and nonlethal capabilities.
14	Time	To obtain additional resources to improve existing perimeter control measures.
15	Time	To pass personnel and vehicles through the guard post.
16	Percent	Of perimeter penetrations detected and reported.
17	Percent	Of mission-capable perimeter control measures.
18	Percent	Of unit personnel to man existing guard posts.
19	Percent	Of protected site perimeter covered by observation from existing guard posts.
20	Number	Of guard posts established.
21	Number	Of personnel to man existing guard posts.
22	Number	Of surface and subsurface (tunnels) perimeter penetrations taking place or attempted.

ART 6.5.3.2 ESTABLISH CHECKPOINTS

6-60. Establish checkpoints to monitor and control movement, inspect cargo, enforce rules and regulations, and provide information. (FM 3-19.4) (USAMPS)

No.	Scale	Measure
01	Yes/No	Personnel manning checkpoints took appropriate actions per rules of engagement and special orders to control movement, inspect cargo, and enforce rules and regulations.
02	Yes/No	Checkpoint hardened against terrorist or level I attack.
03	Yes/No	Personnel manning checkpoint allowed only authorized persons and vehicles to pass through the checkpoint.
04	Yes/No	Checkpoints placed at unanticipated locations and located so the checkpoint cannot be seen more than a short distance away to prevent it being avoided.
05	Yes/No	Checkpoints communicated with response forces, base defense operations center, and unit command post.
06	Yes/No	Method of operating checkpoint was per doctrine, regulations, status-of-forces agreements, and unit standing operating procedures.
07	Yes/No	Male and female search teams were available.
08	Yes/No	Units dismantled roadblocks and established checkpoints.
09	Time	To assess the checkpoint site—identify threat and vulnerabilities, review existing security arrangements, coordinate with facility commander, and conduct reconnaissance of the area.
10	Time	To develop special instructions for checkpoints.
11	Time	To establish communications with response forces, base defense operations center, and unit command post.
12	Time	To emplace checkpoint control measures to include concertina wire, mines, trenches, barricades, fences, and nonlethal capabilities.
13	Time	To establish barriers around checkpoint using available materials.
14	Time	To obtain additional resources to improve existing perimeter control measures.
15	Time	To pass personnel and vehicles through the checkpoint.
16	Percent	Of contraband detected and reported.
17	Percent	Of mission-capable checkpoint control measures.
18	Percent	Of unit personnel to man existing checkpoint.

No.	Scale	Measure
19	Percent	Of personnel and vehicles that initiate fires against the checkpoint killed, destroyed, or captured.
20	Percent	Of personnel aware of rules of engagement and limitations regarding search, arrest, and use of force.
21	Number	Of checkpoints established.
22	Number	Of personnel to man existing checkpoints.
23	Number	Of personnel or vehicles attempting to flee or breach the checkpoint.
24	Number	And types of contraband seized at checkpoints.

ART 6.5.3.3 ESTABLISH PERIMETER SECURITY

6-61. Employ defensive measures to protect a unit, facility, or location from attack, unauthorized access, theft, or sabotage. Measures may include physical barriers, clear zones, lighting, guards or sentries, reaction forces, intrusion detection devices, and defensive positions. (FM 3-90) (USACAC)

No.	Scale	Measure
01	Yes/No	Effective perimeter security existed in a 360-degree arc around the unit.
02	Yes/No	Commander adjusted unit levels of alert based on the factors of mission, enemy, terrain and weather, troops and support available, time available and civil considerations.
03	Yes/No	The perimeter took advantage of the natural defensive characteristics of the terrain.
04	Yes/No	The unit controlled the area surrounding the perimeter to a range beyond that of enemy mortars and rockets.
05	Yes/No	Unit used smoke and deception during the conduct of perimeter security.
06	Yes/No	Lethal and nonlethal systems and munitions were available and sufficient to support the perimeter security plan.
07	Time	To plan for perimeter security.
08	Time	To prepare for the conduct of perimeter security.
09	Time	That perimeter security will be maintained.
10	Time	To establish observation and guard posts.
11	Time	To conduct patrols of the local area.
12	Time	To emplace camouflage.
13	Time	Between observation and surveillance of dead space within direct fire range of the unit's perimeter.
14	Time	For all unit personnel to occupy fighting and survivability positions on receipt of warning of enemy attack or operation order.
15	Time	For unit reaction force to respond to enemy penetration of unit perimeter.
16	Time	To site and emplace protective obstacles, such as concertina wire and command detonated antipersonnel mines.
17	Time	To adjust local security measures in reaction to changes in environmental conditions, such as fog, rain, and nightfall.
18	Time	To establish ambushes to provide local protection under limited visibility conditions.
19	Time	To emplace security measures that are not protective obstacles, such as sally ports, guard towers, intrusion detector sensors, and exterior lights.
20	Percent	Of unit observing stand-to time and procedures as outlined in unit standing operating procedures.
21	Percent	Of unit observing movement control restrictions.
22	Percent	Of unit observing unit noise and light discipline protocols.
23	Percent	Of available ground sensors, night vision devices, and daylight sights in operating condition.

No.	Scale	Measure
24	Percent	Of area around the unit perimeter under continuous observation or surveillance.
25	Percent	Of unit to provide perimeter security.
26	Percent	Of decreased sustainment unit functional capabilities due to the requirement for those units to provide their own perimeter security.
27	Number	And types of ground sensors, night vision devices, and daylight sights in operating condition.
28	Number	Of observation and guard posts established.
29	Number	Of patrols operating at any given time.
30	Number	Of ambushes operating at any given time.
31	Number	Of instances enemy surveillance and reconnaissance attempts disrupted by friendly security activities.
32	Number	Of level I and terrorist attacks directed against the unit perimeter.
33	Number	Of level II attacks directed against the unit perimeter.

ART 6.5.3.4 ESTABLISH OBSERVATION POSTS

6-62. Establish and maintain observation posts to prevent surprise to a protected force or to ensure observation of a designated area. (FM 3-90) (USACAC)

No.	Scale	Measure
01	Yes/No	Observation post position allowed personnel to observe assigned area, such as likely enemy avenues of approach and named areas of interest.
02	Yes/No	Observation post personnel provided early warning in event of enemy activity.
03	Yes/No	Personnel manning observation post engaged and destroyed enemy reconnaissance elements within organic and available supporting capabilities.
04	Yes/No	Observation post was operational not later than the time the operation order specifies.
05	Yes/No	Observation posts communicated with higher headquarters.
06	Yes/No	Minimum of two personnel in observation post. Observation duties rotated on a given schedule.
07	Time	To plan and prepare to establish the observation post.
08	Time	To move from current position to proposed site of the observation point.
09	Time	To assess the proposed site for the observation post and move it to a more suitable location as necessary.
10	Time	To establish communications with higher headquarters.
11	Time	To establish local security including the selection of fighting and hide positions for combat vehicles; preparation of range cards; emplacing chemical agent alarms; and installing camouflage, concertina wire, and protective mines.
12	Time	To orient personnel manning observation posts to terrain and mission control graphics, such as target reference points and trigger points.
13	Percent	Of enemy or civilian activity detected and reported.
14	Percent	Of serviceable, on-hand equipment (such as map with control graphics, compass, communications equipment, and observation devices) to conduct observation mission.
15	Percent	Of unit personnel to man existing observation posts.
16	Percent	Of named area of operations covered by observation from existing observation posts.
17	Number	Of observation posts established.
18	Number	Of personnel to man existing observation posts.
19	Number	Of observation posts detected by enemy reconnaissance assets.

ART 6.5.3.5 CONTROL ACCESS TO EQUIPMENT, INSTALLATIONS, MATERIEL, AND DOCUMENTS

6-63. Establish a system of complementary, overlapping security measures to control access to critical resources and information. Measures may include physical barriers, clear zones, lighting, access and key control, the use of security badges, intrusion detection devices, defensive positions, and nonlethal capabilities. (FM 3-19.30) (USAMPS)

No.	Scale	Measure
01	Yes/No	Unit, base, or installation physical security program protected personnel, information, and critical resources from unauthorized access.
02	Time	To review and refine unit physical security standing operating procedures per factors of mission, enemy, terrain and weather, troops and support available, time available and civil considerations.
03	Time	To refine physical security regulations for installations and major combat formation in an area of operations.
04	Time	To complete a threat analysis.
05	Time	To design, procure, emplace, and activate protective measures, such as physical barriers, clear zones, exterior lighting, access and key control, intrusion detection devices, and defensive positions.
06	Percent	Of units, bases, or installations in the area of operations that have active integrated physical security programs.
07	Percent	Of guidance in unit and base physical security programs actually followed.
08	Percent	Of planned physical barriers, clear zones, exterior lighting, access and key control, intrusion detection devices, and defensive positions operational.
09	Number	Of successful attempts to gain unauthorized access to friendly forces, installations, information, equipment, and supplies.

ART 6.5.3.6 EMPLOY INTRUSION DETECTION DEVICES

6-64. Conduct site surveys. Install and operate intrusion detection systems to protect Army installations, personnel, operations, and critical resources in both tactical and nontactical situations. (FM 3-19.30) (USAMPS)

No.	Scale	Measure
01	Yes/No	Unit, base, or installation intrusion detection devices protected sites from unauthorized access.
02	Yes/No	Intrusion detection device was an integrated system that encompassed interior and exterior sensors; close-circuit television systems for assessing alarm conditions; electronic entry-control systems; data-transmission media; and alarm reporting systems for monitoring, controlling, and displaying various alarm and system information.
03	Yes/No	The intrusion detection device deployed in and around barriers served as a boundary demarcation and means to keep animals and people from causing nuisance alarms by inadvertently straying into controlled area.
04	Yes/No	Voice communications linked (radio, intercom, and telephone) with the response force were located in the security center.
05	Time	To complete a threat analysis.
06	Time	To design, obtain, emplace, and activate intrusion detection device.
07	Time	For response force to respond to report of activity by the intrusion detection device.
08	Percent	Of alerts by intrusion detection device that are incorrect.
09	Number	Of intrusions into protected site that the intrusion detection device does not detect.

ART 6.5.3.7 CONDUCT COMMAND POST SECURITY

6-65. Prevent command and control (C2) disruption due to enemy forces penetrating the perimeter around a command post or the rapid forced displacement of the command post due to the presence of enemy forces. Security of command posts at all levels is essential to the continuity and successful exercise of C2. Security is achieved by using security forces, air defense, camouflage, traffic control, electronic countermeasures, and frequent displacements. (FM 3-19.4) (USAMPS)

No.	Scale	Measure
01	Yes/No	Command post secured without degradation of command post operations.
02	Yes/No	Traffic control points were near the intersection of main supply routes and access roads to the command post. They ensured traffic flowed freely, congestion was avoided, and traffic entering access roads was screened.
03	Yes/No	Unit provided personal security for the commander.
04	Yes/No	Unit established dismount point near the command post entrance.
05	Yes/No	Unit enforced noise, light, and litter discipline.
06	Yes/No	Traffic control points controlled entrance to command post by access rosters.
07	Yes/No	Security force communicated with headquarters commandant.
08	Time	To conduct reconnaissance of routes to the command post and areas around the command post.
09	Time	To conduct troop leading procedures.
10	Time	To establish a screen line around the command post.
11	Time	To conduct patrols around the command post.
12	Time	To establish defensive positions, deploy camouflage and concealment systems, deploy lethal and nonlethal measures, and protective obstacles around the command post.
13	Time	To establish a challenge and password or duress system.
14	Time	To establish communications with headquarters commandant.
15	Time	Between movements, command post displacements, or jumps.
16	Percent	Of available forces to provide command post security.
17	Percent	Of personnel who become casualties due to a level I, level II, or terrorist attack on the command post.
18	Number	Of level I, level II, and terrorist attacks against the command post.
19	Number	Of friendly force actions that disrupt enemy intelligence collection efforts.

ART 6.5.4 PROVIDE PROTECTIVE SERVICES FOR SELECTED INDIVIDUALS

6-66. Protect designated high-risk individuals from assassination, kidnapping, injury, or embarrassment. ART 6.5.4 includes planning, preparing, executing, and assessing close-in protection; coordinating external security with supporting law enforcement and security agencies; and providing technical advice on protective service operations to commanders, to include conducting vulnerability assessments. (FM 3-19.12) (USAMPS)

No.	Scale	Measure
01	Yes/No	Unit protected principal, including key political and societal leaders, accomplished mission while receiving protective services.
02	Yes/No	Unit protected principal was not injured, killed, or captured during time in the unit or installation area of operations (AO).
03	Yes/No	Forces providing protective services received useable antiterrorism intelligence information from host-nation agencies.
04	Yes/No	Unit or installation operations security concerning measures taken to protect the principal were not deliberately or accidentally disclosed to terrorist organizations.

No.	Scale	Measure
05	Yes/No	Local population supported principal's presence in the AO.
06	Yes/No	Effective response or reaction forces existed in the event of an attempted or successful attack on the principal.
07	Time	To refine security plans to reflect changes in protected individual's itinerary.
08	Time	For internal or external reaction and response forces to respond to an attack on the protected individual.
09	Time	To identify facilities scheduled for visit by the protected individual, inspect to determine if safeguards are adequate, and prioritize protection.
10	Time	To identify activities by the protected individual that increases vulnerability to terrorist acts.
11	Time	To collect critical personal data (medical history, likes, and dislikes) on protected individual.
12	Time	To conduct a threat assessment to include the threat level in the AO and the protected individual's history to determine previous threats targeted at the individual and conduct site visits.
13	Time	To determine and obtain any special equipment—such as military working dogs and devices—for detecting the presence of unauthorized listening devices.
14	Percent	Of friendly force and noncombatant casualties due to terrorist actions directed against the principal.
15	Percent	Of reduced unit mission capabilities or installation support capabilities due to security measures designed to protect the principal.
16	Number	And types of groups and enemy forces operating in unit AO or in the vicinity of friendly installations likely to attack the protected individual.
17	Number	And types of individuals and units needed to provide protective services in the AO.
18	Number	And types of counterterrorism and antiterrorism activities conducted to protect the principal in the AO.
19	Number	Of terrorist attacks attempted against the protected individual in the unit AO.
20	Number	Of friendly force and noncombatant casualties due to terrorist actions directed against the protected principal.
21	Cost	To provide protective services in the AO.

ART 6.5.5 CONDUCT RESPONSE FORCE OPERATIONS

6-67. Response force operations include the planning for defeat of levels I and II threats and the shaping of level III threats until the designated tactical combat force arrives for decisive operations. Response force operations use a mobile force with appropriate fire support to deal with level II threats in the area of operations. (FM 3-19.1) (USAMPS)

ART 6.5.5.1 CONDUCT BATTLE HANDOVER FROM BASE AND BASE CLUSTER SECURITY FORCES TO RESPONSE FORCES

6-68. Transfer responsibility for fighting an enemy from the base or base cluster commander to the commander of the response force. (FM 3-90) (USACAC)

No.	Scale	Measure
01	Yes/No	Battle handover occurred before the enemy penetrated base perimeter or base cluster security area.
02	Yes/No	Main bodies of units conducting battle handover were not surprised by the enemy.
03	Time	To prepare and exchange plans.
04	Time	To provide supporting fires.
05	Time	To establish conditions allowing battle handover.

No.	Scale	Measure
06	Time	Difference between when contact at contact point was planned and when actually made.
07	Percent	Of time that participating forces are in contact with each other.
08	Percent	Of previous plans still applicable at time of battle handover.
09	Percent	Of casualties incurred by either force due to fratricide.
10	Number	Of fratricide incidents.
11	Number	Of casualties due to fratricide.
12	Kilometers	Distance between planned and actual unit contact points where linkup occurs.

ART 6.5.5.2 CONDUCT BATTLE HANDOVER FROM RESPONSE FORCES TO TACTICAL COMBAT FORCES

6-69. Transfer responsibility for fighting an enemy from the commander of the response force to the commander of a tactical combat force. A tactical combat force is a combat unit, with appropriate sustainment assets, that is assigned the mission of defeating level III threats. (FM 3-90) (USACAC)

No.	Scale	Measure
01	Yes/No	Battle handover occurred before the enemy penetrated base perimeter or base cluster security area.
02	Yes/No	Main bodies of units conducting battle handover were not surprised by the enemy.
03	Yes/No	Lethal and nonlethal systems and munitions were available and sufficient to support operational requirements.
04	Time	To prepare and exchange plans.
05	Time	To provide supporting fires.
06	Time	To establish conditions allowing battle handover.
07	Time	Difference between when contact at contact point was planned and when actually made.
08	Percent	Of time that participating forces are in contact with each other.
09	Percent	Of previous plans still applicable at time of battle handover.
10	Percent	Of casualties incurred by either force due to fratricide.
11	Number	Of fratricide incidents.
12	Number	Of casualties due to fratricide.
13	Kilometers	Distance between planned and actual unit contact points where linkup occurs.

ART 6.5.6 SECURE SUPPLY ROUTES AND CONVOYS

6-70. Security and protection of supply routes and convoys in the area of operations is critical to military operations since most traffic supporting military operations moves along these routes. Plans to provide mains supply route security may include designating units for convoy security, providing guidance for units to provide their own security during convoys, or establishing protection and security requirements for convoys carrying critical assets. (FM 4-01.45) (CASCOM)

ART 6.5.6.1 CONDUCT CONVOY SECURITY OPERATIONS

6-71. Convoy security operations protect convoys. Units conduct convoy security operations any time there are not enough friendly forces to continuously secure lines of communications in an area of operations, and there is a danger of enemy ground action against the convoy. Convoy security operations are defensive in nature and orient on the protected force. (FM 3-90) (USACAC)

No.	Scale	Measure
01	Yes/No	Operations of the convoy security forces provided the protected convoy with sufficient reaction time and maneuver space to avoid contact with significant enemy forces.
02	Yes/No	Convoy crossed start point and release point at the times indicated in the operation order.
03	Yes/No	Fratricide did not occur.
04	Yes/No	The convoy escort oriented its operations on the movement of the protected convoy.
05	Yes/No	Collateral damage due to the convoy escort operations was within acceptable limits.
06	Yes/No	Convoy screening elements provided early and accurate warning of enemy forces located along the route used by the convoy or moving toward the convoy's route.
07	Time	To conduct coordination with escorted unit and conduct troop leading procedures.
08	Time	To obtain route information.
09	Time	To designate reconnaissance, screen, escort, and reaction force elements, and move these elements into position.
10	Time	To conduct convoy security operation.
11	Time	For reaction force elements to respond.
12	Percent	Of convoy element casualties.
13	Percent	Of convoy escort casualties.
14	Percent	Of convoys provided convoy escorts.
15	Percent	Of available combat power in area used to provide convoy escorts including reaction forces.
16	Percent	Of decreased support capability of sustainment units due to the requirement to provide convoy escort forces from internal assets.
17	Percent	Of increased supply amounts transported along a line of communications or main supply route due to the presence of convoy escorts.
18	Number	And types of forces used to provide convoy escorts.
19	Number	And types of enemy forces operating in the echelon sustainment area.
20	Number	Of convoy escort casualties during the operation.
21	Number	Of obstacles encountered, bypassed, and breached during the conduct of convoy security operations.
22	Number	And types of enemy forces destroyed during the conduct of convoy security operations.
23	Kilometers	Length of the route traveled by the escorted convoy.

ART 6.5.6.2 CONDUCT ROUTE SECURITY OPERATIONS

6-72. Route (including highway, pipeline, rail, and water) security operations protect lines of communications and friendly forces moving along them. Units conduct route security missions to prevent enemy ground forces from moving into direct fire range of the protected route. Route security operations are defensive in nature and terrain-oriented. (FM 3-90) (USACAC)

No.	Scale	Measure
01	Yes/No	Route security elements provided early and accurate warning of enemy forces located along the route or moving toward the secured route.
02	Yes/No	Route security was established not later than the time indicated in the operation order.
03	Yes/No	Fratricide did not occur.
04	Yes/No	The route security force oriented its operations on the secured route.
05	Yes/No	Collateral damage due to providing route security was within acceptable limits.
06	Time	To plan route security operations.
07	Time	To prepare to conduct route security operations.

No.	Scale	Measure
08	Time	To designate reconnaissance, screen, escort, and reaction force elements, and move these elements into position.
09	Time	That security force route provides route security.
10	Time	For reaction force elements to respond to incidents along the route.
11	Time	That the protected route was unavailable for use by friendly forces because of enemy activities.
12	Percent	Of enemy attacks that succeed in closing the protected route.
13	Percent	Of casualties incurred by elements using the protected route.
14	Percent	Of security force casualties.
15	Percent	Of lines of communications or main supply routes in the area of operations secured.
16	Percent	Of available combat power used to provide route security including reaction forces.
17	Percent	Of increased amount of supplies successfully transported along a line of communications or main supply route due to route security.
18	Number	And types of forces used to provide route security.
19	Number	And types of enemy forces operating near the secured route.
20	Number	Of route security force casualties during the operation.
21	Number	Of obstacles encountered, bypassed, and breached during the conduct of route security operations.
22	Number	Of attacks that succeed in closing the protected route.
23	Number	And types of enemy forces destroyed during route security operations.
24	Kilometers	Distance of the route secured.

SECTION VI – ART 6.6: APPLY ANTITERRORISM MEASURES

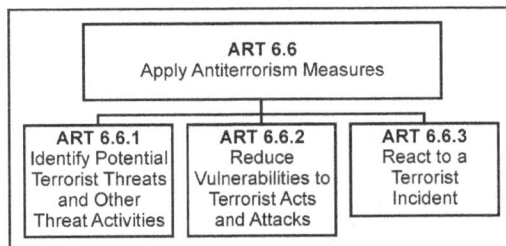

```
                    ┌──────────────────────────┐
                    │         ART 6.6          │
                    │ Apply Antiterrorism Measures │
                    └──────────────────────────┘
          ┌──────────────┬───────────────┬──────────────┐
  ┌───────────────┐ ┌───────────────┐ ┌──────────────┐
  │   ART 6.6.1   │ │   ART 6.6.2   │ │  ART 6.6.3   │
  │Identify Potential│ │    Reduce    │ │  React to a  │
  │Terrorist Threats│ │Vulnerabilities to│ │  Terrorist   │
  │   and Other   │ │ Terrorist Acts │ │   Incident   │
  │Threat Activities│ │  and Attacks  │ │              │
  └───────────────┘ └───────────────┘ └──────────────┘
```

6-73. Antiterrorism consists of defensive measures used to reduce the vulnerability of individuals and property to terrorist acts, to include limited response and containment by local military and civilian forces. It is an element of protection. Antiterrorism is a consideration for all forces during all military operations. (FM 3-19.30) (USAMPS)

Note: This task branch only addresses antiterrorism operations. ART 7.6.4.4 (Combat Terrorism) addresses counterterrorism measures.

No.	Scale	Measure
01	Yes/No	Unit or installation continued its mission while taking actions to combat terrorism in the area of operations (AO).
02	Yes/No	Threat and vulnerability assessments for unit and installation were completed.
03	Yes/No	Unit completed antiterrorism awareness training.
04	Yes/No	Incident response plans included managing the force protection condition system.

No.	Scale	Measure
05	Yes/No	Unit or installation received useable antiterrorism intelligence information from host-nation agencies.
06	Yes/No	Local population supported unit or installation presence in the AO.
07	Yes/No	Effective response or reaction forces existed in the event of an attempted or successful terrorist penetration of unit or installation perimeter.
08	Yes/No	Unit or installation measures designed to combat terrorism—such as operations security—were not deliberately or accidentally disclosed to terrorist organizations.
09	Time	To refine installation or unit security plans and standing operating procedures.
10	Time	For internal or external reaction and response forces to reach individual, installation, or facility under attack.
11	Time	To identify critical facilities, key terrain, and significant sites (cultural, historical, or natural) and prioritize protection.
12	Time	To identify mission essential activities vulnerable to terrorist acts and inspect to determine if safeguards are adequate.
13	Percent	Of terrorist attacks that degrade, delay, or modify friendly force operations.
14	Percent	Of units that have active counterterrorism protocols.
15	Percent	Of successful terrorist attacks.
16	Percent	Of friendly force and noncombatant casualties due to terrorist actions.
17	Percent	Of information systems in the AO hardened against unauthorized access.
18	Percent	Of information systems in the AO with multiple pathways over which to transmit data.
19	Percent	Of critical installations and facilities hardened or protected against terrorist acts.
20	Percent	Of reduced unit mission capabilities or installation support capabilities due to security measures designed to combat terrorism in the AO.
21	Percent	Of reduced line of communications or main supply route through capabilities due to security measures designed to combat terrorism in the AO.
22	Number	And types of terrorists groups operating in unit AO or in the vicinity of friendly installations.
23	Number	Of counterterrorism activities unit supports in the AO.
24	Number	Of terrorist attacks attempted in unit AO.
25	Number	Of successful terrorist attacks.
26	Number	Of friendly force actions to disrupt enemy terrorist activities.
27	Number	Of friendly force and noncombatant casualties due to terrorist actions.
28	Cost	To implement protective measures against terrorism in the AO.

ART 6.6.1 IDENTIFY POTENTIAL TERRORIST THREATS AND OTHER THREAT ACTIVITIES

6-74. Enhance freedom of action by identifying and reducing friendly vulnerability to terrorist threats, acts, influence, or surprise. This includes measures to protect from surprise, observation, detection, interference, espionage, terrorism, and sabotage. (FM 3-19.30) (USAMPS)

No.	Scale	Measure
01	Percent	Of lines of communications secured.
02	Percent	Of total troops used to secure critical facilities and lines of communications.
03	Number	Of incidents by enemy troops or partisans affecting security of force and means in the area of operations.
04	Number	Of security plans not including smoke and obscuration.
05	Hours	To coordinate for additional assets for the area of operations.

ART 6.6.2 REDUCE VULNERABILITIES TO TERRORIST ACTS AND ATTACKS

6-75. Reduce personnel vulnerability to terrorism by understanding the nature of terrorism, knowing current threats, identifying vulnerabilities to terrorist acts, and by implementing protective measures against terrorist acts and attacks. (FM 3-19.30) (USAMPS)

No.	Scale	Measure
01	Yes/No	Actions deterred hostile actions against Soldiers, Army civilians, family members, facilities, information, and equipment; when deterrence failed, actions mitigated the consequences of terrorist attacks against these potential targets.
02	Yes/No	Commander applied judgment to every situation and combined it with available technologies to manage risk.
03	Yes/No	Commander retained freedom of action by reducing friendly force vulnerability to terrorist actions.
04	Yes/No	Unit or installation antiterrorism program included planned and integrated antiterrorism measures, counterterrorism, physical security, operations security, and personal protective services supported by counterintelligence and other security programs.
05	Yes/No	Force protection planning was a continuous process.
06	Yes/No	Force protection assets focused on protecting the most critical assets.
07	Time	To understand how potential terrorists operate.
08	Time	To prioritize unit force protection efforts based upon criticality and vulnerability assessments.
09	Time	In advance that a force is warned of attack.
10	Percent	Of unit force protection activities integrated with those of other Services and nations.
11	Percent	Of reduced enemy targeting effectiveness due to the implementation of force protection measures.
12	Percent	Of reduced the effectiveness of enemy action due to friendly measures to harden units and facilities from enemy attack.
13	Percent	Of enhanced personnel, equipment, and facility survivability because of measures taken to harden them from enemy attack.
14	Number	Of friendly and noncombatant casualties due to terrorist attack.
15	Cost	Of measures to protect the unit or installation from terrorist attack.

ART 6.6.3 REACT TO A TERRORIST INCIDENT

6-76. Implement measures to treat casualties, minimize property damage, restore operations, and expedite the criminal investigation and collection of lessons learned from a terrorist incident. (FM 19-10) (USAMPS)

Note: ART 5.5.1.2.3 (Conduct Criminal Investigations) addresses the conduct of crime analysis. ART 5.3.5.2 (Collect Relevant Information) addresses the collection of lessons learned.

No.	Scale	Measure
01	Yes/No	Response to terrorist incident did not prevent unit or installation from accomplishing the missions.
02	Time	To conduct reconnaissance or criminal investigation of site where terrorist incident occurred.
03	Time	To establish or restore security around site where terrorist incident occurred.
04	Time	To conduct area damage control activities, such as firefighting, power restoration and production, rubble clearance, removal of downed trees, and repair of critical damaged facilities and installations.

No.	Scale	Measure
05	Time	To report the occurrence of terrorist incident to appropriate headquarters and agencies.
06	Time	For response forces or teams to arrive at site of terrorist incident.
07	Time	To search for, collect, identify, and treat injured survivors of terrorist incident.
08	Time	To search for, collect, identify, and process the remains of individuals killed in terrorist incident.
09	Time	To restore damaged facilities to desired level of functionality.
10	Percent	Of decreased attacked facilities capabilities to perform designed function.
11	Percent	Of response forces or teams arriving at terrorist incident site within desired response times.
12	Number	Of friendly force and noncombatant casualties due to terrorist incident.
13	Cost	To provide forces and supplies to provide local security and humanitarian aid and comfort, conduct area damage control, and restore damaged facility in response to a given terrorist incident.

SECTION VII – ART 6.7: CONDUCT SURVIVABILITY OPERATIONS

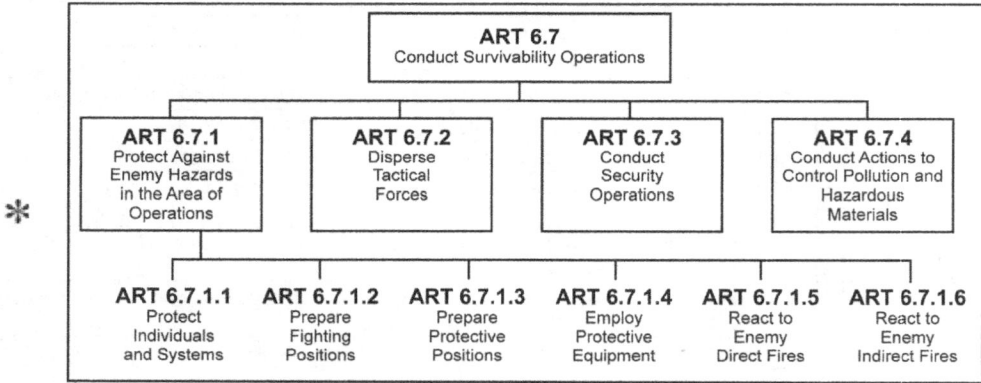

6-77. ART 6.7 is a concept that includes all aspects of protecting personnel, weapons, and supplies while simultaneously deceiving the enemy. Survivability tactics include building a good defense; employing frequent movement; using concealment, deception, and camouflage; and constructing fighting and protective positions for both individuals and equipment. Included are those assessments and surveys completed as a part of focused engineer reconnaissance support that includes infrastructure reconnaissance. (See FM 3-34.170.) Survivability operations are the development and construction of protective positions, such as earth berms, dug-in positions, overhead protection, and countersurveillance means, to reduce the effectiveness of enemy weapon systems. (FM 5-103) (USAES)

Note: This task branch is supported by ART 2.2.3 (Provide Intelligence Support to Protection).

No.	Scale	Measure
01	Yes/No	Unit could continue to conduct operations.
02	Yes/No	Unit losses from hazards were at acceptable levels.
03	Yes/No	Unit could determine when contaminated area was at a level with acceptable risk.
04	Yes/No	Unit performed risk assessment of all areas in the area of operations (AO) that underwent chemical, biological, radiological, and nuclear (CBRN) weapons attack.

No.	Scale	Measure
05	Time	Required to conduct an area reconnaissance.
06	Time	To determine that unit has been attacked by CBRN weapons.
07	Time	To predict downwind hazard from the use of CBRN weapons.
08	Time	To disseminate hazard information to all units in the AO and appropriate headquarters and agencies outside the AO.
09	Time	To harden unit equipment, facilities, and positions.
10	Time	To acquire equipment and supplies necessary to harden a unit or installation.
11	Percent	Of U.S. military and civilian casualties.
12	Percent	Of increased time it takes the unit to conduct its operations due to the need to protect against identified hazards.
13	Percent	Of unit that has completed risk management and safety training.
14	Percent	Of friendly casualties due to failure to report the existence of hazards.
15	Percent	Of on-hand, mission-capable equipment necessary to protect the unit against hazards.
16	Percent	Of on-hand supplies necessary to protect the unit against hazards.
17	Percent	Of planned fighting positions completed.
18	Percent	Of planned protective positions completed.
19	Number	Of planned fighting positions completed.
20	Number	Of planned protective positions completed.
21	Number	And types of on-hand equipment necessary to protect the unit against hazards.
22	Number	And types of friendly equipment destroyed or damaged by enemy action.
23	Number	Of incidents of damage to units and facilities in the AO that impact the commander's concept of operations.
24	Number	Of U.S. military and civilian casualties due to enemy hazards.

ART 6.7.1 PROTECT AGAINST ENEMY HAZARDS IN THE AREA OF OPERATIONS

6-78. Protect the friendly force in an area of operations (AO) by reducing or avoiding the effects of enemy weapon systems. (FM 3-0) (USACAC)

No.	Scale	Measure
01	Yes/No	Unit could continue to conduct operations.
02	Yes/No	Unit losses from hazards were at acceptable levels.
03	Time	To conduct an area reconnaissance to identify hazards.
04	Time	To disseminate hazard data to all elements operating in the AO.
05	Time	To analyze the impact of identified hazards.
06	Time	To obtain necessary equipment and supplies to protect against hazards.
07	Time	To protect the unit and its facilities, equipment, and supplies against hazards.
08	Percent	Of increased time it takes the unit to conduct its operations because of the need to protect against identified hazards.
09	Percent	Of unit that has completed risk management and safety training.
10	Percent	Of friendly casualties due to failure to report existence of hazards.
11	Percent	Of on-hand, mission-capable equipment that is necessary to protect the unit against hazards.
12	Percent	Of U.S. military and civilian casualties.
13	Percent	Of on-hand supplies (chemoprophylaxis, pretreatments, and barrier creams) necessary to protect the unit against hazards.

No.	Scale	Measure
14	Percent	Of planned protective positions completed.
15	Number	Of planned protective positions completed.
16	Number	And types of on-hand equipment necessary to protect against hazards.
17	Number	And types of friendly equipment destroyed or damaged by enemy action.
18	Number	Of incidents of damage to units and facilities in the AO that impact the concept of operations.
19	Number	Of U.S. military and civilian casualties due to enemy hazards.

ART 6.7.1.1 PROTECT INDIVIDUALS AND SYSTEMS

6-79. Use protective positions (natural or artificial), measures, or equipment—such as armor, detection equipment, mission-oriented protective posture gear (MOPP), and collective protective equipment—to reduce effects of enemy weapon systems. ART 6.7.1.1 includes construction of fighting and survivability positions, conduct of chemical, biological, radiological, and nuclear defense, and responding to enemy fires. (FM 3-0) (USACAC)

No.	Scale	Measure
01	Yes/No	Unit losses from effects of enemy weapon systems did not keep the unit from accomplishing its mission.
02	Yes/No	Unit losses from enemy weapons systems effects were at acceptable levels.
03	Yes/No	Unit used the terrain to protect itself from effects of enemy weapons.
04	Time	To conduct an area reconnaissance to identify areas where risk of enemy attack exists.
05	Time	To analyze the impact of identified hazards in the area of operations (AO).
06	Time	To revise the plan based on results of the area reconnaissance and environmental considerations.
07	Time	To disseminate area reconnaissance data and revised execution instructions to all elements operating in the AO.
08	Time	To establish the necessary degree of local security for construction efforts.
09	Time	To obtain the equipment and supplies necessary to protect the unit from the effects of enemy weapon systems.
10	Time	To protect the unit, its facilities, equipment, and supplies from the effects of enemy weapons systems.
11	Percent	Of U.S. military and civilian casualties due to enemy hazards.
12	Percent	Of increased time it takes the unit to conduct operations because of the need to protect itself from the effects of identified enemy weapons systems.
13	Percent	Of unit personnel trained to use available detection and protection equipment.
14	Percent	Of friendly casualties due to failure to use existing protective equipment and structures.
15	Percent	Of on-hand, mission-capable equipment, such as MOPP gear, necessary to protect the unit from the effects of enemy weapon systems.
16	Percent	Of on-hand supplies necessary to protect the unit from effects of enemy weapon systems.
17	Percent	Of unit personnel who become casualties due to the faulty use of terrain to protect themselves from effects of enemy weapon systems.
18	Percent	Of unit personnel who become casualties of effects of enemy weapon systems due to the faulty use of detection equipment.
19	Percent	Of planned protective positions completed.
20	Number	Of planned protective positions completed.
21	Number	And types of on-hand equipment necessary to protect the unit against the effects of enemy weapon systems.
22	Number	And types of friendly equipment destroyed or damaged by enemy action.

No.	Scale	Measure
23	Number	Of incidents of damage to units and facilities in the AO that impact the concept of operations.
24	Number	Of U.S. military and civilian casualties from effects of enemy weapons.

ART 6.7.1.2 Prepare Fighting Positions

6-80. Prepare primary, alternate, and supplementary fighting positions that provide cover, concealment, and protection from the effects of enemy fires for occupants and systems, and allow for fields of fire and maneuver space for combat systems and units engaging the enemy. (FM 5-103) (USAES)

No.	Scale	Measure
01	Yes/No	Unit could accomplish its mission using its fighting positions.
02	Time	To conduct an area reconnaissance including environmental considerations.
03	Time	To design primary, alternate, and supplementary fighting positions, and decide correct placement to maximize terrain effectiveness.
04	Time	To establish the degree of local security necessary for the construction effort.
05	Time	To obtain the equipment and supplies necessary to construct fighting positions.
06	Time	To construct primary, alternate, and supplementary fighting positions with engineer support.
07	Time	To construct primary, alternate, and supplementary fighting positions without engineer support.
08	Percent	Of on-hand equipment and supplies needed to construct fighting positions.
09	Percent	Of personnel trained to properly construct fighting positions.
10	Percent	Of planned fighting positions completed.
11	Percent	Of completed fighting positions constructed to standard.
12	Percent	Of friendly casualties due to improperly constructed or sited fighting positions.
13	Percent	Of planned fighting positions completed.
14	Number	And types of fighting positions completed.
15	Number	Of friendly casualties due to improperly constructed or sited fighting positions.

ART 6.7.1.2.1 Construct Vehicle Fighting Positions

6-81. Construct fighting positions that provide cover, concealment, and protection from direct and indirect fires for combat vehicles, yet allow direct or indirect engagement of enemy forces. (FM 5-103) (USAES)

Note: ART 6.7.1.3.2 (Construct Vehicle, Information Systems, Equipment, and Material Protective Positions) addresses vehicles that do not provide or conduct direct fire.

No.	Scale	Measure
01	Yes/No	Unit could accomplish its mission using its vehicle fighting positions.
02	Time	To conduct area reconnaissance including environmental considerations.
03	Time	To design primary, alternate, and supplementary vehicle fighting positions, and decide correct placement to maximize terrain effectiveness.
04	Time	To establish the local security necessary for the construction effort.
05	Time	To obtain the equipment and supplies necessary to construct vehicle fighting positions.
06	Time	To construct primary, alternate, and supplementary vehicle fighting positions with engineer support.
07	Time	To construct primary, alternate, and supplementary vehicle fighting positions without engineer support.
08	Percent	Of on-hand equipment and supplies necessary to construct vehicle fighting positions.

No.	Scale	Measure
09	Percent	Of personnel trained to properly construct vehicle fighting positions.
10	Percent	Of planned vehicle fighting positions completed.
11	Percent	Of completed vehicle fighting positions constructed to standard.
12	Percent	Of friendly casualties due to improperly constructed or sited vehicle fighting positions.
13	Number	Of planning vehicle fighting positions completed.
14	Number	Of friendly casualties due to improperly constructed or sited vehicle fighting positions.

ART 6.7.1.2.2 Construct Crew-Served Weapon Fighting Positions

6-82. Construct fighting positions and or bunkers for crew-served weapons remaining in defensive positions for extended times. Fighting positions provide cover and concealment from direct and indirect fires while allowing coverage of primary and secondary sectors of fire. (FM 5-103) (USAES)

No.	Scale	Measure
01	Yes/No	Unit could accomplish its mission using crew-served weapon fighting positions.
02	Time	To conduct area reconnaissance including environmental considerations.
03	Time	To design primary, alternate, and supplementary crew-served weapon fighting positions, and decide the correct placement to maximize terrain effectiveness.
04	Time	To establish the degree of local security necessary for the construction effort.
05	Time	To obtain the necessary equipment and supplies to construct fighting positions.
06	Time	To construct primary, alternate, and supplementary crew-served weapon fighting positions with engineer support.
07	Time	To construct primary, alternate, and supplementary crew-served weapon fighting positions without engineer support.
08	Percent	Of on-hand equipment and supplies needed to construct crew-served weapon fighting positions.
09	Percent	Of personnel trained to properly construct crew-served weapon fighting positions.
10	Percent	Of planned crew-served weapon fighting positions completed.
11	Percent	Of completed crew-served weapon fighting positions constructed to standard.
12	Percent	Of friendly casualties due to improperly constructed or sited crew-served weapon fighting positions.
13	Number	And types of planned crew-served weapon fighting positions completed.
14	Number	Of friendly casualties due to improperly constructed or sited crew-served weapon fighting positions.

ART 6.7.1.2.3 Construct Individual Fighting Positions

6-83. Construct individual fighting positions that provide cover and concealment from observation and direct and indirect fires. Properly sited individual fighting positions allow Soldiers to engage the enemy with their assigned weapons while providing observation and fields of fire that overlap those of other positions. (FM 3-21.75) (USAIS)

No.	Scale	Measure
01	Yes/No	Unit could accomplish its mission using individual fighting positions.
02	Time	To conduct an area reconnaissance including environmental considerations.
03	Time	To design primary, alternate, and supplementary individual fighting positions, and decide correct placement to maximize terrain effectiveness.
04	Time	To establish the degree of local security for the construction effort.
05	Time	To obtain the equipment and supplies to construct fighting positions.
06	Time	To construct primary, alternate, and supplementary individual fighting positions with engineer support.

No.	Scale	Measure
07	Time	To construct primary, alternate, and supplementary individual fighting positions without engineer support.
08	Percent	Of on-hand equipment and supplies available to construct individual fighting positions.
09	Percent	Of personnel trained to properly construct individual fighting positions.
10	Percent	Of planned individual fighting positions completed.
11	Percent	Of completed individual fighting positions constructed to standard.
12	Percent	Of completed individual fighting positions that can support unit defensive positions, such as vehicle fighting positions and crew-served weapon fighting positions.
13	Percent	Of friendly casualties due to improperly constructed or sited individual fighting positions.
14	Number	Of planned individual fighting positions completed.
15	Number	Of friendly casualties due to improperly constructed or sited individual fighting positions.

ART 6.7.1.3 PREPARE PROTECTIVE POSITIONS

6-84. Provide cover and concealment for personnel, systems, equipment, supplies, and other materiel not directly involved in fighting. This includes medical patients. These positions reduce risks associated with all forms of enemy contact, such as direct and indirect fires, enemy observations, and employment of chemical, biological, radiological and nuclear weapons. (FM 5-103) (USAES)

No.	Scale	Measure
01	Yes/No	The use of protective positions preserved the unit's personnel, equipment, and supplies for future missions.
02	Time	To conduct an area reconnaissance including environmental considerations.
03	Time	To design protective positions for personnel, systems, equipment, supplies, and other materiel not directly involved in fighting, and to decide correct placement to maximize terrain effectiveness.
04	Time	To establish the degree of local security for the construction effort.
05	Time	To obtain the equipment and supplies to construct protective positions.
06	Time	To construct protective positions with engineer support.
07	Time	To construct protective positions without engineer support.
08	Percent	Of on-hand equipment and supplies to construct protective positions.
09	Percent	Of personnel trained to properly construct protective positions.
10	Percent	Of planned protective positions completed.
11	Percent	Of completed protective positions constructed to standard.
12	Percent	Of unit facilities not hardened.
13	Percent	Of friendly casualties due to improperly constructed or sited protective positions.
14	Percent	Of personnel casualties or equipment and supplies lost due to the nonavailability of protective positions.
15	Number	And types of planned protective positions completed.
16	Number	Of friendly casualties due to improperly constructed or sited protective positions.
17	Number	Of personnel casualties or equipment and supplies lost due to the nonavailability of protective positions.

ART 6.7.1.3.1 Construct Protective Earth Walls, Berms, and Revetments

6-85. Provide cover, concealment, and protection against direct and indirect fires without restricting the operational capability of systems. (FM 5-103) (USAES)

No.	Scale	Measure
01	Yes/No	The use of protective earth walls, berms, and revetments preserved unit personnel, equipment, and supplies for future missions.
02	Time	To conduct an area reconnaissance including environmental considerations.
03	Time	To design protective earth walls, berms, and revetments for personnel, systems, equipment, supplies, and other materiel not directly involved in fighting, and to decide correct placement to maximize terrain effectiveness.
04	Time	To obtain the equipment and supplies to construct protective earth walls, berms, and revetments.
05	Time	To establish degree of local security for the construction effort.
06	Time	To construct protective earth walls, berms, and revetments with engineer support.
07	Time	To construct protective earth walls, berms, and revetments without engineer support.
08	Percent	Of on-hand equipment and supplies to construct protective earth walls, berms, and revetments.
09	Percent	Of personnel trained to construct protective earth walls, berms, and revetments.
10	Percent	Of planned protective earth walls, berms, and revetments completed.
11	Percent	Of completed protective earth walls, berms, and revetments constructed to standard.
12	Percent	Of unit facilities not hardened.
13	Percent	Of friendly casualties due to improperly constructed or sited protective earth walls, berms, and revetments.
14	Percent	Of personnel casualties or equipment and supplies lost to enemy attack due to the nonavailability of protective earth walls, berms, and revetments.
15	Number	Of planned protective earth walls, berms, and revetments completed.
16	Number	Of friendly casualties due to improperly constructed or sited protective earth walls, berms, and revetments.
17	Number	Of friendly casualties or equipment and supplies lost due to the nonavailability of protective earth walls, berms, and revetments.

ART 6.7.1.3.2 Construct Vehicle, Information Systems, Equipment, and Material Protective Positions

6-86. Provide cover and concealment for vehicles, information systems nodes, equipment, supplies, and other materials that do not provide or conduct direct fire. (FM 5-103) (USAES)

No.	Scale	Measure
01	Yes/No	The use of vehicle, information systems, equipment, and materials protective positions preserved unit personnel, equipment, and supplies for future missions.
02	Time	To conduct an area reconnaissance including environmental considerations.
03	Time	To design protective positions for vehicle, information systems, equipment, and materiel not directly involved in fighting, and decide correct placement to maximize terrain effectiveness.
04	Time	To establish the degree of local security for the construction effort.
05	Time	To obtain the equipment and supplies to construct vehicle, information systems, equipment, and materials protective positions.
06	Time	To construct vehicle, information systems, equipment, and materials protective positions with engineer support.
07	Percent	Of on-hand equipment and supplies available to construct vehicle, information systems, equipment, and materials protective positions.
08	Percent	Of personnel trained to construct vehicle, information systems, equipment, and materials protective positions.
09	Percent	Of planned vehicle, information systems, equipment, and materials protective positions completed.

No.	Scale	Measure
10	Percent	Of completed vehicle, information systems, equipment, and materials protective positions constructed to standard.
11	Percent	Of unit vehicles, information systems, equipment, and materials supplies not protected by protective positions.
12	Percent	Of friendly casualties due to improperly constructed or sited vehicle, information systems, equipment, and or materials protective positions.
13	Percent	Of casualties or equipment and supplies lost due to the nonavailability of vehicle, information systems, equipment, and or materials protective positions.
14	Number	Of planned vehicle, information systems, equipment, and materials protective positions completed.
15	Number	Of friendly casualties due to improperly constructed or sited vehicle, information systems, equipment, and materials protective positions.
16	Number	Of friendly casualties or equipment and supplies lost due to the nonavailability of vehicle, information systems, equipment, and materials protective positions.

ART 6.7.1.4 EMPLOY PROTECTIVE EQUIPMENT

6-87. Employ individual and collective equipment to protect personnel, systems, and facilities against hazards caused by enemy action. Protective equipment includes individual and collective chemical, biological, radiological, and nuclear (CBRN) detection and protective systems. ART 6.7.1.4 includes using other items such as bullet-resistant glazing, hydraulically or manually operated vehicle crash barriers, personnel gates that limit the number of personnel passing through at one time, intrusion detection systems, security lighting, and security fences. (FM 5-103) (USAES)

No.	Scale	Measure
01	Yes/No	Using protective equipment preserved unit personnel, equipment, and supplies for future missions.
02	Time	To conduct an area reconnaissance including environmental considerations.
03	Time	To plan or revise the plan for employing protective equipment to account for the existing factors of mission, enemy, terrain and weather, troops and support available, time available, and civil considerations.
04	Time	To establish the degree of local security necessary for installing or constructing protective equipment.
05	Time	To employ protective equipment to harden individuals and facilities.
06	Time	To assume the necessary mission-oriented protective posture level in response to enemy action, given previous mission-oriented protective posture level.
07	Time	To employ protective equipment to harden supply stocks with engineer support.
08	Time	To employ protective equipment to harden supply stocks without engineer support.
09	Time	To emplace warning signs marking the edges of areas contaminated by enemy action such as the use of CBRN weapons.
10	Time	To obtain the protective equipment and systems needed to complete the hardening process.
11	Percent	Of personnel, systems, and facilities hardened by protective equipment and systems.
12	Percent	Of personnel trained to use protective equipment.
13	Percent	Of necessary protective equipment and supplies on hand.
14	Percent	Of protective equipment employed to standard.
15	Percent	Of protective equipment system that is mission capable.
16	Percent	Of friendly casualties due to improperly used protective equipment.
17	Percent	Of casualties or equipment and supplies lost due to the nonavailability of protective equipment.
18	Number	Of mission capable protective equipment systems.

No.	Scale	Measure
19	Number	Of friendly casualties due to improperly used protective equipment.
20	Number	Of personnel casualties or equipment and supplies lost due to the nonavailability of protective equipment.

ART 6.7.1.4.1 Install Bridge Protective Devices

6-88. Provide protective systems for an existing floating bridge or river-crossing site. Protect the bridge or site from waterborne demolition teams, floating mines, or floating debris. The three types of floating protective systems are anti-mine booms, impact booms, and anti-swimmer nets. (FM 5-34) (USAES)

No.	Scale	Measure
01	Yes/No	The use of bridge protective equipment preserved the functionality of the bridge for current and future missions.
02	Time	To conduct an area reconnaissance of the river approaches to the bridge including environmental considerations.
03	Time	To plan or revise the plan to employ bridge protective equipment to account for the existing factors of mission, enemy, terrain and weather, troops and support available, time available, and civil considerations.
04	Time	To establish the degree of local security for installation or construction of the bridge protective equipment.
05	Time	To employ bridge protective equipment.
06	Time	To widen the bridge or remove interior bridge bays to account for flood condition or heavy debris (for floating bridges).
07	Time	To obtain the bridge protective equipment and systems through the supply system or by local procurement.
08	Percent	Of personnel trained to use bridge protective equipment.
09	Percent	Of on-hand bridge protective equipment and supplies.
10	Percent	Of bridge protective equipment employed to standard.
11	Percent	Of mission-capable bridge protective equipment or systems.
12	Percent	Of bridges damaged due to improperly used bridge protective equipment.
13	Number	Of bridges protected by the proper protective device for the situation.
14	Number	Of mission-capable bridge protective equipment systems.
15	Number	Of friendly casualties due to improperly used protective equipment.
16	Number	Of casualties due to accidents while installing bridge protective equipment.

ART 6.7.1.4.2 Install or Remove Protective Obstacles

6-89. Provide friendly forces close-in protection with protective obstacles as part of their force protection plan. ART 6.7.1.4.2 includes employing temporary or permanent protective obstacles and removal or turnover of obstacles to relieving units. (FM 3-34.210) (USAES)

No.	Scale	Measure
01	Yes/No	The use of protective obstacles preserved unit personnel, equipment, and supplies for future missions.
02	Yes/No	The protective obstacles were properly turned over to the relieving unit. This includes transfer of intelligence; maneuver; fires; and mobility, countermobility, and survivability information such as local enemy, friendly, and civilian situations; direct and indirect fire control measures; minefield composition; marking; and layout.
03	Time	To conduct an area reconnaissance including environmental considerations.
04	Time	To plan or revise the plan to employ protective obstacles to account for existing factors of mission, enemy, terrain and weather, troops and support available, time available, and civil considerations.

No.	Scale	Measure
05	Time	To establish the degree of local security necessary for installation or construction of the protective obstacles.
06	Time	To install or remove protective obstacles to include proper marking with engineer support.
07	Time	To install or remove protective obstacles to include proper marking without engineer support.
08	Time	To turn over protective obstacles.
09	Time	To properly record and report protective obstacles.
10	Time	To obtain obstacle emplacing equipment and class IV and class V to install protective obstacles.
11	Percent	Of personnel, systems, unit positions, and facilities protected by protective obstacles.
12	Percent	Of personnel trained to install, maintain, and remove protective obstacles.
13	Percent	Of on-hand protective obstacle installation and removal equipment and class IV and class V.
14	Percent	Of protective obstacles installed and removed to standard.
15	Percent	Of protective obstacles properly turned over.
16	Percent	Of friendly casualties due to improperly installed, marked, and removed protective obstacles.
17	Number	Of mission-capable protective obstacle installation and removal systems.
18	Number	Of friendly casualties due to improperly installed, marked, and removed protective obstacles.

ART 6.7.1.5 REACT TO ENEMY DIRECT FIRES

6-90. Return fire at known or suspected enemy positions and take evasive action upon detecting enemy direct fires. (FM 3-21.75) (USAIS)

No.	Scale	Measure
01	Yes/No	Unit reaction to enemy direct fire allowed the unit to complete its mission.
02	Yes/No	Unit retained its cohesion.
03	Yes/No	Collateral damage due to friendly response to enemy direct fires did not result from violations of the law of war or rules of engagement.
04	Time	That unit was delayed from accomplishing its mission due to enemy direct fire.
05	Time	That unit stayed within the enemy's engagement area before it can suppress the enemy's weapon systems, find cover from which to engage the enemy, or extract itself from the engagement area.
06	Percent	Of enemy casualties inflicted.
07	Percent	Of friendly casualties.
08	Number	Of friendly and noncombatant casualties.
09	Number	And types of friendly systems rendered non-mission capable by enemy direct fires.

ART 6.7.1.6 REACT TO ENEMY INDIRECT FIRES

6-91. Seek protection under the overhead cover of fighting or protective positions or move rapidly out of the impact area in the direction the unit commander orders. If armored vehicles are available, personnel mount, and the vehicles move out of the impact area in the direction and designated distance ordered by the unit commander. (FM 3-21.75) (USAIS)

No.	Scale	Measure
01	Yes/No	Unit reaction to enemy indirect fires allowed the unit to complete its mission.
02	Yes/No	Unit retained its cohesion.

No.	Scale	Measure
03	Yes/No	Collateral damage due to the friendly response to enemy direct fires did not result from violations of the law of war or rules of engagement.
04	Time	To report contact to the higher commander.
05	Time	For personnel to either close hatches on the combat vehicles in which they are riding, seek shelter in positions with suitable overhead cover, or seek shelter offered by the terrain immediately around them.
06	Time	For vehicles to move out of the impact area.
07	Time	For dismounted individuals caught without suitable shelter in the impact area to improve their chances of surviving by digging in using resources immediately available to them.
08	Time	To conduct counterbattery or countermortar fires.
09	Percent	Of Soldiers performing immediate action drill correctly.
10	Percent	Of enemy casualties due to friendly counterbattery or countermortar fires.
11	Percent	Of friendly casualties.
12	Number	Of friendly and noncombatant casualties.
13	Number	And types of friendly systems rendered non-mission capable by enemy indirect fires.

*ART 6.7.1.7 CONDUCT IMPROVISED EXPLOSIVE DEVICE DEFEAT OPERATIONS

6-92. ART 6.7.1.7 (Conduct Improvised Explosive Device Defeat Operations) and ART 6.7.1.7.1 (Plan for Possible Improvised Explosive Devise Threats) has moved to ART 6.12.3.1 and ART 6.12.3.2. ARTs 6.7.1.7.2, 6.7.1.7.3, and 6.7.1.7.4 are rescinded.

*ART 6.7.1.7 PROVIDE FIRE AND EMERGENCY SERVICES

6-93. Provide fire and emergency services (F&ES) in an area of operations (AO) including fire prevention and fire suppression of facilities, equipment, munitions and aviation firefighting/aircraft crash rescue services. Specific capabilities include provide fire prevention programs, inspections, tactical firefighting, technical rescue and first aid capabilities. A fire protection program can protect logistics support areas, intermediate staging bases, forward operating bases, and major facilities. Facilities can include petroleum tank farms, petroleum distribution sites, open and closed warehouse facilities, general warehouses, detainee facilities, and civilian resettlement sites. ART 6.7.1.7 includes initial hazardous material response, aviation firefighting, extrication of personnel and equipment from crashed aircraft, rescuing sick or entrapped personnel from buildings, equipment, vehicles, water, ice, confined space, and high angles. This also includes firefighting protection against grass or brush fires within assigned area when augmented with combat or construction engineer Soldiers or units. (FM 5-415) (USAES)

No.	Scale	Measure
01	Yes/No	F&ES limited the damage caused by fires in the area of operations (AO) so that fires do not disrupt, cancel, or require the modification of the unit's course of action.
02	Yes/No	F&ES provided first-responder level medical response and assistance to victims.
03	Yes/No	F&ES provided an initial response to hazardous material incidents with environmental considerations.
04	Yes/No	F&ES used command and control of nonfirefighting assets when supporting brush firefighting operations.
05	Yes/No	Environmental considerations planning and procedures were present and being followed.
06	Yes/No	F&ES limited the damage caused by aircraft fires and provided aircraft crash rescue services to protect the lives of aviation crewmembers.
07	Yes/No	Assist in training of host nation firefighting assets.
08	Time	To develop a firefighting plan for the AO.
09	Time	To develop mutual aid agreements.

No.	Scale	Measure
10	Time	To respond to reports of fires, medical emergencies, and hazardous material incidents.
11	Time	To establish and maintain 24/7 fire department communications network.
1 2	Time	To reconnoiter water-supply points.
13	Time	To provide water resupply to firefighting teams.
14	Time	To establish local security from external assets for firefighting operations.
15	Time	To provide additional manpower support to firefighting teams from supported units.
16	Time	To train personnel so they remain qualified to fight fires and respond to medical emergencies and hazardous materials incidents.
17	Time	To practice fire drills by units in the AO.
18	Time	To complete fire prevention measures, such as inspections and preventive maintenance checks and services on firefighting equipment.
19	Time	To procure necessary personal protective equipment, firefighting equipment, and fire trucks to protect the AO.
20	Time	To investigate fires.
21	Time	To implement aircraft crash rescue services and to respond to aircraft emergencies.
22	Percent	Of firefighting operations that are petroleum, oils, and lubricants fires.
23	Percent	Of high-value assets protected by firefighting teams.
24	Percent	Of difference between planned level of firefighting support and the required level.
25	Percent	Of qualified personnel assigned to firefighting teams in the AO.
26	Percent	Of firefighting support provided by host nation.
27	Number	Of firefighting teams found in the AO.
28	Number	And types of mission capable fire trucks in the AO.
29	Number	Of crash or rescue operations conducted within a given time.
30	Number	Of normal flight and maintenance operations supported within a given time.
31	Number	Of medical evacuation operations supported within a given time.
32	Number	Of fire prevention inspections performed in a given time.
33	Number	Of emergency water-supply points maintained.

*ART 6.7.1.7.1 Provide General Firefighting

6-94. Provide response to fires with the AO. Provide crash rescue support to medical evacuation (MEDEVAC) and normal flight operations. Provide initial response for hazardous material (HAZMAT) and medical assistance. Provide fire prevention measures, such as, inspections, fire drills, and training. (FM 5-415) (USAES)

No.	Scale	Measure
1	Yes/No	Provide crash-rescue support for MEDEVAC and normal flight or stand-by operations
2	Yes/No	Firefighting team provided first-responder level medical response and assistance to victims.
3	Yes/No	Firefighting team provided an initial response to HAZMAT incidents with environmental considerations.
4	Time	To respond to reports of fires, medical emergencies, and HAZMAT incidents.
5	Time	To reconnoiter water-supply points.
6	Time	To provide water resupply to firefighting teams.
7	Time	To provide additional manpower support to firefighting teams from supported units.
8	Time	To practice fire drills with other units in the AO.
9	Time	To complete fire prevention measures, such as inspections and preventive maintenance checks and services on firefighting equipment.

No.	Scale	Measure
10	Time	To investigate fires.
11	Number	Of emergency water-supply points maintained.

*ART 6.7.1.7.2 Provide Technical Rescue Services

6-95. Provide technical rescue support and extrication of personnel and equipment from crashed aircraft, rescuing sick or entrapped personnel from buildings, equipment, vehicles, water, ice, confined space, and high angles. Provide emergency lifesaving care for victims of accident or sudden illness. (FM 5-415) (USAES)

No.	Scale	Measure
1	Yes/No	Provide initial first aid.
2	Yes/No	Provide initial response to hazardous material.
3	Yes/No	Rescue entrapped, sick, and injured personnel from buildings.
4	Yes/No	Rescue entrapped, sick, and injured personnel from equipment.
5	Yes/No	Rescue entrapped, sick, and injured personnel from vehicles.
6	Yes/No	Rescue entrapped, sick, and injured personnel from water.
7	Yes/No	Rescue entrapped, sick, and injured personnel from confined space.
8	Yes/No	Rescue entrapped, sick, and injured personnel from high angles.

ART 6.7.2 DISPERSE TACTICAL FORCES

6-96. Relocate forces and spread or separate troops, materiel, or activities following concentration and maneuver to enhance survivability. The lethality of modern weaponry significantly increases the threat to concentrated formations. Attacking commanders manipulate their own and the enemy's concentration of forces by a combination of dispersion, concentration, deception, and attack. Dispersion stretches the enemy's defenses and denies lucrative targets to enemy long-range fires. (FM 3-90) (USACAC)

No.	Scale	Measure
01	Yes/No	Unit accomplished mission while tactically dispersed.
02	Time	To refine operation plan or order to reflect risk management assessment.
03	Time	To relocate friendly forces to minimize risks from battlefield hazards.
04	Percent	Of friendly casualties due to failure to disperse.
05	Percent	Of friendly casualties due to an enemy inability to mass combat power because assets are too dispersed.

ART 6.7.3 CONDUCT SECURITY OPERATIONS

6-97. Security operations are those operations undertaken by a commander to provide early and accurate warning of enemy operations, to provide the force being protected with time and maneuver space within which to react to the enemy, and to develop the situation to allow the commander to effectively use the protected forces. Commanders continually conduct some form of security operations. (FM 3-90) (USACAC)

Note: The information obtained on the enemy in conducting this task also pertains to ART 2.0 (The Intelligence Warfighting Function).

No.	Scale	Measure
01	Yes/No	The operations of the security force provided the protected force or installation with sufficient reaction time and maneuver space to conduct defensive operations.
02	Yes/No	Security force was in place not later than time specified in operation order.
03	Yes/No	Security force prevented enemy ground observation of protected force or installation.
04	Yes/No	Collateral damage from security operation was within acceptable limits.
05	Yes/No	Security force provided early and accurate warning of enemy approach.
06	Yes/No	Security force oriented its operations of the force or facility to be secured.
07	Yes/No	Security force performed continuous reconnaissance.
08	Yes/No	Security force maintained contact with enemy forces.
09	Yes/No	Commander developed criteria for ending the security operation.
10	Yes/No	Commander directed that contingency plans be developed for security operations.
11	Time	To conduct reconnaissance of the area surrounding the secured force or installation.
12	Time	To plan security operations.
13	Time	To prepare for the security operations including movement into security area.
14	Time	To execute security operations.
15	Time	To report enemy activities to appropriate headquarters.
16	Time	That the secured force or installation has to prepare prior to its encounter with the enemy.
17	Time	To integrate host-nation or third-nation security forces and means into friendly security operations.
18	Percent	Of security force casualties during the security operation.
19	Percent	Of secured force or installation casualties during the security operation.
20	Percent	Of unit combat power used to provide desired degree of security.
21	Percent	O f decreased support capability of sustainment units due to the requirement to provide security forces from internal assets.
22	Percent	Of increased availability of combat forces through use of host-nation or third-nation security forces.
23	Percent	Of enemy reconnaissance elements within security force capabilities destroyed or repelled.
24	Percent	Of friendly operations judged as not compromised prior to or during execution.
25	Percent	Of operations not compromised (based on enemy prisoner of war interrogations or captured documents).
26	Percent	Of critical facilities hardened or protected by security forces.
27	Percent	Of the AO or security area that can be observed by visual observation or covered by sensors at any given time.
28	Number	Of incidents where enemy forces affect the security of friendly units and facilities.
29	Number	Of incidents where enemy reconnaissance forces compromise friendly course of action causing them to be delayed, disrupted, canceled, or modified.
30	Number	Of security force casualties during the security operation.
31	Number	Of secured force or installation casualties during the security operation.
32	Number	Of mobility corridors or avenues of approach that can be observed by the security force.
33	Number	Of observation posts that can be established by the security force.
34	Number	Of enemy reconnaissance elements destroyed during security operation.
35	Square Kilometers	Size of security area or area of operation.

ART 6.7.4 CONDUCT ACTIONS TO CONTROL POLLUTION AND HAZARDOUS MATERIALS

6-98. Develop actions to prevent pollution generation and hazardous substance releases to avoid exposing friendly personnel to human health hazards, disrupting operations, adversely affecting indigenous or refugee populations and local economies; and to avoid damaging the natural or cultural environment. Conduct the environmental compliance program while appropriately considering the effect on the environment per applicable U.S. and host-nation agreements, environmental laws, policies, and regulations. Promptly report and clean up hazardous substance releases while avoiding tactical interference and ensuring adequate protection of the environment. Manage hazardous wastes correctly prior to transporting them to a permitted treatment, storage, or disposal facility. (FM 3-100.4) (USAES)

No.	Scale	Measure
01	Time	To provide the commander with the technical expertise relating to releases of hazardous materials or petroleum, oils, and lubricants.
02	Time	Of delay in the operation.
03	Time	River closed as a source of drinking water.
04	Time	River closed to traffic.
05	Time	To provide training guidance to the field as required.
06	Time	Of training lost due to release.
07	Percent	Of operations cancelled or delayed.
08	Percent	Of population with newly polluted drinking water.
09	Percent	Of wildlife killed as a result of pollution or a release.
10	Number	Of people with newly polluted drinking water.
11	Number	Of spills reported per week.
12	Number	Of wildlife killed as a result of pollution or a release.
13	Number	Of personnel injured or sickened as a result of pollution or a release.
14	Cost	For hazardous material removal or disposal.
15	Cost	To complete release recovery.
16	Cubic Yards	Of earth cleaned, removed, or replaced.
17	Pounds	Of hazardous materials released.

*SECTION VIII – ART 6.8: PROVIDE FORCE HEALTH PROTECTION

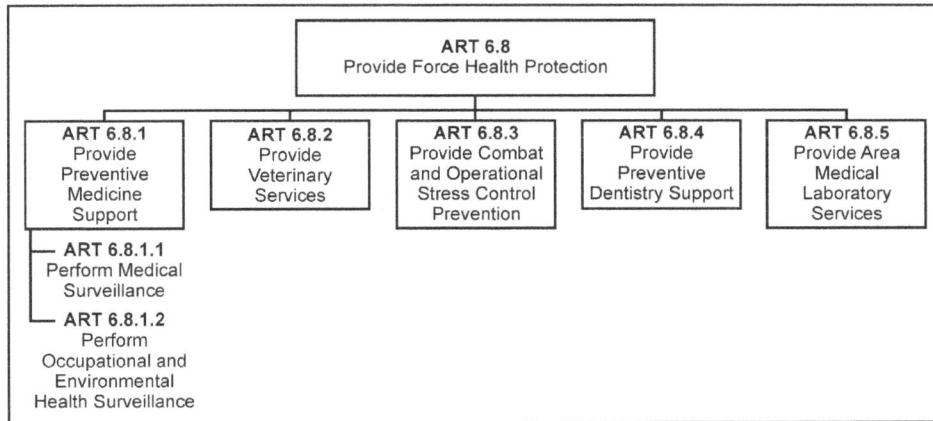

```
                        ART 6.8
              Provide Force Health Protection

   ┌──────────┬──────────┬──────────┬──────────┬──────────┐
  ART 6.8.1   ART 6.8.2   ART 6.8.3   ART 6.8.4   ART 6.8.5
  Provide     Provide     Provide Combat  Provide    Provide Area
  Preventive  Veterinary  and Operational Preventive  Medical
  Medicine    Services    Stress Control  Dentistry Support  Laboratory
  Support                 Prevention                  Services

  ─ ART 6.8.1.1
    Perform Medical
    Surveillance
  ─ ART 6.8.1.2
    Perform
    Occupational and
    Environmental
    Health Surveillance
```

6-99. Force health protection encompasses measures to promote, improve, or conserve the mental and physical well-being of Soldiers. These measures enable a healthy and fit force, prevent injury and illness, and protect the force from health hazards and include the prevention aspects of a number of Army Medical Department functions such as preventive medicine, including medical surveillance and occupational and environmental health (OEH) surveillance; veterinary, services, including the food inspection and animal care missions, and the prevention of zoonotic disease transmissible to man; combat and operational stress control (COSC); dental services (preventive dentistry); and laboratory services (area medical laboratory support). (FM 4-02) (USAMEDDC&S)

No.	Scale	Measure
01	Yes/No	Force health protection programs established in the area of operations (AO) kept casualty rates from disease and nonbattle injuries; chemical, biological, radiological, and nuclear exposures; OEH hazards; and combat operational stress below established thresholds.
02	Time	To refine force health protection in the AO after receipt of warning order.
03	Time	To establish force health protection on activation of the AO.
04	Percent	Of difference between force health protection requirements and actual requirements in the AO.
05	Percent	Of difference between planned area medical laboratory support requirements and actual requirements in the AO.
06	Percent	Of difference between planned preventive medicine services requirements and actual requirements in the AO.
07	Percent	Of difference between planned veterinary services requirements and actual requirements in the AO.
08	Percent	Of planned force health protection support achieved in the AO.
09	Percent	Of difference between planned COSC prevention support requirements and actual requirements in the AO.
10	Percent	Of difference between planned preventive dentistry support requirements and actual requirements in the AO.

*ART 6.8.1 PROVIDE PREVENTIVE MEDICINE SUPPORT

6-100. Prevent disease and nonbattle injuries by establishing preventive medicine programs such as, field hygiene and sanitation, disease surveillance, immunizations, chemoprophylaxis, and education in personal protective measures. (FM 4-02.17) (USAMEDDC&S)

No.	Scale	Measure
01	Yes/No	Preventive medicine programs established in the area of operations (AO) kept disease and nonbattle injury rates below established thresholds.
02	Yes/No	Units communicated the health risks to the at risk population.
03	Yes/No	Units conducted health hazard assessments.
04	Yes/No	Unit implemented the Vision Conservation Program.
05	Time	To refine preventive medicine program for AO after receipt of warning order.
06	Time	To establish preventive medicine plan (to include immunizations, pretreatment, chemoprophylaxis, and barrier creams) in the AO.
07	Time	Required to provide 100-percent immunizations to all Soldiers in the AO.
08	Percent	Of Soldiers who have all of the prescribed predeployment immunizations.
09	Percent	Of planned preventive medicine support achieved in AO.
10	Percent	Of personnel who received all required immunizations.
11	Percent	Of personnel who received required chemoprophylaxis.
12	Percent	Of personnel in AO briefed on health threats and trained in personal and unit-level protective measures and preventive medicine measures.
13	Percent	Of water points inspected for potability.
14	Percent	Of unit field sanitation teams trained.
15	Percent	Of personnel in the AO who have required and serviceable optical devices (such as spectacles, mask inserts, and protective devices).
16	Percent	Of bivouac sites inspected for disease and occupational and environmental health hazards.
17	Number	Of units with all required field sanitation team equipment and supplies.
18	Number	Of aerial spray missions conducted.
19	Number	Of Soldiers not available for duty due to hearing loss.
20	Number	Of consultations provided on preventive medicine measures.
21	Number	Of food service facilities inspected requiring immediate corrective actions.

*ART 6.8.1.1 PERFORM MEDICAL SURVEILLANCE

6-101. Perform medical surveillance, to include the collection and analysis of health status and health threat information before, during, and following deployment. Ensure common awareness of potential health threats and monitor implementation of preventive medicine measures. (FM 4-02.17) (USAMEDDC&S)

No.	Scale	Measure
01	Yes/No	Unit performed health threat assessment was for all areas of the area of operations (AO) and briefed to all personnel.
02	Yes/No	Unit conducted all required epidemiological investigations appropriately and completed them in a timely manner.
03	Yes/No	Unit collected disease and nonbattle injury data daily and categorized it correctly per Joint Chiefs of Staff and theater-specific standards.
04	Yes/No	All personnel completed pre- and postdeployment health assessments within required timeframes.
05	Yes/No	Unit analyzed disease and nonbattle injury data weekly for trends.

No.	Scale	Measure
06	Time	To refine medical surveillance programs for AO after receipt of warning order.
07	Time	To survey operational environment to detect and identify health threats and formulate means for minimizing effects.
08	Percent	Of difference between planned medical surveillance requirements and actual requirements in AO.
09	Percent	Of planned medical surveillance support achieved in AO.
10	Percent	Of Soldiers identified with a measured environmental or occupational exposure that have the exposure noted in their individual health records.
11	Percent	Of recorded medical treatment episodes in individual health records and/or electronic medical records.
12	Percent	Of disease and nonbattle injury reports submitted on time per theater policy.
13	Percent	Of personnel compliant with required personal protective measures.
14	Number	Of epidemiological investigations conducted in AO.
15	Number	Of health threats to the deployed force not identified in the predeployment medical threat assessment.

*ART 6.8.1.2 PERFORM OCCUPATIONAL AND ENVIRONMENTAL HEALTH SURVEILLANCE

6-102. Perform occupational and environmental health (OEH) surveillance. Develop and update the environmental health site assessment. (FM 4-02.17) (USAMEDDC&S)

No.	Scale	Measure
01	Yes/No	Occupational and environmental health surveillance established in the area of operations (AO) prevented or reduced the number and percent of personnel who became exposed to OEH hazards.
02	Yes/No	Significant OEH hazards were identified in the AO.
03	Time	To refine OEH hazard surveillance program for AO after receipt of warning order.
04	Time	To detect, identify, quantify, and evaluate OEH hazards; develop controls and communicate risk to minimize health risks.
05	Percent	Of difference between planned OEH surveillance requirements and actual requirements in the AO.
06	Percent	Of personnel in AO without health threat education and training provided.
07	Percent	Of personnel in AO without appropriate personal protective equipment or engineering controls to minimize health risks of identified OEH hazards.
08	Percent	Of identified OEH hazards in the AO evaluated using risk assessment.
09	Number	Of identified OEH exposures recorded in individual health records and/or electronic medical records.

*ART 6.8.2 PROVIDE VETERINARY SERVICES

6-103. Serve as the Department of Defense (DOD) executive agent for veterinary services for all services with the exception of the food inspection mission on U.S. Air Force installations. Perform food safety and bottled water surveillance—which includes food hygiene and quality assurance, inspection of class I sources, microbial analysis of food, and temperature monitoring of transported and stored food supplies—and assess potential health hazards in the area of operations (AO); identify, evaluate, and assess animal diseases of military significance; and provide complete veterinary health care to DOD military working dogs and any other government-owned animals in the AO. (FM 4-02.18) (USAMEDDC&S)

No.	Scale	Measure
01	Yes/No	Nonavailability of veterinary services did not degrade, delay, or disrupt unit operations.
02	Time	To refine veterinary services program for AO after receipt of warning order.

No.	Scale	Measure
03	Time	To establish comprehensive veterinary plan on activation of the AO.
04	Percent	Of difference between planned veterinary service requirements and actual requirements in AO.
05	Percent	Of planned veterinary support achieved in AO.
06	Percent	Of required food inspections meeting food safety standards in the AO.
07	Percent	Of government-owned animals treated and returned to duty in the AO.
08	Percent	Of veterinary capacity in use per day in AO.
09	Number	Of local food procurement establishment inspections performed in AO per month.
10	Number	Of animal diseases of military significance in the AO.
11	Number	Of military working dogs and other government-owned animals in AO requiring periodic veterinary support.
12	Number	Of approved bottled water and ice plants.
13	Percent	Of animals exposed to chemical, biological, radiological, and nuclear or toxic industrial materials.
14	Yes/No	Published food and beverage procurement source list.
15	Number	Of animals requiring evacuation for additional treatment in theater.

*ART 6.8.3 PROVIDE COMBAT AND OPERATIONAL STRESS CONTROL PREVENTION

6-104. Provide combat and operational stress control (COSC) prevention by establishing behavioral health prevention programs, conducting traumatic event management, and providing consultation and educational services. (FM 4-02.51) (USAMEDDC&S)

No.	Scale	Measure
01	Yes/No	Absence of command personnel from stress-related causes did not degrade, delay, or disrupt unit operations.
02	Time	To refine COSC prevention program for the area of operations (AO) after receipt of warning order.
03	Percent	Of critical incident debriefings planned and actual requirements.
04	Percent	Of required COSC personnel at activation in AO.
05	Percent	Of decrease in number of stress-related casualties after establishing and implementing COSC prevention plan or program in AO.
06	Number	Of consultations on COSC prevention techniques with Soldiers or groups.
07	Number	Of consultations on COSC prevention techniques with unit leaders.
08	Number	Of education and training events on COSC prevention techniques.
09	Number	Of combat and operational stress control cases requiring evacuation from AO.
10	Percent	Of personnel that have received Warrior resilience training.

*ART 6.8.4 PROVIDE PREVENTIVE DENTISTRY SUPPORT

6-105. Military preventive dentistry incorporates primary, secondary, and tertiary preventive measures taken to reduce or eliminate oral conditions that decrease a Soldier's fitness to perform the mission and cause absence from duty. (FM 4-02.19) (USAMEDDC&S)

No.	Scale	Measure
01	Yes/No	Preventive dentistry programs established in the area of operations (AO) prevented or reduced the number and percent of command personnel who became casualties as a result of dental disease and injury.
02	Time	To establish dental combat effectiveness program upon activation of the AO.
03	Time	To refine preventive dentistry plan after receipt of warning order.
04	Percent	Of difference between planned preventive dentistry requirements and actual requirements within the AO.
05	Percent	Of Soldiers receiving prophylaxis treatment.
06	Percent	Of Soldiers receiving fluoride varnish treatments.
07	Percent	Of Soldiers receiving prescriptions for fluoride supplement tablets.
08	Percent	Of Soldiers receiving training or education in field oral hygiene information program.

*ART 6.8.5 PROVIDE AREA MEDICAL LABORATORY SERVICES

6-106. Identify, evaluate, and assess health hazards in the area of operations (AO). This task includes providing chemical, biological, radiological, and nuclear (CBRN) laboratory services, endemic disease laboratory services, and environmental laboratory services. (FM 4-02) (USAMEDDC&S)

No.	Scale	Measure
01	Yes/No	Nonavailability of laboratory services did not degrade, delay, or disrupt unit operations or endanger the health of unit personnel.
02	Time	To refine area medical laboratory services program for AO after receipt of warning order.
03	Time	To establish comprehensive area medical laboratory service plan on activation of the AO.
04	Time	Of turnaround for technical labaratory testing results.
05	Percent	Of difference between planned area medical laboratory requirements and actual requirements in AO.
06	Percent	Of planned laboratory support achieved in AO.
07	Percent	Of required laboratories at activation of AO.
08	Percent	Of required laboratory personnel available at activation of AO.
09	Percent	Of laboratory capacity in use per day in AO.
10	Number	Of toxic industrial materials identified or confirmed through laboratory testing in the AO.
11	Number	Of CBRN warfare agents identified or confirmed through laboratory testing in the AO.
12	Number	Of endemic diseases identified through labaratory testing in AO in support of diagnostic treatment at medical treatment facilities .

SECTION IX – ART 6.9: CONDUCT CHEMICAL, BIOLOGICAL, RADIOLOGICAL, NUCLEAR, AND HIGH-YIELD EXPLOSIVES OPERATIONS

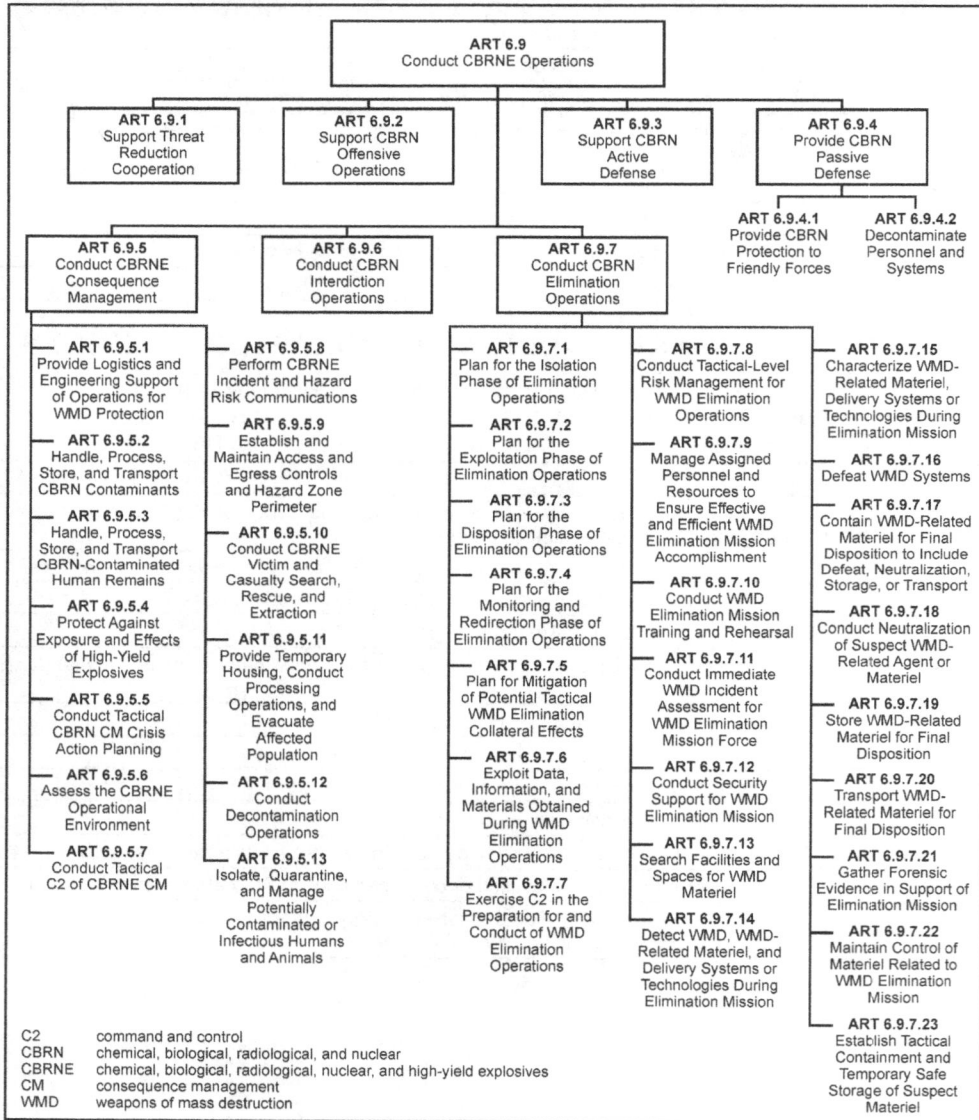

ART 6.9
Conduct CBRNE Operations

- **ART 6.9.1** Support Threat Reduction Cooperation
- **ART 6.9.2** Support CBRN Offensive Operations
- **ART 6.9.3** Support CBRN Active Defense
- **ART 6.9.4** Provide CBRN Passive Defense
  - **ART 6.9.4.1** Provide CBRN Protection to Friendly Forces
  - **ART 6.9.4.2** Decontaminate Personnel and Systems

- **ART 6.9.5** Conduct CBRNE Consequence Management
- **ART 6.9.6** Conduct CBRN Interdiction Operations
- **ART 6.9.7** Conduct CBRN Elimination Operations

ART 6.9.5:
- ART 6.9.5.1 Provide Logistics and Engineering Support of Operations for WMD Protection
- ART 6.9.5.2 Handle, Process, Store, and Transport CBRN Contaminants
- ART 6.9.5.3 Handle, Process, Store, and Transport CBRN-Contaminated Human Remains
- ART 6.9.5.4 Protect Against Exposure and Effects of High-Yield Explosives
- ART 6.9.5.5 Conduct Tactical CBRN CM Crisis Action Planning
- ART 6.9.5.6 Assess the CBRNE Operational Environment
- ART 6.9.5.7 Conduct Tactical C2 of CBRNE CM
- ART 6.9.5.8 Perform CBRN Incident and Hazard Risk Communications
- ART 6.9.5.9 Establish and Maintain Access and Egress Controls and Hazard Zone Perimeter
- ART 6.9.5.10 Conduct CBRNE Victim and Casualty Search, Rescue, and Extraction
- ART 6.9.5.11 Provide Temporary Housing, Conduct Processing Operations, and Evacuate Affected Population
- ART 6.9.5.12 Conduct Decontamination Operations
- ART 6.9.5.13 Isolate, Quarantine, and Manage Potentially Contaminated or Infectious Humans and Animals

ART 6.9.7:
- ART 6.9.7.1 Plan for the Isolation Phase of Elimination Operations
- ART 6.9.7.2 Plan for the Exploitation Phase of Elimination Operations
- ART 6.9.7.3 Plan for the Disposition Phase of Elimination Operations
- ART 6.9.7.4 Plan for the Monitoring and Redirection Phase of Elimination Operations
- ART 6.9.7.5 Plan for Mitigation of Potential Tactical WMD Elimination Collateral Effects
- ART 6.9.7.6 Exploit Data, Information, and Materials Obtained During WMD Elimination Operations
- ART 6.9.7.7 Exercise C2 in the Preparation for and Conduct of WMD Elimination Operations
- ART 6.9.7.8 Conduct Tactical-Level Risk Management for WMD Elimination Operations
- ART 6.9.7.9 Manage Assigned Personnel and Resources to Ensure Effective and Efficient WMD Elimination Mission Accomplishment
- ART 6.9.7.10 Conduct WMD Elimination Mission Training and Rehearsal
- ART 6.9.7.11 Conduct Immediate WMD Incident Assessment for WMD Elimination Mission Force
- ART 6.9.7.12 Conduct Security Support for WMD Elimination Mission
- ART 6.9.7.13 Search Facilities and Spaces for WMD Materiel
- ART 6.9.7.14 Detect WMD, WMD-Related Materiel, and Delivery Systems or Technologies During Elimination Mission
- ART 6.9.7.15 Characterize WMD-Related Materiel, Delivery Systems or Technologies During Elimination Mission
- ART 6.9.7.16 Defeat WMD Systems
- ART 6.9.7.17 Contain WMD-Related Materiel for Final Disposition to Include Defeat, Neutralization, Storage, or Transport
- ART 6.9.7.18 Conduct Neutralization of Suspect WMD-Related Agent or Materiel
- ART 6.9.7.19 Store WMD-Related Materiel for Final Disposition
- ART 6.9.7.20 Transport WMD-Related Materiel for Final Disposition
- ART 6.9.7.21 Gather Forensic Evidence in Support of Elimination Mission
- ART 6.9.7.22 Maintain Control of Materiel Related to WMD Elimination Mission
- ART 6.9.7.23 Establish Tactical Containment and Temporary Safe Storage of Suspect Materiel

C2 command and control
CBRN chemical, biological, radiological, and nuclear
CBRNE chemical, biological, radiological, nuclear, and high-yield explosives
CM consequence management
WMD weapons of mass destruction

6-107. Defend against chemical, biological, radiological, nuclear, and high-yield explosives (CBRNE) weapons using the principles of avoidance, protection, and decontamination. ART 6.9 includes protection from agents deliberately or accidentally released. An example of an accidentally released agent is toxic chemicals leaking from factory storage containers due to collateral damage. (FM 3-11) (USACRBNS)

No.	Scale	Measure
01	Yes/No	Unit could continue its mission when attacked by enemy CBRNE weapons.
02	Time	To conduct area or route reconnaissance to identify the limits of CBRNE weapons effects.
03	Time	To refine annex J to the operation order.
04	Time	To deploy and employ CBRNE monitoring equipment.
05	Time	To identify the CBRNE hazard.
06	Time	To detect the use of CBRNE weapons in the area of operations (AO).
07	Time	To issue downwind hazard warnings of a CBRNE attack in the AO.
08	Time	To conduct area damage control after using CBRNE weapons.
09	Time	To recover unit operational capability after a CBRNE attack.
10	Time	To give and understand CBRNE contamination alarms and signals.
11	Time	To assume appropriate mission-oriented protective posture after warning of the use of CBRNE weapons in the AO.
12	Time	To reconstitute unit to designated level of combat power after exposure to the effects of CBRNE weapons.
13	Time	To coordinate for additional CBRNE reconnaissance, monitoring, and decontamination assets.
14	Time	To administer chemoprophylaxis, immunizations, pretreatments, and barrier creams for protection against CBRNE warfare agents.
15	Percent	Of incidents of the use of CBRNE weapons detected.
16	Percent	Of enemy delivery systems for CBRNE weapons in AO identified, targeted, and destroyed.
17	Percent	Of CBRNE contaminated sites in the AO that have decontamination operations initiated or completed.
18	Percent	Of friendly units in the AO that have CBRNE monitoring, protective, and decontamination equipment.
19	Percent	Of on-hand CBRNE equipment, necessary to protect the unit against hazards, that is mission-capable.
20	Percent	Of CBRNE monitoring, protective and decontamination equipment positioned and operated correctly.
21	Percent	Of CBRNE hazards correctly identified.
22	Percent	Of friendly units in the AO without adequate supplies of individual and collective monitoring and protective equipment, and decontamination materials.
23	Percent	Of reduced unit combat power from the need to defend against the use of CBRNE weapons.
24	Percent	Of friendly and civilian casualties in AO from the use of CBRNE weapons.
25	Number	Of instances where CBRNE weapons are employed.
26	Number	And types of on-hand CBRNE monitoring, protective, and decontamination equipment.
27	Number	And types of friendly systems destroyed, damaged, or rendered inoperable resulting from the use of CBRNE weapons.
28	Number	Of instances where units and facilities are affected by using CBRNE weapons without warning of their use.
29	Number	Of false alarms relating to using CBRNE weapons.

ART 6.9.1 SUPPORT THREAT REDUCTION COOPERATION

6-108. Take action that allows Soldiers to survive and continue the mission under chemical, biological, radiological, and nuclear (CBRN) conditions. (FM 3-11.4) (USACRBNS)

No.	Scale	Measure
01	Yes/No	The use of CBRN protective equipment preserved unit personnel, equipment, and supplies for future missions.
02	Time	To conduct area reconnaissance to detect the use of CBRN weapons.
03	Time	To plan or revise the plan to employ protective equipment to take into account existing factors of mission, enemy, terrain and weather, troops and support available, time available, and civil considerations.
04	Time	To establish the degree of local security for installing collective CBRN protective equipment.
05	Time	To employ additional CBRN protective equipment to harden individuals and facilities from effects of CBRN weapons.
06	Time	To assume mission-oriented protective posture in response to the employment of CBRN weapons given previous mission-oriented protective posture.
07	Time	To employ CBRN protective equipment to harden supply stocks with engineer support.
08	Time	To employ CBRN protective equipment to harden supply stocks without engineer support.
09	Time	To emplace warning signs marking edges of areas contaminated by CBRN weapons.
10	Time	To obtain the CBRN protective equipment and systems needed to complete hardening process.
11	Percent	Of personnel, systems, and facilities hardened with CBRN protective equipment and systems.
12	Percent	Of personnel trained to use CBRN protective equipment.
13	Percent	Of required CBRN protective equipment and supplies available.
14	Percent	Of CBRN individual and collective protective equipment employed to standard.
15	Percent	Of on-hand CBRN equipment, necessary to protect the unit against hazards, that is mission-capable.
16	Percent	Of friendly casualties due to improperly used CBRN protective equipment.
17	Percent	Of casualties or equipment and supplies lost to enemy attack due to the nonavailability of CBRN protective equipment.
18	Number	Of mission-capable individual and collective CBRN protective equipment systems.
19	Number	Of friendly casualties due to improperly used CBRN protective equipment or slow reaction to the use of CBRN weapons.
20	Number	Of casualties or equipment and supplies lost due to the nonavailability of CBRN protective equipment.

ART 6.9.2 SUPPORT CHEMICAL, BIOLOGICAL, RADIOLOGICAL, AND NUCLEAR OFFENSIVE OPERATIONS

6-109. Deter and respond to the acquisition, facility preparation, production, weaponization, exportation, deployment, threat, and use of chemical, biological, radiological, and nuclear (CBRN) weapons. Counterforce capabilities include conventional, unconventional (such as special operations forces), and nuclear. For deterrence to succeed and for Secretary of Defense to possess flexible response options, counterforce must include highly destructive and lethal options, as well as nonlethal options that discriminate or minimize collateral damage and loss of life. (FM 3-11) (USACRBNS)

No.	Scale	Measure
01	Yes/No	Units had available weapon systems designed to destroy, disrupt, or deny access to CBRN weapon targets while minimizing negative collateral effects.
02	Yes/No	Capability to model and predict collateral effects was prior to a strike on CBRN weapon targets.
03	Time	To acquire, positively identify, select, and prioritize CBRN weapon targets as well as other high-value targets.
04	Time	To apportion resources to attack CBRN weapon targets as well as other high-value targets in either a deliberate or adaptive planning mode.

No.	Scale	Measure
05	Percent	Of targets reviewed for collateral damage or effects, damage expectancy, casualties, and political ramifications or sensitivities.
06	Percent	Of targets exceeding guidance.
07	Percent	Of planned targets hit on time.
08	Percent	Of friendly or neutral forces or noncombatants influenced by collateral effects from friendly attacks on CBRN-weapon targets.
09	Percent	Of known or suspected enemy CBRN targets that have been preplanned with the joint targeting cycle process.

ART 6.9.3 SUPPORT CHEMICAL, BIOLOGICAL, RADIOLOGICAL, AND NUCLEAR ACTIVE DEFENSE

6-110. Protect all assets from attack by chemical, biological, radiological, and nuclear (CBRN) weapons by using assets to detect, divert or intercept, and counter or destroy delivery systems. Integrate surveillance, detection, identification, tracking, and interception systems. Includes use of aircraft, air defense missiles, air defense artillery, and nonair defense systems. Protects critical nodes and facilities. (FM 3-11) (USACRBNS)

No.	Scale	Measure
01	Time	To issue threat warning after launch of ballistic missile.
02	Percent	Of enemy CBRN-weapon attacks reached target.
03	Percent	Of enemy CBRN attacks intercepted.
04	Number	Of U.S. casualties both combatant and noncombatant.
05	Number	Of false alarms.

ART 6.9.4 PROVIDE CHEMICAL, BIOLOGICAL, RADIOLOGICAL, AND NUCLEAR PASSIVE DEFENSE

6-111. Passive defense seeks to deter and deny the use of chemical, biological, radiological, and nuclear (CBRN) weapons by ensuring that U.S. forces succeed in a CBRN environment. The highest priorities for passive defense are force survivability and successful mission accomplishment. Passive-defense operations focus on protecting assets, sustaining mission operations, and minimizing casualties. The elements of passive defense against a CBRN attack consist of contamination avoidance, protection, and decontamination. Actions undertaken include measures to provide essential individual and collective protection for friendly forces and critical assets. Passive-defense measures are planned whenever U.S. forces could face a threat with a CBRN capability. (FM 3-11) (USACRBNS)

ART 6.9.4.1 PROVIDE CHEMICAL, BIOLOGICAL, RADIOLOGICAL, AND NUCLEAR PROTECTION TO FRIENDLY FORCES

6-112. Employ detecting, identifying, marking, warning, and reporting methods and equipment to protect personnel, units, and equipment from chemical, biological, radiological, and nuclear (CBRN) hazards. (FM 3-11.4) (USACRBNS)

No.	Scale	Measure
01	Yes/No	Unit could continue its mission when attacked by enemy CBRN weapons.
02	Time	To conduct area reconnaissance to determine limits of the effects of the use of CBRN weapons.
03	Time	To refine annex J to the operation order.
04	Time	To detect the use of CBRN weapons in the area of operations (AO).
05	Time	To issue downwind hazard warnings of a CBRN attack in the AO.

No.	Scale	Measure
06	Time	To conduct area damage control after the use of CBRN weapons.
07	Time	To recover unit operational capability after a CBRN attack.
08	Time	To assume appropriate mission-oriented protective posture after warning of use of CBRN weapons in the AO.
09	Percent	Of enemy delivery systems for CBRN weapons in AO identified, targeted, and destroyed.
10	Percent	Of CBRN-contaminated sites in the AO that have decontamination operations initiated or completed.
11	Percent	Of units in the AO that have CBRN monitoring equipment.
12	Percent	Of on-hand CBRN equipment, necessary to protect the unit against hazards, that is mission-capable.
13	Percent	Of CBRN monitoring equipment positioned and operated correctly.
14	Percent	Of friendly units in the AO lacking supplies of individual and collective protective equipment and decontamination materials.
15	Percent	Of friendly and civilian casualties in AO as a result of the use of CBRN weapons.
16	Number	And types of friendly systems destroyed, damaged, or rendered inoperable as a result of the use of CBRN weapons.

ART 6.9.4.1.1 Employ Contamination Avoidance

6-113. Take measures to avoid or minimize the effects of chemical, biological, radiological, and nuclear (CBRN) attacks and reduce the effects of CBRN hazards. By taking measures to avoid the effects of CBRN attacks, units can reduce their protective posture and decrease the likelihood and extent of decontamination required. (FM 3-11.3) (USACRBNS)

No.	Scale	Measure
01	Yes/No	Unit could continue its mission when attacked by enemy CBRN weapons.
02	Time	To detect the use of CBRN weapons in the area of operations (AO).
03	Time	To conduct route reconnaissance to determine locations where effects of CBRN weapons are present and the degree of contamination along selected routes.
04	Time	To conduct area reconnaissance to determine limits of the effects of CBRN weapons.
05	Time	To refine the operation order to avoid or limit contact with contaminated areas.
06	Time	To use the CBRN warning and reporting system to send reports of CBRN attacks, such as to issue downwind hazard warnings.
07	Time	To employ CBRN monitoring equipment.
08	Time	To identify CBRN hazards.
09	Time	To mark likely entry points into contaminated areas.
10	Time	To conduct contamination control—bypassing, exposing only the absolute minimum number of personnel and equipment, encapsulating personnel and equipment, covering equipment and supplies, and relocating.
11	Percent	Of CBRN contamination in the AO detected and correctly identified.
12	Percent	Of friendly units in the AO that have CBRN monitoring equipment.
13	Percent	Of on-hand CBRN equipment, necessary to protect the unit against hazards, that is mission-capable.
14	Percent	Of CBRN monitoring equipment positioned and operated correctly.
15	Percent	Of unit courses of action modified due to the presence of CBRN contamination.
16	Number	And types of on-hand CBRN monitoring equipment.
17	Number	And types of friendly systems destroyed, damaged, or rendered inoperable resulting from contact with CBRN contamination.
18	Number	Of friendly or civilian casualties in AO resulting from contact with CBRN contamination.

ART 6.9.4.1.2 Identify Chemical, Biological, Radiological, and Nuclear Hazards

6-114.   Obtain information about the chemical, biological, radiological, and nuclear (CBRN) activities and resources of an enemy by visual observation or other detection methods. Detect and identify CBRN hazards, to include finding gaps and detours around CBRN-contaminated areas. CBRN reconnaissance, which provides the information for identifying CBRN hazards, is part of the overall intelligence collection effort. (FM 3-11.19) (USACRBNS)

Note: This task branch is supported by ART 2.3.3 (Conduct Reconnaissance).

No.	Scale	Measure
01	Yes/No	Unit continued its mission when attacked by enemy CBRN weapons.
02	Time	To collect CBRN hazard samples.
03	Time	To identify CBRN hazard samples.
04	Time	Required to obtain medical specimens for suspected biological or chemical hazards.
05	Time	Required to identify CBRN warfare agents from medical specimens.
06	Percent	Of instances in which a CBRN hazard is correctly identified.
07	Percent	Of instances in which a CBRN hazard is incorrectly identified as harmless.
08	Percent	Of instances in which a harmless sample is incorrectly identified as a CBRN hazard.
09	Number	Of casualties due to incorrect identification of CBRN hazards.

ART 6.9.4.1.3 Warn Personnel and Units of Contaminated Areas

6-115.   Alert units and personnel concerning contaminated areas so they can retain freedom of maneuver, orient on the threat, report all information rapidly and accurately, and develop the situation rapidly. (FM 3-11.4) (USACRBNS)

No.	Scale	Measure
01	Yes/No	Personnel and units were warned of the presence and limits of contaminated areas in their area of operations (AO) so they could retain freedom of maneuver.
02	Time	To detect the use of chemical, biological, radiological, and nuclear (CBRN) weapons in the AO.
03	Time	To conduct area reconnaissance to determine limits of the effects of CBRN weapons.
04	Time	To conduct route reconnaissance to determine locations where effects of CBRN weapons are and the degree of contamination.
05	Time	To refine operation order to avoid or limit contact with contaminated areas.
06	Time	To use the CBRN warning and reporting system to send reports of CBRN attacks, such as to issue downwind hazard warnings.
07	Time	To employ CBRN monitoring equipment.
08	Time	To identify CBRN hazards.
09	Time	To mark likely entry points into contaminated areas.
10	Time	To give and understand CBRN contamination alarms and signals.
11	Percent	Of CBRN contamination in the AO detected and correctly identified.
12	Percent	Of friendly units in the AO that have CBRN monitoring equipment.
13	Percent	Of on-hand chemical, biological, radiological, nuclear, and high-yield explosives equipment, necessary to protect the unit against hazards, that is mission-capable.
14	Percent	Of CBRN monitoring equipment positioned and operated correctly.
15	Percent	Of personnel trained to operate in a CBRN environment.
16	Number	And types of on-hand CBRN monitoring equipment.

No.	Scale	Measure
17	Number	And types of friendly systems destroyed, damaged, or rendered inoperable due to unanticipated contact with CBRN contamination.
18	Number	Of friendly and civilian casualties in the AO due to unanticipated contact with CBRN contamination.

ART 6.9.4.1.4 Report Chemical, Biological, Radiological, and Nuclear Hazards Throughout the Area of Operations

6-116. Provide chemical, biological, radiological, and nuclear (CBRN) hazards information to support decisionmaking and permit units and individuals to avoid contaminated areas. (FM 3-11.4) (USACRBNS)

No.	Scale	Measure
01	Yes/No	Unit accomplished its mission.
02	Time	To detect the use of CBRN weapons in the area of operations (AO).
03	Time	To conduct area reconnaissance to determine contaminated locations and the degree of that contamination resulting from the use of CBRN weapons.
04	Time	To conduct route reconnaissance to determine contaminated locations and the degree of that contamination resulting from the use of CBRN weapons.
05	Time	To refine the operation order to avoid or limit contact with contaminated areas.
06	Time	To use the CBRN warning and reporting system to send reports of CBRN attacks, such as to issue downwind hazard warnings.
07	Time	To identify CBRN hazards.
08	Time	To give and understand CBRN contamination alarms and signals.
09	Percent	Of friendly units in the AO that have information systems capable of receiving CBRN warnings.
10	Percent	Of unit courses of action that must be abandoned, changed, or modified due to the warning of the presence of contaminated areas.
11	Number	Of locations contaminated by CBRN detected, correctly identified, and reported in the AO.
12	Number	And types of friendly systems destroyed, damaged, or rendered inoperable due to contact with CBRN contamination.
13	Number	Of friendly or civilian casualties in the AO due to contact with CBRN contamination.

ART 6.9.4.1.5 Prepare for a Nuclear Strike

6-117. Take preparatory actions to warn personnel, harden positions, protect equipment, and conduct periodic monitoring when warned that a nuclear strike is imminent. (FM 3-11) (USACRBNS)

No.	Scale	Measure
01	Yes/No	Unit continued its mission after the nuclear strike.
02	Time	To prepare for the nuclear strike. This includes the time it takes to cover and secure loose, flammable, and explosive items, zero radiation-monitoring equipment, close sights and optics, shut down information systems, disconnect power and antenna leads, and take protective measures to prevent dazzle. The time will also vary with the need to take additional preparatory measures, depending on the unit, installation, or facility's closeness to the predicted detonation point.
03	Time	To move the unit or system to the minimum safe distance (MSD) from the predicted ground zero.
04	Time	To prepare and transmit effective downwind messages (USMTF #C503).
05	Time	To transmit, receive, and understand a nuclear strike warning (USMTF #C505).
06	Time	To conduct surveillance and reconnaissance to detect a nuclear strike and determine ground zero.
07	Percent	Of unit casualties due to the effects—blast, thermal radiation, residual radiation, and electromagnetic pulse—of a nuclear strike.

No.	Scale	Measure
08	Percent	Of nuclear strike preparatory measures completed before a nuclear strike.
09	Percent	Of unit caught within MSD 1, MSD 2, and MSD 3 of ground zero.
10	Percent	Of systems redundancy existing before a nuclear strike.
11	Percent	Of systems in the unit designed to survive the thermal, radiation, and electromagnetic effects of a nuclear strike.
12	Percent	Of unit personnel and equipment not prepared for the nuclear strike.
13	Percent	Of reduced unit combat power due to the need to prepare for a nuclear strike.
14	Percent	Of personnel trained to prepare for a nuclear strike.
15	Number	And types of systems inoperable due to effects of the nuclear strike.
16	Number	Of casualties due to effects of the nuclear strike.
17	Number	Of casualties due to improperly used chemical, biological, radiological, and nuclear (CBRN) protective equipment.
18	Number	Of casualties or equipment and supplies lost due to the nonavailability of CBRN protective equipment.
19	Number	Of casualties attributed to slow reaction to effects of the nuclear strike.

ART 6.9.4.2 DECONTAMINATE PERSONNEL AND SYSTEMS

6-118. Make any person, object, or area safe by absorbing, destroying, neutralizing, making harmless, or removing chemical, biological, radiological, and nuclear materials or agents clinging to or around it. This task encompasses environmental considerations. (FM 3-11.5) (USACRBNS)

Note: ART 6.9.4.2.1 (Perform Immediate Decontamination) addresses immediate decontamination.

ART 6.9.4.2.2 (Perform Operational Decontamination) addresses operational decontamination.

ART 6.9.4.2.3 (Perform Thorough Decontamination) addresses thorough decontamination.

No.	Scale	Measure
01	Yes/No	Unit continued its mission after decontaminating its personnel and systems.
02	Time	That execution of the unit concept of operations is delayed by decontamination procedures.
03	Time	To determine an appropriate decontamination site incorporating environmental considerations.
04	Time	To determine the extent of contamination.
05	Time	To move the required decontamination equipment to the decontamination site and obtain the necessary decontamination supplies.
06	Time	To decontaminate individuals.
07	Time	To decontaminate vehicles and systems.
08	Percent	Of unit personnel and equipment requiring decontamination.
09	Percent	Of unit personnel proficient in conducting decontamination operations.
10	Percent	Of on-hand decontamination equipment and supplies.
11	Percent	Of mission-capable, on-hand decontamination equipment.
12	Number	Of personnel and equipment requiring decontamination.
13	Number	And types of mission-capable, on-hand decontamination equipment.
14	Number	Of casualties due to improper and incomplete decontamination.

ART 6.9.4.2.1 Perform Immediate Decontamination

6-119. Minimize casualties, save lives, and limit the spread of contamination by contaminated individuals. Individuals or crews conduct immediate decontamination by skin decontamination, personal wipe down, and operator's spray down to stop agent from penetrating the equipment. (FM 3-11.5) (USACRBNS)

No.	Scale	Measure
01	Yes/No	Unit continued its mission after conducting immediate decontamination.
02	Time	To complete skin decontamination.
03	Time	To conduct personal equipment wipe down.
04	Time	For equipment spray down.
05	Time	For unit personnel to exchange mission-oriented protective posture gear.
06	Time	To conduct unmasking procedures.
07	Time	To discard contaminated articles incorporating environmental considerations.
08	Percent	Of personnel and equipment requiring immediate decontamination.
09	Percent	Of personnel proficient in conducting immediate decontamination operations.
10	Percent	Of on-hand immediate decontamination equipment and supplies.
11	Percent	Of mission-capable, on-hand immediate decontamination equipment.
12	Number	Of personnel and equipment requiring immediate decontamination.
13	Number	And types of mission-capable, on-hand immediate decontamination equipment and supplies.
14	Number	Of casualties due to improper or incomplete immediate decontamination.

ART 6.9.4.2.2 Perform Operational Decontamination

6-120. Sustain operations, reduce the contact hazard, and limit the spread of contamination to eliminate the necessity for, or reduce the duration of, wearing mission-oriented protective posture (MOPP) gear. Affected units and battalion, crew, or chemical corps decontamination platoons perform operational decontamination. (FM 3-11.5) (USACRBNS)

No.	Scale	Measure
01	Yes/No	Unit continued its mission after conducting operational decontamination.
02	Time	To find a site to perform operational decontamination.
03	Time	To initiate operational decontamination after exposure.
04	Time	To obtain equipment and supplies to conduct operational decontamination.
05	Time	To complete operational decontamination of unit equipment.
06	Time	For unit personnel to exchange MOPP gear.
07	Time	To conduct unmasking procedures.
08	Time	To discard contaminated articles incorporating environmental considerations.
09	Percent	Of unit equipment requiring operational decontamination.
10	Percent	Of unit personnel proficient in conducting operational decontamination operations.
11	Percent	Of on-hand operational decontamination equipment and supplies.
12	Percent	Of mission-capable, on-hand operational decontamination equipment.
13	Percent	Of operations degraded, delayed, or modified due to the inability to perform operational decontamination.
14	Number	And types of equipment requiring operational decontamination.
15	Number	And types of mission-capable, on-hand operational decontamination equipment and supplies.
16	Number	Of casualties due to improper or incomplete operational decontamination.

ART 6.9.4.2.3 Perform Thorough Decontamination

6-121. Reduce contamination on personnel, equipment or materiel, and working areas to the lowest possible level (negligible risk) to permit reducing or removing individual protective equipment and to maintain operations with minimal degradation. There are three thorough decontamination techniques: detailed troop decontamination, detailed equipment decontamination, and detailed aircraft decontamination. To reduce or eliminate the need of individual protective clothing, units can carry out decontamination with assistance from chemical units. (FM 3-11.5) (USACRBNS)

No.	Scale	Measure
01	Yes/No	Unit continued its mission after conducting a thorough decontamination.
02	Time	To find a site to perform thorough decontamination incorporating environmental considerations.
03	Time	To plan and coordinate a thorough decontamination operation, including the time to prioritize decontamination efforts.
04	Time	To obtain equipment and supplies to conduct a thorough decontamination.
05	Time	To move to the decontamination site.
06	Time	To initiate a thorough decontamination.
07	Time	To complete thorough decontamination of unit equipment.
08	Time	For unit personnel to exchange mission-oriented protective posture gear.
09	Time	To conduct unmasking procedures.
10	Time	To discard contaminated articles.
11	Percent	Of unit equipment requiring thorough decontamination.
12	Percent	Of unit personnel proficient in conducting thorough decontamination operations.
13	Percent	Of personnel and equipment completing immediate decontamination before leaving the site of initial contamination.
14	Percent	Of on-hand necessary thorough decontamination equipment and supplies.
15	Percent	Of mission-capable, on-hand thorough decontamination equipment.
16	Percent	Of operations degraded, delayed, or modified due to the inability to perform thorough decontamination.
17	Number	And types of equipment requiring thorough decontamination.
18	Number	And types of mission-capable, on-hand thorough decontamination equipment and supplies.
19	Number	Of casualties due to improper or incomplete decontamination.

ART 6.9.4.2.4 Perform Area Decontamination

6-122. Decontaminate fixed sites and terrain to restore the area to an acceptable level of readiness and effectiveness while conducting the mission. Limit the spread and transfer of contamination, restore mission essential functioning, and open accessibility for entry and exit to key facilities. Fixed sites include command posts, signal facilities, supply installations and points, depots, pre-positioned materiel, airfields, and port facilities. (FM 3-11.5) (USACRBNS)

No.	Scale	Measure
01	Yes/No	Units and unprotected personnel maneuvered through or used the decontaminated area without hindrance from contamination after area decontamination procedures were completed.
02	Time	To perform reconnaissance of the area designated for decontamination in conjunction with environmental considerations.
03	Time	To plan and coordinate the area decontamination, including the time to prioritize decontamination efforts.
04	Time	To obtain equipment and supplies to conduct area decontamination.
05	Time	To move to the decontamination area.

No.	Scale	Measure
06	Time	To initiate the area decontamination after exposure to contaminates.
07	Time	To complete area decontamination of fixed sites and key terrain.
08	Time	To move contaminated soil and hazardous waste generated by the area decontamination to hazardous waste dumps.
09	Percent	Of fixed sites and key terrain requiring area decontamination.
10	Percent	Of unit personnel proficient in conducting area decontamination operations.
11	Percent	Of on-hand area decontamination equipment and supplies.
12	Percent	Of mission-capable, on-hand area decontamination equipment.
13	Percent	Of operations degraded, delayed, or modified due to the inability to perform area decontamination of fixed sites and key terrain.
14	Number	Of fixed sites requiring area decontamination.
15	Number	And types of mission-capable, on-hand area decontamination equipment and supplies.
16	Number	Of casualties due to improper or incomplete area decontamination.

ART 6.9.4.2.5 Perform Patient Decontamination

6-123. Decontaminate patients who are unable to decontaminate themselves through the systematic removal of clothing and contaminants. A patient decontamination team consisting of nonmedical personnel from the supported unit performs patient decontamination. The patient decontamination team operates under the supervision of medical personnel to ensure the decontamination process causes no further injury to the patient. (FM 4-02.7) (USAMEDDC&S)

No.	Scale	Measure
01	Yes/No	Patient decontamination did not result in detrimental effects on the patient.
02	Time	To prepare patient chemical, biological, radiological, and nuclear decontamination equipment and supplies.
03	Time	To decontaminate a litter patient. This includes decontaminating the patient's mask and hood; removing the field medical card; removing gross contamination; removing the patient's protective overgarment, uniform, and personal effects; transferring the patient to a decontamination litter; conducting spot skin decontamination; and transferring the patient through the shuffle pit to the clean treatment area.
04	Time	To decontaminate an ambulatory patient. This includes removing load-bearing equipment, decontaminating the patient's mask and hood, removing the field medical card, removing all gross contamination, removing the patient's protective overgarment and personal effects, checking the patient for contamination, conducting spot skin decontamination, removing bandages and tourniquets (medical personnel perform this action), and moving the patient through the shuffle pit to the clean treatment area.
05	Time	To train the patient decontamination team.
06	Time	To establish clean and dirty patient treatment facilities.
07	Time	To obtain equipment and supplies needed to conduct patient decontamination.
08	Time	To initiate patient decontamination.
09	Time	To discard contaminated articles in conjunction with environmental considerations.
10	Percent	Of patients requiring decontamination before receiving medical treatment.
11	Percent	Of on-hand patient decontamination equipment and supplies.
12	Percent	Of mission-capable, on-hand patient decontamination equipment.
13	Percent	Of medical treatments degraded, delayed, or modified due to the inability to perform patient decontamination.
14	Number	Of patients decontaminated.

ART 6.9.5 CONDUCT CHEMICAL, BIOLIGICAL, RADIOLOGICAL, NUCLEAR, AND HIGH-YIELD EXPLOSIVES CONSEQUENCE MANAGEMENT

6-124. Coordinate support for essential services and activities required to manage and mitigate damage resulting from the use of chemical, biological, radiological, nuclear, and high-yield explosives (CBRNE) weapons or the release of toxic industrial materials or contaminants. Services and activities can include population evacuation, decontamination, transportation, communications, public works and engineering, firefighting, information and planning, mass care, resource support, health and medical services, urban search and rescue, hazardous materials, and food and energy. This task addresses processes for sharing information on stockpiles and local or referral surge capacities. In addition, establishment of alert and notification mechanisms, media management plans, and coordination of a public awareness and education effort are included. Population at risk and CBRNE casualties are estimated. If required, organizations are requested and military units are deployed to support consequence management. (FM 3-11.21) (USACRBNS)

No.	Scale	Measure
01	Yes/No	Department of Defense plans and policies for consequence management operations were in place.
02	Time	To coordinate emergency response plan with civil authorities.
03	Time	To identify relevant participants and determine roles and responsibilities via approved exercise.
04	Time	To develop options for decision makers.
05	Time	To assess consequences and facilitate follow-on support.
06	Time	To deploy reaction teams.
07	Time	To establish plan to distribute medicines and medical supplies when protectively isolation, fearful of the public, and concerned for security.
08	Time	To develop specific entry and exit plans in concert with local, state, and federal response plans.
09	Time	To establish coordination, communication, and contingency plans for joint and government or industry via approved exercises.
10	Time	To establish mechanisms and processes for sharing information on stockpiles and surge capacities via approved exercises.
11	Time	To establish alert and notification mechanisms via approved exercises.
12	Percent	Of actions forwarded with developed protections against failure.
13	Percent	Of medical consequence management teams available.

ART 6.9.5.1 PROVIDE LOGISTICS AND ENGINEERING SUPPORT OF OPERATIONS FOR WEAPONS OF MASS DESTRUCTION PROTECTION

6-125. Provide and use logistics and engineering operations to support weapons of mass destruction (WMD) protection. This involves identifying, allocating, and providing resources necessary to protect people, critical infrastructure, and equipment. Includes engineering activities for protection operations. ART 6.9.5.1 includes logistics activities including the application of critical resources within the prioritization framework. It also encompasses the dissemination of consequence management guidance and program materials. Involves the use of engineering capabilities to clear routes of entry and to remove secondary hazards. (FM 3-11.21) (USACRBNS)

No.	Scale	Measure
01	Yes/No	Unit identified resources to protect personnel.
02	Yes/No	Unit allocated resources to protect personnel.
03	Yes/No	Unit provided resources to protect personnel.

No.	Scale	Measure
04	Yes/No	Unit provided resources to protect critical infrastructure and equipment.
05	Yes/No	Unit identified resources to protect critical infrastructure and equipment.
06	Yes/No	Unit allocated resources to protect critical infrastructure and equipment.
07	Yes/No	Unit conducted engineering activities in support of WMD consequence management activities.
08	Yes/No	Unit supported emplacement of collective protection systems.
09	Yes/No	Consequence management guidance and program materials were provided to non-English-speaking populations.
10	Yes/No	Engineering capabilities were available to clear routes of entry.
11	Yes/No	Engineering capabilities were available to remove secondary hazards.
12	Yes/No	Procedures adequately addressed use of logistics and engineering operations supporting WMD protection.

ART 6.9.5.2 HANDLE, PROCESS, STORE, AND TRANSPORT CHEMICAL, BIOLOGICAL, RADIOLOGICAL, AND NUCLEAR CONTAMINANTS

6-126. The handling, processing, storage, and transport of contaminated equipment, materials, samples, residues, animal remains, and waste. It involves the chain of actions from collection or identification, through processing and storage, to disposition actions as outlined in procedures and higher authority guidance. It includes gathering samples, adhering to chain of evidence procedures, and recovering animal remains. Includes the collection and processing of personal effects of deceased, missing, and medically evacuated personnel. (FM 3-11.21) (USACRBNS)

No	Scale	Measure
01	Yes/No	Unit handled and processed contaminated and infectious equipment, samples, residues, animal remains, and waste.
02	Yes/No	Unit contained contaminated and infectious equipment, samples, residues, animal remains, and waste.
03	Yes/No	Unit safely stored contaminated and infectious equipment, samples, residues, animal remains, and waste.
04	Yes/No	Unit safely transported contaminated and infectious equipment, samples, residues, animal remains, and waste.
05	Yes/No	Procedures outlined the chain of actions and associated processes from collection or identification, through processing and storage, to disposition actions.
06	Yes/No	Procedures outlined how to gather samples and adhere to chain of custody of contaminated and infectious evidence.
07	Yes/No	Procedures addressed how to transfer custody of contaminated and infectious evidence.
08	Yes/No	Personnel were qualified to wear personal protective equipment.
09	Yes/No	Personnel wore appropriate personal protective equipment.
10	Yes/No	A safety officer was appointed to monitor operations in weapons of mass destruction (WMD) environments.
11	Yes/No	Unit identified and used appropriately contaminated and noncontaminated transportation routes.
12	Yes/No	Unit followed procedures to prevent secondary exposure and threat of WMD.

ART 6.9.5.3 HANDLE, PROCESS, STORE, AND TRANSPORT CHEMICAL, BIOLOGICAL, RADIOLOGICAL, AND NUCLEAR-CONTAMINATED HUMAN REMAINS

6-127. Establish mass mortuary operations and collect human remains. Address the performance of marking, handling, decontamination, processing, temporary storage, and preparation for transport of contaminated human remains. This task involves the chain of actions from collection and identification, through processing and storage, to transport from the joint operations area or the conduct of final disposition action in the joint operations area. Encompass requirements addressing the safety of personnel, protection of resources, and the safe containment of remains. (FM 3-11.21) (USACRBNS)

Note: For information concerning decontamination of human remains, refer to JP 4-06.

No	Scale	Measure
01	Yes/No	Unit established mass mortuary operations in a weapons of mass destruction environment.
02	Yes/No	Unit collected contaminated human remains.
03	Yes/No	Procedures existed for chain of actions associated with contaminated human remains processing.
04	Yes/No	Procedures adequately addressed the safe recovery and handling, packaging, marking, decontamination, processing, storage, and preparation for transport of remains and personal effects.
05	Yes/No	Personnel were capable of safely handling, processing, storing, and transporting human remains contaminated with chemical, biological, radiological, and nuclear materials.
06	Yes/No	Personnel were qualified to wear personal protective equipment.
07	Yes/No	Personnel wore appropriate personal protective equipment.
08	Yes/No	Unit could decontaminate human remains.

ART 6.9.5.4 PROTECT AGAINST EXPOSURE AND EFFECTS OF HIGH-YIELD EXPLOSIVES

6-128. Protecting forces and population from exposure and effects related to high-yield explosives. ART 6.9.5.4 includes the application of guidelines for operations involving possible high-yield explosives and their residues, implementation and maintenance of safety requirements, the application of personal protective equipment, marking of explosives materiel and hazard locations, issuance of appropriate protective material, and the implementation of movement controls. Involves ensuring all personnel operating in the hazardous zone are qualified to conduct operations involving high-yield explosives and are equipped with protective equipment. Includes protection against secondary hazards. (FM 3-11) (USACRBNS)

No.	Scale	Measure
01	Yes/No	The high-yield explosives hazardous condition was continually reassessed.
02	Yes/No	Processes existed to designate the required level of protection.
03	Yes/No	Shelter-in-place plans were implemented for appropriate affected populations.
04	Yes/No	Evacuation plans were implemented for appropriate affected populations.
05	Yes/No	Protective equipment was available for identified secondary hazards.
06	Yes/No	Guidance was implemented regarding the provisioning of Department of Defense (DOD) protective materiel to non-DOD entities.
07	Yes/No	Explosive environment safe personal protective equipment was available.
08	Yes/No	Radio frequency exclusion zone was established.
09	Yes/No	Bomb suit protected personnel from chemical, biological, and radiological threat.
10	Yes/No	Robotics were available to reduce human exposure to hazards.
11	Yes/No	Stand-off distances were established.
12	Yes/No	Capability existed to execute radio frequency jamming.
13	Yes/No	Worker safety and health risk assessments were performed.
14	Time	That on-hand protective resources can sustain operations.
15	Percent	Of forces and responders in hazardous areas equipped with appropriate personal protective equipment.
16	Percent	Of military working animals issued personal protective equipment.

No.	Scale	Measure
17	Percent	Of population in the hazardous zone successfully evacuated.
18	Percent	Of potentially affected population protected.

ART 6.9.5.5 CONDUCT TACTICAL CHEMICAL, BIOLOGICAL, RADIOLOGICAL, AND NUCLEAR CONSEQUENCE MANAGEMENT CRISIS ACTION PLANNING

6-129. Applies and adapts contingency plans and procedures to determine forces and capabilities that are required to conduct tasked activities and operations in support of chemical, biological, radiological, and nuclear consequence management. Integrates attached forces, both Department of Defense (DOD) and non-DOD, and defines responsibilities and roles. Specifies subordinate unit tasks and activities to accomplish the mission. Includes contingency planning for potential circumstances. Also includes intelligence information and data provided by intelligence agencies, the meteorological and oceanographic community, other resources, and collaborating with partner entities. (FM 3-11.21) (USACRBNS)

No.	Scale	Measure
01	Yes/No	Procedures adequately addressed weapons of mass destruction (WMD) consequence management crisis action planning at the tactical level.
02	Yes/No	Procedures adequately addressed planning considerations for use of dual-purpose units during WMD consequence management.
03	Yes/No	Procedures adequately addressed planning considerations for use of total-force units during WMD consequence management.
04	Yes/No	Unit provided planning guidance in areas of noninteroperability between participating responding units and activities.
05	Yes/No	Units integrated local incident response planners.
06	Yes/No	Planning procedures were consistent with the National Response Plan; National Incident Management System; and appropriate state, local, and tribal procedures for domestic incidents.
07	Yes/No	Staff met National Incident Management System incident command structure requirements to perform core planning functions.
08	Yes/No	Staff effectively anticipated resource and operational requirements to support anticipated taskings.
09	Yes/No	Unit tracked activities and conducted contingency planning for potential WMD consequence management circumstances.
10	Yes/No	Staffs effectively tracked activities and planned for multiple WMD decontamination operations.
11	Yes/No	Staffs effectively tracked activities and planned for multiple WMD victim and casualty search, rescue, and extraction operations.
12	Yes/No	Staffs effectively tracked activities and planned for temporary housing, processing operations, and evacuation of the affected population.
13	Yes/No	Staffs effectively tracked activities and planned for multiple WMD reconnaissance operations.
14	Yes/No	All planning staffs had adequate WMD subject matter expertise.

ART 6.9.5.6 ASSESS THE CHEMICAL, BIOLOGICAL, RADIOLOGICAL, NUCLEAR, AND HIGH-YIELD EXPLOSIVES OPERATIONAL ENVIRONMENT

6-130. Assess and characterize the operational area, including the integration of information and data, to determine the location and source of chemical, biological, radiological, nuclear, and high-yield explosives and secondary hazards associated with the task. Includes the application of intelligence, surveillance, and reconnaissance (ISR) assets to determine the incident zone and to perform hazard detection. identification, and quantification. (FM 3-11.21) (USACRBNS)

No.	Scale	Measure
01	Yes/No	Unit assessed and characterized the weapons of mass destruction (WMD) threat in the operational area.
02	Yes/No	Unit assessed potential implications and impacts to support of WMD consequence management in the operational area.
03	Yes/No	Tactical intelligence collection, analysis, and dissemination processes were in place.
04	Yes/No	Unit collaborated and shared tactical situational awareness.
05	Yes/No	Unit disseminated information in near-real time.
06	Yes/No	Unit assessed support levels of selected tactical activities.
07	Yes/No	Tactical WMD intelligence guidance identified information requirements and processes.
08	Yes/No	Staff assessed the WMD operations environment.
09	Yes/No	WMD hazard predictive modeling information was available to the commander.
10	Yes/No	Staffs assessed and characterized the impact of WMD hazard on tactical consequence management operations.
11	Yes/No	Staff ISR asset assessed WMD and secondary hazards.
12	Yes/No	Unit processed WMD threat intelligence rapidly to meet needs of commanders.
13	Yes/No	WMD information was disseminated in near-real time.
14	Yes/No	WMD information was available to all units and teams in the operational area.
15	Yes/No	Unit checked the atmosphere for WMD hazards.
16	Yes/No	Unit checked surfaces for WMD hazards.
17	Yes/No	Unit checked water for WMD hazards.
18	Yes/No	Unit checked the presence WMD hazards on humans, animals, and human remains.
19	Yes/No	Dual-purpose units were trained and prepared to conduct WMD reconnaissance missions.

ART 6.9.5.7 CONDUCT TACTICAL COMMAND AND CONTROL OF CHEMICAL, BIOLOGICAL, RADIOLOGICAL, NUCLEAR, AND HIGH-YIELD EXPLOSIVES CONSEQUENCE MANAGEMENT

6-131.   Providing incident zone command and control (C2) of chemical, biological, radiological, nuclear, and high-yield explosives consequence management support operations and activities includes providing direction, guidance, oversight, and management of forces executing the mission. Includes the integration of networked detectors to monitor the weapons of mass destruction (WMD) incident zone. Encompasses the establishment and operation of a WMD consequence management response C2 structure that facilitates WMD consequence management operations and activities and is interoperable with participating forces. (FM 3-11.21) (USACRBNS)

No.	Scale	Measure
01	Yes/No	Unit provided C2 of WMD consequence management operations for Department of Defense (DOD) and non-DOD forces.
02	Yes/No	Procedures existed to provide guidance for WMD consequence management operations.
03	Yes/No	Unit synchronized C2 activities with other WMD C2 operations and higher headquarters for information sharing and decision support.
04	Yes/No	Processes existed to integrate networked detectors to monitor the WMD incident zone.
05	Yes/No	Procedures existed to establish and operate a WMD consequence management response C2 structure.
06	Yes/No	WMD consequence management C2 operations and activities were interoperable with participating forces.
07	Yes/No	C2 structure was established for WMD consequence management response.
08	Yes/No	Staff provided accurate WMD consequence management information to decisionmakers.
09	Yes/No	Staff provided appropriate WMD consequence management recommendations to decisionmakers.
10	Yes/No	Mechanism existed to capture lessons learned.

ART 6.9.5.8 PERFORM CHEMICAL, BIOLOGICAL, RADIOLOGICAL, NUCLEAR, AND HIGH-YIELD EXPLOSIVES INCIDENT AND HAZARD RISK COMMUNICATIONS

6-132. Conduct and maintain communications regarding the weapons of mass destruction (WMD) incident and all identified or projected hazards. Includes establishing a mutually supporting tactical and field-level WMD warning and reporting system to collect and disseminate prompt, accurate WMD information. This information goes to the public, joint forces, host nations, and intergovernmental, nongovernmental, and private volunteer organizations. Disseminate appropriate intelligence information regarding continuing or potential WMD and secondary hazards. (FM 3-11.21) (USACRBNS)

No.	Scale	Measure
01	Yes/No	Unit effectively managed the WMD incident information at the tactical level.
02	Yes/No	Standardized WMD hazard alarm, warning, and reporting formats and procedures were interoperable with interagency partners.
03	Yes/No	WMD incident data were integrated into the common operational picture.
04	Yes/No	Integrated detection network provided automated warning and reporting of the WMD hazard.
05	Yes/No	Mechanism existed to rapidly disseminate WMD incident threat intelligence.
06	Yes/No	Information on the operational impact of WMD events was integrated into the common operational picture.
07	Yes/No	Staff operated a mutually supporting tactical- and field-level warning and reporting system.
08	Yes/No	Staff briefed participating forces on risks, hazards, and appropriate safety measures.
09	Yes/No	Staff debriefed participants upon leaving the hazard zone to collect information and identify potential secondary hazards.
10	Yes/No	Procedures existed to conduct communications regarding the WMD incident and identified and projected hazards.
11	Yes/No	Procedures existed to establish mutually supporting tactical- and field-level WMD reporting system that disseminates appropriate WMD consequence management intelligence and information, including potential and secondary hazards.
12	Yes/No	Unit promptly disseminated hazard information to the affected population.
13	Yes/No	Unit communicated hazards at daily briefs, situation reports, and debriefs.
14	Yes/No	Unit communicated hazards that developed during the operational cycle to all key personnel at the hazard site.
15	Yes/No	Process existed to establish nuclear, biological, and chemical warning and reporting.

ART 6.9.5.9 ESTABLISH AND MAINTAIN ACCESS AND EGRESS CONTROLS AND HAZARD ZONE PERIMETER

6-133. Implement measures to contain and control contamination, include marking and identifying hazard areas, maintaining clear access to the incident site, and controlling access to and from hazard areas. Encompass the application of directed movement controls of the affected population and the enforcement of required isolation or quarantine restrictions. (FM 3-11.21) (USACRBNS)

No.	Scale	Measure
01	Yes/No	Procedures existed to contain and control contamination.
02	Yes/No	Procedures existed to standardize the marking of hazard areas.
03	Yes/No	Procedures existed to direct the identifying of hazard areas.
04	Yes/No	Procedures existed to direct uncontaminated access to the incident site.
05	Yes/No	Procedures existed to control access to and from the hazard areas.
06	Yes/No	Procedures existed to provide movement control of the affected population.
07	Yes/No	Measures existed to enforce isolation and quarantine events.
08	Yes/No	Unit conducted crowd control.

ART 6.9.5.10 CONDUCT CHEMICAL, BIOLOGICAL, RADIOLOGICAL, NUCLEAR, AND HIGH-YIELD EXPLOSIVES VICTIM AND CASUALTY SEARCH, RESCUE, AND EXTRACTION

6-134. Find, rescue, and extract victims and casualties from the hazardous environment. Includes support for people with special needs (physical or mental disabilities) requiring medical attention or personal care beyond basic first aid, managing and conducting immediate lifesaving measures, evacuating casualties to hot-zone egress points and into decontamination site, and evacuating noncontaminated victims to appropriate medical or mass care facilities. (FM 3-11.21) (USACRBNS)

No.	Scale	Measure
01	Yes/No	Procedures existed to locate, rescue, and extract victims and casualties in a hazardous environment.
02	Yes/No	Procedures adequately addressed weapons of mass destruction victim and casualty search, rescue, and extraction operations.
03	Yes/No	Procedures coordinated pet rescue, care, and handling.
04	Yes/No	Search and rescue centers were established and operational.
05	Yes/No	Support requirements for special-needs individuals were addressed.
06	Yes/No	Measures existed to protect weapons of mass destruction victims and casualties from the effects of weapons of mass destruction and secondary hazards while in the hazardous zone.
07	Yes/No	Procedures and guidance on animal rescue were disseminated to operating personnel.
08	Yes/No	Personnel protected victims and casualties from further contamination, exposure, or injury.
09	Yes/No	Personnel provided medical treatment en route.

ART 6.9.5.11 PROVIDE TEMPORARY HOUSING, CONDUCT PROCESSING OPERATIONS, AND EVACUATE AFFECTED POPULATION

6-135. Conduct operations to temporarily house, clothe, feed, and care for affected populations until evacuation or release; process affected populations for evacuation; and manage evacuation operations. Encompasses separate operations for potentially contaminated or infectious populations. Apply quarantine and isolation measures. ART 6.9.5.11 includes activities identifying support requirements, including supporting transport capabilities. Includes coordination with medical authorities for the screening of incoming persons, the provision of medical care, and the conduct of evaluations for the safety and health of the housed population. Addresses the establishment and operation of processing centers to receive and process the affected population, including screening for medical needs; collect relevant information; and facilitate evacuation. (FM 3-11.21) (USACRBNS)

No.	Scale	Measure
01	Yes/No	Unit could temporarily house, clothe, and feed affected populations until evacuation or release.
02	Yes/No	Unit processed affected populations for evacuation (including special-needs populations).
03	Yes/No	Processes existed to manage evacuation operations.
04	Yes/No	Unit operated evacuee processing centers in support of evacuation operations.
05	Yes/No	Procedures supported evacuation of affected population from incident site to reception and processing centers.
06	Yes/No	Quarantine and isolation measures were disseminated to operating personnel.
07	Yes/No	Unit could support operations with temporary utilities.
08	Yes/No	Unit could support establishment with temporary feeding operations.
09	Yes/No	Unit coordinated with medical authorities to evaluate the health of the housed population.
10	Yes/No	Unit collected relevant information from evacuee population.
11	Yes/No	Unit processed affected population for relocation.
12	Yes/No	Unit identified and tracked members of the affected population.

No.	Scale	Measure
13	Yes/No	Unit allocated evacuation transportation.
14	Yes/No	Unit identified and reunited family members and caretakers.
15	Yes/No	Unit retained a record of activities associated with affected individuals.
16	Yes/No	Unit provided temporary housing and conducted processing and evacuation operations of affected populations.
17	Yes/No	Unit conducted crowd control.

ART 6.9.5.12 CONDUCT DECONTAMINATION OPERATIONS

6-136. Conduct decontamination operations associated with chemical, biological, radiological, nuclear, and high-yield explosives consequence management. It includes the performance of decontamination operations of civilian and military personnel, equipment, and assets. Encompasses both ambulatory and nonambulatory personnel, including casualties. Decontamination may include a range of activities such as removal, sealing, weathering, neutralization, and other means of mitigation or eliminating contamination. (FM 3-11) (USACRBNS)

No.	Scale	Measure
01	Yes/No	Unit could conduct weapons of mass destruction consequence management decontamination operations.
02	Yes/No	Unit could decontaminate personnel and military working animals.
03	Yes/No	Unit could decontaminate equipment and assets.
04	Yes/No	Unit could decontaminate critical and key infrastructure.
05	Yes/No	Unit could decontaminate local hospitals.
06	Yes/No	Unit could decontaminate operational terrain.
07	Yes/No	Unit could establish joint decontamination and egress sites.
08	Yes/No	Unit could control contaminated waste and runoff.

ART 6.9.5.13 ISOLATE, QUARANTINE, AND MANAGE POTENTIALLY CONTAMINATED OR INFECTIOUS HUMANS AND ANIMALS

6-137. Contain the spread of disease by isolating or quarantining potentially contagious humans and animals. Includes screening individuals and animals for exposure patterns and symptoms, establishing isolation and quarantine enforcement requirements, and implementing movement controls from isolation and quarantine areas. Includes feeding, clothing, and providing shelter to isolated and quarantined individuals. Includes providing response personnel working in isolation areas with personal protection to prevent disease transmission. (FM 3-11.21) (USACRBNS)

No.	Scale	Measure
01	Yes/No	Isolation and quarantine plans and procedures were implemented.
02	Yes/No	Quarantine and isolation guidelines were strictly enforced.
03	Yes/No	Unit enforced isolation and quarantine operations.
04	Yes/No	Unit conducted crowd control.
05	Percent	Of units capable of conducting operations requiring the isolation and quarantining of contaminated or infectious human populations.
06	Percent	Of units capable of conducting operations requiring the isolation and quarantining of contaminated or infectious animal populations.

ART 6.9.6 CONDUCT CHEMICAL, BIOLOGICAL, RADIOLOGICAL, AND NUCLEAR INTERDICTION OPERATIONS

6-138. Integrate intelligence, surveillance, and reconnaissance (ISR) with chemical, biological, radiological, and nuclear (CBRN) situation. Integrate the CBRN weapons situation into the command, control, communications, and computer for ISR systems in the area of operations (AO). Include processing information from strategic, operational, and tactical sources on CBRN weapon delivery systems, enemy intent, and possible courses of action. Also include the characterization of any toxic industrial materials in the AO, tracking the operational situation for CBRN weapon hazards, directing counterforce responses, alerting active defenses, activating CBRN defense procedures, or initiating consequence management activities. Exploit captured technical data on CBRN weapons and assesses enemy capabilities, intent, and probable courses of action. Recommend friendly actions. Provide links between proactive and passive defenses. This task also includes medical surveillance. (FM 3-11.19) (USACRBNS)

No.	Scale	Measure
01	Yes/No	CBRN warning system was established with all forces in the AO.
02	Yes/No	CBRN warning system was established with all necessary non-Department of Defense organizations in the AO.
03	Time	To provide unambiguous attack warning.
04	Time	To provide accurate attack assessment.
05	Time	Of friendly or neutral forces or noncombatants influenced by collateral effects from friendly attacks on CBRN weapon targets.
06	Time	To establish plan to distribute medicines and medical supplies.
07	Time	To send or receive messages.
08	Percent	Of CBRN weapon capabilities that are detected and identified.
09	Percent	Of intelligence on enemy CBRN weapon systems that prove to be accurate.
10	Percent	Of command and control nodes surviving a CBRN weapons attack.
11	Percent	Of in-place plans and policies for consequence management operations in continental United States (CONUS) and outside CONUS locations.

ART 6.9.7 CONDUCT CHEMICAL, BIOLOGICAL, RADIOLOGICAL, AND NUCLEAR ELIMINATION OPERATIONS

6-139. Combine, sequence, and integrate operations of the unit's assigned, attached, or supporting forces to eliminate weapons of mass destruction (WMD), WMD-related materiel, delivery systems, and related materiel, technologies, and expertise. Conducting elimination operations may include multiple sites of varying sizes with different kinds of WMD. This task therefore involves vertical and horizontal integration of tasks in time and space to include supporting operations. Preparatory activities are coordinated with affected joint, interagency, and multinational organizations. This task may require facilitating the reception of site assessment and exploitation teams and integrating them with tactical units. (FMI 3-90.10) (USACRBNS)

No.	Scale	Measure
01	Yes/No	Forces were properly trained, equipped, and available to conduct WMD elimination missions.
02	Yes/No	Staff processes and expertise were in place to coordinate WMD elimination mission elements.

ART 6.9.7.1 PLAN FOR THE ISOLATION PHASE OF ELIMINATION OPERATIONS

6-140. Develop a plan to isolate a weapons of mass destruction (WMD) site. Isolation involves preventing interaction with and movement of tangible components, as well as physical access to the site. It also means establishing conditions that allow site assessment teams to enter and inspect the site. Plans for the isolation of a WMD site may include coordinating maneuver forces, special operations forces, and intelligence; identifying and locating the site; erecting barriers and establishing perimeters; identifying and using personnel to prevent unwanted use or interference with systems located at a WMD site (such as unauthorized access to computers and networks); using security forces; and transitioning the exploitation phase. (FMI 3-90.10) (USACRBNS)

No.	Scale	Measure
01	Yes/No	Doctrine existed to plan the isolation phase of tactical-level WMD elimination operations.
02	Yes/No	Doctrine existed to plan physical security of sensitive WMD sites.
03	Yes/No	Training and exercises provided feedback on efficacy of the planning process.
04	Yes/No	Security plan provided site security assets, augmentation, communication, reinforcement, and logistics.
05	Yes/No	Security plan defined mission, equipment, time, troops, terrain, movement, stability, and support requirements and also contained an exit strategy.
06	Yes/No	Required WMD intelligence, operational, and technical expertise was available during the synchronization planning process.

ART 6.9.7.2 PLAN FOR THE EXPLOITATION PHASE OF ELIMINATION OPERATIONS

6-141. Develop a plan to exploit a weapons of mass destruction (WMD) site. This task includes plans for collection and initial characterization of a site's WMD materiel, weapons, equipment, personnel, data, and infrastructure. This task also includes planning for the limited destruction, rendering safe, and dismantling of WMD materiel or weapons that may be required to facilitate exploitation. Plans include the possibility for more explicit and detailed exploitation such as the enhanced security of documents, electronic media, personnel, materials, weapons, and equipment. Furthermore, plans should provide for the gathering of forensic evidence leading to attribution and prosecution. (FMI 3-90.10) (USACRBNS)

No.	Scale	Measure
01	Yes/No	Doctrine existed for the planning of the exploitation phase of tactical-level WMD elimination operations.
02	Yes/No	Required information was available to aid in developing all parts of the exploitation.
03	Yes/No	A process existed to assess potential tactical WMD elimination effects.
04	Yes/No	Required personnel with WMD intelligence, operational, and technical expertise was available during the planning process.
05	Yes/No	Training and exercises provided feedback on efficacy of tactical planning for the mitigation of WMD hazards.

ART 6.9.7.3 PLAN FOR THE DISPOSITION PHASE OF ELIMINATION OPERATIONS

6-142. Develop a plan for the disposition of weapons of mass destruction (WMD) materiel and weapons. This task includes planning for the employment of engineers, explosive ordnance disposal units, and personnel with requisite WMD-related skills. Plans may include not only physical destruction of WMD materiel and weapons, but also means to package items for the secure transportation and the transfer of formal custody to permanent storage or other sites for final disposition. (FMI 3-90.10) (USACRBNS)

No.	Scale	Measure
01	Yes/No	Doctrine existed to plan the exploitation phase of tactical-level WMD elimination operations.
02	Yes/No	Training and exercises provided feedback on efficacy of tactical planning for the disposition of WMD-related materiel.
03	Yes/No	Doctrine existed to plan for physical security of sensitive WMD sites.
04	Yes/No	Required personnel with WMD intelligence, operational, and technical expertise was available during the planning process.
05	Yes/No	A process existed to assess potential tactical WMD elimination effects.

ART 6.9.7.4 PLAN FOR THE MONITORING AND REDIRECTION PHASE OF ELIMINATION OPERATIONS

6-143.   Develop a plan for the monitoring or redirection phase of elimination operations. This task focuses on establishing and maintaining a permissive tactical environment in which monitoring and redirection can be carried out. Therefore, this task includes the following subtasks: planning to facilitate and maintain positive surveillance of former or potential weapons of mass destruction (WMD) programs, planning to facilitate the receiving and sustaining of specialist or expert personnel, and planning for the continued maintenance of perimeters and barriers by means of which interaction with the WMD site can be controlled. (FMI 3-90.10) (USACRBNS)

No.	Scale	Measure
01	Yes/No	Doctrine existed for the planning of the monitoring or redirection phase of tactical-level WMD elimination operations.
02	Yes/No	Training and exercises provided feedback on efficacy of the planning process.
03	Yes/No	Appropriate personnel were available for the planning of redirection and monitoring of WMD personnel, sites, and facilities.
04	Yes/No	Appropriate legal means were in place to allow for redirection and monitoring of WMD personnel, sites, and facilities.

ART 6.9.7.5 PLAN FOR MITIGATION OF POTENTIAL TACTICAL WEAPONS OF MASS DESTRUCTION ELIMINATION COLLATERAL EFFECTS

6-144.   Plan for the mitigation of potential collateral damage and effects during tactical weapons of mass destruction (WMD) elimination operations. Probable WMD sites are reviewed for the possibility of collateral damage, personnel and civilian casualties, environmental impact, and political sensitivity. It is important that planners have access to accurate data that provide the range of potential WMD collateral effects that might be encountered in the WMD elimination mission. Additionally, this task includes consideration of individual protective equipment requirements and responses (in accordance with doctrine and standing operating procedures) to possible contingencies. Planners must also ensure a methodology is in place to allow recommendations and options of appropriate resources and techniques—to include all available WMD intelligence, operational, and technical expertise—are available during the planning process. (FMI 3-90.10) (USACRBNS)

No.	Scale	Measure
01	Yes/No	A process existed to assess potential tactical WMD elimination collateral effects.
02	Yes/No	Required WMD intelligence, operational, and technical expertise was available during the planning process.
03	Yes/No	Training and exercises provided feedback on efficacy of tactical planning for the mitigation of WMD hazards.

ART 6.9.7.6 EXPLOIT DATA, INFORMATION, AND MATERIALS OBTAINED DURING WEAPONS OF MASS DESTRUCTION ELIMINATION OPERATIONS

6-145.   Unit personnel must be prepared to identify and report critical intelligence from the information gained during weapons of mass destruction (WMD) elimination operations. The goal is to rapidly collect, identify, and report intelligence related to identified intelligence priorities. This task may include a wide range of WMD proliferation information related to technical expertise and personnel, programs, facilities, infrastructure, materials, agents, delivery means, and the procurement, transfer of WMD-related materiel. ART 6.9.7.6 involves collecting and filtering relevant information and reporting or forwarding raw intelligence to higher headquarters in accordance with established intelligence priorities and reporting procedures. This task also includes obtaining information and data from all sources that provide information about the WMD threat situation in the joint operations area. This task requires guidance on priority intelligence requirements from higher headquarters, as well as personnel at the tactical level with the ability to identify critical information and data quickly that support these requirements. (FMI 3-90.10) (USACRBNS)

No.	Scale	Measure
01	Yes/No	WMD elimination tactical units were aware of and understood all established priority intelligence requirements.
02	Yes/No	Sufficient qualified personnel were available to collect and filter information and data to allow rapid reporting of time-sensitive priority intelligence.
03	Yes/No	WMD elimination tactical units were capable of rapidly converting captured documents to electronic files suitable for transmission to the joint task force and the intelligence community.
04	Yes/No	Procedures and communications channels were available to report intelligence gained during WMD elimination operations quickly.

ART 6.9.7.7 EXERCISE COMMAND AND CONTROL IN THE PREPARATION FOR AND CONDUCT OF WEAPONS OF MASS DESTRUCTION ELIMINATION OPERATIONS

6-146. Provide tactical command and control (C2) of assigned forces for weapons of mass destruction (WMD) elimination operations as directed by the joint task force. Tactical units must be capable of exercising C2 with or over multinational forces supporting WMD elimination operations. C2 requires direct communication and coordination with assigned, supporting, and supported forces and other friendly forces operating in and around the area of operations. Command and control includes maintaining situational awareness of the operational environment and situation, mission taskings, and status of assigned, supporting, and supported forces during the planning, execution, and redeployment phases of WMD elimination operations. (FMI 3-90.10) (USACRBNS)

No.	Scale	Measure
01	Yes/No	Command and support relationships and reporting requirements between tactical commands, multinational forces, and supporting agencies were clearly defined.
02	Yes/No	Procedures were established to ensure the timely dissemination of intelligence to lower, lateral, and higher echelons, including multinational forces.
03	Yes/No	Communication channels and capabilities can support timely coordination and execution of WMD elimination operations.
04	Yes/No	Tactical elimination units can maintain situational awareness via access to the common operational picture.

ART 6.9.7.8 CONDUCT TACTICAL-LEVEL RISK MANAGEMENT FOR WEAPONS OF MASS DESTRUCTION ELIMINATION OPERATIONS

6-147. Tactical commanders and leaders apply applicable risk management tools and procedures to continuously identify and assess operational risks associated with weapons of mass destruction (WMD) elimination operations and develop appropriate countermeasures, direct appropriate courses of action for assigned forces, and monitor and adjust those courses of action. (FMI 3-90.10) (USACRBNS)

No.	Scale	Measure
01	Yes/No	WMD elimination tactical unit leaders had sufficient training and experience to manage the risks associated with WMD elimination operations.
02	Yes/No	Risk management for units tasked to isolate WMD elimination sites was supported by WMD elimination forces.
03	Yes/No	Unit risk mitigation procedures addressed the risks associated with WMD elimination operations.
04	Yes/No	WMD elimination tactical command staffs continuously applied the risk management process during WMD elimination operations.

ART 6.9.7.9 MANAGE ASSIGNED PERSONNEL AND RESOURCES TO ENSURE EFFECTIVE AND EFFICIENT WEAPONS OF MASS DESTRUCTION ELIMINATION MISSION ACCOMPLISHMENT

6-148.   Determine best use of assigned resources to complete required elimination mission tasks. This task involves allocating unit resources to perform weapons of mass destruction (WMD) elimination tasks at preidentified (potential) WMD sites and reallocate as needed to address changes in the mission, including discovery of new (potential) WMD sites during the course of operations. This task includes providing and coordinating support to assigned forces during all phases of elimination operations, from isolation through exploitation and disposition. (FMI 3-90.10) (USACRBNS)

No.	Scale	Measure
01	Yes/No	Tactical command staffs were trained to determine best use of available resources to complete assigned elimination mission tasks.
02	Yes/No	Tactical command staffs were organized and manned to manage and support assigned units and forces during all phases of elimination operations.
03	Yes/No	Tactical command staffs could rapidly adjust allocation of assigned forces and unit resources to address changes effectively in the mission and the operational environment.

ART 6.9.7.10 CONDUCT WEAPONS OF MASS DESTRUCTION ELIMINATION MISSION TRAINING AND REHEARSAL

6-149.   Train units, staffs, leaders, and personnel for tactical weapons of mass destruction (WMD) elimination operations. In stability operations, this activity includes training assistance for friendly nations and groups. This task includes conducting rehearsals for tactical-level operations or commander and staff rehearsals within the headquarters. It also includes making sure dedicated information technology and learning tools are available. Conduct training and rehearsals by unit staff cells under conditions that simulate those expected in actual operations. Standards and doctrine should exist to facilitate the conduct of training. Feedback will be obtained to ensure continued improvement and development. (FMI 3-90.10) (USACRBNS)

No.	Scale	Measure
01	Yes/No	WMD elimination organization was tasked to conduct training and rehearsals.
02	Yes/No	Unit training and rehearsals were conducted under simulated WMD-related conditions.
03	Yes/No	Adequate training support such as dedicated information technology was available.
04	Yes/No	Personnel task organized to support the operation were trained to perform their specific WMD elimination task.

ART 6.9.7.11 CONDUCT IMMEDIATE WEAPONS OF MASS DESTRUCTION INCIDENT ASSESSMENT FOR WEAPONS OF MASS DESTRUCTION ELIMINATION MISSION FORCE

6-150.   Conduct, for purposes of determining the safe and responsible mitigation of any possible effects, an immediate incident assessment of damage to or effects on the elimination force from released weapons of mass destruction (WMD)-related hazards. This task requires that personnel participating in tactical WMD elimination operations be trained to conduct an assessment and to be well versed in standing operating procedures. Organizations tasked to support elimination operations require good external communications; therefore, units need to have good contact with leaders and may require access to situation-critical data such as local meteorological data. (FMI 3-90.10) (USACRBNS)

No.	Scale	Measure
01	Yes/No	Capabilities and procedures existed to conduct an immediate assessment.
02	Yes/No	There were sufficient doctrine for effective assessments.
03	Yes/No	All necessary communication was available for units to communicate with leaders.
04	Yes/No	Accurate local meteorological data were available.
05	Yes/No	Training and exercises provided feedback on the efficacy of immediate WMD incident assessments.

ART 6.9.7.12 CONDUCT SECURITY SUPPORT FOR WEAPONS OF MASS DESTRUCTION ELIMINATION MISSION

6-151. Conduct security support for weapons of mass destruction (WMD) elimination operations. This task addresses the physical security of a WMD site. Physical security involves preventing both penetration of the site and the theft or smuggling of materials from the site. Security support includes preservation of the site to prevent accidental release of WMD or WMD materiel. It also includes maintaining an environment that allows safely carrying out assessment and collection activities. Security support may include both maneuver and nonmaneuver forces such as military police. (FMI 3-90.10) (USACRBNS)

No.	Scale	Measure
01	Yes/No	Security forces were available and effectively provided force protection.
02	Yes/No	Forces were trained and equipped to provide strict access control to suspected WMD facilities and spaces.
03	Yes/No	Security forces had necessary individual protective equipment.
04	Time	For security forces to set up strict access control to suspected WMD facilities and spaces.

ART 6.9.7.13 SEARCH FACILITIES AND SPACES FOR WEAPONS OF MASS DESTRUCTION MATERIEL

6-152. Thoroughly search and inspect facilities at a weapons of mass destruction (WMD) site. This task includes the designation, description, and collection of WMD materiel for later assessment and disposition. Also included in this task is the interrogation of personnel found at the site. Designation, description, collection, and interrogation are done to facilitate exploitation and to ensure that no WMD materiel or intelligence is overlooked. This task therefore requires adequate time and personnel trained to identify items and facilities that are found at a site. It also requires personnel who can interrogate site personnel. The legal facets of searching facilities must be determined as a prerequisite to this task. (FMI 3-90.10) (USACRBNS)

No.	Scale	Measure
01	Yes/No	Personnel searched all types of facilities and spaces successfully.
02	Yes/No	Plans existed for searching facilities and spaces during WMD elimination mission.
03	Yes/No	Legal facets of searching facilities and spaces during WMD elimination existed at mission outset.
04	Yes/No	Operational tempo degradation and conflict existed due to time necessary to search facilities and spaces.
05	Yes/No	Adequate number of personnel was trained to search and inspect a WMD site and interrogate personnel.

ART 6.9.7.14 DETECT WEAPONS OF MASS DESTRUCTION (WMD), WMD-RELATED MATERIEL, AND DELIVERY SYSTEMS OR TECHNOLOGIES DURING ELIMINATION MISSION

6-153. Detect suspected WMD or WMD-related materiel for the purposes of further identification and characterization. This task will involve both organic and specialized personnel who have experience in explosive ordnance disposal, use of bomb dogs, determination of sample areas, sample collecting procedures, use of radiological detection equipment, and use of chemical, biological, and radiological detection equipment. This task may also require reachback capability to technical experts to assist in detection. (FMI 3-90.10) (USACRBNS)

No.	Scale	Measure
01	Yes/No	Unit detected all chemical warfare agents.
02	Yes/No	Unit detected all high-priority toxic industrial chemicals.
03	Yes/No	Unit detected all biological warfare agents.
04	Yes/No	Unit detected all types of radiation.

No.	Scale	Measure
05	Yes/No	Unit detected all types and components of WMD-related materiel, delivery systems, and technologies.
06	Yes/No	Personnel with appropriate detection skills were available.
07	Yes/No	Reachback communications and procedures were coordinated, in place, and exercised.

ART 6.9.7.15 CHARACTERIZE WEAPONS OF MASS DESTRUCTION-RELATED MATERIEL, DELIVERY SYSTEMS, OR TECHNOLOGIES DURING ELIMINATION MISSION

6-154. Identify and characterize weapons of mass destruction (WMD), WMD-related materiel, delivery systems, and related materiel and technologies discovered or detected during search. After establishing the general type of site (such as a production, storage, or research site), characterization begins by determining the specific kinds of WMD materiel present. Such characterization also determines whether the site and material can be useful for non-WMD purposes. Characterization will also include collecting forensic evidence. This task will require the presence of personnel trained in characterization of different types of WMD materiel and their delivery systems and evidence collection. It will require the intelligence community, commercial, and private-sector nonprofit support. It may also require the support and cooperation of foreign nations. (FMI 3-90.10) (USACRBNS)

No.	Scale	Measure
01	Yes/No	Unit identified and characterized all chemical warfare agents.
02	Yes/No	Unit identified and characterized all radiological materials.
03	Yes/No	Unit detected all biological warfare agents.
04	Yes/No	Unit identified and characterized all types and components of WMD-related material, delivery systems, and technologies.
05	Yes/No	Reachback existed for characterization of suspected WMD-related materiel, delivery systems, and technologies.
06	Yes/No	Personnel with appropriate detection skills were available.

ART 6.9.7.16 DEFEAT WEAPONS OF MASS DESTRUCTION SYSTEMS

6-155. Defeat a weapons of mass destruction (WMD) system by rendering safe or inoperable explosives, triggering systems, guidance systems, and other WMD-related materiel. This task presupposes previous identification and characterization of WMD sites and materiel. All types of WMD-related systems on the site must be capable of being defeated. This task includes providing means and procedures to report chemical, biological, radiological, nuclear, and high-yield explosives (CBRNE) incidents over a specified period. Also included are means and procedures to ensure collateral forwarding to intelligence agencies. This task requires enough time to identify safety requirements, coordinate with the reporting agency for site support assistance (engineer, medical, security, and transportation), and clear munitions and improvised devices. (FMI 3-90.10) (USACRBNS)

No.	Scale	Measure
01	Yes/No	All types of WMD-related delivery systems or weapon triggering mechanisms were defeated.
02	Yes/No	Personnel identified CBRNE and conventional ordnance incidents that can be responded to within a given period.
03	Yes/No	Personnel determined munitions for which intelligence information (what, when, where, how delivered, and type).
04	Yes/No	Personnel identified safety requirements and considerations.
05	Yes/No	Site support assistance—such as engineer, medical, security, and transportation—was coordinated with reporting agency.
06	Yes/No	Unit cleared munitions and improvised devices.

No.	Scale	Measure
07	Yes/No	Facilities and delivery systems were disabled or eliminated. Includes production facilities and equipment, research and development facilities and equipment, parts fabrication and equipment, chemical processing equipment and materials, weapons delivery systems and facilities to build delivery systems, and biological agent material development and equipment.
08	Yes/No	Explosive ordnance disposal support was identified (including contact procedures) and practiced.
09	Yes/No	Reporting requirements and formats were identified and in place.

ART 6.9.7.17 CONTAIN WEAPONS OF MASS DESTRUCTION-RELATED MATERIEL FOR FINAL DISPOSITION TO INCLUDE DEFEAT, NEUTRALIZATION, STORAGE, OR TRANSPORT

6-156. Contain weapons of mass destruction (WMD)-related material for final disposition. This task requires the use of protective packaging for the safe containment and transportation of WMD-related materiel. This task requires accessible storage facilities to safely store materiel awaiting final disposition. Arrangements and agreements must be made for transportation and maintaining control of WMD-related materiel across various regions. This task should include, where necessary, interagency, contractor, and foreign nation approvals, and technical advice and skills. Finally, this task requires obtaining trained personnel who can perform the functions listed above. (FMI 3-90.10) (USACRBNS)

No.	Scale	Measure
01	Yes/No	All chemical warfare agents were safely contained for further disposition.
02	Yes/No	All biological warfare agents were safely contained for further disposition.
03	Yes/No	Radiological materials were safely contained for further disposition.
04	Yes/No	All types and components of WMD-related materiel, delivery systems, and technologies were contained for further disposition.
05	Yes/No	Interagency, contractor, and foreign nation support were available.
06	Yes/No	Personnel required to perform containment and storage functions were trained and available.
07	Yes/No	Labeling and tracking requirements were identified and in place.

ART 6.9.7.18 CONDUCT NEUTRALIZATION OF SUSPECT WEAPONS OF MASS DESTRUCTION-RELATED AGENT OR MATERIEL

6-157. Render weapons of mass destruction (WMD)-related agents, materials, precursors, and residual products ineffective, unusable, or harmless. This task presupposes previous identification and characterization of WMD sites and materials, and can include safe storage. All types of chemical and biological agents determined to be at the site must be capable of being neutralized. Part of this task is to carry out the neutralization process in a way that minimizes the release of WMD-related hazards into the environment as well as to minimize exposure to personnel, particularly exposure to or release that can cause casualties or fatalities. Adequate time must be available for safe neutralization. (FMI 3-90.10) (USACRBNS)

No.	Scale	Measure
01	Yes/No	Unit neutralized all types of chemical and biological agents and materials.
02	Yes/No	Available neutralization processes minimized release of WMD-related hazards into the environment.
03	Yes/No	Available neutralization processes were conducted safely to prevent personnel casualties and fatalities.
04	Yes/No	There was time to safely neutralize WMD agent without degrading or conflicting with operational tempo.
05	Percent	Of found WMD agents that are safely neutralized.

ART 6.9.7.19 Store Weapons of Mass Destruction-Related Materiel for Final Disposition

6-158.   Provide safe storage for weapons of mass destruction (WMD)-related materiel. This task includes WMD-related agents, materials, precursors, and residual products. This task presupposes previous identification and characterization of WMD sites and materials. All types of WMD-related materials on the site must be capable of being safely stored. Trained personnel must be available for controlling stored materials. The task should include, where necessary, interagency, contractor, and foreign nation approvals and technical advice and skills. (FMI 3-90.10) (USACRBNS)

No.	Scale	Measure
01	Yes/No	Chemical agents and high-priority toxic industrial chemicals were stored for further disposition.
02	Yes/No	Biological warfare agents were stored for further disposition.
03	Yes/No	Radiological materials were stored for further disposition.
04	Yes/No	Trained personnel for storing materials were available.
05	Yes/No	Interagency, contractor, and foreign nation support were available.
06	Yes/No	Labeling and tracking requirements and formats were identified and in place.

ART 6.9.7.20 Transport Weapons of Mass Destruction-Related Materiel for Final Disposition

6-159.   Provide safe transportation for weapons of mass destruction (WMD)-related agents, materials, precursors, and residual products while awaiting final disposition. (FMI 3-90.10) (USACRBNS)

No.	Scale	Measure
01	Yes/No	Protective packages and containers were available to transport contained WMD-related materials safely.
02	Yes/No	National support and agreements were in place for transporting across spaces and maintaining control of materials.
03	Yes/No	Platforms certified for transportation of hazardous materials were available.
04	Yes/No	Trained personnel for transporting materials were available.
05	Yes/No	Labeling and tracking requirements and formats were identified and in place.

ART 6.9.7.21 Gather Forensic Evidence in Support of Elimination Mission

6-160.   Provide safe, efficient, and consistent collection and handling of evidence in support of the weapons of mass destruction elimination mission. Evidence will be used to attribute proliferation activities to culpable state or nonstate actors. This task includes accurately identifying, marking, and transporting samples with care to ensure no contamination is inadvertently spread. (FMI 3-90.10) (USACRBNS)

No.	Scale	Measure
01	Yes/No	Appropriate collection and handling procedures existed.
02	Yes/No	Elimination force included personnel trained in collection and handling procedures.
03	Yes/No	Protective packages and containers were available to safely package and transport materials.

ART 6.9.7.22 Maintain Control of Materiel Related to Weapons of Mass Destruction Elimination Mission

6-161.   Maintain control of hazardous materials used in defeat and neutralization processes and maintain control of hazardous by-products of defeat and neutralization processes. (FMI 3-90.10) (USACRBNS)

No.	Scale	Measure
01	Yes/No	National support and agreements were in place for transporting across spaces and maintaining control of materials.
02	Yes/No	Platforms certified for transportation of hazardous materials were available.
03	Yes/No	Trained personnel for transporting and controlling hazardous materials were available.

ART 6.9.7.23 ESTABLISH TACTICAL CONTAINMENT AND TEMPORARY SAFE STORAGE OF SUSPECT MATERIEL

6-162. Assess the ability to collect, control, and monitor all weapons of mass destruction (WMD)-related agents, precursors, materiel, and by-products retrieved during or resulting from elimination operations. This task includes the ability to oversee and manage temporarily stored agents or materials per theater requirements and guidelines. (FMI 3-90.10) (USACRBNS)

No.	Scale	Measure
01	Yes/No	Units had trained personnel to establish and operate temporary storage sites for captured WMD-related materials.
02	Yes/No	Devices capable of monitoring hazardous WMD-related materials in temporary storage sites were available.

SECTION X – ART 6.10: EMPLOY SAFETY TECHNIQUES

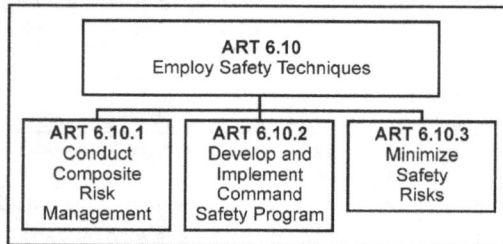

6-163. Safety in the protection warfighting function identifies and assesses hazards to the force and makes recommendations on ways to mitigate those hazards. Responsibility for safety starts with the commander and continues through the chain of command to individuals. All staffs understand and factor into their analysis how their execution recommendations could adversely affect Soldiers. (FM 5-19) (CRC)

ART 6.10.1 CONDUCT COMPOSITE RISK MANAGEMENT

6-164. Identify and control hazards to protect the force and increase the chance of mission accomplishment. Use risk management throughout the conduct (planning, preparing, executing, and assessing) of missions to recommend how to reduce tactical and accidental risk. ART 6.10.1 includes the requirement to establish, communicate, and enforce controls that reduce the risk of tactical and accident hazards. (FM 5-19) (CRC)

No.	Scale	Measure
01	Yes/No	Unit established, communicated, and enforced tactical and accident hazard controls.
02	Yes/No	Probability of successful mission accomplishment increased because of risk management.
03	Yes/No	Casualties and vehicle or building damage reduced because of the use of risk management techniques.
04	Yes/No	Environmental considerations planning and procedures were present and being followed.
05	Time	To employ risk management.

No.	Scale	Measure
06	Time	To communicate controls or changes to force.
07	Percent	Of identified significant risks to mission accomplishment and force protection.
08	Percent	Of identified risks that are determined to be acceptable by the commander.
09	Percent	Of force affected by identified accident hazards.
10	Percent	Of hazards identified and mitigated to include environmental hazards.
11	Percent	Of residual risk accepted.
12	Percent	Of force affected by unidentified accident hazards.
13	Number	And types of hazards not identified affecting operation or casualties.
14	Number	Of controls averting identified accident hazards.
15	Number	Of controls averting unidentified accident hazards.

ART 6.10.2 DEVELOP AND IMPLEMENT COMMAND SAFETY PROGRAM

6-165.   Develop and implement command safety and occupational health, risk management, and accident prevention programs. (FM 5-19) (CRC)

No.	Scale	Measure
01	Yes/No	The command's published safety program incorporated the commander's safety philosophy and had realistic safety goals, objectives, and priorities.
02	Time	To modify command safety program to include new safety hazards.
03	Number	Of measures introduced to mitigate risk as a result of the risk assessment.
04	Number	Of violations of command safety program occurring within a given time.
05	Number	Of safety hazards not covered initially by command safety program.

ART 6.10.3 MINIMIZE SAFETY RISKS

6-166.   Ensure that programs are in place to identify potential safety threats, to apply risk management, and to take action to abate such risks. (FM 5-19) (CRC)

No.	Scale	Measure
01	Time	To complete accident investigation and report.
02	Percent	Of accidents attributed to human error in last 12 months.
03	Number	Of fatalities in last 12 months.
04	Number	Of man-hours lost because of accidents in last 6 months.
05	Number	Of people with lost time because of contact with hazardous materials.
06	Number	Of people with lost time because of nonwork-related accidents.
07	Number	Of people with lost time because of work-related accidents

SECTION XI – ART 6.11: IMPLEMENT OPERATIONS SECURITY

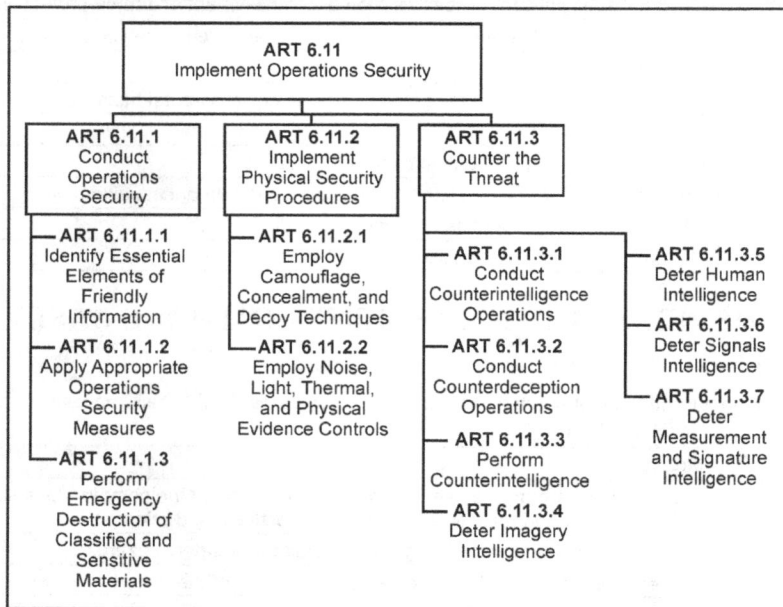

6-167. Operations security is a process of identifying essential elements of friendly information and subsequently analyzing friendly actions attendant to military operations and other activities to identify those actions that can be ob-served by adversary intelligence systems; determine indicators that hostile intelligence systems might obtain that could be interpreted or pieced together to derive critical information in time to be useful to adversaries; and select and execute measures that eliminate or reduce to an acceptable level the vulnerabilities of friendly actions to adversary exploitation. (FM 3-13) (USACAC)

ART 6.11.1 CONDUCT OPERATIONS SECURITY

6-168. Identify essential elements of friendly information (EEFI) and subsequently analyze friendly actions attendant to military operations and other activities. ART 6.11.1 identifies actions that can be observed by adversary intelligence systems and determines indicators adversary intelligence systems might obtain that could be interpreted or pieced together to derive EEFI in time to be useful to adversaries. ART 6.11.1 also involves selecting and executing measures that eliminate or reduce to an acceptable level the vulnerabilities of friendly actions to adversary exploitation. (FM 3-13) (USACAC)

No.	Scale	Measure
01	Yes/No	Operations security (OPSEC) compromised degraded, delayed, or modified unit operation.
02	Time	To refine OPSEC appendix to the operation order.
03	Time	To complete OPSEC assessment in the area of operations (AO).
04	Time	To identify possible compromises of EEFI in AO.
05	Time	To identify EEFI for an operation.
06	Percent	Of increased or decreased number of security violations on combat net radios in the AO within a given period.
07	Percent	Of enemy sensor coverage in AO known to friendly forces.

No.	Scale	Measure
08	Percent	Of successful enemy attempted penetration of friendly information systems.
09	Percent	Of information systems administrators and operators who have current OPSEC training.
10	Percent	Of identified friendly vulnerabilities in AO exploited by enemy actions.
11	Percent	Of friendly troop movements conducted without the possibility of enemy overhead surveillance (satellite and manned and unmanned aerial reconnaissance platforms).
12	Percent	Of units, facilities, and installations protected from enemy observation or surveillance.
13	Percent	Of electronic communications in AO encrypted or secured.
14	Percent	Of message traffic in AO exploited by enemy.
15	Percent	Of friendly emitters in AO exploited by enemy.
16	Percent	Of EEFI items covered by two or more measures.
17	Percent	Of enemy capabilities not covered by OPSEC measures covered by other elements such as deception, and electronic warfare.
18	Percent	Of friendly plan determined from self-monitoring of EEFI.
19	Percent	Of OPSEC measures previously assessed unsatisfactory that have improved based on assessment.
20	Percent	Of OPSEC measures selected tied to vulnerability analysis.
21	Percent	Of OPSEC planners who accommodate measures required to protect trusted agent planning such as given access.
22	Percent	Of OPSEC planners who have access to compartmented planning efforts.
23	Percent	Of OPSEC planners who have input to and receive guidance and results from higher headquarters OPSEC plans and surveys.
24	Percent	Of OPSEC surveys reflected in OPSEC plans.
25	Percent	Of routine actions with timing or location changed at least weekly.
26	Percent	Of units equipped with antisurveillance sensor and sensor jamming devices.
27	Percent	Of vulnerabilities tied to specific enemy capabilities by planners.
28	Number	And types of information needed by the commander to make decisions listed as EEFI.
29	Number	Of security violations on combat net radios in the AO.
30	Number	Of instances of friendly force operational patterns repeated in the AO.

ART 6.11.1.1 IDENTIFY ESSENTIAL ELEMENTS OF FRIENDLY INFORMATION

6-169. Identify friendly vulnerabilities that are exploitable by enemies and potential adversaries. Include recommendations concerning countermeasures and corrective action. (FM 3-13) (USACAC)

No.	Scale	Measure
01	Yes/No	Commander and staff identified friendly vulnerabilities that can be exploited by an enemy.
02	Time	To develop essential elements of friendly information (EEFI).
03	Time	To disseminate initial and subsequent EEFI requirements to subordinate elements of the force.
04	Time	Between updates of priority information requirements.
05	Time	To disseminate to all force elements and agencies information obtained due to the answering of EEFI.
06	Time	In advance of collection that EEFI are identified.
07	Time	Since most current information regarding EEFI was last collected.
08	Time	Of turnaround to process new EEFI.
09	Percent	Of friendly activities and resource expenditures accurately predicted by friendly reference materials, checklists, and other previously prepared documents and databases.
10	Percent	Of total EEFI identified only during execution.
11	Percent	Of EEFI collected in time to meet current operational needs.

No.	Scale	Measure
12	Percent	Of EEFI included in collection plan.
13	Number	Of EEFI not identified during planning.

ART 6.11.1.2 APPLY APPROPRIATE OPERATIONS SECURITY MEASURES

6-170. Deny adversaries information about friendly capabilities and intentions by identifying, controlling, and protecting indicators associated with planning and conducting military operations. (FM 3-13) (USACAC)

No.	Scale	Measure
01	Yes/No	Unit application of operations security (OPSEC) measures prevented the enemy from detecting the correct indicators of friendly operations until it was too late for the enemy to react.
02	Yes/No	Units changed patterns of operation on an irregular basis.
03	Time	To apply appropriate OPSEC measures.
04	Time	To brief unit information engagement cell and unit plans cell on OPSEC requirements.
05	Time	To identify target sets and desired effect, by priority.
06	Percent	Of OPSEC surveys reflected in OPSEC plans.
07	Percent	Of routine actions with timing or location changed at least weekly.
08	Percent	Of favorable signal security assessments.
09	Number	Of public media disclosures.
10	Number	Of critical essential elements of friendly information that must be concealed from the enemy.

ART 6.11.1.3 PERFORM EMERGENCY DESTRUCTION OF CLASSIFIED AND SENSITIVE MATERIALS

6-171. Establish and execute procedures for the emergency destruction of classified materials. (AR 380-5) (USAIC&FH)

No.	Scale	Measure
01	Yes/No	Emergency destruction of classified and sensitive documents and other materials was accomplished without compromising classified and sensitive information.
02	Time	To identify classified and sensitive materials for destruction.
03	Time	To destroy classified and sensitive materials.
04	Time	To develop and rehearse emergency destruction procedures.
05	Percent	Of identified classified and sensitive s identified for destruction that is actually destroyed.
06	Percent	Of classified and sensitive materials accountability procedures followed during the emergency destruction process.
07	Percent	Of mission-capable, on-hand equipment to perform emergency destruction of classified and sensitive materials.
08	Number	And types of paper shredders, thermal grenades, burn barrels, and magnets used to perform emergency destruction of classified and sensitive materials.
09	Number	And types of classified and sensitive documents and other materials destroyed.

ART 6.11.2 IMPLEMENT PHYSICAL SECURITY PROCEDURES

6-172. Protect personnel, information, and critical resources in all locations and situations against various threats by developing and implementing effective security policies and procedures. This total system approach is based on the continuing analysis and employment of protective measures, to include physical barriers, clear zones, lighting, access and key control, intrusion detection devices, defensive positions, and nonlethal capabilities. (FM 3-19.30) (USAMPS)

No.	Scale	Measure
01	Yes/No	Unit, base, and installation physical security program protected personnel, information, and critical resources from unauthorized access.
02	Time	To review and refine unit physical security standing operating procedures in accordance with the factors of mission, enemy, terrain and weather, troops and support available, time available, civil considerations.
03	Time	To refine physical security regulations for installations and major combat formation in an area of operations.
04	Time	To complete a threat analysis.
05	Time	To design, procure, emplace and activate protective measures, such as physical barriers, clear zones, exterior lighting, access and key control, intrusion detection devices, defensive positions, and nonlethal capabilities.
06	Percent	Of units, bases, and installations in the area of operations that have active integrated physical security programs.
07	Percent	Of guidance in unit and base physical security programs actually followed.
08	Percent	Of decreased crime rate.
09	Percent	Of increased reported crimes cleared.
10	Percent	Of perpetrators arrested or killed.
11	Percent	Of decreased serious crimes reported.
12	Percent	Of decreased fear of crime by unit personnel.
13	Percent	Of planned physical barriers, clear zones, exterior lighting, access and key control, intrusion detection devices, defensive positions and nonlethal capabilities operational.
14	Number	Of successful attempts to gain unauthorized access to friendly forces, installations, information, equipment, and supplies.

ART 6.11.2.1 EMPLOY CAMOUFLAGE, CONCEALMENT, AND DECOY TECHNIQUES

6-173. Protect friendly forces, personnel, materiel, equipment, and information system nodes from observation and surveillance by using natural or artificial materials. Employ an imitation in any sense of a person, object, or phenomenon with the intentions of deceiving enemy surveillance devices or misleading enemy evaluation. (FM 20-3) (USAES)

No.	Scale	Measure
01	Yes/No	The unit's use of camouflage, concealment, and decoy techniques enhanced unit survivability.
02	Time	To employ camouflage, concealment, and decoy techniques.
03	Time	To conduct a preliminary assessment of camouflage, concealment, and decoy effort in the area of operations.
04	Time	To obtain additional operational camouflage and decoy systems as required by the factors of mission, enemy, terrain and weather, troops and support available, time available, civil considerations.
05	Percent	Of unit concealed from enemy observation and sensor detection.
06	Percent	Of unit personnel trained to correctly employ camouflage and concealment and to use decoys.
07	Percent	Of casualties due to improper use of camouflage and concealment and decoys.
08	Percent	Of on-hand camouflage and decoy systems that are serviceable against enemy observation and sensors.
09	Number	And types of on-hand, serviceable camouflage and decoy systems.

ART 6.11.2.2 EMPLOY NOISE, LIGHT, THERMAL, AND PHYSICAL EVIDENCE CONTROLS

6-174. Reduce friendly indicators by controlling personnel and equipment sounds, light emissions, and physical evidence of occupying a position. (FM 3-21.75) (USAIS)

No.	Scale	Measure
01	Yes/No	Unit personnel did not compromise unit course of action by violations of noise, light, thermal, and physical evidence controls.
02	Time	To assess unit noise, light, thermal, and physical evidence controls.
03	Time	To employ noise, light, thermal, and physical evidence controls.
04	Percent	Of increased time to conduct operations required by the need to maintain noise, light, thermal, and physical evidence controls.
05	Percent	Of unit maintaining noise, light, thermal, and physical evidence controls.
06	Percent	Of unit personnel trained in noise, light, thermal, and physical evidence controls.
07	Percent	Of units, installations, and facilities that have recorded violations of noise, light, thermal, and physical evidence controls.
08	Number	Of friendly casualties due to violations of unit noise, light, thermal, and physical evidence controls.

ART 6.11.3 COUNTER THE THREAT

6-175. The task of providing the commander information and intelligence support for targeting the threat forces through lethal and nonlethal fires including electronic warfare and information engagement. Intelligence support to force protection and counterintelligence including the tactics, techniques, and procedures to deny or degrade threat intelligence, surveillance, and reconnaissance capabilities to access and collect information and intelligence on friendly forces. (FM 2-0) (USAIC&FH)

No.	Scale	Measure
01	Yes/No	Unit identified threat capabilities and limitations.
02	Yes/No	Unit identified friendly force vulnerabilities to threat forces.
03	Yes/No	Unit developed countermeasures to deny or degrade threat capabilities.
04	Yes/No	Unit developed countermeasures to mitigate friendly force vulnerability.
05	Yes/No	Unit identified threat capabilities and friendly forces countermeasures in sufficient time to integrate into the plan, prepare, execute, and assess operations process.
06	Yes/No	Unit disseminated countermeasures to friendly forces.
07	Time	To develop threat databases and templates.
08	Time	To develop countermeasures.
09	Percent	Of threat capabilities correctly identified.
10	Percent	Of countermeasures that effectively denied or degraded threat's ability.

ART 6.11.3.1 CONDUCT COUNTERINTELLIGENCE OPERATIONS

6-176. Counterintelligence is designed to defeat or degrade threat intelligence collection capabilities. The intelligence staff provides counterintelligence to the commander outlining the capabilities and limitations of threat intelligence services to limit or eliminate these capabilities. (FM 34-60) (USAIC&FH)

No.	Scale	Measure
01	Yes/No	Enemy intelligence operations directed against the unit and its personnel degraded, delayed, or modified unit operations.
02	Yes/No	Control element provided counterintelligence guidance.
03	Time	To conduct an area reconnaissance to identify hazards.
04	Time	To review counterintelligence plans for major tactical formations in the area of operations (AO).
05	Time	To conduct counterintelligence collection, operations, and investigations; create and maintain counterintelligence databases; analyze, assess counterintelligence information; and publish, disseminate, and technically support counterintelligence products and reports.

No.	Scale	Measure
06	Time	To conduct counterintelligence screening and collection at enemy prisoner of war collection points or dislocated civilian checkpoints.
07	Time	To conduct counterintelligence liaison with host-nation agencies and multinational forces.
08	Time	To conduct the counterintelligence portion of a vulnerability assessment and recommend countermeasures.
09	Percent	Of friendly force operations in the AO degraded, delayed, or modified due to successful enemy intelligence activities.
10	Percent	Of friendly force units in the AO that have current counterintelligence training.
11	Percent	Of friendly force action designed to mislead or prevent enemy intelligence efforts.
12	Percent	Of units in the AO that have active counterintelligence plans.
13	Number	Of enemy intelligence efforts detected by counterintelligence activities.
14	Number	Of counterintelligence teams available for use in the AO.
15	Number	And quality of sources developed in the AO.

ART 6.11.3.2 CONDUCT COUNTERDECEPTION OPERATIONS

6-177. Conduct activities that preclude the commander from being deceived by enemy deception operations. (FM 3-13) (USACAC)

No.	Scale	Measure
01	Yes/No	Friendly course of action was not affected by enemy deception except as desired to deceive the enemy when the friendly force accepted the enemy deception story.
02	Time	To identify adversary attempts to deceive friendly forces.
03	Time	To develop counterdeception operations options as required.
04	Percent	Of enemy deception activities detected.
05	Percent	Of enemy deception activities using multiple sources to transmit deception story.

ART 6.11.3.3 PERFORM COUNTERINTELLIGENCE

6-178. Gather information and conduct activities to protect against espionage, other intelligence activities, sabotage, or assassinations conducted by or on behalf of foreign governments or elements thereof, foreign organizations, or foreign persons, or international terrorist activities. (FM 34-60) (USAIC&FH)

No.	Scale	Measure
01	Yes/No	Unit identified threat intelligence collection capabilities and limitations.
02	Yes/No	Unit identified friendly forces vulnerabilities to threat intelligence collection.
03	Yes/No	Unit developed countermeasures to deny or degrade threat's ability to collect on friendly forces.
04	Yes/No	Unit developed countermeasures to mitigate friendly forces vulnerability to threat collection.
05	Yes/No	Unit identified threat capabilities and friendly forces countermeasures in sufficient time for implementation.
06	Time	To develop threat databases and templates.
07	Time	To develop countermeasures.
08	Percent	Of threat intelligence collection capabilities correctly identified.
09	Percent	Of countermeasures that effectively denied or degraded threat's ability to collect on friendly forces.

ART 6.11.3.4 DETER IMAGERY INTELLIGENCE

6-179. Defeat or degrade threat imagery intelligence collection capabilities. The intelligence staff will provide counterintelligence to the commander outlining the capabilities and limitation of threat imagery intelligence services to limit or eliminate these imagery capabilities. (FM 2-0) (USAIC&FH)

No.	Scale	Measure
01	Yes/No	Unit identified threat imagery intelligence collection capabilities and limitations.
02	Yes/No	Unit identified friendly forces vulnerabilities to threat imagery intelligence collection.
03	Yes/No	Unit developed countermeasures to deny or degrade threat's ability to collect on friendly forces with imagery assets.
04	Yes/No	Unit developed countermeasures to mitigate friendly forces vulnerability to threat collection.
05	Yes/No	Unit identified threat capabilities and friendly forces countermeasures in sufficient time for implementation.
06	Time	To develop threat databases and templates.
07	Time	To develop countermeasures.
08	Percent	Of threat imagery intelligence collection capabilities correctly identified.
09	Percent	Of countermeasures that effectively denied or degraded threat's ability to collect on friendly forces with imagery assets.

ART 6.11.3.5 DETER HUMAN INTELLIGENCE

6-180. Defeat or degrade threat human intelligence collection capabilities. The intelligence staff will provide counterintelligence to the commander outlining the capabilities and limitation of threat human intelligence services to limit or eliminate these human intelligence capabilities. (FM 2-0) (USAIC&FH)

No.	Scale	Measure
01	Yes/No	Unit identified threat human intelligence collection capabilities and limitations.
02	Yes/No	Unit identified friendly force vulnerabilities to threat human intelligence collection.
03	Yes/No	Unit developed countermeasures to deny or degrade threat's ability to collect on friendly forces.
04	Yes/No	Unit developed countermeasures to mitigate friendly forces vulnerability to threat human collection.
05	Yes/No	Unit identified threat capabilities and friendly forces countermeasures in sufficient time for implementation.
06	Time	To develop threat databases and templates.
07	Time	To develop countermeasures.
08	Percent	Of threat human intelligence collection capabilities correctly identified.
09	Percent	Of countermeasures that effectively denied or degraded threat's ability to collect on friendly forces using human intelligence.

ART 6.11.3.6 DETER SIGNALS INTELLIGENCE

6-181. Defeat or degrade threat signal intelligence collection capabilities. The intelligence staff will provide counterintelligence to the commander outlining the capabilities and limitation of threat signal intelligence services to limit or eliminate these signal intelligence capabilities. (FM 2-0) (USAIC&FH)

No.	Scale	Measure
01	Yes/No	Unit identified threat signal intelligence collection capabilities and limitations.
02	Yes/No	Unit identified friendly forces vulnerabilities to threat signal intelligence collection.

No.	Scale	Measure
03	Yes/No	Unit developed countermeasures to deny or degrade threat's ability to collect on friendly force use of the electromagnetic spectrum.
04	Yes/No	Unit developed countermeasures to mitigate friendly force vulnerability to threat signal collection.
05	Yes/No	Unit identified threat capabilities and friendly force countermeasures in sufficient time for implementation.
06	Time	To develop threat databases and templates.
07	Time	To develop counter measures.
08	Percent	Of threat signals intelligence collection capabilities correctly identified.
09	Percent	Of countermeasures that effectively denied or degraded threat's ability to collect on friendly force using signals intelligence.

ART 6.11.3.7 DETER MEASUREMENT AND SIGNATURE INTELLIGENCE

6-182. Defeat or degrade threat measurement and signature intelligence collection capabilities. The intelligence staff will provide counterintelligence to the commander outlining the capabilities and limitation of threat measurement and signature intelligence services to limit or eliminate these measurement and signature intelligence capabilities. (FM 2-0) (USAIC&FH)

No.	Scale	Measure
01	Yes/No	Unit identified threat measurement and signature intelligence collection capabilities and limitations.
02	Yes/No	Unit identified friendly force vulnerabilities to threat measurement and signature intelligence collection.
03	Yes/No	Unit developed countermeasures to deny or degrade threat's ability to collect on friendly force.
04	Yes/No	Unit developed countermeasures to mitigate friendly force vulnerability to threat measurement and signature collection.
05	Yes/No	Unit identified threat capabilities and friendly force countermeasures in sufficient time for implementation.
06	Time	To develop threat databases and templates.
07	Time	To develop countermeasures.
08	Percent	Of threat measurement and signature intelligence collection capabilities correctly identified.
09	Percent	Of countermeasures that effectively denied or degraded threat's ability to collect on friendly forces using measurement and signature intelligence.

*SECTION XII – ART 6.12: PROVIDE EXPLOSIVE ORDNANCE DISPOSAL PROTECTION SUPPORT

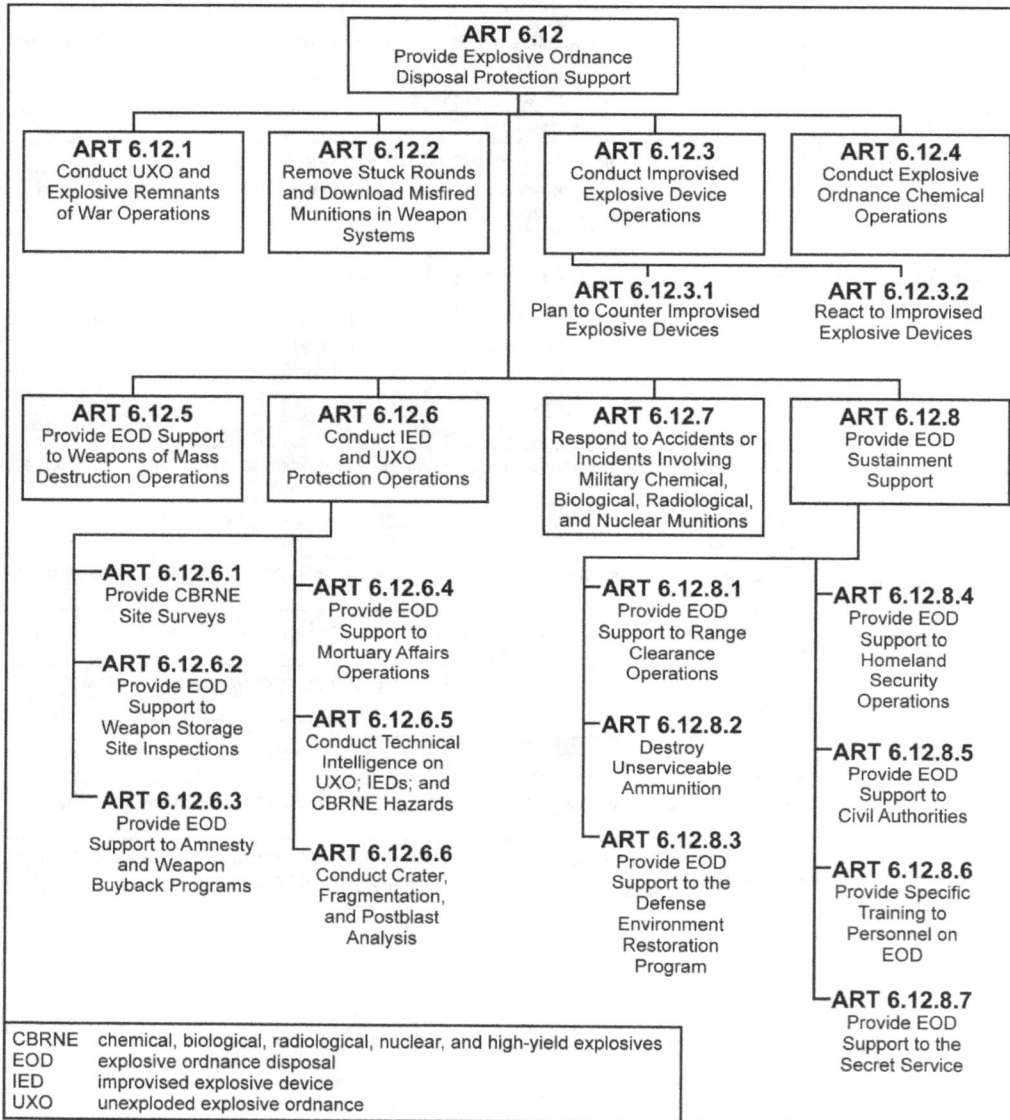

ART 6.12
Provide Explosive Ordnance Disposal Protection Support

ART 6.12.1 Conduct UXO and Explosive Remnants of War Operations

ART 6.12.2 Remove Stuck Rounds and Download Misfired Munitions in Weapon Systems

ART 6.12.3 Conduct Improvised Explosive Device Operations

ART 6.12.4 Conduct Explosive Ordnance Chemical Operations

ART 6.12.3.1 Plan to Counter Improvised Explosive Devices

ART 6.12.3.2 React to Improvised Explosive Devices

ART 6.12.5 Provide EOD Support to Weapons of Mass Destruction Operations

ART 6.12.6 Conduct IED and UXO Protection Operations

ART 6.12.7 Respond to Accidents or Incidents Involving Military Chemical, Biological, Radiological, and Nuclear Munitions

ART 6.12.8 Provide EOD Sustainment Support

ART 6.12.6.1 Provide CBRNE Site Surveys

ART 6.12.6.2 Provide EOD Support to Weapon Storage Site Inspections

ART 6.12.6.3 Provide EOD Support to Amnesty and Weapon Buyback Programs

ART 6.12.6.4 Provide EOD Support to Mortuary Affairs Operations

ART 6.12.6.5 Conduct Technical Intelligence on UXO; IEDs; and CBRNE Hazards

ART 6.12.6.6 Conduct Crater, Fragmentation, and Postblast Analysis

ART 6.12.8.1 Provide EOD Support to Range Clearance Operations

ART 6.12.8.2 Destroy Unserviceable Ammunition

ART 6.12.8.3 Provide EOD Support to the Defense Environment Restoration Program

ART 6.12.8.4 Provide EOD Support to Homeland Security Operations

ART 6.12.8.5 Provide EOD Support to Civil Authorities

ART 6.12.8.6 Provide Specific Training to Personnel on EOD

ART 6.12.8.7 Provide EOD Support to the Secret Service

CBRNE chemical, biological, radiological, nuclear, and high-yield explosives
EOD explosive ordnance disposal
IED improvised explosive device
UXO unexploded explosive ordnance

6-183. Eliminate or reduce the effects of unexploded explosive ordnance (UXO), improvised explosive devices (IEDs), and chemical, biological, radiological, and nuclear (CBRN) hazards to protect the commander's combat power. Explosive ordnance hazards limit battlefield mobility, deny use of critical assets, and threaten to reduce the commander's combat power significantly. Neutralize domestic or foreign conventional, chemical, biological, and nuclear munitions; enhanced explosive devices; and IEDs that present a threat to civilian facilities, materiel, and personnel, regardless of location. The Departments of

Justice, State, and Energy may receive this support per current agreements and directives. (FM 3-34.214) (CASCOM)

> *Note:* ART 6.12.8.6 includes conducting bomb and sabotage device recognition and safety precaution training. ART 6.12.8.7 includes providing explosive ordnance disposal (EOD) support to the U.S. Secret Service, Department of State, and the Department of Defense to protect the President and other designated persons.
>
> Tasks identified in ART 6.12 are tasks that can be provided in support of stability operations as well as in civil support operations.

No.	Scale	Measure
01	Yes/No	EOD support allowed the unit to accomplish its mission.
02	Yes/No	Collateral damage incurred during the EOD operation was within acceptable limits.
03	Yes/No	Unit safeguarded classified materials and publications during the EOD operation.
04	Yes/No	Unit forwarded items and components of technical intelligence value to appropriate headquarters or agency.
05	Time	To provide EOD input to annex I to the operation plan or order.
06	Time	To respond to a request for EOD support.
07	Time	Delay in executing the concept of operations due to the presence of UXOs and IEDs.
08	Time	To gather intelligence information (what, when, where, how delivered, and type) regarding munitions.
09	Time	To identify safety requirements and considerations.
10	Time	To identify personnel, equipment, procedures, and additional support requirements.
11	Time	To coordinate with reporting agency for site support assistance, such as engineer, medical, security, and transportation.
12	Time	To clear munitions and improvised devices.
13	Time	To document render-safe procedures, as conducted, for unknown ordnance, if technical intelligence data does not exist.
14	Percent	Of safety precautions enforced during EOD operations.
15	Percent	Of reported munitions and improvised devices rendered safe.
16	Percent	Of reported munitions and improvised devices rendered safe per commander's intent.
17	Percent	Of available EOD support expended on conducting bomb and sabotage device recognition and safety training.
18	Percent	Of patients received at medical treatment facilities with UXO in their wounds.
19	Number	Of casualties during the EOD operation.
20	Number	And types of ordnance located and destroyed by EOD personnel.
21	Number	Of chemical, biological, radiological, nuclear, and high-yield explosives incidents responded to within a given period.

ART 6.12.1 CONDUCT UNEXPLODED EXPLOSIVE ORDNANCE AND EXPLOSIVE REMNANTS OF WAR OPERATIONS

6-184. Explosive ordnance disposal (EOD) units identify, render safe, recover, evaluate, dispose of, mitigate the threat of, and report (EOD and intelligence) U.S. and foreign unexploded explosive ordnance (UXO) to eliminate or reduce hazards and to protect the commander's combat power. EOD units maintain the capability to render safe, identify, destroy, and move UXO hazards to safe holding or disposal areas within the limitations of their organic transportation assets, perform technical intelligence on new or first seen ordnance items, and perform postblast forensic analysis. EOD performs initial assessment of explosive remnants of war sites, eliminates immediate hazards, conducts technical intelligence procedures, and recommends disposition to capturing unit. (FM 3-34.214) (CASCOM)

No.	Scale	Measure
01	Yes/No	EOD support allowed the requesting unit to accomplish its mission.
02	Yes/No	Collateral damage incurred during the conduct of the EOD operation was in acceptable limits.
03	Yes/No	Unit safeguarded classified materials and publications during EOD operations.
04	Yes/No	Unit forwarded items and components of technical intelligence value to appropriate headquarters or agency.
05	Time	To respond to a request for EOD support to conduct UXO and explosive remnants of war operations.
06	Time	Delay to execute operations due to the presence of UXO and explosive remnants of war.
07	Time	To gather intelligence information (what, when, where, how delivered, and type) regarding UXO and explosive remnants of war.
08	Time	To identify safety requirements and considerations concerned with the destruction or elimination of UXO or rendering safe explosive remnants of war.
09	Time	To identify personnel, equipment, procedures, and additional support requirements.
10	Time	To coordinate with the reporting agency for site support assistance such as engineer, medical, security, and transportation.
11	Time	To clear UXO and explosive remnants of war.
12	Time	Time to move UXOs to safe holding or dispersal areas.
13	Time	To document render-safe procedures, as conducted, for unknown UXO or explosive remnants of war if technical intelligence data does not exist.
14	Time	To provide disposition instructions of explosive remnants of war to capturing unit.
15	Percent	Of safety precautions enforced during the EOD operation.
16	Percent	Of reported UXO and explosive remnants of war rendered safe.
17	Percent	Of reported UXO and explosive remnants of war destroyed or rendered safe per the commander's priorities.
18	Number	Of casualties during the conduct of the EOD operation.
19	Number	And types of UXO and explosive remnants of war destroyed by EOD personnel.

ART 6.12.2 REMOVE STUCK ROUNDS AND DOWNLOAD MISFIRED MUNITIONS IN WEAPON SYSTEMS

6-185.   Removal of stuck rounds is a routine explosive ordnance disposal (EOD) operation. In a removal of stuck rounds mission, the EOD unit removes stuck rounds in mortars, artillery tubes, and other weapon systems; inspects possible unsafe ammunition removed during EOD or operator procedures to determine if it is safe for storage or transport; and destroys ammunition that EOD personnel determine to be unsafe for storage or transport. A battle damaged vehicle with ammunition on board requires EOD to download munitions and clear the vehicle before it can be repaired. EOD also supports the postblast investigation and performs technical intelligence on enemy weapons effects on U.S. vehicles through appropriate channels. (FM 3-34.214) (CASCOM)

No.	Scale	Measure
01	Yes/No	EOD support allowed the unit to accomplish its mission.
02	Yes/No	Collateral damage incurred during the conduct of the EOD operation was in acceptable limits.
03	Yes/No	Unit safeguarded classified materials and publications during EOD operations.
04	Yes/No	Unit forwarded items and components of technical intelligence value to appropriate headquarters or agency.
05	Time	To identify safety requirements and considerations concerned with removal of stuck rounds or download misfired munitions.
06	Time	To identify personnel, equipment, procedures, and additional support requirements.

No.	Scale	Measure
07	Time	To move from the current location to the work site.
08	Time	To coordinate with the requesting agency for site support assistance in areas such as engineer, medical, security, and transportation.
09	Time	To remove the stuck rounds or download misfired munitions.
10	Time	To gather intelligence information (what, when, where, how delivered, and type) regarding munitions.
11	Time	To coordinate with reporting agency for site support assistance such as engineer, medical, security, and transportation.
12	Time	To develop and attempt render-safe procedures or conduct analysis for forensic evidence.
13	Percent	Of available EOD support assets expended on developing render-safe procedures or conducting postblast analysis.
14	Percent	Of safety precautions enforced during the EOD operation.
15	Percent	Of stuck rounds or misfired munitions identified as safe for storage or transport.
16	Percent	Of stuck rounds or misfired munitions identified as unsafe for storage or transport and destroyed.
17	Number	Of casualties during the conduct of the EOD operation.
18	Number	And types of stuck rounds removed or misfired munitions downloaded.
19	Number	And types of appropriate intelligence reports prepared within a given period.

*ART 6.12.3 CONDUCT IMPROVISED EXPLOSIVE DEVICE OPERATIONS

6-186. Explosive ordnance disposal (EOD) teams identify, render safe, and dispose of improvised explosive devices (IEDs) and conduct postblast analysis and intelligence reporting. EOD teams maintain capabilities for remote investigation, identification, and movement of IEDs and emplacement or operation of disruption tools and disposal methods. IEDs with potential chemical, biological, radiological, nuclear, and high-yield explosives may require additional EOD personnel and equipment to augment the initial EOD response team. (FM 3-34.214) (CASCOM)

No.	Scale	Measure
01	Yes/No	EOD support allowed the requesting unit to accomplish its mission.
02	Yes/No	Collateral damage incurred during the conduct of the EOD operation was in acceptable limits.
03	Yes/No	Unit safeguarded classified materials and publications during EOD operations.
04	Yes/No	Unit forwarded items and components of technical intelligence value to appropriate headquarters or agency.
05	Time	To respond to a request for EOD support to conduct counter-IED operations.
06	Time	Delay in executing the concept of operations due to the presence of IEDs.
07	Time	To gather intelligence information (what, when, where, how delivered, and type) regarding IEDs.
08	Time	To identify safety requirements and considerations concerned with the destruction or elimination of IEDs.
09	Time	To identify personnel, equipment, procedures, and additional support requirements.
10	Time	To coordinate with the reporting agency for site support assistance such as engineer, medical, security, decontamination, and transportation.
11	Time	To gain access to and identify IEDs.
12	Time	To emplace or operate disruption tools.
13	Time	To render safe and or dispose of IEDs.
14	Time	To conduct postblast analysis

No.	Scale	Measure
15	Percent	Of safety precautions enforced during the EOD operation.
16	Percent	Of IEDs destroyed or rendered safe.
17	Number	Of casualties while disposing of explosive ordnance.
18	Number	And types of IEDs destroyed by EOD personnel.
19	Number	Of IED components retrieved by EOD personnel for exploitation.

*ART 6.12.3.1 PLAN TO COUNTER IMPROVISED EXPLOSIVE DEVICES

6-187.   Conduct improvised explosive device (IED) defeat operations to defeat asymmetric attacks against U.S. forces. (FM 3-90.119) (USAES)

No.	Scale	Measure
01	Yes/No	Unit planned for possible IED threats in a counterinsurgency environment.
02	Yes/No	Unit prepared for IED defeat using the tenets of IED defeat (predict, detect, prevent, neutralize, and mitigate).
03	Yes/No	Element prepared for a suspected vehicle borne, suicide vehicle borne, and person borne IED attack against static positions.
04	Yes/No	Element reacted to a possible IED, vehicle borne, suicide vehicle borne or person borne IED by using the 5Cs (confirm, clear, cordon, check, and control).

*ART 6.12.3.2 REACT TO IMPROVISED EXPLOSIVE DEVICES

6-188.   Planning elements are proactive actions taken by friendly forces to predict, detect, prevent, avoid, neutralize, and protect against improvised explosive device events. The process and the products of mission analysis help the commander and staffs develop and refine their situational understanding and develop effective plans. By having a thorough understanding of the mission, enemy, terrain and weather, troops and support available, time available, civil considerations factors, the commander and staff are better equipped to develop effective plans to accomplish the mission. (FM 3-90.119) (USAES)

No.	Scale	Measure
01	Yes/No	Unit predicted actions and circumstances that could affect the ability of the force to maintain movement and maneuver.
02	Yes/No	Unit prevented potential impediments to movement and maneuver from affecting the mobility of the force by acting early.
03	Yes/No	Unit detected early indicators of impediments to battlefield mobility and identify solutions through the use of intelligence, surveillance, and reconnaissance assets.
04	Yes/No	Unit avoided detected impediments to movement and maneuver if prevention fails.
05	Yes/No	Unit neutralized, reduced, or overcame impediments to movement and maneuver that could not be prevented or avoided.
06	Yes/No	Unit protected against enemy countermobility efforts.

ART 6.12.4 CONDUCT EXPLOSIVE ORDNANCE DISPOSAL CHEMICAL OPERATIONS

6-189.   Explosive ordnance disposal (EOD) units respond to U.S. and foreign chemical weapons incidents to identify, render safe, perform preliminary packaging, and limit the spread of chemical contamination. EOD units maintain the capability to presume identification of chemical agents, perform preliminary packaging of chemical munitions, perform emergency personnel decontamination station operations (EOD personnel and limited equipment only), mark and perform hasty decontamination of the immediate chemical incident site, perform render-safe procedures, collect samples for additional identification and testing, and dispose of chemical munitions. (FM 3-34.214) (CASCOM)

No.	Scale	Measure
01	Yes/No	EOD support allowed the unit to accomplish its mission.
02	Yes/No	Collateral damage incurred during EOD operation was within acceptable limits.
03	Yes/No	Unit safeguarded classified materials and publications during the EOD operation.
04	Yes/No	Unit forwarded items and components of technical intelligence value to appropriate headquarters or agency.
05	Time	To provide EOD input to annex I to the operation plan or order.
06	Time	To respond to a request for EOD support.
07	Time	Delay in executing the concept of operations due to the presence of unexploded explosive ordnance and improvised explosive device.
08	Time	To gather intelligence information (what, when, where, how delivered, and type) regarding munitions.
09	Time	To identify safety requirements and considerations.
10	Time	To identify personnel, equipment, procedures, and additional support requirements.
11	Time	To coordinate with reporting agency for site support assistance, such as engineer, medical, security, and transportation.
12	Time	To clear munitions and improvised devices.
13	Time	To document render-safe procedures, as conducted, for unknown ordnance, if technical intelligence data does not exist.
14	Percent	Of safety precautions enforced during EOD operations.
15	Percent	Of reported munitions and improvised devices rendered safe.
16	Percent	Of reported munitions and improvised devices rendered safe per commander's intent.
17	Percent	Of available EOD support expended on conducting bomb and sabotage device recognition and safety training.
18	Percent	Of patients received at medical treatment facilities with unexploded explosive ordnance in their wounds.
19	Percent	Of casualties during the EOD operation.
20	Number	And types of ordnance located and destroyed by EOD personnel.
21	Number	Of chemical, biological, radiological, nuclear, and high-yield explosives incidents responded to within a given period.

ART 6.12.5 PROVIDE EXPLOSIVE ORDNANCE DISPOSAL SUPPORT TO WEAPONS OF MASS DESTRUCTION OPERATIONS

6-190.   Reduce the threat of arms and weapons of mass destruction to regional security; conduct disposal operations of hazardous U.S. and foreign munitions; conduct clearance and destruction operations of ammunition storage areas and caches that may have hazardous munitions or booby traps; assist with weapons or ammunition storage site inspections to satisfy treaty obligations or agreements and policies enforcement; and detect, identify, and respond to chemical, biological, radiological, nuclear, and high-yield explosives threat. Explosive ordnance disposal (EOD) units have the capability to presume identification of biological agents, perform preliminary packaging of biological munitions and agents, perform emergency personnel decontamination station operations (EOD personnel only), mark and perform limited decontamination of incident site, perform render-safe procedures, and collect samples for additional identification and testing. (FM 3-34.214) (CASCOM)

No.	Scale	Measure
01	Yes/No	EOD support allowed the requesting unit to accomplish its mission.
02	Yes/No	Collateral damage incurred during the conduct of the EOD operation was in acceptable limits.
03	Yes/No	Unit safeguarded classified materials and publications during EOD operations.

No.	Scale	Measure
04	Yes/No	Unit forwarded items and components of technical intelligence value to appropriate headquarters or agency.
05	Time	To move from current location to the weapons of mass destruction (WMD) site.
06	Time	To locate WMD.
07	Time	To gain access to WMD
08	Time	To conduct advanced diagnostics on WMD.
09	Time	To gather intelligence information (what, when, where, how delivered, and type) regarding WMD.
10	Time	To identify safety requirements and considerations concerned with the destruction or elimination of WMD.
11	Time	To identify personnel, equipment, procedures, and additional support requirements.
12	Time	To coordinate with the reporting agency for site support assistance such as engineer, medical, security, decontamination, and transportation.
13	Time	To conduct clearance and destruction operations of ammunition storage areas and caches that may contain WMD.
14	Time	To document render-safe procedures, as conducted, for unknown ordnance, if technical intelligence data does not exist.
15	Percent	Of safety precautions enforced during the EOD operation.
16	Percent	Of reported munitions rendered safe.
17	Number	Of casualties during the conduct of the EOD operation.
18	Number	And types of ordnance or WMD destroyed by EOD personnel.

ART 6.12.6 CONDUCT IMPROVED EXPLOSIVE DEVICE AND UNEXPLODED EXPLOSIVE ORDNANCE PROTECTION OPERATIONS

6-191. Assist commanders with protection of the force. This includes reviewing base defense plans and bomb threat or search procedures, assisting in facility explosive hazard site surveys, and developing and implementing explosive ordnance disposal (EOD) emergency response plans and the antiterrorism or force protection plan. Provide training to emergency preparedness personnel on bomb threat search, improvised explosive device (IED) defeat techniques, unexploded explosive ordnance (UXO) identification, and marking procedures. Provide chemical, biological, radiological, nuclear, and high-yield explosives emergency response plans and facility site surveys to commanders. Provide EOD support to weapon storage site inspections. Provide EOD support to amnesty and weapon buyback programs. (FM 3-34.214) (CASCOM)

No.	Scale	Measure
01	Yes/No	EOD support allowed the requesting unit to accomplish its mission.
02	Yes/No	Collateral damage incurred during the conduct of the EOD operation was in acceptable limits.
03	Yes/No	Unit safeguarded classified materials and publications during EOD operations.
04	Yes/No	Unit forwarded items and components of technical intelligence value to appropriate headquarters or agency.
05	Time	To move from current location to the work site.
06	Time	To gather intelligence information (what, when, where, how delivered, and type) regarding munitions.
07	Time	To identify safety requirements and considerations concerned with force protection operations.
08	Time	To identify personnel, equipment, procedures, and additional support requirements.

No.	Scale	Measure
09	Time	To coordinate with the reporting agency for site support assistance such as engineer, medical, security, and transportation.
10	Time	To document training, response preparedness, and review of defense plans, UXO identification and marking procedures, site surveys, and bomb threat or search procedures conducted to assist commanders with force protection.
11	Time	To clear munitions and improvised devices found during the conduct of force protection operations.
12	Percent	Of safety precautions enforced during the EOD operation.
13	Percent	Of reported munitions and IEDs rendered safe.
14	Percent	Of available EOD support expended on conducting bomb threat or search procedures, device recognition, and safety training.
15	Number	Of casualties during EOD operations.
16	Number	And types of ordnance located and destroyed by EOD personnel.
17	Number	Of force protection requests responded to within a given period.

ART 6.12.6.1 PROVIDE CHEMICAL, BIOLOGICAL, RADIOLOGICAL, NUCLEAR, AND HIGH-YIELD EXPLOSIVES SITE SURVEYS

6-192. Assist commanders with protection of the force by providing explosive ordnance disposal (EOD) support to review base defense plans and bomb threat or search procedures, develop and implement base defense plans, and assist in facility site surveys. (FM 3-34.214) (CASCOM)

No.	Scale	Measure
01	Yes/No	EOD support allowed the unit to accomplish its mission.
02	Yes/No	Collateral damage incurred during the conduct of the EOD operation was in acceptable limits.
03	Yes/No	Unit safeguarded classified materials and publications during EOD operations.
04	Time	To identify safety requirements and considerations concerned with emergency response plans and facility explosive hazard site surveys.
05	Time	To identify personnel, equipment, procedures, and additional support requirements.
06	Time	To move from the current location to the work site.
07	Time	To coordinate with the requesting agency for site support assistance in areas such as engineer, medical, security, and transportation.
08	Time	To document render-safe procedures that relate to emergency response plans and facility site surveys.
09	Percent	Of safety precautions enforced during the EOD operation.
10	Percent	Of reported munitions rendered safe.
11	Percent	Of reported munitions rendered safe per the requesting agency's priorities.
12	Number	Of casualties during the conduct of the EOD operation.
13	Number	Of requests for assistance responded to in a given period.

ART 6.12.6.2 PROVIDE EXPLOSIVE ORDNANCE DISPOSAL SUPPORT TO WEAPON STORAGE SITE INSPECTIONS

6-193. Sensitive site exploitation operations focus on locating, characterizing, seizing, securing, and searching facilities, supplies, weapons, equipment, personnel, and infrastructures. Exploitation operations may additionally include disablement operations to render chemical, biological, radiological, nuclear, and high-yield explosives (CBRNE) materiel, systems, and equipment ineffective for use against the joint, interagency, and multinational forces. Disablement operations include neutralization, incineration, destruction, confiscation, and evacuation of CBRNE and associated materials. (FM 3-34.214) (CASCOM)

No.	Scale	Measure
01	Yes/No	Explosive ordnance disposal (EOD) support allowed the unit to accomplish its mission.
02	Yes/No	Collateral damage incurred during the conduct of the EOD operation was in acceptable limits.
03	Yes/No	Unit safeguarded classified materials and publications during EOD operations.
04	Time	To identify safety requirements and considerations concerned with weapon storage site inspections.
05	Time	To identify personnel, equipment, procedures, and additional support requirements.
06	Time	To move from the current location to the work site.
07	Time	To coordinate with the requesting agency for site support assistance in areas such as engineer, medical, security, decontamination, and transportation.
08	Time	To document render-safe procedures related to weapon storage site inspections.
09	Percent	Of safety precautions enforced during the EOD operation.
10	Percent	Of reported munitions rendered harmless.
11	Percent	Of reported munitions cleared per the requesting agency's priorities.
12	Number	Of casualties during the conduct of the EOD operation.
13	Number	Of requests for assistance responded to in a given period.

ART 6.12.6.3 PROVIDE EXPLOSIVE ORDNANCE DISPOSAL SUPPORT TO AMNESTY AND WEAPON BUYBACK PROGRAMS

6-194. Assist with and respond to amnesty collection points to ensure armed or unsafe ordnance items are disposed of properly. (FM 3-34.214) (CASCOM)

No.	Scale	Measure
01	Yes/No	Explosive ordnance disposal (EOD) support allowed the unit to accomplish its mission.
02	Yes/No	Collateral damage incurred during the conduct of the EOD operation was in acceptable limits.
03	Yes/No	Unit safeguarded classified materials and publications during EOD operations.
04	Time	To identify safety requirements and considerations concerned with amnesty and weapon buyback programs.
05	Time	To identify personnel, equipment, procedures, and additional support requirements.
06	Time	To move from the current location to the work site.
07	Time	To coordinate with the requesting agency for site support assistance in areas such as engineer, medical, security, and transportation.
08	Time	To document render-safe procedures related to amnesty and weapon buyback programs.
09	Percent	Of safety precautions enforced during the EOD operation.
10	Percent	Of reported munitions rendered harmless.
11	Percent	Of reported munitions cleared per the requesting agency's priorities.
12	Percent	Of munitions destroyed.
13	Number	Of casualties during the conduct of the EOD operation.
14	Number	Of requests for assistance responded to in a given period.

ART 6.12.6.4 PROVIDE EXPLOSIVE ORDNANCE DISPOSAL SUPPORT TO MORTUARY AFFAIRS OPERATIONS

6-195. Explosive ordnance disposal (EOD) involvement in recovery and processing of deceased personnel supports the immediate recovery and clearance of deceased persons, an Army priority. (FM 3-34.214) (CASCOM)

No.	Scale	Measure
01	Yes/No	EOD support allowed the unit to accomplish its mission.
02	Yes/No	Collateral damage incurred during the conduct of the EOD operation was in acceptable limits.
03	Yes/No	Unit safeguarded classified materials and publications during EOD operations.
04	Time	To identify safety requirements and considerations concerned with recovery and clearance of deceased persons.
05	Time	To identify personnel, equipment, procedures, and additional support requirements.
06	Time	To move from the current location to the work site.
07	Time	To coordinate with the requesting agency for site support assistance in areas such as engineer, medical, security, and transportation.
08	Time	To document render-safe procedures related to recovery and clearance of deceased persons.
09	Percent	Of safety precautions enforced during the EOD operation.
10	Percent	Of reported munitions rendered safe.
11	Percent	Of reported munitions rendered safe per the requesting agency's priorities.
12	Number	Of casualties during the conduct of the EOD operation.
13	Number	Of requests for assistance responded to in a given period.

ART 6.12.6.5 CONDUCT TECHNICAL INTELLIGENCE ON UNEXPLODED EXPLOSIVE ORDNANCE; IMPROVISED EXPLOSIVE DEVICES; AND CHEMICAL, BIOLOGICAL, RADIOLOGICAL, NUCLEAR, AND HIGH-YIELD EXPLOSIVES HAZARDS

6-196. Explosive ordnance disposal (EOD) personnel provide technical intelligence support to maneuver units by identifying and requesting disposition of first seen ordnance and explosive remnants of war, improvised explosive devices (IEDs), and chemical, biological, radiological, nuclear, and high-yield explosives (CBRNE) hazards of intelligence value. EOD personnel also prepare and submit an appropriate intelligence report based upon type of ordnance and type of function. They determine if items are safe for shipment and storage. EOD personnel develop and attempt render-safe procedures. EOD personnel conduct postblast analysis for forensic evidence, and if the need exists, EOD personnel can collect chemical and biological samples for analysis. EOD personnel can recognize and test for CBRNE hazards. (FM 3-34.214) (CASCOM)

No.	Scale	Measure
01	Yes/No	EOD support allowed the unit to accomplish its mission.
02	Yes/No	Collateral damage incurred during EOD operation was within acceptable limits.
03	Yes/No	Unit safeguarded classified materials and publications during the EOD operation.
04	Yes/No	Unit forwarded items and components of technical intelligence value to appropriate headquarters or agency.
05	Time	To respond to a request for EOD support.
06	Time	Delay to execute operations due to the conduct of technical intelligence procedures on unexploded explosive ordnance, IEDs, and CBRNE hazards.
07	Time	To gather intelligence information (what, when, where, how delivered, and type) regarding munitions.
08	Time	To identify personnel, equipment, procedures, and additional support requirements.
09	Time	To coordinate with reporting agency for site support assistance such as engineer, medical, security, and transportation.
10	Time	To develop and attempt render-safe procedures or conduct postblast analysis for forensic evidence.
11	Time	To document render-safe procedures, as conducted, for unknown ordnance if technical intelligence data does not exist.

No.	Scale	Measure
12	Time	To determine if items are safe for shipment or storage.
13	Time	To collect chemical and biological samples for analysis.
14	Percent	Of safety precautions enforced during EOD operations.
15	Percent	Of available EOD personnel expended on developing render-safe procedures or conducting postblast analysis.
16	Number	And types of appropriate intelligence reports prepared within a given period.
17	Number	Of battle damaged vehicles, buildings, or the like analyzed for postblast damage.

ART 6.12.6.6 CONDUCT CRATER, FRAGMENTATION, AND POSTBLAST ANALYSIS

6-197. Explosive ordnance disposal (EOD) units conduct postblast analysis for forensic evidence. And, if the need exists, EOD personnel can collect chemical and biological samples for analysis. (FM 3-34.214) (CASCOM)

No.	Scale	Measure
01	Yes/No	EOD support allowed the unit to accomplish its mission.
02	Yes/No	Collateral damage incurred during the conduct of the EOD operation was in acceptable limits.
03	Yes/No	Unit safeguarded classified materials and publications during EOD operations.
04	Yes/No	Unit forwarded items and components of technical intelligence value to appropriate headquarters or agency.
05	Time	To respond to a request for EOD support.
06	Time	Delay to execute operations due to the conduct of technical intelligence procedures on unexploded explosive ordnance, improvised explosive devices, and chemical, biological, radiological, nuclear, and high-yield explosives hazards.
07	Time	To gather intelligence information (what, when, where, how delivered, and type) regarding munitions.
08	Time	To identify personnel, equipment, procedures, and additional support requirements.
09	Time	To coordinate with reporting agency for site support assistance such as engineer, medical, security, and transportation.
10	Time	To develop and attempt render-safe procedures and conduct postblast analysis for forensic evidence.
11	Time	To document render-safe procedures, as conducted, for unknown ordnance if technical intelligence data does not exist.
12	Time	To determine if items are safe for shipment or storage.
13	Time	To collect chemical land or biological samples for analysis.
14	Percent	Of safety precautions enforced during EOD operations.
15	Percent	Of available EOD support expended on developing render-safe procedures and postblast analysis.
16	Number	And types of appropriate intelligence reports prepared within a given period.

ART 6.12.7 RESPOND TO ACCIDENTS OR INCIDENTS INVOLVING MILITARY CHEMICAL, BIOLOGICAL, RADIOLOGICAL, AND NUCLEAR MUNITIONS

6-198. Explosive ordnance disposal (EOD) units respond to U.S. and foreign chemical, biological, radiological, and nuclear (CBRN) weapons incidents to identify, render safe, perform preliminary packaging, and limit the spread of CBRN contamination. Three identifiable operational responses to a CBRN accident or event are initial response, emergency actions, and site remediation. (FM 3-34.214) (CASCOM)

No.	Scale	Measure
01	Yes/No	EOD support accomplished the purpose of the requesting agency.
02	Yes/No	Collateral damage incurred during the EOD operation was within acceptable limits.
03	Yes/No	Unit safeguarded classified materials and publications during the EOD operation.
04	Yes/No	Unit forwarded items and components of technical intelligence value to appropriate headquarters or agency.
05	Yes/No	Spread of contamination was within acceptable limits.
06	Time	To move from the current location to the work site.
07	Time	To identify safety requirements and considerations concerned with destruction of the CBRN materiel.
08	Time	To gather information about the CBRN munitions (what, when, where, how delivered, and type).
09	Time	To perform initial response, emergency actions, and site remediation.
10	Time	To coordinate with the reporting agency for site support assistance in areas such as engineer, medial, security and transportation.
11	Time	To identify personnel, equipment, procedures, and additional support requirements.
12	Time	To identify, render safe, package, and limit the spread of CBRN contamination.
13	Percent	Of safety precautions enforced during the EOD operation.
14	Percent	Of CBRN munitions destroyed or rendered safe.
15	Number	Of casualties during the conduct of the EOD operation.
16	Number	And types of CBRN munitions destroyed or rendered harmless.

ART 6.12.8 PROVIDE EXPLOSIVE ORDNANCE DISPOSAL SUSTAINMENT SUPPORT

6-199. Neutralize domestic or foreign munitions and improvised devices that present a threat to military operations and military and civilian facilities, materiel, and personnel, regardless of location. The Departments of Justice, State, and Energy may receive this support per current agreements and directives. ART 6.12.8 includes providing explosive ordnance disposal (EOD) support to the U.S. Secret Service, Department of State, and Department of Defense to protect the President and other designated persons. ART 6.12.8 also includes performing bomb and sabotage device recognition and safety precaution training. (FM 3-34.214) (CASCOM)

No.	Scale	Measure
01	Yes/No	EOD support allowed the unit to accomplish its mission
02	Yes/No	Collateral damage incurred during the EOD operation was within the acceptable limits.
03	Yes/No	Unit safeguarded classified materials and publications during the EOD operation.
04	Yes/No	Unit forwarded items and components of technical intelligence value to appropriate headquarters or agency.
05	Time	To respond to a request for EOD support
06	Time	To identify safety requirements and considerations
07	Time	To identify personnel, equipment, procedures, and additional support requirements.
08	Time	To coordinate with reporting agency for site support assistance, such as engineer, medical, security, and transportation.
09	Time	To clear domestic or foreign munitions and improvised devices that present a threat to military operations and military and civilian facilities, materiel, and personnel.
10	Time	To document render-safe procedures, as conducted, for unknown ordnance, if technical intelligence data does not exist.
11	Percent	Of safety precautions enforced during EOD operations.
12	Percent	Of reported munitions and improvised devices rendered safe.

No.	Scale	Measure
13	Percent	Of available EOD support expended on conducting bomb and sabotage device recognition and safety training.
14	Number	Of casualties during the EOD operation.
15	Number	And types of ordnance located and destroyed by EOD personnel.

ART 6.12.8.1 PROVIDE EXPLOSIVE ORDNANCE DISPOSAL SUPPORT TO RANGE CLEARANCE OPERATIONS

6-200. Range clearance operations require long-range planning. Range clearance operations are complex, have specific requirements, and vary for each range and installation. Range clearance operations and planning considerations should be covered in local plans and directives. (FM 3-34.214) (CASCOM)

No	Scale	Measure
01	Yes/No	Explosive ordnance disposal (EOD) support met the aim of the requesting agency.
02	Yes/No	Collateral damage incurred during the EOD operation was within acceptable limits.
03	Yes/No	Unit safeguarded classified materials and publications during the EOD operation.
04	Yes/No	Unit forwarded items and components of technical intelligence value to appropriate headquarters or agency.
05	Time	To move from the current location to the work site.
06	Time	To identify safety requirements and considerations concerned with ordnance found during range clearance.
07	Time	To identify personnel, equipment, procedures, and additional support requirements.
08	Time	To clear ordnance from the range clearance site.
09	Time	To coordinate with the reporting agency for site support assistance in areas such as engineer, medial, security, and transportation.
10	Time	To document render-safe procedures, as completed, for unknown ordnance if technical intelligence data does not exist.
11	Percent	Of safety precautions enforced during the EOD operation.
12	Percent	Of reported munitions destroyed or rendered safe.
13	Percent	Of reported munitions rendered safe per the requesting agency's priorities.
14	Number	Of casualties during the conduct of the EOD operation.
15	Number	And types of chemical, biological, radiological, and nuclear munitions destroyed or rendered harmless.
16	Number	Of requests for assistance responded to in a given period.

ART 6.12.8.2 DESTROY UNSERVICEABLE AMMUNITION

6-201. Supervise or assist in the routine destruction of unserviceable or surplus ammunition upon the request of an accountable agency. (FM 3-34.214) (CASCOM)

No.	Scale	Measure
01	Yes/No	Explosive ordnance disposal (EOD) support accomplishes the intent of the requesting agency.
02	Yes/No	Collateral damage incurred during the conduct of the EOD operation was in acceptable limits.
03	Yes/No	Unit safeguarded classified materials and publications during EOD operation.
04	Time	To identify safety requirements and considerations concerned with destruction of the unserviceable or surplus ordnance.
05	Time	To identify personnel, equipment, procedures, and additional support requirements.
06	Time	To move from the current location to the work site.

No.	Scale	Measure
07	Time	To coordinate with the reporting agency for site support assistance, in areas such as engineer, medical, security, and transportation.
08	Time	To destroy the unserviceable and surplus ordnance.
09	Percent	Of safety precautions enforced during the EOD operation.
10	Percent	Of nominated ordnance destroyed or rendered safe.
11	Number	Of nominated ordnance destroyed or rendered safe per the requesting agencies priorities.
12	Number	Of casualties during the execution of the EOD operation.
13	Number	And types of ordnance destroyed by EOD personnel.

ART 6.12.8.3 PROVIDE EXPLOSIVE ORDNANCE DISPOSAL SUPPORT TO THE DEFENSE ENVIRONMENT RESTORATION PROGRAM

6-202. Provide explosive ordnance disposal (EOD) support to agencies restoring the environment at military installations. (FM 3-34.214) (CASCOM)

No	Scale	Measure
01	Yes/No	EOD support met the aim of the requesting authorities.
02	Yes/No	Collateral damage incurred during the conduct of the EOD operation was in acceptable limits.
03	Yes/No	Unit safeguarded classified materials and publications during EOD operation.
04	Yes/No	Unit forwarded items and components of technical intelligence value to appropriate headquarters or agency.
05	Time	To move from the current location to the work site.
06	Time	To identify safety requirements and considerations concerned with the ordnance found during environment restoration projects.
07	Time	To identify personnel, equipment, procedures, and additional support requirements.
08	Time	To clear ordnance from the environmental restoration project site.
09	Time	To coordinate with the reporting agency for site support assistance, such as engineer, medical, security, and transportation.
10	Time	To document render-safe procedures, as completed, for unknown ordnance if technical intelligence data does not exist.
11	Percent	Of safety precautions enforced during the EOD operation.
12	Percent	Of reported munitions rendered safe.
13	Number	Of casualities during the conduct of the EOD operation.
14	Number	And types of ordnance located and destroyed by EOD personnel.
15	Number	Of requests for assistance responded to in a given period.

ART 6.12.8.4 PROVIDE EXPLOSIVE ORDNANCE DISPOSAL SUPPORT TO HOMELAND SECURITY OPERATIONS

6-203. Explosive ordnance disposal (EOD) supports homeland security operations by protecting its citizens and infrastructure from conventional and unconventional threats. Homeland security has two components. The first component is homeland defense. If the United States comes under direct attack or is threatened by hostile armed forces, Army forces under joint command conduct offensive and defensive missions as part of homeland defense. The second component is civil support. Army EOD forces conduct civil support operations, when requested, providing EOD expertise and capabilities to lead agency authorities. (FM 3-34.214) (CASCOM)

No	Scale	Measure
01	Yes/No	EOD support allowed the unit to accomplish its mission.
02	Yes/No	Collateral damage incurred during the EOD operation was within acceptable limits.
03	Yes/No	Unit safeguarded classified materials and publications during the EOD operation.
04	Yes/No	Unit forwarded items and components of technical intelligence value to appropriate headquarters or agency.
05	Time	To respond to a request for EOD support.
06	Time	Delay in executing the concept of operations due to the presence of unexploded ordnance and improvised devices.
07	Time	To gather intelligence information (what, when, where, how delivered, and type) regarding munitions.
08	Time	To identify safety requirements and considerations.
9	Time	To identify personnel, equipment, procedures, as completed, for unknown ordnance if technical intelligence data does not exist.
10	Time	To coordinate with reporting agency for site support assistance, such as engineer, medical, security, and transportation.
11	Time	To clear explosive ordnance and improvised devices.
12	Time	To document render-safe procedures, as completed, for unknown ordnance if technical intelligence data does not exist.
13	Percent	Of safety precautions enforced during the EOD operation.
14	Percent	Of reported munitions and improvised devices rendered safe.
15	Percent	Of available EOD support expended on performing bomb and sabotage device recognition and safety training.
16	Percent	Of patients received at medical treatment facilities who have unexploded ordnance with unexploded explosive ordnance in the wound.
17	Number	Of causalities during the EOD operation.
18	Number	And types of ordnance located and destroyed by EOD personnel.
19	Number	Of chemical, biological, radiological, nuclear, and high-yield explosives incidents responded to within a given period.

ART 6.12.8.5 PROVIDE EXPLOSIVE ORDNANCE DISPOSAL SUPPORT TO CIVIL AUTHORITIES

6-204. Provide assistance to include training to public safety and law enforcement agencies to address improvised explosive devices (IEDs). Provide explosive ordnance disposal (EOD) service when requested by local, state, or federal authorities in the interest of public safety. ART 6.12.8.5 includes assisting law enforcement personnel with war souvenir collection campaigns and the disposition of the explosive ordnance collected. (FM 3-34.214) (CASCOM)

No	Scale	Measure
01	Yes/No	The EOD support met the aim of the requesting civil authorities.
02	Yes/No	Collateral damage during the EOD operation was within acceptable limits.
03	Yes/No	Unit safeguarded classified materials and publications during the EOD operation.
04	Yes/No	Unit forwarded items and components of technical intelligence value to the appropriate headquarters or agency.
05	Time	To determine whether the EOD support requested by a civil authority is authorized under current laws and regulations.
06	Time	To move from the current location to the work site.
07	Time	To gather information about munitions (what, when, where, how delivered, and type).
08	Time	To identify safety requirements and considerations.
09	Time	To identify personnel, equipment, procedures, and additional support requirements.

No	Scale	Measure
10	Time	To coordinate with the reporting agency for site support assistance such as engineer, medical, security, and transportation.
11	Time	To clear munitions and IEDs.
12	Time	To document render-safe procedures, as completed, for unknown ordnance if technical intelligence data does not exist.
13	Time	Spent in developing and coordinating public or Department of Defense awareness campaigns on the dangers that war souvenirs pose to civilians.
14	Percent	Of safety precautions enforced during the EOD operation.
15	Percent	Of reported munitions and IEDs rendered harmless.
16	Percent	Of reported munitions and IEDs cleared per the requesting agencies priorities.
17	Percent	Of available EOD support expended on public safety training, including mine recognition training.
18	Number	Of casualties during the EOD operation.
19	Number	And types of ordnance located and destroyed by EOD personnel.
20	Number	Of requests for assistance from civil authorities responded to within a given period.

ART 6.12.8.6 PROVIDE SPECIFIC TRAINING TO PERSONNEL ON EXPLOSIVE ORDNANCE DISPOSAL

6-205. Explosive ordnance disposal (EOD) units provide training for military personnel, federal agencies, and public officials involved with civil emergency preparedness, law enforcement, and fire protection as requested. EOD units also provide training to Department of Defense, law enforcement, nongovernmental organizations, and emergency preparedness personnel on bomb threat search, improvised explosive devices defeat techniques, and unexploded explosive ordnance identification and marking procedures. (FM 3-34.214) (CASCOM)

No	Scale	Measure
01	Yes/No	EOD support met the aim of the requesting agency.
02	Yes/No	Unit safeguarded classified materials and publications during the EOD operation.
03	Time	To determine if current laws and regulations authorize the provision of the requested EOD support.
04	Time	To move from the current location to the work site.
05	Time	To gather intelligence information (what, when, where, how delivered, and type). regarding requested training.
06	Time	To identify safety requirements and considerations.
07	Time	To identify personnel, equipment, procedures, and additional support requirements.
08	Time	To coordinate with the reporting agency for site support assistance in areas such as engineer, medial, security, and transportation.
09	Time	To document render-safe procedures, as completed, for unknown ordnance if technical intelligence data does not exist.
10	Time	Spent in training personnel on recognizing and practicing immediate action drills when confronted by conventional or improvised explosive devices.
11	Percent	Of safety precautions enforced during the EOD operation.
12	Percent	Of available EOD support personnel expended on training.
13	Number	Of requests for assistance from civil authorities responded to within a given period.

ART 6.12.8.7 PROVIDE EXPLOSIVE ORDNANCE DISPOSAL SUPPORT TO THE SECRET SERVICE

6-206. Provide support to the U.S. Secret Service, Department of State, and Department of Defense to protect the President and other designated high-risk personnel. (FM 3-34.214) (CASCOM)

No	Scale	Measure
01	Yes/No	Explosive ordnance disposal (EOD) support met the aim of the requesting authorities.
02	Yes/No	Collateral damage incurred during the conduct of the EOD operation was in acceptable limits.
03	Yes/No	Unit safeguarded classified materials and publications during the EOD operation.
04	Yes/No	Unit forwarded items and components of technical intelligence value to appropriate headquarters or agency.
05	Time	To move from the current location to the work site.
06	Time	To identify safety requirements and considerations concerned with the ordnance found during environment restoration projects.
07	Time	To gather intelligence information (what, when, where, how delivered, and type) regarding munitions and improvised explosive devices.
08	Time	To identify safety requirements and considerations.
09	Time	To identify personnel, equipment, procedures, and additional support requirements.
10	Time	To coordinate with reporting and other agencies for additional site support assistance such as engineer, medical, security, and transportation.
11	Time	To clear munitions and improvised explosive devices (IEDs).
12	Time	To document render-safe procedures, as completed, for unknown ordnance if technical intelligence data does not exist.
13	Time	To train personnel providing executive protection services on recognizing and practicing immediate action drills when confronted by conventional or IEDs.
14	Percent	Of safety precautions enforce during EOD operations.
15	Percent	Of reported munitions and IEDs rendered safe.
16	Percent	Of reported munitions and IEDs rendered safe per the requesting agencies priorities.
17	Percent	Of available EOD support expended on training executive protection services personnel.
18	Percent	Of available EOD support expended on assisting the U.S. Secret Service and other government agencies that provide executive protection services.
19	Number	Of casualties during the conduct of the EOD operation.
20	Number	And types of ordnance located and destroyed by EOD personnel.
21	Number	Of requests for assistance from civil authorities responded to within a given period.

Chapter 7

ART 7.0: Full Spectrum Operations, Tactical Mission Tasks, and Operational Themes

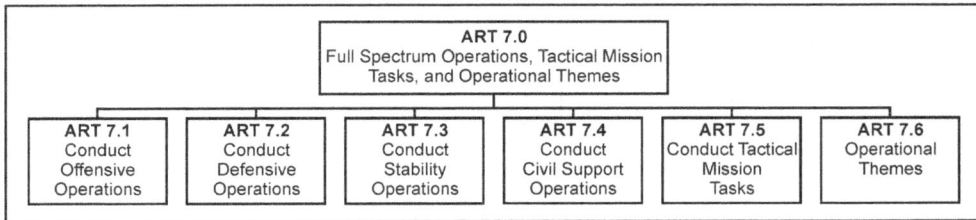

```
┌─────────────────────────────────────────────────────────────────────────────┐
│                              ART 7.0                                          │
│              Full Spectrum Operations, Tactical Mission                       │
│                   Tasks, and Operational Themes                               │
│                                                                               │
│  ┌──────────┐ ┌──────────┐ ┌──────────┐ ┌──────────┐ ┌──────────┐ ┌────────┐ │
│  │ ART 7.1  │ │ ART 7.2  │ │ ART 7.3  │ │ ART 7.4  │ │ ART 7.5  │ │ART 7.6 │ │
│  │ Conduct  │ │ Conduct  │ │ Conduct  │ │ Conduct  │ │ Conduct  │ │Operat- │ │
│  │Offensive │ │Defensive │ │Stability │ │Civil     │ │Tactical  │ │ional   │ │
│  │Operations│ │Operations│ │Operations│ │Support   │ │Mission   │ │Themes  │ │
│  │          │ │          │ │          │ │Operations│ │Tasks     │ │        │ │
│  └──────────┘ └──────────┘ └──────────┘ └──────────┘ └──────────┘ └────────┘ │
└─────────────────────────────────────────────────────────────────────────────┘
```

Tactical proficiency is not defined by mastery of written doctrine, but by the ability to employ available means to win battles and engagements. A tactical solution may not match any previous example; however, the language used to communicate that concept must be technically precise and doctrinally consistent, using commonly understood and accepted terms and concepts. This chapter sustains this doctrinal consistency by delineating the tactical missions and operations conducted by Army tactical forces. **A *tactical mission task* is a specific activity performed by a unit while executing a form of tactical operation or form of maneuver. It may be expressed as either an action by a friendly force or effects on an enemy force.** The tactical mission tasks provide commanders, their staffs, combat developers, training developers, and doctrine analysts a resource to assist in identifying missions units are to accomplish. Training developers use this chapter to base collective training plans on one or more of these missions.

7-1. Measures of any mission accomplishment are both objective and subjective in their evaluation by commanders. The most critical measure for all Army tactical missions and tasks is if the mission is accomplished. There is no other measure above this one. To measure mission accomplishment, commanders also consider if mission accomplishment was within the—

- Higher commander's intent of what the force must do and the conditions.
- Specified timeline.
- Commander's risk assessment for fratricide avoidance and collateral damage.
- Minimum expenditure of resources.
- Unit's capability of continuing or being assigned future missions and operations.

SECTION I – ART 7.1: CONDUCT OFFENSIVE OPERATIONS

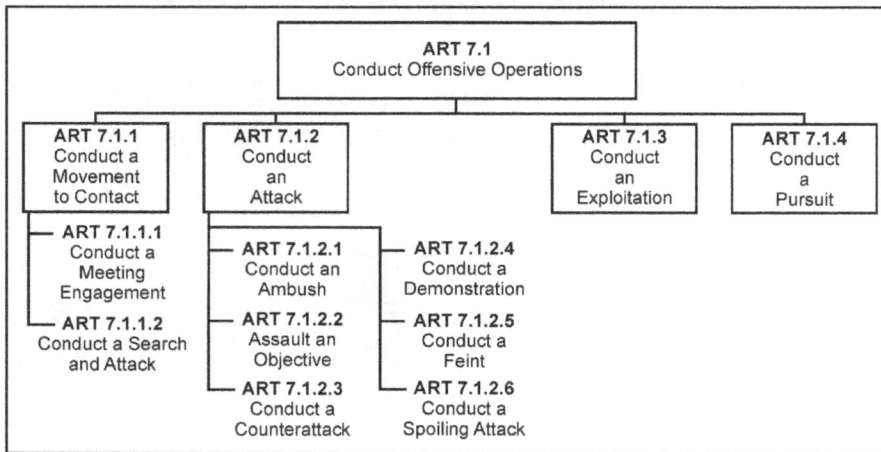

```
                        ART 7.1
              Conduct Offensive Operations

    ART 7.1.1          ART 7.1.2          ART 7.1.3        ART 7.1.4
    Conduct a          Conduct            Conduct          Conduct
    Movement           an                 an               a
    to Contact         Attack             Exploitation     Pursuit

    ART 7.1.1.1     ART 7.1.2.1      ART 7.1.2.4
    Conduct a       Conduct an       Conduct a
    Meeting         Ambush           Demonstration
    Engagement
    ART 7.1.1.2     ART 7.1.2.2      ART 7.1.2.5
    Conduct a Search Assault an       Conduct a
    and Attack      Objective        Feint

                    ART 7.1.2.3      ART 7.1.2.6
                    Conduct a        Conduct a
                    Counterattack    Spoiling Attack
```

7-2.   Offensive operations are operations conducted to defeat and destroy enemy forces and seize terrain, resources, and population centers. They impose the commander's will on the enemy. In combat operations, the offense is the decisive element of full spectrum operations. Against an adaptive and capable enemy, offensive operations are the most direct and sure means of seizing, retaining, and exploiting the initiative. Executing offensive operations compels the enemy to react, creating or revealing weaknesses that attacking forces can exploit. Successful offensive operations place tremendous pressure on defenders, creating a cycle of deterioration that can lead to their disintegration. Army forces conduct four tactical offensive tasks: movement to contact, attack, exploitation, and pursuit. (FM 3-0) (USACAC)

No.	Scale	Measure
01	Yes/No	Mission accomplished.
02	Yes/No	Mission accomplished within existing rules of engagement.
03	Yes/No	Mission accomplished without significant collateral damage or loss of noncombatants.
04	Time	To accomplish mission.
05	Percent	Of friendly forces available to continue mission.
06	Percent	Combat effectiveness of enemy force.

ART 7.1.1 CONDUCT A MOVEMENT TO CONTACT

7-3.   The movement to contact is an offensive action designed to develop the situation and establish or regain contact. Forces conducting a movement to contact seek to make contact with the smallest forces feasible. A movement to contact may result in a meeting engagement. (FM 3-90) (USACAC)

ART 7.1.1.1 CONDUCT A MEETING ENGAGEMENT

7-4.   A meeting engagement is a combat action that occurs when a moving force engages an enemy at an unexpected time and place. Such encounters normally occur by chance in small-unit operations, typically when two moving forces collide. They may result in brigade or larger unit operations when intelligence, surveillance, and reconnaissance operations have been ineffective. Meeting engagements can also occur when opposing forces are aware of the general presence, but not the exact location, of each other, and both decide to attack immediately. (FM 3-90) (USACAC)

ART 7.1.1.2 CONDUCT A SEARCH AND ATTACK

7-5.   Search and attack is a technique of conducting a movement to contact that shares many of the characteristics of an area security mission. (FM 3-90) (USACAC)

ART 7.1.2 CONDUCT AN ATTACK

7-6.   An attack is an offensive operation that destroys or defeats enemy forces, seizes and secures terrain, or both. Attacks incorporate coordinated movement supported by direct and indirect fires. They may be either decisive or shaping operations. Attacks may be hasty or deliberate, depending on the time available for assessing the situation, planning, and preparing. (FM 3-90) (USACAC)

ART 7.1.2.1 CONDUCT AN AMBUSH

7-7.   An ambush is an attack by lethal or nonlethal means from concealed positions on a moving or temporarily halted enemy. An ambush stops, denies, or destroys enemy forces by maximizing the element of surprise. Ambushes can employ direct fire systems as well as other destructive means, such as command-detonated mines, indirect fires, and supporting nonlethal fires. They may include an assault to close with and destroy the enemy forces. (FM 3-90) (USACAC)

ART 7.1.2.1.1 Conduct a Point Ambush

7-8.   A point ambush is a surprise attack by lethal or nonlethal capabilities from concealed positions on a moving or temporarily halted enemy near a given point. It may include an assault to close with and destroy enemy forces. (FM 3-21.10) (USAIS)

ART 7.1.2.1.2 Conduct an Area Ambush

7-9.   An area ambush consists of a series of surprise attacks by lethal or nonlethal capabilities from concealed positions on a moving or temporarily halted enemy in a specific area. It may include an assault to close with and destroy enemy forces. (FM 3-21.20) (USAIS)

ART 7.1.2.1.3 Conduct an Antiarmor Ambush

7-10.   An antiarmor ambush is a surprise attack by fire or other destructive means from concealed positions on moving or temporarily halted enemy armored vehicles. The ambush ends with the displacement of the antiarmor system to preclude its destruction by enemy counterambush actions. (FM 3-21.91) (USAIS)

ART 7.1.2.2 ASSAULT AN OBJECTIVE

7-11.   The assault on an objective is a short, violent, but well-ordered attack against an objective. It is the climax of an attack and involves the act of closing with the enemy, including the possible conduct of hand-to-hand fighting. (FM 3-90) (USACAC)

ART 7.1.2.3 CONDUCT A COUNTERATTACK

7-12.   A counterattack is an attack by part or all of a defending force against an attacking force with the general objective of denying attackers their goals. Commanders normally conduct counterattacks from a defensive posture. They direct them to defeat or destroy enemy forces or to regain control of terrain and facilities after enemy successes. They counterattack after enemies launch an attack, reveal their main effort, or offer an assailable flank. (FM 3-90) (USACAC)

ART 7.1.2.4 CONDUCT A DEMONSTRATION

7-13.   A demonstration is an attack designed to deceive enemies as to the location or time of the decisive operation by a display of force. Forces conducting a demonstration do not seek contact. Demonstrations are shaping operations. They seek to mislead enemies concerning the attacker's true intentions. They facilitate decisive operations by fixing enemies or diverting their attention from the decisive operation. Commanders allow enemies to detect a demonstration. (FM 3-90) (USACAC)

ART 7.1.2.5 CONDUCT A FEINT

7-14. A feint is a form of attack used to deceive the enemy as to the location or time of the actual decisive operation. Forces conducting a feint seek direct fire contact with the enemy but avoid decisive engagement. Feints divert attention from the decisive operation and prevent the enemy from focusing elements of combat power against it. They are usually shallow, limited-objective attacks conducted before or during the decisive operation. (FM 3-90) (USACAC)

ART 7.1.2.6 CONDUCT A SPOILING ATTACK

7-15. A spoiling attack is an attack that preempts or seriously impairs an enemy attack while the enemy is planning or preparing for it. Normally conducted from a defensive posture, spoiling attacks strike where and when enemies are most vulnerable—during preparations for attack in assembly areas and attack positions or while they are moving toward their line of departure. Therefore, proper timing and coordination with higher headquarters are critical requirements for spoiling attacks. A spoiling attack depends on accurate information on enemy dispositions. (FM 3-90) (USACAC)

ART 7.1.3 CONDUCT AN EXPLOITATION

7-16. Exploitation is an offensive operation that usually follows a successful attack and is designed to disorganize the enemy in depth. Exploitations seek to disintegrate enemy forces to the point where they have no alternative but surrender or take flight. Exploitations take advantage of tactical opportunities, foreseen or unforeseen. Division and higher headquarters normally plan exploitations as branches or sequels. (FM 3-90) (USACAC)

ART 7.1.4 CONDUCT A PURSUIT

7-17. A pursuit is an offensive operation designed to catch or cut off a hostile force attempting to escape, with the aim of destroying it. Pursuits are decisive operations that follow successful attacks or exploitations. They occur when enemies fail to organize a defense and attempt to disengage. If it becomes apparent that resistance has broken down entirely and the enemy is fleeing, a force can transition to a pursuit from any type of offensive operation. Pursuits entail rapid movement and decentralized control. (FM 3-90) (USACAC)

SECTION II – ART 7.2: CONDUCT DEFENSIVE OPERATIONS

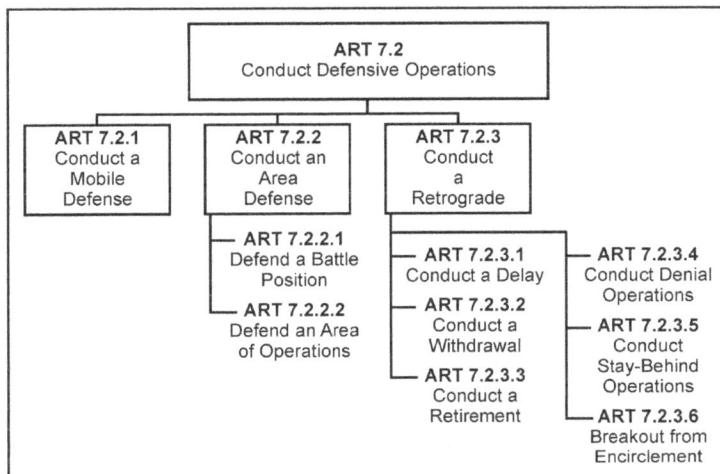

```
                        ART 7.2
                Conduct Defensive Operations

   ART 7.2.1        ART 7.2.2           ART 7.2.3
   Conduct a        Conduct an          Conduct
    Mobile            Area                 a
    Defense          Defense           Retrograde

                   ART 7.2.2.1        ART 7.2.3.1        ART 7.2.3.4
                   Defend a Battle    Conduct a Delay    Conduct Denial
                   Position                              Operations
                   ART 7.2.2.2        ART 7.2.3.2        ART 7.2.3.5
                   Defend an Area     Conduct a          Conduct
                   of Operations      Withdrawal         Stay-Behind
                                                         Operations
                                      ART 7.2.3.3
                                      Conduct a          ART 7.2.3.6
                                      Retirement         Breakout from
                                                         Encirclement
```

7-18. *Defensive operations* are combat operations conducted to defeat an enemy attack, gain time, economize forces, and develop conditions favorable for offensive or stability operations. The defense alone normally cannot achieve overarching mission objectives. However, it can create conditions for a counteroffensive that allows Army forces to regain the initiative. Defensive operations can also establish a protective barrier behind which stability operations can progress. Defensive operations counter enemy offensive operations. They defeat attacks, destroying as much of the attacking enemy force as possible. They also preserve control over land, resources, and populations. Defensive operations retain terrain, guard populations, and protect critical capabilities. They can be used to gain valuable time and economize forces to allow execution of offensive tasks elsewhere. Three tasks are associated with the defense: mobile defense, area defense, and retrograde. Defending commanders combine the three tasks to fit the situation. (FM 3-0) (USACAC)

ART 7.2.1 CONDUCT A MOBILE DEFENSE

7-19. The mobile defense concentrates on the destruction or defeat of the enemy through a decisive attack by a striking force. A mobile defense requires defenders to have greater mobility than attackers. Defenders combine offensive, defensive, and delaying actions to lure attackers into positions where attackers are vulnerable to counterattack. (FM 3-90) (USACAC)

ART 7.2.2 CONDUCT AN AREA DEFENSE

7-20. The area defense concentrates on denying an enemy force access to designated terrain for a specific time rather than destroying the enemy outright. The bulk of the defending force combines static defensive positions, engagement areas, and small, mobile reserves to retain ground. Keys to successful area defenses include effective and flexible control, synchronization, and distribution of fires. Area defenses can also be part of a larger mobile defense. (FM 3-90) (USACAC)

ART 7.2.2.1 DEFEND A BATTLE POSITION

7-21. This task involves denying an enemy force access to the terrain encompassed by a specific battle position. The battle position is a tactical control graphic that depicts the location and general orientation of the majority of the defending forces. Five kinds of battle positions exist to include the strong point. (FM 3-90) (USACAC)

ART 7.2.2.2 Defend an Area of Operations

7-22. This task involves denying an enemy force access across a designated area of operations. (FM 3-90) (USACAC)

ART 7.2.3 CONDUCT A RETROGRADE

7-23. The retrograde is a type of defensive operation that involves organized movement away from the enemy. The three forms of retrograde operations are delays, withdrawals, and retirements. A commander can direct the conduct of denial operations and stay-behind operations as subordinate activities within the context of a retrograde operation. Commanders use retrogrades as part of a larger concept of operations to create conditions to regain the initiative and defeat the enemy. Retrogrades improve the current situation or prevent a worse situation from occurring. (FM 3-90) (USACAC)

ART 7.2.3.1 Conduct a Delay

7-24. A delay is a form of retrograde in which a force under pressure trades space for time by slowing the enemy's momentum and inflicting maximum damage on the enemy without, in principle, becoming decisively engaged. Delays gain time for friendly forces to establish defenses, cover defending or withdrawing units, protect friendly unit flanks, contribute to economy of force, draw the enemy into unfavorable positions, and determine the enemy main effort. (FM 3-90) (USACAC)

ART 7.2.3.2 Conduct a Withdrawal

7-25. A withdrawal is a planned operation in which a force in contact disengages from an enemy force. Withdrawals may involve all or part of a committed force. Commanders conduct withdrawals to preserve the force, release it for a new mission, avoid combat under undesirable conditions, or reposition forces. Enemy pressure may or may not be present during withdrawals. Withdrawing forces may be unassisted or assisted by another friendly force. (FM 3-90) (USACAC)

ART 7.2.3.3 Conduct a Retirement

7-26. A retirement is a retrograde in which a force not in contact with the enemy moves away from the enemy. (FM 3-90) (USACAC)

ART 7.2.3.4 Conduct Denial Operations

7-27. Denial operations hinder or deny the enemy the use of space, personnel, supplies, or facilities. It may include destroying, removing, or contaminating those supplies and facilities, or erecting obstacles. (FM 3-90) (USACAC)

ART 7.2.3.5 Conduct Stay-Behind Operations

7-28. A stay-behind operation occurs when a commander leaves a unit in position to conduct a specified mission while the remainder of the force withdraws or retires from an area. (FM 3-90) (USACAC)

ART 7.2.3.6 Breakout from Encirclement

7-29. A breakout is an offensive and a defensive operation. An encircled force normally attempts to conduct breakout operations when one of the following four conditions exist: the commander directs the breakout or the breakout falls within the intent of a higher commander; the encircled force does not have sufficient relative combat power to defend itself against enemy forces attempting to reduce the encirclement; the encircled force lacks adequate terrain available to conduct its defense; the encircled force cannot sustain itself long enough to be relieved by forces outside the encirclement. (FM 3-90) (USACAC)

SECTION III – ART 7.3: CONDUCT STABILITY OPERATIONS

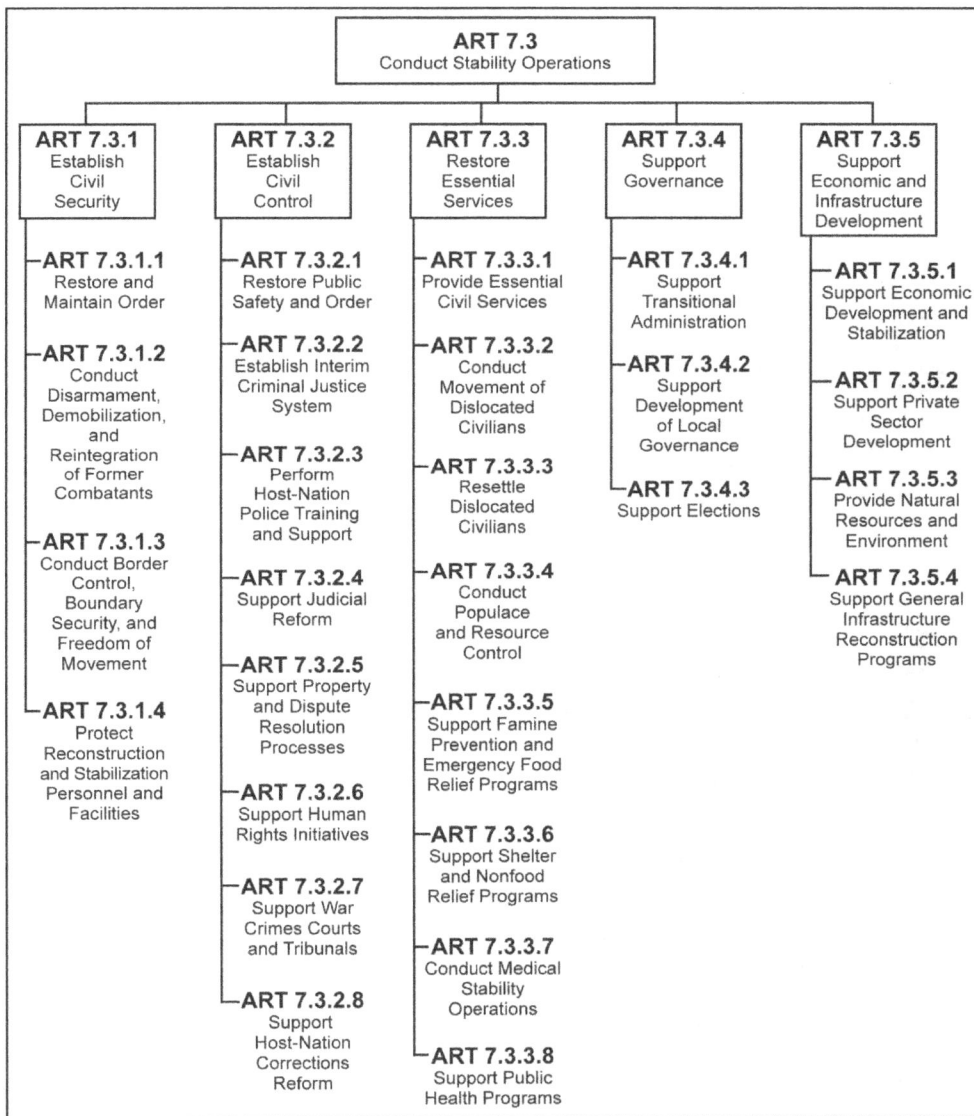

ART 7.3
Conduct Stability Operations

ART 7.3.1
Establish Civil Security

- **ART 7.3.1.1** Restore and Maintain Order
- **ART 7.3.1.2** Conduct Disarmament, Demobilization, and Reintegration of Former Combatants
- **ART 7.3.1.3** Conduct Border Control, Boundary Security, and Freedom of Movement
- **ART 7.3.1.4** Protect Reconstruction and Stabilization Personnel and Facilities

ART 7.3.2
Establish Civil Control

- **ART 7.3.2.1** Restore Public Safety and Order
- **ART 7.3.2.2** Establish Interim Criminal Justice System
- **ART 7.3.2.3** Perform Host-Nation Police Training and Support
- **ART 7.3.2.4** Support Judicial Reform
- **ART 7.3.2.5** Support Property and Dispute Resolution Processes
- **ART 7.3.2.6** Support Human Rights Initiatives
- **ART 7.3.2.7** Support War Crimes Courts and Tribunals
- **ART 7.3.2.8** Support Host-Nation Corrections Reform

ART 7.3.3
Restore Essential Services

- **ART 7.3.3.1** Provide Essential Civil Services
- **ART 7.3.3.2** Conduct Movement of Dislocated Civilians
- **ART 7.3.3.3** Resettle Dislocated Civilians
- **ART 7.3.3.4** Conduct Populace and Resource Control
- **ART 7.3.3.5** Support Famine Prevention and Emergency Food Relief Programs
- **ART 7.3.3.6** Support Shelter and Nonfood Relief Programs
- **ART 7.3.3.7** Conduct Medical Stability Operations
- **ART 7.3.3.8** Support Public Health Programs

ART 7.3.4
Support Governance

- **ART 7.3.4.1** Support Transitional Administration
- **ART 7.3.4.2** Support Development of Local Governance
- **ART 7.3.4.3** Support Elections

ART 7.3.5
Support Economic and Infrastructure Development

- **ART 7.3.5.1** Support Economic Development and Stabilization
- **ART 7.3.5.2** Support Private Sector Development
- **ART 7.3.5.3** Provide Natural Resources and Environment
- **ART 7.3.5.4** Support General Infrastructure Reconstruction Programs

7-30. Stability operations encompass various military missions, tasks, and activities conducted outside the United States in coordination with elements of combat to maintain or reestablish a safe and secure environment, provide essential government services, emergency infrastructure reconstruction, and humanitarian relief. Stability operations can be conducted in support of a host-nation or interim government or as part of an occupation when no government exists. Stability operations involve both coercive and constructive actions by the military force. They are designed to establish a safe and secure environment; facilitate reconciliation among local or regional adversaries; establish political, legal, social, and economic institutions; and facilitate the transition to legitimate local governance. Army forces conduct five stability tasks: civil security, civil control, restore essential services, support to governance, and support to economic and infrastructure development. (FM 3-0) (USACAC)

> *Note:* This task branch only addresses those tasks that support the conduct of Army stability operations. Other tasks that support stability operations are addressed elsewhere in the AUTL:
> ART 6.5.4 (Provide Protective Services for Selected Individuals) addresses the protection of personnel.
> ART 6.5.2 (Conduct Critical Installations and Facilities Security) addresses the protection of facilities.
> ART 6.12 (Provide Explosive Ordnance Disposal Protection Support) addresses the conduct of unexploded ordnance disposal.

ART 7.3.1 ESTABLISH CIVIL SECURITY

7-31. Civil security involves providing for the safety of the state and its population, including protection from internal and external threats. Civil security includes a diverse set of activities, ranging from enforcing peace agreements to executing disarmament, demobilization, and reintegration. For the other stability tasks to be effective, civil security is required. Establishing a safe, secure, and stable environment for the local populace is key to obtaining their support for the overall stability operation. To help establish civil security, military forces enforce ceasefires, supervise disengagement of belligerent forces, and neutralize potential adversaries. As soon the host-nation security forces can safely perform this task, Army forces transition civil security responsibilities to them. (FM 3-0) (USACAC)

> *Note:* ART 7.6.3.4 (Conduct Peace Enforcement Operations) addresses the measures that help military forces establish a sustained peace.

No.	Scale	Measure
01	Yes/No	Unit protected vulnerable elements of population—refugees, internally displaced persons, women, children.
02	Yes/No	Unit provided interim security programs for at-risk populations.
03	Yes/No	Unit established and maintained order in refugee camps and population centers.
04	Yes/No	Unit ensured safety of quartered personnel and families.
05	Yes/No	Unit ensured adequate health, food provisions, and security for demobilized belligerents.
06	Yes/No	Unit protected and secured places of religious worship and cultural sites.
07	Yes/No	Unit protected and secured critical infrastructure, natural resources, civil registries, and property ownership documents.
08	Yes/No	Unit identified, secured, and protected stockpiles of conventional, biological, radiological, nuclear, and chemical materials.
09	Yes/No	Unit secured military depots, equipment, ammunition depots, and means of communication.
10	Yes/No	Unit protected and secured strategically important institutions such as government buildings, museums, religious sites, courthouses, and communications.
11	Yes/No	Unit secured records, storage, equipment, and funds related to criminal justice and security institutions.
12	Yes/No	Unit provided security for negotiations among host-nation belligerents.
13	Yes/No	Unit protected private property and factories.
14	Time	To begin ceasefire enforcement.
15	Time	To reintegrate combatants and promote civilian control.
16	Time	To establish public order and safety
17	Time	To develop integrated command, control, intelligence and information sharing arrangements between international military, constabulary, and civilian police forces.
18	Time	To create host-nation capacity to protect government sponsored civilian stabilization and reconstruction personnel.
19	Time	To create host-nation capacity to protect host-nation individuals, infrastructure, and institutions.

ART 7.3.1.1 RESTORE AND MAINTAIN ORDER

7-32. Operations to restore order are conducted to halt violence and to support, reinstate, or establish civil authorities. These operations provide security and stability after a conflict while setting the preconditions for beginning disarmament, demobilization, and reintegration. This is critical to providing effective security for the local populace by reducing their exposure to the threat of violent conflict and help military forces establish a sustained peace by focusing on processes and activities fundamental to conflict transformation. (FM 3-07) (USACAC)

No.	Scale	Measure
01	Yes/No	Vulnerable elements of population (dislocated civilians) were protected.
02	Yes/No	Security forces ensured humanitarian aid and access to endangered populations and refugee camps.
03	Yes/No	Civilian police functions included investigating crimes and making arrests.
04	Yes/No	Unit supervised incarceration processes and transfers to prison facilities.
05	Yes/No	Unit transferred public security responsibilities to host-nation police forces.
06	Yes/No	Procedures were in place to control crowds, prevent looting, and manage civil disturbances.
07	Yes/No	Unit controlled crowds, prevented looting, and managed civil disturbances.
08	Yes/No	Unit maintained positive relations with the local populace while conducting crowd and disturbance control.
09	Yes/No	Unit maintained positive relations with the local populace while conducting interim police operations.
10	Time	To transfer police operations responsibility to host-nation police.
11	Time	To transfer civil disturbance responsibility to host-nation police.

ART 7.3.1.2 CONDUCT DISARMAMENT, DEMOBILIZATION, AND REINTEGRATION OF FORMER COMBATANTS

7-33. Disarmament, demobilization, and reintegration of former combatants is fundamental to most efforts to establish stability and lasting peace. It includes physically disbanding armed groups, removing the means of combat from former combatants and belligerents, and reintegrating them into society. Disarmament may include seizing ammunition, collecting and destroying weapons and supplies, closing weapons and ammunition factories, and preventing resupply. (FM 3-07) (USACAC)

No.	Scale	Measure
01	Yes/No	Unit implemented plans for disposition on host-nation forces and other national security institutions.
02	Yes/No	Unit identified, gathered, and disbanded structural elements of belligerent groups.
03	Yes/No	Unit implemented reintegration strategy including assessment of absorptive capacity of economic and social sectors.
04	Yes/No	Unit coordinated with overall political and economic recovery plans.
05	Yes/No	Unit reintegrated former combatants into society.
06	Yes/No	Unit established and enforced weapons control programs, including collection and destruction.
07	Yes/No	Unit disarmed former combatants and belligerents.
08	Yes/No	Unit reduced availability of unauthorized weapons.
09	Yes/No	Unit provided reassurances and incentives for disarmed factions.
10	Yes/No	Unit ensured safety of quartered personnel and families.
11	Yes/No	Unit established monitoring program.
12	Yes/No	Unit ensured adequate health, food provisions, and security for belligerents including follow-up services for reintegration.

No.	Scale	Measure
13	Time	To implement plans to identify future roles, missions, and structure of host-nation forces and other national security institutions.
14	Time	To vet senior officers and other individuals for past abuses.
15	Time	To coordinate and integrate actions with disarmament, demobilization, and reintegration plans.
16	Time	To implement plans establishing transparent entry, promotion, and retirement systems.
17	Time	To implement plans establishing programs to support civilian oversight of military.
18	Time	To implement plans to provide jobs, pensions, or other material support for demobilized forces.
19	Time	To provide job training, health screening, education, and employment assistance for demobilized forces.

ART 7.3.1.2.1 Forcibly Separate Belligerents

7-34. Forcible separation may involve reducing the combat capability of one or more of the belligerent parties by disarming and demobilizing them. The peace enforcement force normally retains the right of first use of force. Forces conducting forcible separation require extensive offensive combat capability. The goal is to force the belligerent parties to disengage, withdraw, and subsequently establish a buffer zone or demilitarized zone. Security operations—such as screening, combat and reconnaissance patrolling, performing cordon and search, and establishing checkpoints and roadblocks to control movement into and within the buffer zone or demilitarized zone—may be conducted to maintain the separation of belligerent parties. (FM 3-07) (USACAC)

No.	Scale	Measure
01	Yes/No	Unit established military technical agreement that enumerates the agreed separation parameters.
02	Yes/No	Unit established collaboration mechanisms to negotiate and verify withdrawal procedures.
03	Yes/No	Unit established and controlled buffer zone or demilitarized zone in three dimensions.
04	Yes/No	Unit observed and reported on the disputing parties' compliance with a cease-fire.
05	Yes/No	Unit established liaison officers, with appropriate transportation and communications, with the headquarters of the opposing sides.
06	Yes/No	Unit established a quick reaction force.
07	Yes/No	Unit planned to support civilians in the buffer zone or demilitarized zone.
08	Yes/No	Unit planned for uncooperative local officials, demonstrations, and other forms of civil disturbance preventing mission accomplishment.
09	Yes/No	Unit assisted in moving personnel and equipment through the buffer zone or demilitarized zone.

ART 7.3.1.2.2 Disarm Belligerents

7-35. The mandate may require the peace enforcement force to disarm or demobilize the belligerent parties. These tasks are complex, difficult, and often dangerous. The peace enforcement force demonstrates a clear resolve and intent to disarm or demobilize designated belligerent parties according to the agreement. (FM 3-07) (USACAC)

No.	Scale	Measure
01	Yes/No	Unit developed integrated plan outlining the parameters for disarmament and weapons control that accounts for joint and interagency approaches.
02	Yes/No	Unit established collaboration mechanisms to negotiate and verify disarmament.
03	Yes/No	Unit determined the inventory of weapons and munitions.
04	Yes/No	Unit identified and prepared the areas for weapons turn in, disposal, and unit demobilization.

No.	Scale	Measure
05	Yes/No	Unit conducted information engagement.
06	Yes/No	Unit provided technical expertise on weapons and munitions.
07	Yes/No	Unit established accountability mechanisms.
08	Yes/No	Unit ensured that resources were allocated.

ART 7.3.1.2.3 Demobilize Belligerents

7-36. Demobilization is the process of transitioning a conflict or wartime military establishment and defense-based civilian economy to a peacetime configuration while maintaining national security and economic vitality. Following demobilization, Army forces may support integrating military and paramilitary forces into society by providing training, advice, and assistance for the new defense or security force. (FM 3-07) (USACAC)

No.	Scale	Measure
01	Yes/No	Unit provided intelligence support and overall security during the demobilization process.
02	Yes/No	Unit provided incentives and disincentives for the forming, arming, and training of the new defense force.
03	Yes/No	Unit was supported with information operations.
04	Yes/No	Unit provided liaison coordination teams to local commanders as a confidence-building measure during the disarmament and demobilization.
05	Yes/No	Unit developed and coordinated training plans.
06	Yes/No	Unit outfitted the new force per plan.
07	Yes/No	Unit trained the new force.

ART 7.3.1.2.4 Establish Protected Areas

7-37. The requirement to establish and supervise a protected or safe area can arise when any community is at risk from persistent attack. Unless those in the safe area are disarmed, it may be used as a base from which to conduct raids and attacks. Commanders must be clear on what is expected of the force tasked to establish and maintain a protected or safe area. (FM 3-07) (USACAC)

No.	Scale	Measure
01	Yes/No	Unit surveyed and identified area.
02	Yes/No	Unit developed the rules of engagement and the memorandum of understanding for the safe area.
03	Yes/No	Unit disarmed and demobilized the safe area.
04	Yes/No	Unit was supported with information operations.
05	Yes/No	Unit established checkpoints and control measures.
06	Yes/No	Unit established a quick reaction force.
07	Yes/No	Unit planned for extraction or reinforcement.
08	Yes/No	Unit supported humanitarian efforts in safe area.

ART 7.3.1.3 CONDUCT BORDER CONTROL, BOUNDARY SECURITY, AND FREEDOM OF MOVEMENT

7-38. Border controls are necessary to regulate immigration, control the movements of the local populace, collect excise taxes or duties, limit smuggling, and control the spread of disease vectors through quarantine. Generally, border and coast guard forces secure national boundaries while customs officials regulate the flow of people, animals, and goods across state borders. (FM 3-07) (USACAC)

No.	Scale	Measure
01	Yes/No	Unit established border security, including customs regime to prevent arms smuggling, interdict contraband (such as drugs and natural resources), prevent trafficking of persons, regulate immigration and emigration, and establish control over major points of entry.
02	Yes/No	Unit trained and equipped border control and boundary security personnel.
03	Yes/No	Unit established and disseminated rules pertaining to populace movement and curfews.
04	Yes/No	Unit assisted in establishing a policy to regulate air and overland movement.
05	Yes/No	Unit transferred responsibility for movement regulations to the indigenous populations and institutions, nongovernmental organizations, or other government agencies as appropriate.
06	Yes/No	Unit provided full freedom of movement.
07	Yes/No	Unit dismantled roadblocks and established checkpoints.
08	Time	To establish identification rules and protocol relating to personal identification, property ownership, court records, voter registries, birth certificates, and driving licenses, including securing documents.

ART 7.3.1.4 PROTECT RECONSTRUCTION AND STABILIZATION PERSONNEL AND FACILITIES

7-39. It may be necessary for Army forces to extend protection and support to key civilian personnel to ensure their continued contribution to the overall stability operation. In the interest of transparency, military forces specifically request and carefully negotiate this protection. Additionally, forces may provide transportation and security for nongovernmental organizations and other government agencies. (FM 3-07) (USACAC)

No.	Scale	Measure
01	Yes/No	Unit protected government-sponsored civilian reconstruction and stabilization personnel.
02	Yes/No	Unit protected contractor and civilian reconstruction and stabilization personnel and resources.
03	Yes/No	Unit provided emergency logistic support, as required.
04	Yes/No	Unit completed emergency de-mining and unexploded explosive ordnance removal.
05	Yes/No	Unit completed mapping, surveys, and marked minefields.

ART 7.3.2 ESTABLISH CIVIL CONTROL

7-40. Civil control regulates selected behavior and activities of individuals and groups. This control reduces risk to individuals or groups and promotes security. Civil control centers on rule of law, supported by efforts to rebuild the host-nation judiciary and corrections systems. Civil control channels the population's activities to allow provision of security and essential services while coexisting with a military force conducting operations. (FM 3-0) (USACAC)

No.	Scale	Measure
01	Yes/No	Unit established broad public information programs to promote efforts for reconciliation.
02	Time	To create and strengthen legal aid and nongovernmental organization groups.
03	Time	To foster support for or establish mechanisms and local capacity to protect human rights and resolve conflict; support citizen advocacy organizations.
04	Time	To train existing host-nation police in international policing standards.
05	Time	To initiate training programs based upon institutional reforms and new laws, consisting of established mentoring programs with both international and local professionals.
06	Time	To deploy interim justice personnel (judges, prosecutors, defense advocates, court administrators, corrections staff, police investigators) to supplement host-nation criminal justice system.
07	Time	To enhance participation through public outreach.

No.	Scale	Measure
08	Time	To rebuild correctional institutions, including administrative and rehabilitative capacities.
09	Time	To implement mechanisms to prevent unauthorized seizures of land or property.

ART 7.3.2.1 RESTORE PUBLIC SAFETY AND ORDER

7-41. The military may provide a broad range of activities to protect the civilian populace, provide interim policing and crowd control, and secure critical infrastructure. ART 7.3.2.1 represents actions that must occur during and after direct armed conflict to ensure the long-term sustainability of any reform efforts. (FM 3-07) (USACAC)

No.	Scale	Measure
01	Yes/No	Unit protected vulnerable elements of the population.
02	Yes/No	Unit ensured humanitarian aid and security forces had access to endangered populations and refugee camps.
03	Yes/No	Unit performed civil police functions, including investigating crimes and making arrests.
04	Yes/No	Unit located and safeguarded key witnesses, documents, and other evidence related to key ongoing or potential investigations and prosecutions.
05	Yes/No	Unit controlled crowds, prevented looting, and managed civil disturbances.
05	Yes/No	Unit protected and secured places of religious worship and cultural sites.
06	Yes/No	Unit protected and secured critical infrastructure, natural resources, civil registries, and property ownership documents.
07	Yes/No	Unit protected and secured strategically important institutions (such as government buildings, museums, religious sites, and courthouses).
08	Yes/No	Unit secured records, storage equipment, and funds related to criminal justice and security institutions.
09	Yes/No	Unit identified, secured, and protected stockpiles of conventional, nuclear, biological, radiological, and chemical materiel.
10	Yes/No	Unit secured military depots, equipment, ammunition dumps, and means of communication.
11	Yes/No	Unit trained and mentored host-nation police forces.
12	Yes/No	Unit built host-nation capacity to protect infrastructure and public institutions.

ART 7.3.2.2 ESTABLISH INTERIM CRIMINAL JUSTICE SYSTEM

7-42. Army forces may be required assist in the restoration of governance since establishing an interim justice system is a prerequisite. This restoration requires a wide range of skilled professionals, including judges, prosecutors, court administrators, corrections personnel, law enforcement, and investigators. (FM 3-07) (USACAC)

No.	Scale	Measure
01	Yes/No	Unit deployed interim justice personnel to supplement host-nation criminal justice system.
02	Yes/No	Unit enacted interim legal codes and procedures permitted by international law.
03	Yes/No	Unit assessed host-nation capacity to combat organized crime.

+ ART 7.3.2.3 PERFORM HOST-NATION POLICE TRAINING AND SUPPORT

7-43. Support and establish viable host-nation police forces and penal organizations capable of maintaining a safe and secure environment, upholding the rule of law, and reinforcing the legitimacy of the government. Ensure coordination of host-nation police and penal capabilities with the host-nation judicial system to maintain credibility and legitimacy for the government. (FM 3-39) (USAMPS)

No.	Scale	Measure
01	Yes/No	Unit identified and documented existing and required police capabilities.
02	Yes/No	Unit identified and documented existing and required police infrastructure.
03	Yes/No	Efforts coordinated and synchronized with Department of State, Department of Justice, and international police agencies.
04	Yes/No	Unit identified training requirements to include basic skills, entry- through senior-level management, and specialty sections.
05	Yes/No	Civilian police functions included investigating crimes and making arrests.
06	Yes/No	Unit transferred police responsibilities to host-nation police forces.
07	Yes/No	Procedures were in place for recruiting, vetting, training, and employing host-nation police.
08	Time	To establish liaison with Department of State, Department of Justice, and international police agencies.
09	Time	To develop plan for establishment of host-nation police training program.
10	Time	To develop patrol and facilities distribution plan and manning requirements.
11	Number	Of host-nation police trained and supported.

ART 7.3.2.4 SUPPORT JUDICIAL REFORM

7-44. The reform of judicial bodies is integral to rule of law and provides the necessary framework for security reform. The support provided to judicial institutions parallels efforts with police and security forces to enhance the state's capability in the security sector. In a failed state, military forces may initially perform these functions and can be a critical enabler of success over time. (FM 3-07) (USACAC)

No.	Scale	Measure
01	Yes/No	Unit identified host-nation legal professionals.
02	Yes/No	Unit identified actual and potential leaders to incorporate into reform process.
03	Yes/No	Unit established vetting criteria.
04	Yes/No	Unit educated criminal justice personnel on interim legal codes.
05	Yes/No	Unit inventoried and assessed courts, law schools, legal libraries, and bar associations.
06	Yes/No	Unit rehabilitated or constructed necessary facilities.
07	Yes/No	Unit vetted host-nation legal professionals.

ART 7.3.2.5 SUPPORT PROPERTY AND DISPUTE RESOLUTION PROCESSES

7-45. One of the most vital services provided by the judiciary branch is the resolution of property disputes. In a fragile state, long-standing disputes over ownership and control of property are common. Authorities must implement dispute resolution mechanisms. This prevents the escalation of violence that can occur in the absence of law and order as people seek resolution on their own terms. Typically, the military's role in resolving disputes is limited to transitional military authority where these mechanisms are implemented in the absence of a functioning host-nation government. (FM 3-07) (USACAC)

No.	Scale	Measure
01	Yes/No	Unit implemented mechanisms to prevent unauthorized occupation or seizure of land or property.
02	Yes/No	Unit publicized dispute resolution process.
03	Yes/No	Unit coordinated dispute resolution process to deter violence and retribution.
04	Yes/No	Unit developed plans to improve the existing property control system of the host nation.
05	Yes/No	Unit established liaison with civilian and military engineers or property control agencies and evaluated the current property control system.
06	Yes/No	Unit prepared and reviewed property control support plans for compliance with international laws, treaties, and agreements.

No.	Scale	Measure
07	Yes/No	Unit established a claims process for return and compensation for seized property.
08	Yes/No	Property acquisition plans classified the property to be acquired; identified and coordinated the acquisition of property for military use; prescribed a recording system for property acquired by the military from civilian sources; prescribed measures to protect and preserve civilian ownership records; prescribed measures to safeguard and properly managed the acquired property; planned for scheduling the acquisition of property; and planned for controlling negotiable assets and resources of potential military use not under the supervision of other agencies.
09	Time	To establish procedures to resolve property rights for land and subterranean resources.
10	Time	To establish a process to determine land ownership, if disputed (for example, if internally displaced person claims to own land).

ART 7.3.2.6 SUPPORT HUMAN RIGHTS INITIATIVES

7-46. A safe and secure environment maintained by a civilian law enforcement system must exist and operate in accordance with internationally recognized standards and with respect for internationally recognized human rights and freedoms. Army forces may need to provide limited support. (FM 3-07) (USACAC)

No.	Scale	Measure
01	Yes/No	Unit assessed needs of vulnerable population.
02	Yes/No	Unit prevented further abuse of vulnerable populations.
03	Yes/No	Unit established the conditions that enable success of agencies and organizations that provide for the long-term well being of these populations.
04	Yes/No	Unit acted preemptively to deter human rights abuses.
05	Yes/No	Unit provided medical assistance to victims of human trafficking.
06	Yes/No	Unit implemented rape prevention and medical treatment procedures.

ART 7.3.2.7 SUPPORT WAR CRIMES COURTS AND TRIBUNALS

7-47. While the military government operates military commissions and provost courts, the international community oversees the conduct of war crimes courts and tribunals. As part of the broad processes that represent legal system reform, military forces identify, secure, and preserve evidence for courts and tribunals of war crimes and crimes against humanity. Military forces also provide support in other forms, to include helping to establish courts and tribunals, supporting the investigation and arrest of war criminals, and coordinating efforts with other agencies and organizations. (FM 3-07) (USACAC)

No.	Scale	Measure
01	Yes/No	Unit acquired secure facilities.
02	Yes/No	Unit established atrocity reporting system.
03	Yes/No	Unit documented and preserved evidence of mass atrocities.
04	Yes/No	Unit published progress reports.
05	Yes/No	Unit published indictments and statements.
06	Yes/No	Unit assisted in investigation, arrest, and transfer of suspects to international courts.
07	Yes/No	Unit ensured witness protection and media access.

+ ART 7.3.2.8 SUPPORT HOST-NATION CORRECTIONS REFORM

7-48. Support and establish viable host-nation internment/corrections officers capable of maintaining a safe and secure environment, upholding the rule of law, and reinforcing the legitimacy of the government. Ensure coordination of host-nation internment and corrections officers and penal facilities with the host-nation judicial system to maintain credibility and legitimacy for the government. Aligned with the internment and resettlement function. (FM 3-19.40) (USAMPS)

No.	Scale	Measure
01	Yes/No	Unit identified and documented existing and required internment and corrections officers and penal capabilities.
02	Yes/No	Unit identified and documented existing and required internment and corrections officers and penal infrastructure.
03	Yes/No	Efforts coordinated and synchronized with Department of State, Department of Justice, and international prison agencies.
04	Yes/No	Unit identified training requirements to include basic skills, entry- through senior-level management, and specialty sections.
05	Yes/No	Unit supervised incarceration processes and transfer to prison facilities.
06	Yes/No	Unit transferred penal and internment responsibilities to host-nation internment and correction officers.
07	Yes/No	Procedures were in place for recruiting, vetting, training, and employing host-nation internment and corrections officers.
08	Time	To establish liaison with Department of State, Department of Justice, and international internment and corrections agencies.
09	Time	To develop plan for establishment of host-nation internment and corrections officers training program.
10	Time	To develop facilities distribution plan and manning requirements.
11	Number	Of host-nation penal personnel trained and supported.
12	Number	Of host-nation penal facilities supported.
13	Number	Of units supporting host-nation internment and corrections.

ART 7.3.3 RESTORE ESSENTIAL SERVICES

7-49. Army forces establish or restore the most basic services and protect them until a civil authority or the host nation can provide them. Normally, Army forces support other government, intergovernmental, and host-nation agencies. When the host-nation or other agency cannot perform its role, Army forces may provide the basics directly. (FM 3-0) (USACAC)

ART 7.3.3.1 PROVIDE ESSENTIAL CIVIL SERVICES

7-50. Army forces provide essential services, assets, or specialized resources to help civil authorities deal with situations beyond their capabilities. Activities of military forces to provide essential civil services are often defined in terms of the immediate humanitarian needs of the people: providing the food, water, shelter, and medical support necessary to sustain the population until local civil services are restored. (FM 3-0) (USACAC)

No.	Scale	Measure
01	Yes/No	Unit provided for immediate humanitarian needs of the population (food, water, shelter, and medical support).
02	Yes/No	Unit ensured proper sanitation, purification, and distribution of drinking water.
03	Yes/No	Unit provided interim sanitation, wastewater, and waste disposal services.
04	Yes/No	Unit established or reopened accessible clinics to deal with emergency health problems.

ART 7.3.3.2 CONTROL MOVEMENT OF DISLOCATED CIVILIANS

7-51. Assist, direct, or deny the movement of civilians whose location, direction of movement, or actions may hinder operations. Army forces do not assume control of dislocated civilian movement unless requested to do so by the host nation or unless operating in an environment with a hostile government. (FM 3-19.40) (USAMPS)

No.	Scale	Measure
01	Yes/No	Dislocated civilians did not interfere with the conduct of unit operations.
02	Yes/No	Unit developed plans to handle the movement of dislocated civilians in the area of operations.
03	Yes/No	Unit updated higher headquarters on the progress of the operation.
04	Yes/No	Soldiers treated dislocated civilians per the Geneva Convention and agreements between the United States and host nation.
05	Percent	Of subordinate elements assisting civil affairs and host-nation agencies in conducting populace and resources control operations.
06	Number	Of identified populace resources that must be denied contact with the enemy.
07	Number	Of dislocated civilians diverted from routes and areas when they might interfere with U.S. and allied military operations.
08	Number	And location of dislocated civilian collection points and assembly areas.

*ART 7.3.3.3 RESETTLE DISLOCATED CIVILIANS

7-52. Dislocated civilians are a central feature of many stability operations. Army forces may directly provide support to a recovering host nation or population. Specific types of support may include, but are not limited to, temporary support to or repatriation for refugees, return of displaced persons to their original homes, electoral assistance, maintaining public order and security, and maintaining a deterrent presence. (FM 3-07) (USACAC)

No.	Scale	Measure
01	Yes/No	Unit ensured humanitarian access to populations in need.
02	Yes/No	Unit estimated food aid needs for affected populations.
03	Yes/No	Unit assessed the adequacy of local physical transport, distribution, and storage.
04	Yes/No	Unit provided humanitarian assistance, including emergency food, water, sanitation, shelter, and health services.
05	Yes/No	Unit established camps for dislocated civilians.
06	Yes/No	Unit established and maintained order in camps.
07	Yes/No	Unit ensured adequate protection and monitoring.
08	Yes/No	Unit ensured humanitarian aid and security forces access to camp.
09	Time	To develop policy for providing minimum essential support requirements to dislocated civilians.
10	Time	To establish policy for final disposition of dislocated civilians to include guidelines for their release to return to their homes and transition of U.S. control over their camps to host-nation, multinational, and nongovernmental organizations.

ART 7.3.3.4 CONDUCT POPULACE AND RESOURCE CONTROL

7-53. Provide security for a populace, deny personnel and materials to the enemy, mobilize population and material resources, and detect and reduce the effectiveness of enemy agents. Populace controls also include implementing martial law during a complete breakdown of law and order; maintaining curfews; restricting movement, travel permits, and registration cards; and resettling villagers. Resource control measures include licenses, regulations or guidelines, checkpoints (for example, roadblocks), reaction controls, amnesty programs, and inspections of facilities. Two subdivisions of populace and resource control operations are dislocated civilian operations and evacuation of host-nation noncombatants. (FM 3-19.1) (USAMPS)

No.	Scale	Measure
01	Yes/No	The imposition of population and resource control measures met commander's intent of establishing control measures.
02	Yes/No	Unit accounted for and disposed of collected contraband per international and U.S. law, applicable regulations, and status-of-forces agreements.
03	Yes/No	Unit maintained chain of custody for contraband if possession of contraband would result in criminal or administrative trial.

No.	Scale	Measure
04	Time	Required to gain approval for adjustments in population and resource control measures did not detract from effectiveness of program.
05	Time	To coordinate with host-nation agencies for implementing population and resource control measures.
06	Time	To develop plans for imposing populace control by the enforcement of curfews, movement restrictions, travel permits and registration cards, and evacuating noncombatants.
07	Time	To develop plans for resource control, such as establishing roadblocks and checkpoints, inspecting facilities, enforcing local regulations and guidelines, controlling rations, and assisting with amnesty program.
08	Time	To inform local population of new or revised population and resource control measures imposed on them.
09	Time	To determine required population and resource control measures.
10	Time	Restrictions imposed on local civilians.
11	Percent	Of local population supporting population and resource control measures.
12	Percent	Of contraband detected during the conduct of resource control operations.
13	Percent	Of available effort devoted to population and resource control measures.
14	Percent	Of local population affected by population and resource control measures.
15	Number	And types of units involved in populations and resource control operations.
16	Number	Of instances that violations of population and resource control measures units detect.
17	Number	Of tons and types of contraband detected during resource control operations.
18	Number	Of military working dogs needed for patrol or the detection of narcotics and explosives.
19	Cost	To establish and administer population and resource control operation.

ART 7.3.3.5 SUPPORT FAMINE PREVENTION AND EMERGENCY FOOD RELIEF PROGRAMS

7-54. In response to a potential or actual food crisis, relief focuses on lifesaving measures to alleviate the immediate needs of a population in crises. Support may include medical, logistics, transportation, and security. Those activities identified as specifically as humanitarian and civic assistance are specific programs authorized in legislation and controlled per Title 10, U.S. Code. (FM 3-07) (USACAC)

No.	Scale	Measure
01	Yes/No	Unit monitored and analyzed food security and market prices.
02	Yes/No	Unit predicted the effects of conflict on access to food.
03	Yes/No	Unit estimated total food needs.
04	Yes/No	Unit assessed the adequacy of local physical transport, distribution, and storage.
05	Yes/No	Unit delivered emergency food to most vulnerable populations.
06	Yes/No	Unit assessed the effects of weather and climate on transportation networks and storage facilities.
07	Yes/No	Unit secured emergency nonfood relief distribution networks.
08	Yes/No	Unit delivered emergency nonfood items.
09	Yes/No	Unit provided emergency shelter for immediate needs.
10	Yes/No	Unit developed local expertise and ownership through capacity building.
11	Yes/No	Unit coordinated with other donors and humanitarian agencies.

ART 7.3.3.6 SUPPORT SHELTER AND NONFOOD RELIEF PROGRAMS

7-55. Military forces offer significant support capability to the broader effort to provide adequate shelter and nonfood relief during humanitarian crises. The welfare and perceptions of host-nation populations are often central to the mission during stability operations. (FM 3-07) (USACAC)

No.	Scale	Measure
01	Yes/No	Unit provided secure nonfood relief distribution networks.
02	Yes/No	Unit delivered emergency nonfood items.
03	Yes/No	Unit provided emergency shelter for immediate needs.
04	Yes/No	Unit cleared devastated housing and assessed damage.

ART 7.3.3.7 CONDUCT MEDICAL STABILITY OPERATIONS

7-56. Civil-military medicine is a discipline within operational medicine comprising public health and medical issues that involve a civil-military interface (foreign or domestic), including military medical support to civil authorities (domestic), medical elements of security cooperation activities, and medical civil military operations. (JP 4-02) (USJFCOM JWFC)

ART 7.3.3.8 SUPPORT PUBLIC HEALTH PROGRAMS

7-57. The military contribution to the public health sector, especially early in an operation, enables the complementary efforts of local and international aid organizations. The initial efforts of military forces aim to stabilize the public health sector. These efforts may include assessments of the medical and public health system such as infrastructure, medical staff, training and education, medical logistics, and public health programs. Following these initial response tasks, civilian organizations tailor their efforts to reforming the three public health sector through health systems strengthening and other public health capacity-building activities. (FM 3-07) (USACAC)

No.	Scale	Measure
01	Yes/No	Unit repaired and rebuilt clinics and hospitals.
02	Yes/No	Unit evaluated need for new clinics and hospitals.
03	Yes/No	Unit prevented epidemics through immediate vaccinations.
04	Yes/No	Unit assessed chronic and acute malnutrition.
05	Yes/No	Unit assessed emergency physical and psychological care needs.
06	Yes/No	Unit identified, safeguarded, and eliminated most dangerous public health hazards.
07	Yes/No	Unit evaluated water sources.
08	Yes/No	Unit supported host-nation waste and wastewater management capacity.
09	Yes/No	Unit supported public health information campaigns to educate population.
10	Yes/No	Unit identified public health information requirements for commander's critical information requirements consideration.
11	Yes/No	Unit conducted passive collection of medical information and utilized medical intelligence to obtain situational awareness.
12	Yes/No	Unit assists in coordinating U.S. Government, host-nation, and private resources to support public health programs.

ART 7.3.4 SUPPORT GOVERNANCE

7-58. Stability operations establish conditions that enable actions by civilian and host-nation actions to succeed. By establishing security and control, stability operations provide a foundation for transitioning authority to other government or intergovernmental agencies and eventually to the host nation. Once this transition is complete, commanders focus on transferring control to a legitimate civil authority according to the desired end state. (FM 3-0) (USACAC)

ART 7.3.4.1 SUPPORT TRANSITIONAL ADMINISTRATION

7-59. When the host-nation government has collapsed or been deposed, initial response efforts focus on immediately filling the void in governance. In either situation, the reliability and trustworthiness of local officials is suspect; due care and prudence is necessary to avoid empowering officials whose interests and loyalties are inconsistent with those of the force. (FM 3-07) (USACAC)

No.	Scale	Measure
01	Yes/No	Actions of temporary civil administration supported accomplishment of the mission of U.S. forces in the area of operations.
02	Yes/No	Unit vetted host-nation officials.
03	Yes/No	Leadership at multiple levels of government reconstituted.
04	Yes/No	Unit established interim legislative processes.
05	Yes/No	Unit established identification regime including securing documents relating to personal identification, property ownership, court records, voter registries, birth certificates, and driving licenses.
06	Yes/No	Unit developed mechanisms for dealing with claims and disputes relating to property ownership and court records.

ART 7.3.4.2 SUPPORT DEVELOPMENT OF LOCAL GOVERNANCE

7-60. Establishing effective governance at the local level is necessary before developing governance institutions and processes throughout the state. Initially, effective local governance almost depends entirely on the ability to provide essential civil services to the people; restoring these services is also fundamental to humanitarian relief efforts. (FM 3-07) (USACAC)

No.	Scale	Measure
01	Yes/No	Local governance in the area of operations (AO) supported the mission of U.S. forces.
02	Yes/No	Unit established mechanisms for local level participation.
03	Yes/No	Unit provided resources to maintain local public services.
04	Yes/No	Unit identified, secured, rehabilitated, and maintained basic facilities for local government.
05	Yes/No	Unit restored essential local public services.
06	Yes/No	Unit provided legal guidance and assistance to the transitional government to mitigate the near-term effects of corruption.
07	Yes/No	Unit provided advice to local community health authorities and committees.
08	Time	To develop plans to restore civilian authority during and after military operations.
09	Time	To establish civil society and media by strengthening the civil society environment, improving civic education, and strengthening civil capacity and partnerships.
10	Time	To disseminate public information and communication.
11	Time	To create political parties through party formation and training.
12	Time	To begin elections planning and execution at local levels.
13	Time	To conduct legal review of planned changes to public administration system and regulations.
14	Percent	Of local control of public administration in the AO.
15	Percent	Of U.S. forces in AO providing support to local governance.
16	Percent	Of local population satisfied with services provided them in the AO.

ART 7.3.4.3 SUPPORT ELECTIONS

7-61. Military forces may be required to provide security assistance to the host-nation civil authorities or international election commission agencies to prevent the disruption of elections. The primary role of military forces in support of elections is to facilitate civil order by providing a secure environment and should not be perceived as influencing the election outcome. U.S. forces should not be perceived as influencing elections by endorsing political platforms, parties, or candidates. Ideally, host-nation government and institutions will have the lead in developing and implementing elections with U.S. military in a supporting role. This support may include assisting the electoral commission in the planning and coordination of the election execution process and augmenting the host-nation security forces, as directed. Overt U.S. forces involvement in elections may be perceived as detrimental to an open, democratic, and legitimate election and taint the outcome. Planning and execution must be consistent with U.S. law, military guidance, and host-nation government regulations. The integrity of the election is the responsibility of the host-nation government or international election commission. (FM 3-07) (USACAC)

Note: Army forces performing ancillary tasks—such as security patrols, observation posts, and medical support—contribute to the performance of this task. (FM 3-07) (USACAC)

No.	Scale	Measure
01	Yes/No	Unit identified American Embassy and task force commander's guidance on roles, responsibilities, instructions, and parameters for U.S. forces.
02	Yes/No	Unit ensured military support roles and responsibilities complied with U.S. laws, military guidance, and host-nation regulations.
03	Yes/No	Unit coordinated planning with the lead election security agency.
04	Yes/No	Unit coordinated security mission requirements with joint, interagency, and multinational observers; election commission; host-nation government; and indigenous security forces.
05	Yes/No	Unit monitored intelligence and civil information reports for potential security issues and other contingencies.
06	Yes/No	Unit assessed security force capability and vulnerability to accomplish mission requirements.
07	Yes/No	Unit conducted area and route security assessment.
08	Yes/No	Unit augmented host-nation security forces in providing area security and reconnaissance in preelections, during elections, or post-elections, as required.
09	Yes/No	Unit helped develop shaping security plans prior to election.
10	Yes/No	Unit helped develop decisive security plans during conduct of election.
11	Yes/No	Unit helped develop contingency security plans for various election outcome scenarios.
12	Yes/No	Unit helped develop contingency security plans for various election outcome scenarios.

ART 7.3.5 SUPPORT ECONOMIC AND INFRASTRUCTURE DEVELOPMENT

7-62. Support to economic and infrastructure development helps a host nation develop capability and capacity in these areas. It may involve direct and indirect military assistance to local, regional, and national entities. Infrastructure development complements and reinforces efforts to stabilize the economy. It focuses on the society's physical aspects that enable the state's economic viability. These physical aspects of infrastructure include transportation (roads, railways, airports, and ports and waterways), telecommunications, energy (natural resources, the electrical power sector, and energy production and distribution), and general (engineering and construction, and municipal services) infrastructure. (FM 3-0) (USACAC)

No.	Scale	Measure
01	Yes/No	Unit developed partnerships with local organizations to meet community needs and increase local capacity to develop and maintain transportation, telecommunications, energy, and information critical infrastructure.
02	Time	To develop local and regional transportation plans.
03	Time	To develop local and regional telecommunication plans.
04	Time	To develop local and regional energy plans.
05	Time	To construct facilities that restore and promote overall host-nation governance, commerce, and social well-being.

ART 7.3.5.1 SUPPORT ECONOMIC DEVELOPMENT AND STABILIZATION

7-63. Economic recovery begins with an actively engaged labor force. When a military force occupies an operational area, the demand for local goods, services, and labor creates employment opportunities for the local populace. Local projects, such as restoring public services, rebuilding schools, or clearing roads, offer additional opportunities for the local labor pool. Drawing on local goods, services, and labor presents the force with the first opportunity to infuse cash into the local economy, which in turn stimulates market activity. (FM 3-07) (USACAC)

No.	Scale	Measure
01	Yes/No	Unit implemented initiatives to provide immediate employment.
02	Yes/No	Unit created employment opportunities for young males.
03	Yes/No	Unit assessed the labor force for critical skills requirements and deficiencies.
04	Yes/No	Unit established vetting program to ensure the reliability of the workforce.
05	Yes/No	Unit implemented public works projects.
06	Time	To assess the availability of civilian resources for civilian and military use.
07	Time	To reestablish government payment mechanisms to pay recurrent and emergency expenditures.
08	Time	To implement plans for revenue generation, customs taxation.
09	Time	To implement strategy for improved tax audit, collection, and enforcement.
10	Time	To assist national government in formulating recovery plan.
11	Time	To institutionalize regulatory system to govern financial transactions by banks.
12	Time	To foster economic integration through local, regional, and global organizations.

ART 7.3.5.2 SUPPORT PRIVATE SECTOR DEVELOPMENT

7-64. Developing the private sector typically begins with employing large portions of the labor force. In addition to acquiring goods and services from the local economy, the tasks that support private sector development infuse much-needed cash into local markets and initiate additional public investment and development. (FM 3-07) (USACAC)

No.	Scale	Measure
01	Yes/No	Unit assessed the depth of the private sector.
02	Yes/No	Unit identified obstacles to private sector development (such as barriers to entry, high import taxes, import restrictions and lack of business credit, power, transport, or telecommunications).
03	Yes/No	Unit strengthened private sector through contracting and out-sourcing.
04	Yes/No	Unit provided investors with protection and incentives.

ART 7.3.5.3 PROTECT NATURAL RESOURCES AND ENVIRONMENT

7-65. Protecting a nation's natural resources is an extension of the requirement to secure and protect other institutions of the state. Additionally, it preserves the long-term economic development and investment capacity of a fragile state. This capacity includes the revenues generated by the storage, distribution, and trade in natural resources. Rival factions often target these resources to finance illegitimate interests. (FM 3-07) (USACAC)

No.	Scale	Measure
01	Yes/No	Unit assessed and secured access to valuable natural resources.
02	Yes/No	Unit prevented the capture of revenues from natural resources.
03	Yes/No	Unit stopped illicit trade in natural resources.
04	Yes/No	Unit preserved long-term economic development and investment capacity.
05	Yes/No	Unit protected water resources.
06	Yes/No	Unit secured and protected post-harvest storage facilities.
07	Yes/No	Unit established work programs to support agricultural development.
08	Yes/No	Unit identified constraints to agricultural production.
09	Yes/No	Unit assessed health, diversity, and number of animals.
10	Yes/No	Unit kept core reproductive group alive through water and fodder provision.
11	Yes/No	Unit destocked as appropriate.
12	Yes/No	Unit provided veterinary services.

No.	Scale	Measure
13	Yes/No	Unit established sanitary practices and procedures (production and processing).
14	Yes/No	Unit improved food safety systems to facilitate agricultural trade.
15	Time	To identify policy makers in the agricultural, natural resources, and environment areas and discuss their priorities for their respective sectors.

ART 7.3.5.4 SUPPORT GENERAL INFRASTRUCTURE RECONSTRUCTION PROGRAMS

7-66. General infrastructure reconstruction programs focus on rehabilitating the state's ability to produce and distribute fossil fuels, generate electrical power, exercise engineering and construction support, and provide municipal and other services to the populace. As with the restoration of essential services, support to general infrastructure programs requires a thorough understanding of the civil component of the area of operations. Civil affairs personnel support this information collection to inform the prioritization of programs and projects. (FM 3-07) (USACAC)

No.	Scale	Measure
01	Yes/No	Unit assessed overall condition of national transportation infrastructure (airports, roads, bridges, railways, and coastal and inland ports, harbors, and waterways).
02	Yes/No	Unit initiated immediate improvement to the transportation and distribution networks of the host nation.
03	Yes/No	Unit assessed overall condition of national telecommunications infrastructure.
04	Yes/No	Unit assessed overall condition of national energy infrastructure.
05	Yes/No	Unit assessed overall condition of existing power generation and distribution facilities.
06	Yes/No	Unit assessed overall condition of existing natural resources conversion and distribution facilities.
07	Yes/No	Unit assessed overall condition of existing facilities integral for effectively executing essential tasks in other sectors.
08	Yes/No	Unit assessed overall condition of existing local, municipal facilities that provide essential services.
09	Yes/No	Unit determined and prioritized essential infrastructure programs and projects.
10	Yes/No	Unit conducted expedient repairs or built new facilities to support stabilization.
11	Yes/No	Unit conducted expedient repairs or built new facilities to facilitate commercial trade.
12	Yes/No	Unit conducted expedient repairs or built new facilities to support local populace (such as schools, medical clinics, and municipal buildings).

+ ART 7.3.5.4.1 Support Host-Nation Telecommunications Infrastructure Reconstruction Program

7-67. Support to Host-Nation Telecommunications Infrastructure Program focuses on providing initial assessment of the operational readiness and capabilities of the host nation's official public communications systems (and when or if necessary private telecommunications systems.) The systems enable a government to communicate with its populace. This includes mass media systems including radio station equipment and antennae systems, television stations and antenna systems, data network equipment and servers which connect to the wider internet, and wired or wireless telephone central systems and antennae. This requires thoroughly understanding the technical operation of commercial off-the-shelf equipment and systems used in providing mass communications systems. (FM 3-07). (USACAC)

No.	Scale	Measure
01	Yes/No	Unit assessed and prioritized television broadcast telecommunications capabilities available to the host-nation government authority.
02	Yes/No	Unit assessed and prioritized radio broadcast telecommunications capabilities available to the host-nation government authority.
03	Yes/No	Unit assessed and prioritized internet system capabilities available to the host-nation government for internal and external communications.

No.	Scale	Measure
04	Yes/No	Unit assessed and prioritized wired and wireless telephonic telecommunications capabilities available to the host-nation government authority.
05	Percent	Of state or geographic region's government authorities which can complete two-way electronic communications with the host-nation government authority and supporting US nongovernmental organizations.
06	Percent	Of state or geographic region's government authorities with which the host-nation government authority can communicate with one-way communications.

SECTION IV – ART 7.4: CONDUCT CIVIL SUPPORT OPERATIONS

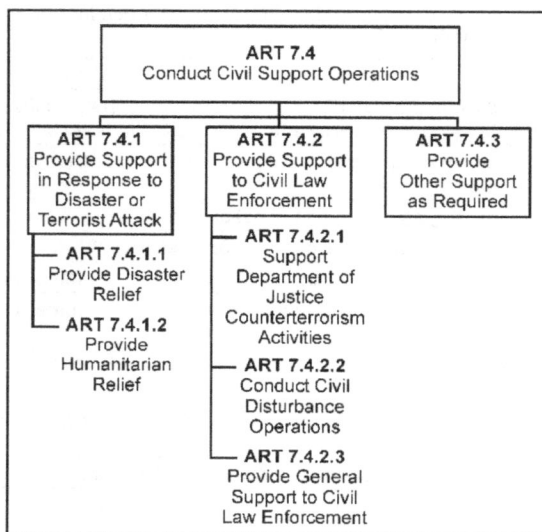

7-68. *Civil support* is Department of Defense support to U.S. civil authorities for domestic emergencies, and for designated law enforcement and other activities. Civil support includes operations that address the consequences of natural or manmade disasters, accidents, and incidents within the United States and its territories. Army forces conduct civil support operations when the size and scope of events exceed the capabilities of domestic civilian agencies. The Army National Guard is often the first military force to respond on behalf of state authorities. Army civil support operations include three tasks: provide support in response to a disaster, support civil law enforcement, and provide other support as required. (FM 3-0) (USACAC)

Note: Chemical, biological, radiological, nuclear, and high-yield explosives weapons and explosive ordnance disposal support provided in civil support are tasks addressed elsewhere in the AUTL:

ART 6.9.5 (Conduct Chemical, Biological, Radiological, Nuclear, and High-Yield Explosives Consequence Management) discusses the coordination of essential services and activities required to manage and mitigate damage resulting from the use of chemical, biological, radiological, and nuclear weapons or the release of toxic industrial materials or contaminants.

ART 6.12.8.5 (Provide Explosive Ordnance Disposal Support to Civil Authorities) addresses assistance to include providing explosive ordnance disposal service when requested by local, state, or federal authorities in the interest of public safety.

ART 6.12.8.6 (Provide Specific Training to Personnel on Explosive Ordnance Disposal) addresses training for military personnel, federal agencies, and public officials involved with civil emergency preparedness, law enforcement, and fire protection as requested.

ART 7.4.1 PROVIDE SUPPORT IN RESPONSE TO DISASTER OR TERRORIST ATTACK

7-69. Disaster relief operations focus on recovery of critical infrastructure after natural or man-made disasters. Humanitarian relief focuses on the well-being of supported populations. Both normally occur simultaneously. In the case of a disaster, state, and local authorities are responsible for restoring essential services. (FM 3-07) (USACAC)

ART 7.4.1.1 PROVIDE DISASTER RELIEF

7-70. Disaster relief restores or recreates essential infrastructure. It includes establishing and maintaining the minimum safe working conditions, less security measures, necessary to protect relief workers and the affected population. Disaster relief allows effective humanitarian relief and creates conditions for long-term recovery. It may involve consultation on and provision of emergency medical treatment and evacuation; repairing or demolishing damaged structures; restoring or building bridges, roads, and airfields; and removing debris from supply routes and relief sites. (FM 3-07) (USACAC)

Note: Disaster relief includes the submissions of military assistance to civilian disaster organizations, natural disaster relief, oil and hazardous substances incident or emergency, public health emergency, as well as technological or man-made disaster relief.

ART 7.4.1.2 PROVIDE HUMANITARIAN RELIEF

7-71. Humanitarian relief focuses on lifesaving measures that alleviate the immediate needs of a population in crisis. It often includes providing medical support, food, water, medicine, clothing, blankets, shelter, and heating or cooking fuels. In some cases, it involves transporting affected people from a disaster area. Civilian relief agencies, government and nongovernmental, are best suited to provide this type of relief. Army forces conducting humanitarian relief usually facilitate civil relief efforts. (FM 3-07) (USACAC)

ART 7.4.2 PROVIDE SUPPORT TO CIVIL LAW ENFORCEMENT

7-72. Support to domestic civil law enforcement involves activities related to the Department of Justice's counterterrorism activities, counterdrug activities, military assistance during civil disturbances, and general support. Army support involves providing resources, training, or augmentation. Federal military forces remain under the military chain of command while supporting civil law enforcement. The supported law enforcement agency coordinates Army forces activities per appropriate civil laws and interagency agreements. Army National Guard units in state status can be a particularly useful military resource. They may be able to provide assistance to civil authorities when federal units cannot due to the Posse Comitatus Act. Title 10 U.S. Code prohibits the military from directly participating in arrests, searches, seizures, or other similar activities unless authorized by law. (FM 3-07) (USACAC)

Note: Support to civil law enforcement includes border protection enforcement, combating terrorism, critical infrastructure protection as well as Department of Defense support to counterdrug operations.

Army missions related to supporting civil law enforcement in counterdrug operations are discussed under ART 7.6.1.5 (Provide Military Support to Counterdrug Efforts).

ART 7.4.2.1 Support Department of Justice Counterterrorism Activities

7-73. When directed by the President, Army forces may provide assistance to the Department of Justice in the areas of transportation, equipment, training, and personnel. When terrorists pose an imminent threat, Army forces may be used to counter these threats. Army forces may also support crisis management. Crisis management of a terrorist incident includes measures to resolve a situation and investigate a criminal case for prosecution under federal law. The Federal Bureau of Investigation is the lead agency and has responsibility for crisis management within the United States. Army forces may provide specialized or technical capabilities to assist in defusing or resolving a terrorist crisis. Support of crisis management includes opening lines of communications for military assistance, evacuating casualties, executing reconnaissance, and decontaminating or assessing the effects of weapons of mass destruction. In the aftermath of a terrorist incident, Army forces may be involved in consequence management activities. These activities include casualty and medical assistance, evaluation and repair of damage to structures and utilities, explosive ordnance disposal, and mortuary affairs. (FM 3-07) (USACAC)

ART 7.4.2.2 Conduct Civil Disturbance Operations

7-74. Army forces assist civil authorities in restoring law and order when state and local law enforcement agencies are unable to control civil disturbances. The Army National Guard is the first military responder during most civil disturbance situations. It usually remains on state active-duty status throughout the operation. When conditions of domestic violence and disorder endanger life and property to the extent that state law enforcement agencies, to include the Army National Guard, cannot suppress violence and restore law and order, the President may federalize Army National Guard units under Title 10, U.S. Code, Chapter 15. The President may use federalized Army National Guard and federal forces to restore law and order. Restrictions may be placed on federal military forces either in the executive order directing their use or through the rules for the use of force outlined in the Department of Defense Civil Disturbance Plan. (FM 3-07) (USACAC)

ART 7.4.2.3 Provide General Support to Civil Law Enforcement

7-75. Provide limited military support to law enforcement agencies. Department of Defense may direct Army forces to provide training to federal, state, and local civilian law enforcement agencies. Such assistance may include training in the operation and maintenance of military equipment. (FM 3-07) (USACAC)

ART 7.4.3 PROVIDE OTHER SUPPORT AS REQUIRED

7-76. Community assistance is a broad range of activities that provide support and maintain a strong connection between the military and civilian communities. Community assistance activities provide effective means of projecting a positive military image, provide training opportunities, and enhance the relationship between Army forces and the American public. They should fulfill community needs that would not otherwise be met. Community assistance activities can enhance individual and unit combat readiness. Army assistance to the community can include air ambulance support, search and rescue activities, firefighting capability, explosive ordnance disposal, emergency or broad-based medical care, wildlife and domestic animal management, assistance in safety and traffic control, emergency snow removal, temporary supplemental housing for the displaced or disadvantaged, and postal augmentation. (FM 3-0) (USACAC)

SECTION V – ART 7.5: CONDUCT TACTICAL MISSION TASKS

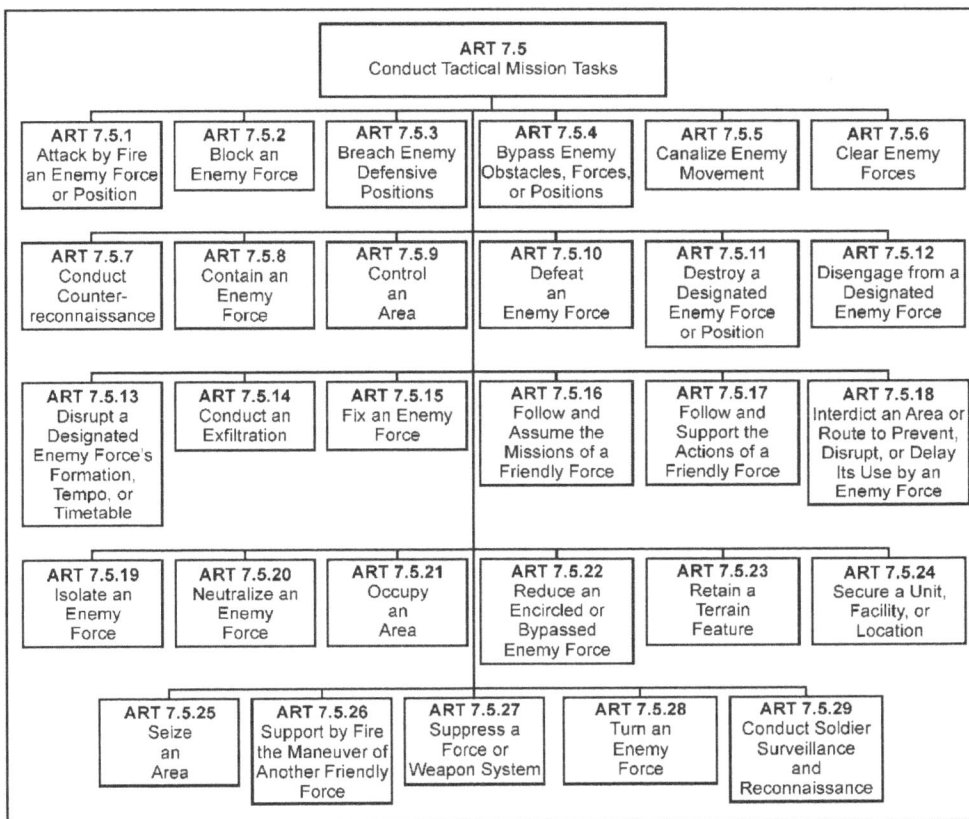

```
                        ART 7.5
               Conduct Tactical Mission Tasks

ART 7.5.1      ART 7.5.2    ART 7.5.3    ART 7.5.4      ART 7.5.5     ART 7.5.6
Attack by Fire Block an     Breach Enemy Bypass Enemy   Canalize Enemy Clear Enemy
an Enemy Force Enemy Force  Defensive    Obstacles, Forces, Movement   Forces
or Position                 Positions    or Positions

ART 7.5.7      ART 7.5.8    ART 7.5.9    ART 7.5.10     ART 7.5.11    ART 7.5.12
Conduct        Contain an   Control      Defeat         Destroy a     Disengage from a
Counter-       Enemy        an           an             Designated    Designated
reconnaissance Force        Area         Enemy Force    Enemy Force   Enemy Force
                                                        or Position

ART 7.5.13     ART 7.5.14   ART 7.5.15   ART 7.5.16     ART 7.5.17    ART 7.5.18
Disrupt a      Conduct an   Fix an Enemy Follow and     Follow and    Interdict an Area or
Designated     Exfiltration Force        Assume the     Support the   Route to Prevent,
Enemy Force's                            Missions of a  Actions of a  Disrupt, or Delay
Formation,                               Friendly Force Friendly Force Its Use by an
Tempo, or                                                             Enemy Force
Timetable

ART 7.5.19     ART 7.5.20   ART 7.5.21   ART 7.5.22     ART 7.5.23    ART 7.5.24
Isolate an     Neutralize an Occupy      Reduce an      Retain a      Secure a Unit,
Enemy          Enemy        an           Encircled or   Terrain       Facility, or
Force          Force        Area         Bypassed       Feature       Location
                                         Enemy Force

        ART 7.5.25     ART 7.5.26     ART 7.5.27     ART 7.5.28    ART 7.5.29
        Seize          Support by Fire Suppress a     Turn an       Conduct Soldier
        an             the Maneuver of Force or       Enemy         Surveillance
        Area           Another Friendly Weapon System Force         and
                       Force                                        Reconnaissance
```

7-77. Tactical mission tasks describe lethal and nonlethal results or effects the commander wants to generate or create to accomplish the mission (they are the what or why of a mission statement). The tasks in this section are often given to units as the tasks or purpose parts of their mission statement. (FM 3-90) (USACAC)

ART 7.5.1 ATTACK BY FIRE AN ENEMY FORCE OR POSITION

7-78. Attack by fire uses direct fires, supported by indirect fires, to engage an enemy without closing to destroy, suppress, fix, or deceive the enemy. (FM 3-90) (USACAC)

ART 7.5.2 BLOCK AN ENEMY FORCE

7-79. Block denies the enemy access to an area or prevents an advance in a direction or along an avenue of approach. (FM 3-90) (USACAC)

ART 7.5.3 BREACH ENEMY DEFENSIVE POSITIONS

7-80. Breach employs all available means to break through or secure a passage through a defense, obstacle, minefield, or fortification. (FM 3-90) (USACAC)

ART 7.5.4 BYPASS ENEMY OBSTACLES, FORCES, OR POSITIONS

7-81. Bypass is when the commander directs the unit to maneuver around an obstacle, position, or enemy force to maintain the momentum of the operation, while deliberately avoiding combat with the enemy force. (FM 3-90) (USACAC)

ART 7.5.5 CANALIZE ENEMY MOVEMENT

7-82. Canalize is when the commander restricts enemy movement to a narrow area by exploiting terrain, obstacles, fires, or friendly maneuver. (FM 3-90) (USACAC)

ART 7.5.6 CLEAR ENEMY FORCES

7-83. Clear requires the commander to remove all enemy forces and eliminate organized resistance within an assigned area. Physical conditions of the area will affect the specific tactics, techniques, and procedures employed. (FM 3-90) (USACAC)

ART 7.5.7 CONDUCT COUNTERRECONNAISSANCE

7-84. Counterreconnaissance encompasses all measures taken by a commander to counter enemy surveillance and reconnaissance efforts. Counterreconnaissance is not a distinct mission but a component of all forms of security operations. See ART 6.7.3 (Conduct Security Operations). (FM 3-90) (USACAC)

ART 7.5.8 CONTAIN AN ENEMY FORCE

7-85. Contain requires the commander stop, hold, or surround enemy forces; or cause them to focus their activity on a given front and prevent them from withdrawing any element for use elsewhere. (FM 3-90) (USACAC)

ART 7.5.9 CONTROL AN AREA

7-86. Control requires the commander to maintain physical influence over a specified area to prevent its use by an enemy. (FM 3-90) (USACAC)

ART 7.5.10 DEFEAT AN ENEMY FORCE

7-87. Defeat occurs when an enemy force has temporarily or permanently lost the physical means or will to fight. The defeated force's commander is unwilling or unable to pursue an adopted course of action, thereby yielding to the friendly commander's will and can no longer interfere to a significant degree with the actions of friendly forces. Defeat can result from the use of force or the threat of its use. (FM 3-90) (USACAC)

ART 7.5.11 DESTROY A DESIGNATED ENEMY FORCE OR POSITION

7-88. Destroy involves the physical rendering of an enemy force to combat ineffectiveness until it is reconstituted. Alternatively, to destroy a combat system is to damage it so badly that it cannot perform its function or be restored to a usable condition without being entirely rebuilt. (FM 3-90) (USACAC)

ART 7.5.12 DISENGAGE FROM A DESIGNATED ENEMY FORCE

7-89. Disengagement is when a commander has the unit break contact with the enemy to allow the conduct of another mission or to avoid decisive engagement. (FM 3-90) (USACAC)

ART 7.5.13 DISRUPT A DESIGNATED ENEMY FORCE'S FORMATION, TEMPO, OR TIMETABLE

7-90. Disrupt is when a commander integrates direct and indirect fires, terrain, and obstacles to upset an enemy formation or tempo, interrupt the timetable, or cause forces to commit prematurely or attack in a piecemeal fashion. (FM 3-90) (USACAC)

ART 7.5.14 CONDUCT AN EXFILTRATION

7-91. Exfiltrate is where a commander removes personnel or units from areas under enemy control by stealth, military deception, surprise, or clandestine means. (FM 3-90) (USACAC)

ART 7.5.15 FIX AN ENEMY FORCE

7-92. Fix is where a commander prevents the enemy from moving any part of the force from a specific location for a specific time. (FM 3-90) (USACAC)

ART 7.5.16 FOLLOW AND ASSUME THE MISSIONS OF A FRIENDLY FORCE

7-93. Follow and assume is when a second committed force follows a force conducting an offensive operation and is prepared to continue the mission of that force when it becomes fixed, attritted, or otherwise unable to continue. (FM 3-90) (USACAC)

ART 7.5.17 FOLLOW AND SUPPORT THE ACTIONS OF A FRIENDLY FORCE

7-94. Follow and support is when a committed force follows and supports the mission accomplishment of a leading force conducting an offensive operation. (FM 3-90) (USACAC)

ART 7.5.18 INTERDICT AN AREA OR ROUTE TO PREVENT, DISRUPT, OR DELAY ITS USE BY AN ENEMY FORCE

7-95. Interdict is where the commander prevents, disrupts, or delays enemy use of an area or route. (FM 3-90) (USACAC)

ART 7.5.19 ISOLATE AN ENEMY FORCE

7-96. Isolate requires a unit to seal off—physically and psychologically—an enemy force from its sources of support, deny it freedom of movement, and prevent it from contacting other enemy forces. (FM 3-90) (USACAC)

ART 7.5.20 NEUTRALIZE AN ENEMY FORCE

7-97. Neutralize results in rendering enemy personnel or materiel incapable of interfering with friendly operations. (FM 3-90) (USACAC)

ART 7.5.21 OCCUPY AN AREA

7-98. Occupy involves moving a force into an area so that it can control the entire area. Both the movement to and occupation of the area occurs without opposition. (FM 3-90) (USACAC)

ART 7.5.22 REDUCE AN ENCIRCLED OR BYPASSED ENEMY FORCE

7-99. Reduce involves the destruction of an encircled or bypassed enemy force. (FM 3-90) (USACAC)

ART 7.5.23 RETAIN A TERRAIN FEATURE

7-100.   Retain is when the commander ensures a terrain feature already controlled by a friendly force remains free of enemy occupation or use. (FM 3-90) (USACAC)

ART 7.5.24 SECURE A UNIT, FACILITY, OR LOCATION

7-101.   Secure involves preventing the enemy from damaging or destroying a unit, facility, or geographical location. (FM 3-90) (USACAC)

ART 7.5.25 SEIZE AN AREA

7-102.   Seize involves taking possession of an area using overwhelming force. (FM 3-90) (USACAC)

ART 7.5.26 SUPPORT BY FIRE THE MANEUVER OF ANOTHER FRIENDLY FORCE

7-103.   Support by fire is when a maneuver force moves to a position where it can engage the enemy by direct fire to support another maneuvering force. (FM 3-90) (USACAC)

ART 7.5.27 SUPPRESS A FORCE OR WEAPON SYSTEM

7-104.   Suppression results in the temporary degradation of the performance of a force or weapon systems below the level needed to accomplish its mission. (FM 3-90) (USACAC)

ART 7.5.28 TURN AN ENEMY FORCE

7-105.   Turn involves forcing an enemy force to move from one avenue of approach or mobility corridor to another. (FM 3-90) (USACAC)

ART 7.5.29 CONDUCT SOLDIER SURVEILLANCE AND RECONNAISSANCE

7-106.   Every Soldier, as a part of a small unit, is a potential information collector and an essential component to answer commander's critical information requirements and facilitate the commander's situational understanding. Each Soldier develops a special level of awareness simply due to exposure to events occurring in the area of operations (AO) and has the opportunity to collect information by observation and interaction with the population. These observations and interactions provide depth and context to information collected through surveillance and reconnaissance. Collecting and reporting this information within an AO is a critical element to achieve situational understanding of the AO. Leaders must create a climate that allows all Soldiers to feel free to report what they see and learn on a mission. (FM 2-0 and FM 2-91.6) (USAIC)

Note: This task is supported by ART 2.3.5.1.1 (Establish a Mission Intelligence Briefing Plan) and direct support ART 2.3.5.1.2 (Establish a Debriefing Plan).

SECTION VI – ART 7.6: OPERATIONAL THEMES

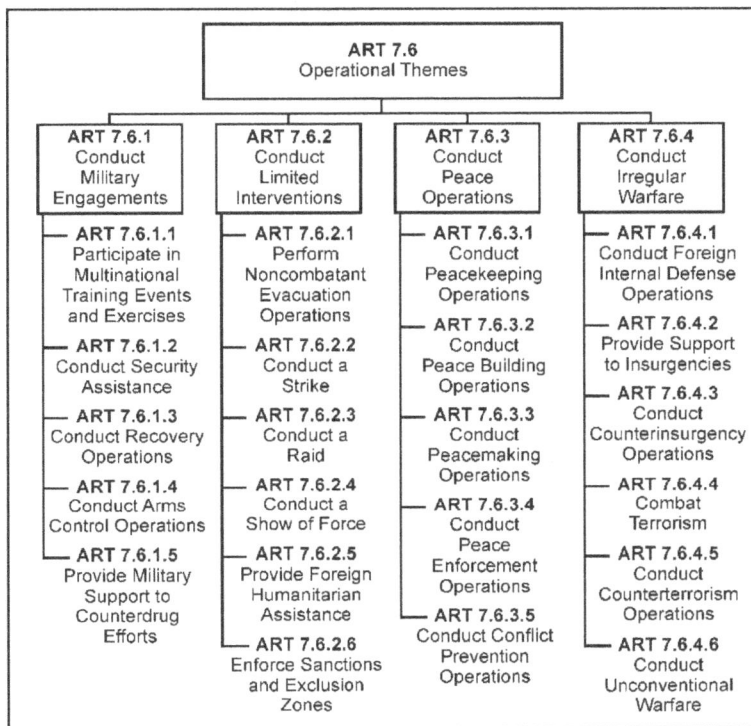

```
                        ┌─────────────────────┐
                        │       ART 7.6        │
                        │  Operational Themes  │
                        └─────────────────────┘
```

ART 7.6.1 Conduct Military Engagements	ART 7.6.2 Conduct Limited Interventions	ART 7.6.3 Conduct Peace Operations	ART 7.6.4 Conduct Irregular Warfare
ART 7.6.1.1 Participate in Multinational Training Events and Exercises	ART 7.6.2.1 Perform Noncombatant Evacuation Operations	ART 7.6.3.1 Conduct Peacekeeping Operations	ART 7.6.4.1 Conduct Foreign Internal Defense Operations
ART 7.6.1.2 Conduct Security Assistance	ART 7.6.2.2 Conduct a Strike	ART 7.6.3.2 Conduct Peace Building Operations	ART 7.6.4.2 Provide Support to Insurgencies
ART 7.6.1.3 Conduct Recovery Operations	ART 7.6.2.3 Conduct a Raid	ART 7.6.3.3 Conduct Peacemaking Operations	ART 7.6.4.3 Conduct Counterinsurgency Operations
ART 7.6.1.4 Conduct Arms Control Operations	ART 7.6.2.4 Conduct a Show of Force	ART 7.6.3.4 Conduct Peace Enforcement Operations	ART 7.6.4.4 Combat Terrorism
ART 7.6.1.5 Provide Military Support to Counterdrug Efforts	ART 7.6.2.5 Provide Foreign Humanitarian Assistance	ART 7.6.3.5 Conduct Conflict Prevention Operations	ART 7.6.4.5 Conduct Counterterrorism Operations
	ART 7.6.2.6 Enforce Sanctions and Exclusion Zones		ART 7.6.4.6 Conduct Unconventional Warfare

7-107.   An operational theme describes the character of the dominant major operation being conducted at any time within a land force commander's area of operations. The operational theme helps convey the nature of the major operation to the force to facilitate common understanding of how the commander broadly intends to operate. (FM 3-0) (USACAC)

ART 7.6.1 CONDUCT MILITARY ENGAGEMENTS

7-108.   Peacetime military engagement comprises all military activities that involve other nations and are intended to shape the security environment in peacetime. It includes programs and exercises that the U.S. military conducts with other nations to shape the international environment, improve mutual understanding, and improve interoperability with treaty partners or potential coalition partners. Peacetime military engagement activities are designed to support a combatant commander's objectives within the theater security cooperation plan. (FM 3-0) (USACAC)

ART 7.6.1.1 PARTICIPATE IN MULTINATIONAL TRAINING EVENTS AND EXERCISES

7-109.   Army forces support the Chairman of the Joint Chiefs of Staff Exercise Program that is the Chairman of the Joint Chiefs of Staff's principal vehicle for performing joint and multinational training. The program provides combatant commanders with their primary means to train staffs and forces in joint and combined operations, to evaluate war plans, and to execute their engagement strategies. (FM 3-07) (USACAC)

ART 7.6.1.2 Conduct Security Assistance

7-110.   Security assistance refers to a group of programs that support U.S. national policies and objectives by providing defense articles, military training, and other defense-related services to host nations by grant, loan, credit, or cash sales. Army forces support security assistance efforts through military training teams, maintenance support personnel and training, and related activities, such as humanitarian mine removal operations. (FM 3-07) (USACAC)

No.	Scale	Measure
01	Yes/No	Unit trained and equipped host-nation military forces.
02	Yes/No	Unit provided conventional military assistance programs.
03	Yes/No	Unit established military-to-military programs with host-nation forces.
04	Yes/No	Unit trained and equipped border security personnel.
05	Yes/No	Unit trained and equipped host-nation de-mining capability.
06	Yes/No	Unit created host-nation capacity to protect private institutions and key leaders.
07	Yes/No	Unit created host-nation capacity to protect critical infrastructure.
08	Yes/No	Unit created host-nation capacity to protect military infrastructure.
09	Yes/No	Unit created host-nation capacity to protect public institutions.
10	Yes/No	Unit created host-nation capacity to protect government-sponsored civilian stabilization and reconstruction personnel.
11	Yes/No	Unit created host-nation capacity to protect contractor and nongovernmental organization stabilization personnel and resources.
12	Yes/No	Unit identified military infrastructure modernization needs and means to achieve them.
13	Yes/No	Unit monitored compliance with and reinforced regional security arrangements.
14	Time	To develop host-nation arms control capacity.
15	Time	To develop host-nation capacity to assure and regulate movement.
16	Time	To transfer de-mining and unexploded explosive ordnance removal operations to host-nation personnel.
17	Time	To establish mechanisms for implementing regional security arrangements.

ART 7.6.1.3 Conduct Recovery Operations

7-111.   Recovery operations are conducted to search for, locate, identify, recover, and return isolated personnel, human remains, sensitive equipment, or items critical to national security. (FM 3-0) (USACAC)

Note: See ART 6.2 (Conduct Personnel Recovery Operations) for tasks as they relate to personnel recovery.

ART 7.6.1.4 Conduct Arms Control Operations

7-112.   Army forces normally conduct arms control operations to support arms control treaties and enforcement agencies. Army forces can help locate, seize, and destroy weapons of mass destruction after hostilities. Other actions include escorting deliveries of weapons and materials (such as enriched uranium) to preclude loss or unauthorized use, inspecting and monitoring production and storage facilities, and training foreign forces to secure weapons and facilities. Army forces may conduct arms control operations to prevent escalation of a conflict and reduce instability. This could include the mandated disarming of belligerents as part of a peace operation. (FM 3-07) (USACAC)

No.	Scale	Measure
01	Yes/No	Unit established and enforced weapons control regimes, including collection and destruction.
02	Yes/No	Unit cooperated with legal authorities to prosecute arms dealers.
03	Yes/No	Unit collaborated with neighboring countries on weapons flows.
04	Yes/No	Unit apprehended illegal arms dealers.

ART 7.6.1.5 PROVIDE MILITARY SUPPORT TO COUNTERDRUG EFFORTS

7-113.   The Army participates in counterdrug operations under provisions of the national drug control strategy. Army forces may be employed in various operations to support other government agencies responsible for detecting, disrupting, interdicting, and destroying illicit drugs and the infrastructure (personnel, materiel, and distribution systems) of illicit drug-trafficking entities. When conducted inside the United States and its territories, they are civil support operations. When conducted outside the United States and its territories, counterdrug operations are considered stability operations. Army forces do not engage in direct action in counterdrug operations. Units that support counterdrug operations comply with U.S. and foreign legal limitations concerning the acquisition of information from civilians and the conduct of law enforcement activities. (FM 3-07) (USACAC)

ART 7.6.1.5.1 Support Detection and Monitoring of Drug Shipments

7-114.   Provide aerial and ground reconnaissance to support counterdrug operations by law enforcement agencies. The goal is to provide early notification to—and, as necessary, prolonged tracking of—aerial and surface targets for appropriate law enforcement agencies. This support enables law enforcement agencies to intercept, search, and arrest traffickers, and seize illegal drugs and illegally obtained property. Aerial reconnaissance may be conducted with fixed- or rotary-wing aircraft, unmanned aircraft systems, or ground-based radars. Land reconnaissance may be executed by observation posts, patrols, ground surveillance radars, and remote ground sensors. (FM 3-07) (USACAC)

ART 7.6.1.5.2 Provide Command, Control, Communication, Computer, and Intelligence Support to Counterdrug Efforts

7-115.   Army personnel and equipment may assist law enforcement agencies and host nations in designing, implementing, and integrating command, control, communication, computer, and intelligence systems. Army personnel support national and departmental drug operations and law enforcement agency analytical centers. In addition, Army forces provide liaison to law enforcement agencies and host nations to facilitate smooth and successful integration of military support. Army units and personnel provide intelligence support targeted at the full range of narcotics traffickers' operations. The principal means of providing this support is through tactical analysis teams. These teams co-locate with the U.S. country team, support law enforcement agencies, and provide focused detection and monitoring of narcotics trafficking activities. (FM 3-07) (USACAC)

ART 7.6.1.5.3 Provide Planning Support to Counterdrug Efforts

7-116.   Army personnel support counterdrug planning of both law enforcement agencies and host nations. Understanding the supported agency or host nation, including its culture and people, is critical. Planning support provided to law enforcement agencies must take into account the organization's mission, current goals, structure or chain of command, measures of success, and relationships with other government agencies or countries. Planning support provided to host nations is similar to that provided to domestic law enforcement agencies. However, the host-nation culture, historical perspectives, political climate, and economic conditions are also considered. (FM 3-07) (USACAC)

ART 7.6.1.5.4 Provide Logistic Support to Counterdrug Efforts

7-117.   Army forces can assist law enforcement agencies or host nations during their conduct of counterdrug operations with logistic management and execution. This includes transportation, maintenance, engineer design and construction, use of facilities, equipment loans, or military working dog support. Army forces can provide supplies and field services directly, if authorized, or assist other agencies in procuring and managing them from other sources. Commanders who assist law enforcement with transportation of evidence, seized property, or contraband ensure a law enforcement officer accompanies the shipment to maintain continuity of the chain of custody. (FM 3-07) (USACAC)

ART 7.6.1.5.5 Provide Training Support to Counterdrug Efforts

7-118.   Training support to law enforcement agencies and host nations includes basic military skills, such as basic marksmanship, patrolling, mission planning, medical, and survival skills. Commanders provide support using a mix of mobile training teams; participation in operational planning groups, joint or combined exercises, institutional training, noninstitutional training, and training simulations; and extended training service specialists. (FM 3-07) (USACAC)

ART 7.6.1.5.6 Provide Manpower Support to Counterdrug Efforts

7-119.   Army forces may provide various individuals or units to support interagency and host-nation counterdrug efforts. Categories of manpower support are eradication, administrative—including staff judge advocate officers, legal specialists, and accounting specialists—linguist, liaison officer, inspection, military police, and intelligence analyst. (FM 3-07) (USACAC)

ART 7.6.1.5.7 Provide Research, Development, and Acquisition Support to Counterdrug Efforts

7-120.   The Army Counterdrug Research, Development, and Acquisition Office makes military research, development, and acquisition efforts available to law enforcement agencies. It informs them of new technical capabilities that have possible law enforcement application. It can also assist them in contracting and procuring technical equipment. (FM 3-07) (USACAC)

ART 7.6.2 CONDUCT LIMITED INTERVENTIONS

7-121.   Limited interventions are executed to achieve a clearly defined end state, limited in scope. Corresponding limitations are imposed on the supporting operations and size of the forces involved. These operations may be phased but are not intended to become campaigns. Although limited interventions are confined in terms of end state and forces, their execution may be lengthy. Joint task forces usually conduct these limited interventions. (FM 3-0) (USACAC)

ART 7.6.2.1 PERFORM NONCOMBATANT EVACUATION OPERATIONS

7-122.   Noncombatant evacuation operations relocate threatened civilian noncombatants from locations in a host nation to secure areas. Normally, these operations involve U.S. citizens whose lives are in danger, either from the threat of hostilities or from a natural disaster. They may also include host-nation citizens and third-country nationals. Army forces, normally as part of a joint task force, conduct noncombatant evacuation operations to assist and support the Department of State. Noncombatant evacuation operations usually involve swift insertions of a force, temporary occupation of an objective, and a planned withdrawal upon accomplishment of the mission. This operation can occur in three environments: permissive, uncertain, or hostile. The environment in which a noncombatant evacuation operation occurs can quickly change from one to another with little warning. The commander prepares to operate in all three environments. This task includes sustainment functions involving emergency medical treatment, transportation, administrative processing, and coordination with the Department of State and other agencies involved in the evacuation. Task organization for noncombatant evacuation operations is based on the operational environment in which the action is performed. However, since the environment can change rapidly, the permissive form of noncombatant evacuation operations can become uncertain or hostile. (FM 3-07) (USACAC)

ART 7.6.2.2 CONDUCT A STRIKE

7-123.   Strikes are attacks conducted to damage or destroy an objective or capability or to compel a hostile government or force to defer from taking hostile actions. Strikes are usually planned and executed as part of or in support of a joint operation. (JP 3-0) (USJFCOM JWFC)

ART 7.6.2.3 CONDUCT A RAID

7-124.   A raid is an attack, usually small scale, involving a swift entry into hostile territory to secure information, confuse the enemy, or destroy installations. It usually ends with a planned withdrawal from the objective area upon mission accomplishment. Raids have narrowly defined purposes. They require both detailed intelligence and deliberate planning. Raids may destroy key enemy installations and facilities, capture or free prisoners, or disrupt enemy command and control or other important systems. (FM 3-90) (USACAC)

ART 7.6.2.4 CONDUCT A SHOW OF FORCE

7-125. Shows of force are flexible deterrence options designed to demonstrate U.S. resolve. They involve increasing the visibility of U.S. forces to defuse a situation that, if allowed to continue, may be detrimental to national interests or objectives. The United States conducts shows of force for three reasons: to bolster or reassure allies, deter potential aggressors, and gain or increase influence. Shows of force demonstrate a credible and specific threat to an aggressor or potential aggressor. They involve increasing the visibility of U.S. forces in the eyes of the target audience through establishing an area presence and performing exercises and demonstrations. Presence patrols executed by small tactical units are one technique of conducting shows of force. Although actual combat is not desired, shows of force can rapidly and unexpectedly escalate. Therefore, units assigned a show of force mission assume that combat is probable and prepare accordingly. All actions ordinarily associated with the projection of a force to conduct combat operations pertain to show of force deployments. (FM 3-07) (USACAC)

ART 7.6.2.5 CONDUCT FOREIGN HUMANITARIAN ASSISTANCE

7-126. Foreign humanitarian assistance operations occur outside the United States and its territories. Army forces usually conduct them to relieve or reduce the results of natural or man-made disasters before, during, or after the event. They also relieve conditions—such as pain, disease, hunger, or privation—that present a serious threat to life or loss of property. Army forces supplement or complement efforts of host-nation civil authorities or other agencies that provide assistance. Army forces participate in foreign humanitarian assistance operations that may be unilateral, multinational, or United Nations-coordinated responses. Foreign humanitarian assistance is limited in scope and duration. It focuses exclusively on prompt aid to resolve an immediate crisis. Long-term activities designed to support full recovery and a return to predisaster conditions normally will become part of a combatant commander's theater engagement plan. In such cases, a foreign humanitarian assistance operation transitions to a stability operation. (FM 3-07) (USACAC)

No.	Scale	Measure
01	Yes/No	Unit surveyed the disaster area; prioritized needs; conducted medical assessments; and provided medical services, communications, shelter, subsistence, water, engineering support, transportation, fire fighting, mass care, urban search and rescue, hazardous materials response, and energy distribution.
02	Percent	Of mission with host-nation health provider participation.
03	Percent	Of missions coordinated with command civil affairs and surgeons guidance.
04	Percent	Of programs which conform to guidance from Department of Defense and National Security Council.
05	Percent	Of subversive acts, lawlessness, or insurgent attacks in target area.
06	Number	Of program mission and patients seen per assistance mission and their medical surveillance.
07	Number	Of patients trained in public health.

ART 7.6.2.6 ENFORCE SANCTIONS AND EXCLUSION ZONES

7-127. Enforcement of sanctions includes a broad range of possible missions. Commanders must understand that actions to enforce sanctions, while endorsed by the United Nations Security Council, have traditionally been considered acts of war and should posture their forces accordingly. (FM 3-07.31) (TRADOC JADD)

No.	Scale	Measure
01	Yes/No	Unit established checkpoints and control measures and occupied key terrain.
02	Yes/No	Unit provided sufficient force to ensure deterrence.
03	Yes/No	Unit established appropriate rules of engagement.
04	Yes/No	Unit supported with information operations.
05	Yes/No	Unit established communications with controlling authorities.
06	Yes/No	Unit established a quick reaction force.
07	Yes/No	Unit ensured coordination and negotiation mechanisms were established.

ART 7.6.3 CONDUCT PEACE OPERATIONS

7-128. *Peace operations* is a broad term that encompasses multiagency and multinational crisis response and limited contingency operations involving all instruments of national power with military missions to contain conflict, redress the peace, and shape the environment to support reconciliation and rebuilding and facilitate the transition to legitimate governance. Peace operations include peacekeeping, peace enforcement, peacemaking, peace building, and conflict prevention efforts. (FM 3-0) (USACAC)

No.	Scale	Measure
01	Yes/No	Unit physically occupied key terrain to establish control over urban and rural areas.
02	Yes/No	Unit separated belligerent forces.
03	Yes/No	Unit disarmed, demobilized, and reintegrated belligerent forces.
04	Yes/No	Unit controlled weapons and borders.
05	Yes/No	Unit regulated movement of persons and goods across borders.
06	Yes/No	Unit secured key sites.
07	Yes/No	Unit established visible control measures and made them known to the local population.
08	Yes/No	Unit established public security and freedom of movement.
09	Yes/No	Unit established protected areas and secure bases.

ART 7.6.3.1 CONDUCT PEACEKEEPING OPERATIONS

7-129. Peacekeeping operations are military operations undertaken with the consent of all major parties to a dispute. They are designed to monitor and facilitate implementation of agreements (cease fire, truce, or other such agreements) and support diplomatic efforts to reach a long-term political settlement. Peacekeeping operations usually involve observing, monitoring, or supervising and assisting parties to a dispute. To achieve their objectives, Army forces conducting peacekeeping operations rely on the legitimacy acknowledged by all major belligerents and international or regional organizations. They use or threaten the use of force only in self-defense or as a last resort. (FM 3-0) (USACAC)

No.	Scale	Measure
01	Yes/No	Unit supervised disengagement of belligerent forces.
02	Yes/No	Unit monitored exchange of prisoners of war.
03	Yes/No	Unit developed confidence-building measures between host-nation belligerents.
04	Yes/No	Unit investigated alleged breaches of agreements.
05	Yes/No	Unit supported and sustained confidence-building measures among belligerents.
06	Time	To transfer monitor requirements to host-nation security institutions.

ART 7.6.3.1.1 Conduct Observation

7-130. Observation missions are performed primarily by unarmed military observers but may also be performed by peacekeeping forces. In either case, observers forces help ensure that parties to the dispute follow the agreements. (FM 3-07.31) (TRADOC JADD)

No.	Scale	Measure
01	Yes/No	Unit established observation posts, mounted and dismounted patrols, and aerial reconnaissance.
02	Yes/No	Unit observed, monitored, verified, and reported any alleged violation of the governing agreements.
03	Yes/No	Unit confirmed or supervised a cease-fire.
04	Yes/No	Unit investigated alleged cease-fire violations, boundary incidents, and complaints. The observer force investigated alleged infractions to gain evidence regarding agreement violations.
05	Yes/No	Unit conducted regular liaison visits in the area of operations.
06	Yes/No	Unit planned for uncooperative local officials, demonstrations, and other forms of civil disturbance preventing mission accomplishment.
07	Time	To verify the storage or destruction of certain categories of military equipment specified in the relevant agreements.

ART 7.6.3.1.2 Supervise Cease-Fires, Withdrawals, and Disengagements

7-131. Lightly armed forces normally perform supervision and assistance missions. The force undertaking these tasks requires large service support organizations, equipment, and finances. In addition to those tasks being performed by military observers in observation missions, peacekeeping forces may perform other tasks when they are within the scope of the military mission, such as supervising cease fires. (FM 3-07.31) (TRADOC JADD)

No.	Scale	Measure
01	Yes/No	Unit established observation posts, mounted and dismounted patrols, and aerial reconnaissance.
02	Yes/No	Unit deployed on the territory of the disputing parties between contending parties.
03	Yes/No	Unit supervised demilitarized zone or buffer zone.
04	Yes/No	Unit observed and reported on the disputing parties' compliance with a cease-fire.
05	Yes/No	Unit investigated alleged cease-fire violations, boundary incidents, and complaints.
06	Yes/No	Unit planned for uncooperative local officials, demonstrations, and other forms of civil disturbance preventing mission accomplishment.
07	Yes/No	Unit verified the storage or destruction of certain categories of military equipment specified in the relevant agreements.
08	Yes/No	Unit conducted regular liaison visits in the area of operations.
09	Yes/No	Unit assisted in prisoner of war exchanges between the parties to include transportation.

ART 7.6.3.2 CONDUCT PEACE BUILDING OPERATIONS

7-132. Peace building operations are post-conflict actions, predominantly diplomatic and economic, that strengthen and rebuild governmental infrastructure and institutions to avoid a relapse into conflict. (FM 3-07) (USACAC)

Note: The tasks in ART 7.3 (Conduct Stability Operations) describe the tasks that pertain to peace building.

ART 7.6.3.3 CONDUCT PEACEMAKING OPERATIONS

7-133. Peacemaking is the process of diplomacy, mediation, negotiation, or other forms of peaceful settlements that arranges an end to a dispute and resolves issues that led to it. (FM 3-07) (USACAC)

Note: Selective tasks in ART 7.6.1 (Conduct Military Engagements) can support peacemaking operations.

ART 7.6.3.4 Conduct Peace Enforcement Operations

7-134. Peace enforcement operations are the application of military force, or the threat of its use, normally pursuant to international authorization, to compel compliance with resolutions or sanctions designed to maintain or restore peace and order. Unlike peacekeeping operations, peace enforcement operations do not require the consent of all parties. Peace enforcement operations maintain or restore peace and support diplomatic efforts to reach a long-term political settlement. Army forces assigned a peace enforcement mission must be able to apply sufficient combat power for self-defense and to perform tasks forcibly. Units must also be prepared to transition to peacekeeping operations. Peace enforcement operations normally include one or more of six subordinate operations: forcible separation of belligerents, establishment and supervision of protected areas, sanction and exclusion zone enforcement, movement denial and guarantee, restoration and maintenance of order, and protection of humanitarian assistance. (FM 3-07) (USACAC)

No.	Scale	Measure
01	Yes/No	Unit enforced ceasefires.
02	Yes/No	Unit identified and neutralized potential spoilers.
03	Yes/No	Unit established and controlled buffers, including demilitarized zones.
04	Yes/No	Unit supported and enforced political, military, and economic terms arrangements.
05	Yes/No	Unit established and enforced weapons control regimes, including collection and destruction.
06	Yes/No	Unit provided reassurances and incentives for disarmed factions.
07	Yes/No	Unit disarmed belligerents.
08	Yes/No	Unit reduced availability of unauthorized weapons.
09	Yes/No	Unit secured, stored, and disposed of weapons.
10	Yes/No	Unit monitored and verified demobilization.
11	Yes/No	Unit ensured full freedom of movement.
12	Time	To transfer enforcement requirements to host-nation authorities.
13	Time	To establish monitoring regime.

ART 7.6.3.5 Conduct Conflict Prevention Operations

7-135. Conflict prevention consists of actions taken before a predictable crisis to prevent or limit violence, deter parties, and reach an agreement before armed hostilities begin. Conflict prevention often involves diplomatic initiatives. It also includes efforts designed to reform a country's security sector and make it more accountable to civilian control. Conflict prevention may require deploying forces to contain a dispute or prevent it from escalating into hostilities. (FM 3-0) (USACAC)

No.	Scale	Measure
01	Yes/No	Military activities were tailored to meet the political and situational demands.
02	Yes/No	Military efforts had the desired result in achieving the mission specifically assigned to the force.
03	Yes/No	Commanders were provided with a means to evaluate the contribution of military efforts.
04	Yes/No	Trouble spots were isolated in time and space from outside influence or interaction.
05	Yes/No	Unit dominated the situation through force presence.
06	Yes/No	Unit maintained situational awareness.
07	Yes/No	Unit used all available resources to influence the outcome.
08	Yes/No	Commander planned for transition and termination before deployment or as soon as possible during the initial phase.

ART 7.6.4 CONDUCT IRREGULAR WARFARE

7-136. Irregular warfare is a violent struggle among state and nonstate actors for legitimacy and influence over a population. U.S. Army forces operations grouped under irregular warfare are foreign internal defense, support for insurgencies, counterinsurgency, combating terrorism, and unconventional warfare. (FM 3-0) (USACAC)

ART 7.6.4.1 CONDUCT FOREIGN INTERNAL DEFENSE OPERATIONS

7-137.   Foreign internal defense is participation by civilian and military agencies of a government in any of the action programs taken by another government or other designated organization to free and protect its society from subversion, lawlessness, and insurgency. Foreign internal defense missions are applicable to a wide variety of operational environments. U.S. policy currently deals with threats through the indirect use of military force in concert with the diplomatic, informational, military, and economic instruments of national power. Direct use of military force is the exception rather than the rule. This approach relies on supporting efforts of the government of the nation in which the problem is developing. (FM 3-05.202) (USAJFKSWCS)

ART 7.6.4.1.1 Provide Indirect Support to Foreign Internal Defense

7-138.   Indirect support builds strong national infrastructures through economic and military capabilities that contribute to self-sufficiency. This can include unit exchange programs, personnel exchange programs, individual exchange programs, and combination programs. (FM 3-05.202) (USAJFKSWCS)

No.	Scale	Measure
01	Yes/No	Commander understood authorities, funding, and restrictions.
02	Yes/No	Unit developed measures of effectiveness to ensure objectives were met.
03	Yes/No	Unit acquired mission approval.
04	Yes/No	Unit coordinated plan with host nation.
05	Yes/No	Unit facilitated transitions to civil authorities or other agency.
06	Time	To identify national objectives.
07	Time	To evaluate host-nation capabilities.
08	Time	To develop a plan to provide required assistance to host nation.

ART 7.6.4.1.2 Provide Direct Support to Foreign Internal Defense (Not Involving Combat Operations)

7-139.   In direct support, U.S. forces provide direct assistance to the host-nation civilians or military. This support can be evaluation, training, limited information exchange, and equipment support. (FM 3-05.202) (USAJFKSWCS)

No.	Scale	Measure
01	Yes/No	Commander understood authorities, funding, and restrictions.
02	Yes/No	Unit developed measures of effectiveness to ensure objectives were met.
03	Yes/No	Unit acquired mission approval.
04	Yes/No	Unit coordinated plan with host nation.
05	Yes/No	Unit facilitated transitions to civil authorities or other agency.
06	Time	To identify national objectives.
07	Time	To evaluate host-nation capabilities.
08	Time	To develop a plan to provide required assistance to host nation.

ART 7.6.4.1.3 Conduct Combat Operations in Support to Foreign Internal Defense

7-140.   The President must approve combat operations. Combat operations are a temporary solution until host-nation forces can stabilize the situation and provide security for the populace. Emphasis should be placed on host-nation forces in the forefront during these operations to maintain host-nation legitimacy with the population. Combat operations can include counterinsurgency operations. (FM 3-05.202) (USAJFKSWCS)

No.	Scale	Measure
01	Yes/No	Commander understood authorities, funding, and restrictions.
02	Yes/No	Unit developed measures of effectiveness to ensure objectives were met.
03	Yes/No	Unit acquired mission approval.
04	Yes/No	Unit coordinated plan with host nation.
05	Yes/No	Unit facilitated transitions to civil authorities or other agency.
06	Time	To identify national objectives.
07	Time	To evaluate host-nation capabilities.
08	Time	To develop a plan to provide required assistance to host nation.

ART 7.6.4.2 PROVIDE SUPPORT TO INSURGENCIES

7-141. Insurgencies are movements organized to overthrow a constituted government through subversion and armed conflict. By order of the President, Army forces support insurgencies that oppose regimes that threaten U.S. interests or regional stability. While any Army force can be tasked to support an insurgency, Army special operations forces usually receive these missions. The training, organization, and regional focus of Army special operations forces make them well suited for these operations. Army forces supporting insurgencies may provide logistic and training support. They can, but normally do not, conduct combat operations. (FM 3-07) (USACAC)

ART 7.6.4.3 CONDUCT COUNTERINSURGENCY OPERATIONS

7-142. *Counterinsurgency* is those military, paramilitary, political, economic, psychological, and civic actions taken by a government to defeat insurgency. In counterinsurgency, host-nation forces and their partners operate to defeat armed resistance, reduce passive opposition, and establish or reestablish the legitimacy of the host-nation government. Counterinsurgency is the predominant joint operation in Operations Iraqi Freedom and Enduring Freedom. (FM 3-24) (USACAC)

ART 7.6.4.4 COMBAT TERRORISM

7-143. Terrorism is the calculated use of unlawful violence or threat of unlawful violence to inculcate fear. It is intended to coerce or intimidate governments or societies in pursuit of goals that are generally political, religious, or ideological. Enemies who cannot compete with Army forces conventionally often turn to terrorist tactics. Terrorist attacks often create a disproportionate effect on even the most capable conventional forces. Tactics used by terrorists range from arson to the use of weapons of mass destruction. Army forces routinely conduct operations to deter or defeat these attacks. Offensively-oriented operations are categorized as; defensively-oriented operations are antiterrorism. (FM 3-07) (USACAC)

Note: This task branch only addresses counterterrorism operations. ART 6.6 (Apply Antiterrorism Measures) addresses antiterrorism measures.

ART 7.6.4.5 CONDUCT COUNTERTERRORISM OPERATIONS

7-144. Counterterrorism are operations that include the offensive measures taken to prevent, deter, preempt, and respond to terrorism. By law, the counterterrorism mission is assigned to designated special operations forces that are organized and trained to combat terrorism. Those forces conduct counterterrorism outside the territory of the United States. Relevant National Security Decision Directives, National Security Directives, contingency plans, and other relevant classified documents address sensitive and compartmentalized counterterrorism programs. Commanders who employ conventional forces against organized terrorist forces operating in their AO are conducting conventional offensive operations, not counterterrorism operations. (FM 3-07) (USACAC)

ART 7.6.4.6 CONDUCT UNCONVENTIONAL WARFARE

7-145.   Unconventional warfare is a broad spectrum of military and paramilitary operations, normally of long duration, predominantly conducted through, with, or by host-nation or surrogate forces that are organized, trained, equipped, supported, and directed in varying degrees by an external source. It includes, but is not limited to, guerrilla warfare, subversion, sabotage, intelligence activities, and unconventional assisted recovery. Unconventional warfare is operations conducted by, with, or through irregular forces in support of a resistance movement, insurgency, or conventional military operations. (FM 3-05.202) (USAJFKSWCS)

This page intentionally left blank.

Glossary

The glossary lists acronyms and terms with Army, multi-Service, or joint definitions, and other selected terms. Where Army and joint definitions are different, (Army) follows the term. Terms for which FM 7-15 is the proponent manual (authority) are marked with an asterisk (*). The proponent manual for other terms is listed in parentheses after the definition.

SECTION I – ACRONYMS AND ABBREVIATIONS

AAFES	Army and Air Force Exchange Service
A/DACG	arrival/departure airfield control group
ALSA	Air Land Sea Application Center
AO	area of operations
APOD	aerial port of debarkation
ARSOF	Army special operations force
ART	Army tactical task
ASCC	Army Service component command
ASCOPE	areas, structures, capabilities, organizations, people, events
AUEL	automated unit equipment list
AUTL	Army Universal Task List
C2	command and control
CA	civil affairs
CAS	close air support
CASCOM	Combined Arms Support Command
CBRN	chemical, biological, radiological, and nuclear
CBRNE	chemical, biological, radiological, nuclear, and high-yield explosives
CCIR	commander's critical information requirement
CI	counterintelligence
CJCSM	Chairman, Joint Chiefs of Staff memorandum
CMO	civil-military operations
COA	course of action
CONUS	continental United States
COP	common operational picture
COSC	combat and operational stress control
CRC	United States Army Combat Readiness Center
CSAR	combat search and rescue
DA	Department of the Army
DD	Department of Defense (forms only)
DOD	Department of Defense
DODD	Department of Defensive directive

EEFI	essential elements of friendly information
EMS	electromagnetic spectrum
EOD	explosive ordnance disposal
EPW	enemy prisoner of war
EW	electronic warfare
FARP	forward arming and refueling point
FM	field manual
FMI	field manual-interim
G-2	assistant chief of staff, intelligence
G-3	assistant chief of staff, operations
G-6	assistant chief of staff, signal
G-7	assistant chief of staff, information
G-9	assistant chief of staff, civil affairs
GCCS-A	Global Command and Control System-Army
HR	human resources
HUMINT	human intelligence
IED	improvised explosive device
IFF	identification, friend or foe
IMDC	isolated, missing, detained, or captured
IMINT	imagery intelligence
IO	information operations
IPB	intelligence preparation of the battlefield
ISR	intelligence, surveillance, and reconnaissance
JADD	Joint and Allied Doctrine Division
JOA	joint operations area
JOPES	Joint Operation Planning and Execution System
JP	joint publication
JPADS	Joint Precision Airdrop System
JRSOI	joint reception, staging, onward movement, and integration
JWFC	Joint Warfighting Center
kph	kilometers per hour
LOC	line of communications
LOGCAP	logistics civilian augmentation program
LOGPAC	logistics package
MASINT	measurement and signature intelligence
MCEB	Military Communications-Electronics Board
MDMP	military decisionmaking process
MEDEVAC	medical evacuation
METL	mission-essential task list
METT-TC	*See* METT-TC under terms.

MI	military intelligence
MOPP	mission-oriented protective posture
MSD	minimum safe distance
MSR	main supply route
MTF	medical treatment facility
OCPA	Office of the Chief of Public Affairs
OEH	occupational and environmental health
OPSEC	operations security
PAM	pamphlet
PIR	priority intelligence requirement
PMCS	preventative maintenance checks and services
POD	port of debarkation
POE	port of embarkation
PR	personnel recovery
PSYOP	psychological operations
R5	reception, replacement, rest and recuperation, redeployment, and return to duty
RAM	rocket, artillery, and mortar
RDD	required delivery date
RFI	request for information
RFID	radio frequency identification
ROE	rules of engagement
RSOI	reception, staging, onward movement, and integration
S-2	intelligence staff officer
S-3	operations staff officer
S-9	civil affairs staff officer
SAR	search and rescue
SEAD	suppression of enemy air defenses
SIGINT	signals intelligence
SOP	standing operating procedure
SPOD	seaport of debarkation
TC-ACCIS	Transportation Coordinator's Automated Command and Control Information System
TC-AIMS II	Transportation Coordinator's Automated Information for Movement System II
TECHINT	technical intelligence
T&EO	training and evaluation outline
TJAGLCS	The Judge Advocate General's Legal Center and School
TOE	table of organization and equipment
TPFDD	time-phased force and deployment data
TPFDL	time-phased force and deployment list
TRADOC	United States Army Training and Doctrine Command
TTP	tactics, techniques, and procedures

U.S.	United States
UJTL	Universal Joint Task List
USAADASCH	United States Army Air Defense Artillery School
USAAGS	United States Army Adjutant General School
USAARMC	United States Army Armor Center
USAAWC	United States Army Aviation Warfighting Center
USACAC	United States Army Combined Arms Center
USACBRNS	United States Army Chemical, Biological, Radiological, and Nuclear School
USACHCS	United States Army Chaplain Center and School
USAES	United States Army Engineer School
USAFAS	United States Army Field Artillery School
USAFMS	United States Army Financial Management School
USAIC&FH	United States Army Intelligence Center and Fort Huachuca
USAIS	United States Army Infantry School
USAJFKSWCS	United States Army John Fitzgerald Kennedy Special Warfare Center and School
USAMEDDC&S	United States Army Medical Department Center and School
USAMPS	United States Army Military Police School
USASC&FG	United States Army Signal Center and Fort Gordon
USASMDC	United States Army Space and Missile Defense Command
USJFCOM	United States Joint Forces Command
USMTF	United States message text format
UXO	unexploded explosive ordnance
WMD	weapons of mass destruction

SECTION II – TERMS

civil support

(joint) Department of Defense support to U.S. civil authorities for domestic emergencies, and for designated law enforcement and other activities. (JP 3-28)

collective task

A unit of work or action requiring interaction between two or more individuals for its accomplishment. It may also be a mission requirement which can be broken down into individual tasks. (TRADOC Pamphlet 350-70-1)

command and control warfare

The integrated use of physical attack, electronic warfare, and computer network operations, supported by intelligence, to degrade, destroy, and exploit an enemy's or adversary's command and control system or to deny information to it. (FM 3-0)

command and control warfighting function

The related tasks and systems that support commanders in exercising authority and direction. (FM 3-0)

condition

(joint) Those variables of an operational environment or situation in which a unit, system, or individual is expected to operate and may affect performance. (JP 1-02)

core mission-essential task list

A list of a unit's core capability mission-essential tasks and general mission-essential tasks. (FM 7-0)

counterinsurgency

(joint) Those military, paramilitary, political, economic, psychological, and civic actions taken by a government to defeat insurgency. (JP 1-02)

counterterrorism

(joint) Operations that include the offensive measures taken to prevent, deter, preempt, and respond to terrorism. (JP 3-05)

criterion

The minimum acceptable level of performance associated with a particular measure of task performance. (CJCSM 3500.04E)

defensive operations

Combat operations conducted to defeat an enemy attack, gain time, economize forces, and develop conditions favorable for offensive or stability operations. (FM 3-0)

directed mission-essential task list

A list of mission-essential tasks that must be performed to accomplish a directed mission. (FM 7-0)

fires warfighting function

The related tasks and systems that provide collective and coordinated Army indirect fires, joint fires, and command and control warfare, including nonlethal fires, through the targeting process. (FM 3-0)

forward passage of lines

When a unit passes through another unit's positions while moving toward the enemy. (FM 3-90)

information engagement

The integrated employment of public affairs to inform U.S. and friendly audiences; psychological operations, combat camera, U.S. Government strategic communication and defense support to public diplomacy, and other means necessary to influence foreign audiences; and, leader and Soldier engagements to support both efforts. (FM 3-0)

information protection

Active or passive measures that protect and defend friendly information and information systems to ensure timely, accurate, and relevant friendly information. It denies enemies, adversaries, and others the opportunity to exploit friendly information and information systems for their own purposes. (FM 3-0)

intelligence, surveillance, and reconnaissance

(Army) An activity that synchronizes and integrates the planning and operation of sensors, assets, and processing, exploitation, and dissemination systems in direct support of current and future operations. This is an integrated intelligence and operations function. For Army forces, this activity is a combined arms operation that focuses on priority intelligence requirements while answering the commander's critical information requirements. (FM 3-0)

intelligence, surveillance, and reconnaissance integration

The task of assigning and controlling a unit's intelligence, surveillance, and reconnaissance assets (in terms of space, time, and purpose) to collect and report information as a concerted and integrated portion of operation plans and orders. (FM 3-0)

intelligence, surveillance, and reconnaissance synchronization

The task that accomplishes the following: analyzes information requirements and intelligence gaps; evaluates available assets internal and external to the organization; determines gaps in the use of those assets; recommends intelligence, surveillance, and reconnaissance assets controlled by the organization to collect on the commander's critical information requirements; and submits requests for information for adjacent and higher collection support. (FM 3-0)

intelligence warfighting function

The related tasks and systems that facilitate understanding of the operational environment, enemy, terrain, and civil considerations. It includes tasks associated with intelligence, surveillance, and reconnaissance (ISR) operations, and is driven by the commander. (FM 3-0)

linkup

(Army) A meeting of friendly ground forces which occurs in a variety of circumstances. (FM 3-90)

maneuver

(joint) Employment of forces in the operational area through movement in combination with fires to achieve a position of advantage in respect to the enemy in order to accomplish the mission. (JP 3-0)

measure

A parameter that provides the basis for describing varying levels of performance of a task. (CJCSM 3500.04E)

measure of performance

(joint) A criterion used to assess friendly actions that are tied to measuring task accomplishment. (JP 3-0)

METT-TC

A memory aid used in two contexts: 1. In the context of information management, the major subject categories into which relevant information is grouped for military operations: mission, enemy, terrain and weather, troops and support available, time available, civil considerations. (FM 6-0) 2. In the context of tactics, major variables considered during mission analysis (mission variables). (FM 3-90)

mission

(joint) The task, together with the purpose, that clearly indicates the action to be taken and the reason therefor. (JP 1-02)

mission-essential task

A collective task a unit must be able to perform successfully in order to accomplish its doctrinal or directed mission. (FM 7-0)

mission-essential task list

A compilation of mission-essential tasks that an organization must perform successfully to accomplish its doctrinal or directed missions. (FM 7-0)

movement and maneuver warfighting function

The related tasks and systems that move forces to achieve a position of advantage in relation to the enemy. Direct fire is inherent in maneuver, as is close combat. (FM 3-0)

operation

(joint) 1. A military action or the carrying out of a strategic, operational, tactical, service, training, or administrative military mission. 2. The process of carrying on combat, including movement, supply, attack, defense, and maneuvers needed to gain the objectives of any battle or campaign. (JP 1-02)

peace operations

(joint) A broad term that encompasses multiagency and multinational crisis response and limited contingency operations involving all instruments of national power with military missions to contain conflict, redress the peace, and shape the environment to support reconciliation and rebuilding and facilitate the transition to legitimate governance. Peace operations include peacekeeping, peace enforcement, peacemaking, peace building, and conflict prevention efforts. (JP 3-07.3)

protection warfighting function

The related tasks and systems that preserve the force so the commander can apply maximum combat power. (FM 3-0)

relief in place

A tactical enabling operation in which, by the direction of higher authority, all or part of a unit is replaced in an area by the incoming unit. (FM 3-90)

situational awareness

Immediate knowledge of the conditions of the operation, constrained geographically and in time. (FM 3-0)

standard

A quantitative or qualitative measure and criterion for specifying the levels of performance of a task. (FM 7-0)

surveillance

(joint) The systematic observation of aerospace, surface, or subsurface areas, places, persons, or things, by visual, aural, electronic, photographic, or other means. (JP 3-0)

sustainment warfighting function

The related tasks and systems that provide support and services to ensure freedom of action, extend operational reach, and prolong endurance. (FM 3-0)

*tactical mission task

A specific activity performed by a unit while executing a form of tactical operation or form of maneuver. It may be expressed as either an action by a friendly force or effects on an enemy force.

task

A clearly defined and measurable activity accomplished by individuals and organizations. (FM 7-0)

warfighting function

A group of tasks and systems (people, organizations, information, and processes) united by a common purpose that commanders use to accomplish missions and training objectives. (FM 3-0)

This page intentionally left blank.

References

Field manuals and selected joint publications are listed by new number followed by old number.

REQUIRED PUBLICATIONS

These documents must be available to intended users of this publication.

FM 1-02. *Operational Terms and Graphics*. 21 September 2004.

JP 1-02. *Department of Defense Dictionary of Military and Associated Terms*. 12 April 2001.

RELATED PUBLICATIONS

These sources contain relevant supplemental information.

JOINT AND DEPARTMENT OF DEFENSE PUBLICATIONS

CJCSM 3500.04E. *Universal Joint Task Manual*. 25 August 2008.

DOD 4140.25-M. *DOD Management of Bulk Petroleum Products, Natural Gas, and Coal*. 22 June 1994.

DODD 2310.01E. *The Department of Defense Detainee Program*. 5 September 2006.

DODD 5100.1. *Functions of the Department of Defense and Its Major Components*. 1 August 2002.

JP 3-0. *Joint Operations*. 17 September 2006.

JP 3-02. *Joint Doctrine for Amphibious Operations*. 19 September 2001.

JP 3-05. *Doctrine for Joint Special Operations*. 17 December 2003.

JP 3-07.3. *Peace Operations*. 17 October 2007.

JP 3-15. *Barriers, Obstacles, and Mine Warfare for Joint Operations*. 26 April 2007.

JP 3-18. *Joint Forcible Entry Operations*. 16 June 2008.

JP 3-28. *Civil Support*. 14 September 2007.

JP 3-35. *Deployment and Redeployment Operations*. 7 May 2007.

JP 3-60. *Joint Targeting*. 13 April 2007.

JP 4-02. *Health Service Support*. 31 October 2006.

JP 4-05. *Joint Mobilization Planning*. 11 January 2006.

JP 4-06. *Mortuary Affairs in Joint Operations*. 5 June 2006.

JP 6-0. *Joint Communications System*. 20 March 2006.

ARMY PUBLICATIONS

AR 5-22. *The Army Force Modernization Proponent System*. 6 February 2009.

AR 15-6. *Procedures for Investigating Officers and Boards of Officers*. 2 October 2006.

AR 25-400-2. *The Army Records Information Management System (ARIMS)*. 2 October 2007.

AR 195-2. *Criminal Investigation Activities*. 30 October 1985.

AR 200-1. *Environmental Protection and Enhancement*. 13 December 2007.

AR 380-5. *Department of the Army Information Security Program*. 29 September 2000.

AR 870-5. *Military History: Responsibilities, Policies, and Procedures*. 21 September 2007.

AR 870-20. *Museums and Historical Artifacts*. 11 January 1999.

DA PAM 710-2-1. *Using Unit Supply System (Manual Procedures)*. 31 December 1997.

FM 1-0. *Human Resources Support*. 21 February 2007.

FM 1-05. *Religious Support*. 18 April 2003.

FM 1-06. *Financial Management Operations*. 21 September 2006.

FM 1-20. *Military History Operations*. 3 February 2003.

FM 2-0. *Intelligence*. 17 May 2004.

FM 2-91.6. *Soldier Surveillance and Reconnaissance: Fundamentals of Tactical Information Collection*. 10 October 2007.

FM 3-0. *Operations*. 27 February 2008.

FM 3-04.104. *Tactics, Techniques, and Procedures for Forward Arming and Refueling Point*. 3 August 2006.

FM 3-04.126. *Attack Reconnaissance Helicopter Operations*. 16 February 2007.

FM 3-05.30. *Psychological Operations*. 15 April 2005.

FM 3-05.40. *Civil Affairs Operations*. 29 September 2006.

FM 3-05.60. *Army Special Operations Forces Aviation Operations*. 30 October 2007.

FM 3-05.202. *Special Forces Foreign Internal Defense Operations*. 2 February 2007.

FM 3-05.231. *Special Forces Personnel Recovery*. 13 June 2003.

FM 3-06.11. *Combined Arms Operations in Urban Terrain*. 28 February 2002.

FM 3-07. *Stability Operations*. 6 October 2008.

FM 3-07.31. *Peace Operations: Multi-Service Tactics, Techniques, and Procedures for Conducting Peace Operations*. 26 October 2003.

FM 3-09.15. *Tactics, Techniques, and Procedures for Field Artillery Meteorology*. 25 October 2007.

FM 3-09.32. *JFIRE: Multi-Service Tactics, Techniques, and Procedures for the Joint Application of Firepower*. 20 December 2007.

FM 3-11. *Multiservice Tactics, Techniques, and Procedures for Nuclear, Biological, and Chemical Defense Operations*. 10 March 2003.

FM 3-11.3. *Multiservice Tactics, Techniques, and Procedures for Chemical, Biological, Radiological, and Nuclear Contamination Avoidance*. 2 February 2006.

FM 3-11.4. *Multiservice Tactics, Techniques, and Procedures for Nuclear, Biological, and Chemical (NBC) Protection*. 2 June 2003.

FM 3-11.5. *Multiservice Tactics, Techniques, and Procedures for Chemical, Biological, Radiological, and Nuclear Decontamination*. 4 April 2006.

FM 3-11.19. *Multiservice Tactics, Techniques, and Procedures for Nuclear, Biological, and Chemical Reconnaissance*. 30 July 2004.

FM 3-11.21. *Multiservice Tactics, Techniques, and Procedures for Chemical, Biological, Radiological, and Nuclear Consequence Management Operations*. 1 April 2008.

FM 3-11.50. *Battlefield Obscuration*. 31 December 2008.

FM 3-13. *Information Operations: Doctrine, Tactics, Techniques, and Procedures*. 28 November 2003.

FM 3-14. *Space Support to Army Operations*. 18 May 2005.

FM 3-19.1. *Military Police Operations*. 22 March 2001.

FM 3-19.4. *Military Police Leaders' Handbook*. 4 March 2002.

FM 3-19.12. *Protective Services*. 11 August 2004.

FM 3-19.13. *Law Enforcement Investigations*. 10 January 2005.

FM 3-19.15. *Civil Disturbance Operations*. 18 April 2005.

FM 3-19.30. *Physical Security*. 8 January 2001.

FM 3-19.40. *Internment/Resettlement Operations.* 4 September 2007. (Incorporating change 1, 17 December 2007.)

+ FM 3-19.50. *Police Intelligence Operations.* 21 July 2006.

FM 3-20.15. *Tank Platoon.* 22 February 2007.

FM 3-21.5 (22-5). *Drill and Ceremonies.* 7 July 2003.

FM 3-21.10 (7-10). *The Infantry Rifle Company.* 27 July 2006.

FM 3-21.20 (7-20). *The Infantry Battalion.* 13 December 2006.

FM 3-21.75 (21-75). *The Warrior Ethos and Soldier Combat Skills.* 28 January 2008.

FM 3-21.91 (7-91). *Tactical Employment of Antiarmor Platoons and Companies.* 26 November 2002.

FM 3-24 (3-07.22). *Counterinsurgency.* 15 December 2006.

FM 3-25.26. *Map Reading and Land Navigation.* 18 January 2005.

FM 3-27.10 (3-26-10). *Army Ground-Based Midcourse Defense (GMD) Systems Operations.* 24 April 2008.

FM 3-34 (5-100 and 5-114). *Engineer Operations.* 2 January 2004.

FM 3-34.170 (5-170). *Engineer Reconnaissance.* 25 March 2008.

FM 3-34.210 (20-32). *Explosive Hazards Operations.* 27 March 2007.

FM 3-34.214 (5-250). *Explosives and Demolitions.* 11 July 2007.

FM 3-34.230 (5-105). *Topographic Operations.* 3 August 2000.

FM 3-34.280 (5-490). *Engineer Diving Operations.* 20 December 2004.

FM 3-34.343 (5-446). *Military Nonstandard Fixed Bridging.* 12 February 2002.

FM 3-34.400 (5-104). *General Engineering.* 9 December 2008.

FM 3-34.480 (5-422). *Engineer Prime Power Operations.* 4 April 2007.

+ FM 3-39. *Military Police Operations.* 16 February 2010.

+ FM 3-39.40. *Internment and Resettlement Operations.* 12 February 2010.

FM 3-50.1. *Army Personnel Recovery.* 10 August 2005.

FM 3-52 (100-103). *Army Airspace Command and Control in a Combat Zone.* 1 August 2002.

FM 3-90. *Tactics.* 4 July 2001.

FM 3-90.12 (90-13). *Combined Arms Gap-Crossing Operations.* 1 July 2008.

FM 3-90.119 (3-34.119). *Combined Arms Improvised Explosive Device Defeat Operations.* 21 September 2007. (Incorporating change 1, 6 August 2008.)

FM 3-100.4. *Environmental Considerations in Military Operations.* 15 June 2000.

FM 4-0 (100-10). *Combat Service Support.* 29 August 2003.

FM 4-01.30 (55-10). *Movement Control.* 1 September 2003.

FM 4-01.41 (55-20). *Army Rail Operations.* 12 December 2003.

FM 4-01.45. *Multi-Service Tactics, Techniques, and Procedures for Tactical Convoy Operations.* 24 March 2005.

FM 4-02 (8-10). *Force Health Protection in a Global Environment.* 13 February 2003.

FM 4-02.1. *Combat Health Logistics.* 28 September 2001.

FM 4-02.2 (8-10-26 and 8-10-6). *Medical Evacuation.* 8 May 2007.

FM 4-02.7 (8-10-7). *Health Service Support in a Nuclear, Biological, and Chemical Environment Tactics, Techniques and Procedures.* 1 October 2002.

FM 4-02.10. *Theater Hospitalization.* 3 January 2005.

FM 4-02.17. *Preventive Medicine Services.* 28 August 2000.

FM 4-02.18 (8-10-18). *Veterinary Service Tactics, Techniques, and Procedures.* 30 December 2004.

FM 4-02.19 (8-10-19). *Dental Service Support in a Theater of Operations.* 1 March 2001.

FM 4-02.51 (8-51). *Combat and Operational Stress Control*. 6 July 2006.

FM 4-20.07 (42-424). *Quartermaster Force Provider Company*. 29 August 2008.

FM 4-20.41 (10-500-1). *Aerial Delivery Distribution in the Theater of Operations*. 29 August 2003.

FM 4-20.64 (10-64). *Mortuary Affairs Operations*. 9 January 2007.

FM 4-30.1 (9-6). *Munitions Distribution in the Theater of Operations*. 16 December 2003.

FM 4-30.3. *Maintenance Operations and Procedures*. 28 July 2004.

FM 4-30.31 (9-43-2). *Recovery and Battle Damage Assessment and Repair*. 19 September 2006.

FM 4-90.7. *Stryker Brigade Combat Team Logistics*. 10 September 2007.

FM 5-0 (101-5). *Army Planning and Orders Production*. 20 January 2005.

FM 5-19 (100-14). *Composite Risk Management*. 21 August 2006.

FM 5-34. *Engineer Field Data*. 19 July 2005.

FM 5-102. *Countermobility*. 14 March 1985.

FM 5-103 (5-15). *Survivability*. 10 June 1985.

FM 5-430-00-2. *Planning and Design of Roads, Airfields, and Heliports in the Theater of Operations—Airfield and Heliport Design*. 29 September 1994.

FM 5-480. *Port Construction and Repair*. 12 December 1990.

FM 5-482. *Military Petroleum Pipeline Systems*. 26 August 1994.

FM 6-0. *Mission Command: Command and Control of Army Forces*. 11 August 2003.

FM 6-02.40 (24-40). *Visual Information Operations*. 24 January 2002.

FM 6-02.72 (11-1). *Tactical Radios: Multiservice Communications Procedures for Tactical Radios in a Joint Environment*. 14 June 2002.

FM 6-2. *Tactics, Techniques, and Procedures for Field Artillery Survey*. 23 September 1993. (Incorporating, change 1, 16 October 1996.)

FM 6-20. *Fire Support in the Airland Battle*. 17 May 1988.

FM 6-20-10. *Tactics, Techniques, and Procedures for the Targeting Process*. 8 May 1996.

FM 6-22 (22-100). *Army Leadership*. 12 October 2006.

FM 6-22.5. *Combat Stress*. 23 June 2000.

FM 7-0. *Training for Full Spectrum Operations*. 22 December 2008.

FM 7-1 (25-101). *Battle Focused Training*. 15 September 2003.

FM 7-85. *Ranger Unit Operations*. 9 June 1987.

FM 10-1. *Quartermaster Principles*. 11 August 1994.

FM 10-16. *General Fabric Repair*. 24 May 2000.

FM 10-23. *Basic Doctrine for Army Field Feeding and Class I Operations Management*. 18 April 1996.

FM 10-27. *General Supply in Theaters of Operations*. 20 April 1993.

FM 10-52. *Water Supply in Theaters of Operations*. 11 July 1990.

FM 10-67. *Petroleum Supply in Theaters of Operations*. 18 February 1983.

FM 10-67-2 (10-70 and 10-72). *Petroleum Laboratory Testing and Operations*. 2 April 1997.

FM 19-10. *Military Police Law and Order Operations*. 20 September 1987.

FM 20-3. *Camouflage, Concealment, and Decoys*. 30 August 1999.

FM 27-100. *Legal Support to Operations*. 1 March 2000.

FM 34-60. *Counterintelligence*. 3 October 1995.

FM 55-30. *Army Motor Transport Units and Operations*. 27 June 1997.

FM 55-50. *Army Water Transport Operations*. 30 September 1993.

FM 55-60. *Army Terminal Operations*. 15 April 1996.

FM 90-4. *Air Assault Operations*. 16 March 1987.

FM 90-7. *Combined Arms Obstacle Integration*. 29 September 1994.

FM 90-26. *Airborne Operations*. 18 December 1990.

FM 100-9. *Reconstitution*. 13 January 1992.

FM 100-10-1. *Theater Distribution*. 1 October 1999.

FM 100-10-2. *Contracting Support on the Battlefield*. 4 August 1999.

FMI 3-01.60. *Counter-Rocket, Artillery, and Mortar (C-RAM) Intercept Operations*. 16 March 2006.

FMI 3-35. *Army Deployment and Redeployment*. 15 June 2007.

FMI 3-90.10. *Chemical, Biological, Radiological, Nuclear, and High Yield Explosives Operational Headquarters*. 24 January 2008.

FMI 4-93.41. *Army Field Support Brigade Tactics, Techniques, and Procedures*. 22 February 2007.

FMI 6-02.60. *Tactics, Techniques, and Procedures (TTPs) for the Joint Network Node-Network (JNN-N)*. 5 September 2006.

TRADOC Pamphlet 350-70-1. *Training: Guide for Developing Collective Training Products*. 17 May 2004. (http://www.tradoc.army.mil/tpubs/pamndx.htm)

TRADOC Regulation 350-70. *Systems Approach to Training Management, Processes, and Products*. 9 March 1999. (http://www.tradoc.army.mil/tpubs/regndx.htm)

OTHER PUBLICATIONS

The United States Code and Acts by their popular names are available at the Library of Congress Web site (http://www.loc.gov/law/help/guide/federal/uscode.php).

Detainee Treatment Act of 2005.

Geneva Conventions of 1949 (http://www.loc.gov/rr/frd/Military_Law/RC-Fin-Rec_Dipl-Conf-1949.html).

Possee Comitatus Act.

Prompt Payment Act of 1982.

REFERENCED FORMS

DA forms are available on the APD Web site (www.apd.army.mil). DD forms are available on the OSD Web site (www.dtic.mil/whs/directives/infomgt/forms/formsprogram.htm).

DA Form 2028. Recommended Changes to Publications and Blank Forms.

DA Form 2627. Record of Proceedings Under Article 15, UCMJ.

DA Form 3953. Purchase Request and Commitment.

DD Form 1494. Application for Equipment Frequency Allocation.

SF 135. Records Transmittal and Receipt.